Classical and Generalized Models of Elastic Rods

CRC Series: Modern Mechanics and Mathematics

Series Editors: David Gao and Martin Ostoja-Starzewski

PUBLISHED TITLES

Beyond Perturbation: Introduction to The Homotopy Analysis Method
by Shijun Liao

Classical and Generalized Models of Elastic Rods
by Dorin Ieşan

Configurational Forces: Thermomechanics, Physics, Mathematics, and Numerics
by Gérard A. Maugin

Continuum Mechanics and Plasticity
by Han-Chin Wu

Hybrid and Incompatible Finite Element Methods
by Theodore H.H. Pian and Chang-Chun Wu

Introduction to Asymptotic Methods
by Jan Awrejcewicz and Vadim A. Krysko

Mechanics of Elastic Composites
by Nicolaie Dan Cristescu, Eduard-Marius Craciun, and Eugen Soós

Modeling and Analytical Methods in Tribology
by Ilya I. Kudish and Michael J. Covitch

Microstructural Randomness and Scaling in Mechanics of Materials
by Martin Ostoja-Starzewski

CRC Series: Modern Mechanics and Mathematics

Classical and Generalized Models of Elastic Rods

Dorin Ieşan
AI. I. Cuza University of Iaşi
Romania

CRC Press
Taylor & Francis Group
Boca Raton London New York
CRC Press is an imprint of the
Taylor & Francis Group, an **informa** business
A CHAPMAN & HALL BOOK

First published 2009 by Chapman and Hall

Published 2019 by CRC Press
Taylor & Francis Group
6000 Broken Sound Parkway NW, Suite 300
Boca Raton, FL 33487-2742

ISBN 13: 978-1-4200-8649-2 (hbk)

Library of Congress Cataloging-in-Publication Data

Ieşan, Dorin.
Classical and generalized models of elastic rods / Dorin Ieşan.
p. cm. -- (CRC series--modern mechanics and mathematics)
Includes bibliographical references and index.
ISBN 978-1-4200-8649-2 (hardback : alk. paper)
1. Elastic rods and wires. 2. Saint-Venant's principle. 3. Elasticity--Mathematical models. I. Title. II. Series.

QA931.I35 2009
620.1'1232--dc22 2008025421

Visit the Taylor & Francis Web site at
http://www.taylorandfrancis.com

and the CRC Press Web site at
http://www.crcpress.com

Contents

Preface

The deformation of elastic cylinders has been a subject of intensive study. In the theory of classical elasticity, the deformation of homogeneous elastic beams has been of interest for many years and has been studied from numerous aspects. In contrast, the case when the material is inhomogeneous has received relatively little attention. Recently, research activity on functionally graded materials, that is, materials with continuum varying material properties designed for specific engineering applications, has stimulated renewed interest in problems of inhomogeneous elasticity. A major part of this book is concerned with the study of inhomogeneous beams. Interest in the construction of a theory for the deformation of elastic cylinders dates back to Coulomb, Navier, and Cauchy. However, only Saint-Venant has been able to give a solution to the problem.

The importance of Saint-Venant's celebrated memoirs [291,292] on what has long since become known as Saint-Venant's problem requires no emphasis. To review the vast literature to which the work contained in Refs. 291 and 292 has given impetus is not our intention. An account of the historical developments as well as references to various contributions may be found in the books and some of the works cited. We recall that Saint-Venant's problem consists of determining the equilibrium of a homogeneous and isotropic linearly elastic cylinder loaded by surface forces distributed over its plane ends. Saint-Venant proposed an approximation to the solution of the three-dimensional problem, which requires only the solution of two-dimensional problems in the cross section of the cylinder. Saint-Venant's formulation leads to the four basic problems of extension, bending, torsion, and flexure. His analysis is founded on physical intuition and elementary beam theory. Saint-Venant's approach to the problem is based on a relaxed statement in which the pointwise assignment of the terminal tractions is replaced by prescribing the corresponding resultant force and resultant moment. Justification of the procedure is twofold. First, it is difficult, in practice, to determine the actual distribution of applied stresses on the ends, although the resultant force and moment can be measured accurately. Second, one invokes Saint-Venant's principle. This principle states, roughly speaking, that if two sets of loadings are statically equivalent at each end, then the difference in stress fields and strain fields is negligible, except possibly near the ends. The precise meaning of Saint-Venant's hypothesis and its justification has been the subject of many studies, almost from the time of the original Saint-Venant's papers. References to some of the early investigations of the question can be found in [211, 313, and 315]. The classic work on linear elasticity is given by Toupin [329] (see also Refs. 90, 91, 182, and 282 for further important developments). For the history of the problem and the detailed analysis of various results on Saint-Venant's principle, we refer to the works of Gurtin [119], Fichera [89], Horgan and Knowles [129], and Horgan [130,131]. Saint-Venant's problem continues to attract attention from both mathematical and technical points of view. Recently, elastic rods have been used as continuum-type model of DNA.

The relaxed statement of the problem fails to characterize the solution uniquely. This fact led various authors to establish characterizations of Saint-Venant's solution. Clebsch [52] proved that Saint-Venant's solution can be derived from the assumption that the stress vector on any plane normal to the cross sections of the cylinder is parallel to its generators. Voigt [342] rediscovered Saint-Venant's solution by using another assumption regarding the structure of the stress field. Thus, Saint-Venant's extension, bending, and torsion solutions are derived from the hypothesis that the stress field is independent of the axial coordinate, and Saint-Venant's flexure solution is obtained if the stress field depends on the axial coordinate at most linearly.

Sternberg and Knowles [322] characterized Saint-Venant's solutions in terms of certain associated minimum strain-energy properties. Other intrinsic criteria that distinguish Saint-Venant's solutions from all the solutions of the relaxed problem were established in the work [159]. The work [159] presents a new method of deriving Saint-Venant's solutions. The advantage of this method is that it does not involve artificial *a priori* assumptions. The method permits construction of a solution of the relaxed Saint-Venant's problem for other kinds of constitutive equations (anisotropic media, Cosserat continua, etc.) where the physical intuition or semi-inverse method cannot be used. The work [159] points out the importance of the plane strain problem in solving Saint-Venant's problem.

Truesdell [331,334,336] proposed a problem which, roughly speaking, consists of the generalization of Saint-Venant's notion of twist which could be applied to any solution of the torsion problem. An elegant solution of Truesdell's problem has been established by Day [62].

A generalization of Saint-Venant's problem consists of determining the equilibrium of an elastic cylinder which, in the presence of body forces, is subjected to surface tractions arbitrarily prescribed over the lateral boundary and to appropriate stress resultants over its ends. Study of this problem was initiated by Almansi [6] and Michell [221] and was developed in various later works [68,163,175,313]. Saint-Venant's results were established within the equilibrium theory of homogeneous and isotropic elastic bodies. A large number of works are concerned with the relaxed Saint-Venant's problem for other kinds of elastic materials [32,85,204,209].

This book attempts to present several results established in the theory of deformation of elastic cylinders from a unified point of view. An effort is made to provide a systematic treatment of the subject. The theory of prestressed cylinders and the case of finite deformations are not considered here. The reader interested in these subjects will find an account in Refs. 7, 108, 164, 217, 222, and 280.

Chapter 1 is concerned mainly with results with which Saint-Venant's solutions are involved. We give a method of construction of these solutions and then we characterize them in terms of certain associated minimum strain-energy properties. A study of Truesdell's problem is presented. This chapter also includes a proof of Saint-Venant's principle and a study of the plane strain problem.

Chapter 2 deals with the generalization of Saint-Venant's problem to the case when the cylinder is subject to body forces and surface tractions on the lateral boundary. We study the problems of Almansi and Michell and present a scheme for deriving a solution of Almansi–Michell problem.

Chapter 3 is concerned with the deformation of nonhomogeneous and isotropic cylinders, where the elastic coefficients are independent of the axial coordinate. First, the plane strain problem is investigated. Then, the Saint-Venant's problem is reduced to the study of certain plane strain problems. The method is used to study the deformation of elastic cylinders composed of different materials. The problems of Almansi and Michell are also investigated.

Chapter 4 is devoted to anisotropic elastic bodies. We first establish a solution of Saint-Venant's problem. The method does not involve artificial *a priori* assumptions and permits a treatment of the problem even for nonhomogeneous bodies. Then, the problem of loaded anisotropic elastic cylinders is studied. The deformation of cylinders composed of different anisotropic materials is also investigated. The results are specialized for orthotropic elastic cylinders.

In Chapter 5, we study the deformation of cylinders within the linearized theory of homogeneous Cosserat elastic solids. We first present some results concerning the plane strain problem. Then, a solution of Saint-Venant's problem is established. A generalization of the problems of Almansi and Michell is also investigated.

Chapter 6 is concerned with the deformation of nonhomogeneous Cosserat cylinders. Saint-Venant's problem and the problem of loaded cylinders are studied.

Chapter 7 is devoted to the study of porous elastic cylinders. In the first part of the chapter, we study the plane strain problem. Then, the solution to the problem of extension, bending, and torsion is expressed in terms of solutions of certain plane strain problems.

The applications included are problems considered relevant to the purpose of the text. By no means can any claim be made with regard to completeness of the coverage. We have tried to maintain the level of rigor now customary in applied mathematics. However, to ease the burden of the reader, many results are stated with hypotheses that are more stringent than necessary. No attempt is made to provide a complete list of works on Saint-Venant's problem. Neither the list of works cited nor the contents is exhaustive. Nevertheless, it is hoped that the developments presented reflect the state of knowledge in the study of the problem.

Chapter 1

Saint-Venant's Problem

1.1 Preliminaries

We consider a body that at some instant occupies the region B of Euclidean three-dimensional space E^3. In what follows, unless specified to the contrary, B will denote a bounded regular region [119]. We let $\overline{B}$ denote the closure of B, call ∂B the boundary of B, and designate by $\mathbf{n}$ the outward unit normal of ∂B. The deformation of the body is referred to the reference configuration B and a fixed cartesian coordinate frame. The cartesian coordinate frame consists of the orthonormal basis $\{\mathbf{e}_1, \mathbf{e}_2, \mathbf{e}_3\}$ and the origin O. We identify a typical particle x of the body B with its position $\mathbf{x}$ in the reference configuration. Letters in boldface stand for tensors of an order $p \geq 1$, and if $\mathbf{v}$ has the order p, we write $v_{ij\ldots k}$ (p subscripts) for the rectangular cartesian components of $\mathbf{v}$. We shall employ the usual summation and differentiation conventions: Greek subscripts are understood to range over the integers $(1, 2)$, whereas Latin subscripts, unless otherwise specified, are confined to the range $(1, 2, 3)$; summation over repeated subscripts is implied and subscripts preceded by a comma denote partial differentiation with respect to the corresponding cartesian coordinate. The inner product of two vectors $\mathbf{a}$ and $\mathbf{b}$ will be designated by $\mathbf{a} \cdot \mathbf{b}$. We denote the vector product of the vectors $\mathbf{a}$ and $\mathbf{b}$ by $\mathbf{a} \times \mathbf{b}$.

We assume that the body occupying B is a linearly elastic material. In what follows, we restrict our attention to the equilibrium theory of elastic bodies. Let $\mathbf{u}$ be a displacement field over B,

$$\mathbf{u} = \mathbf{u}(x_1, x_2, x_3), \quad (x_1, x_2, x_3) \in B$$

The strain field associated with $\mathbf{u}$ is given by

$$e_{ij}(\mathbf{u}) = \frac{1}{2}(u_{i,j} + u_{j,i}) \tag{1.1.1}$$

The stress–strain relations for an anisotropic medium are

$$t_{ij}(\mathbf{u}) = C_{ijrs} e_{rs}(\mathbf{u}) \tag{1.1.2}$$

Here $\mathbf{t}(\mathbf{u})$ is the stress field associated with $\mathbf{u}$, whereas $\mathbf{C}$ stands for the elasticity field. We assume that $\mathbf{C}$ is positive-definite, smooth on $\overline{B}$, and satisfies the symmetry relations

$$C_{ijrs} = C_{jirs} = C_{rsij} \tag{1.1.3}$$

If the body is homogeneous, then $\mathbf{C}$ is independent of $\mathbf{x}$. For the particular case of an isotropic elastic medium, the tensor field $\mathbf{C}$ admits the representation

$$C_{ijrs} = \lambda\delta_{ij}\delta_{rs} + \mu(\delta_{ir}\delta_{js} + \delta_{is}\delta_{jr})$$

where λ and μ are the Lamé moduli and δ_{ij} is the Kronecker delta. In this case, the constitutive equations 1.1.2 reduce to

$$t_{ij}(\mathbf{u}) = \lambda e_{rr}(\mathbf{u})\delta_{ij} + 2\mu e_{ij}(\mathbf{u}) \tag{1.1.4}$$

If the material is isotropic, then the positive definiteness of $\mathbf{C}$ is equivalent to

$$3\lambda + 2\mu > 0, \quad \mu > 0 \tag{1.1.5}$$

The stress–strain relations 1.1.4 may be inverted to give

$$e_{ij}(\mathbf{u}) = \frac{1}{E}[(1+\nu)t_{ij}(\mathbf{u}) - \nu\delta_{ij}t_{ss}(\mathbf{u})] \tag{1.1.6}$$

where

$$E = \frac{\mu(3\lambda + 2\mu)}{\lambda + \mu}, \qquad \nu = \frac{\lambda}{2(\lambda + \mu)} \tag{1.1.7}$$

The constitutive coefficient E is known as Young's modulus and ν is known as Poisson's ratio.

The equations of equilibrium, in the absence of body forces, are

$$t_{ji}(\mathbf{u})_{,j} = 0 \tag{1.1.8}$$

on B. In view of Equations 1.1.1 and 1.1.3, the constitutive equations 1.1.2 can be written in the form

$$t_{ij}(\mathbf{u}) = C_{ijrs}u_{r,s} \tag{1.1.9}$$

Equation 1.1.8 imply the displacement equations of equilibrium

$$(C_{ijrs}u_{r,s})_{,j} = 0 \tag{1.1.10}$$

on B. We call a vector field $\mathbf{u}$ an *equilibrium displacement field* for B if $\mathbf{u} \in C^2(B) \cap C^1(\overline{B})$ and $\mathbf{u}$ satisfies Equations 1.1.10 on B.

Let $\mathbf{s}(\mathbf{u})$ be the surface traction at regular points of ∂B belonging to the stress field $\mathbf{t}(\mathbf{u})$ defined on $\overline{B}$, that is,

$$s_i(\mathbf{u}) = t_{ji}(\mathbf{u})n_j \tag{1.1.11}$$

The strain energy $U(\mathbf{u})$ corresponding to a smooth displacement $\mathbf{u}$ on $\overline{B}$ is

$$U(\mathbf{u}) = \frac{1}{2}\int_B C_{ijrs}e_{ij}(\mathbf{u})e_{rs}(\mathbf{u})dv \tag{1.1.12}$$

In what follows, two displacement fields differing by an infinitesimal rigid displacement will be regarded identical.

The functional $U(\cdot)$ generates the bilinear functional

$$U(\mathbf{u},\mathbf{v}) = \frac{1}{2}\int_B C_{ijrs}e_{ij}(\mathbf{u})e_{rs}(\mathbf{v})dv \tag{1.1.13}$$

The set of smooth vector fields over $\overline{B}$ can be made into a real vector space with the inner product

$$\langle \mathbf{u},\mathbf{v}\rangle = 2U(\mathbf{u},\mathbf{v}) \tag{1.1.14}$$

This inner product generates the energy norm

$$\|\mathbf{u}\|_e^2 = \langle \mathbf{u},\mathbf{v}\rangle \tag{1.1.15}$$

Let $\mathbf{u}$ and $\mathbf{v}$ be any equilibrium displacement fields. It follows from Equations 1.1.10 and the divergence theorem that

$$\langle \mathbf{u},\mathbf{v}\rangle = \int_{\partial B} \mathbf{u}\cdot\mathbf{s}(\mathbf{v})da \tag{1.1.16}$$

In view of Equations 1.1.3, 1.1.13, 1.1.14, and 1.1.16, we get the reciprocity relation

$$\int_{\partial B} \mathbf{u}\cdot\mathbf{s}(\mathbf{v})da = \int_{\partial B} \mathbf{v}\cdot\mathbf{s}(\mathbf{u})da \tag{1.1.17}$$

We note that the strain field $\mathbf{e}(\mathbf{u})$ associated with a class C^3 displacement field over B satisfies the following equations of compatibility

$$\varepsilon_{ipq}\varepsilon_{jrs}e_{pr,qs} = 0 \tag{1.1.18}$$

where ε_{ijk} is the three-dimensional alternator. Conversely, let B be simply-connected, and let $\mathbf{e}$ be a class C^2 symmetric tensor field on B that satisfies the Equations 1.1.18. Then there exists a displacement field $\mathbf{u}$ of class C^3 on B such that $\mathbf{e}$ and $\mathbf{u}$ satisfy the strain–displacement relations 1.1.1 [119,241].

1.2 Formulation of Saint-Venant's Problem

We assume that the region B from here on refers to the interior of a right cylinder of length h, with open cross section Σ and the lateral boundary Π. The rectangular cartesian frame is supposed to be chosen in such a way that the x_3-axis is parallel to the generators of B and the x_1Ox_2 plane contains one of the terminal cross sections. We denote by Σ_1 and Σ_2, respectively, the cross section located at $x_3 = 0$ and $x_3 = h$ (Figure 1.1).

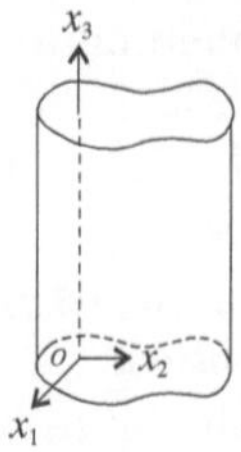

FIGURE 1.1 A prismatic bar.

We assume that the generic cross section Σ is a simply-connected regular region. We denote by Γ the boundary of Σ_1. In view of the foregoing agreements, we have

$$B = \{x : (x_1, x_2) \in \Sigma,\ 0 < x_3 < h\},$$
$$\Pi = \{x : (x_1, x_2) \in L, 0 \leq x_3 \leq h\}$$
$$\Sigma_1 = \{x : (x_1, x_2) \in \Sigma,\ x_3 = 0\}, \qquad \Sigma_2 = \{x : (x_1, x_2) \in \Sigma, x_3 = h\}$$

We consider the equilibrium problem of the cylinder which, in the absence of body forces, is subjected to surface tractions prescribed over its ends and is free from lateral loading. Thus, the problem consists in the determination of an equilibrium displacement field $\mathbf{u}$ on B subjected to the requirements

$$\mathbf{s}(\mathbf{u}) = \mathbf{0} \text{ on } \Pi, \qquad \mathbf{s}(\mathbf{u}) = \mathbf{s}^{(\alpha)} \text{ on } \Sigma_\alpha,\ (\alpha = 1, 2) \tag{1.2.1}$$

where $\mathbf{s}^{(\alpha)}$ is a vector-valued function preassigned to Σ_α. Necessary conditions for the existence of a solution to this problem are given by

$$\int_{\Sigma_1} \mathbf{s}^{(1)} da + \int_{\Sigma_2} \mathbf{s}^{(2)} da = \mathbf{0}, \qquad \int_{\Sigma_1} \mathbf{x} \times \mathbf{s}^{(1)} da + \int_{\Sigma_2} \mathbf{x} \times \mathbf{s}^{(2)} da = \mathbf{0} \tag{1.2.2}$$

where $\mathbf{x}$ is the position vector of a point with respect to O.

Under suitable smoothness hypotheses on Γ and on the given forces, a solution of the problem exists [88].

The importance of Saint-Venant's celebrated memoirs [291,292] in the study of this problem requires no emphasis. Saint-Venant's approach of the problem is based on a relaxed statement in which the pointwise assignment of the terminal tractions is replaced by prescribing the corresponding resultant force and resultant moment. Justification of the procedure is twofold. First, it is difficult in practice to determine the actual distribution of applied stresses on the ends, although the resultant force and moment can be measured accurately. Second, one invokes Saint-Venant's principle. This states, roughly speaking, that if two sets of loadings are statically equivalent at each end, then the difference in stress fields and strain fields are negligible, except possibly near the ends. The precise meaning of Saint-Venant's hypothesis and its justification has been the subject of many studies, almost from the time of the original Saint-Venant's works. A proof of Saint-Venant's principle is presented in Section 1.10.

In the formulation of Saint-Venant, the conditions 1.2.1 are replaced by

$$\mathbf{s}(\mathbf{u}) = \mathbf{0} \text{ on } \Pi, \qquad \mathbf{R}(\mathbf{u}) = \mathbf{F}, \qquad \mathbf{H}(\mathbf{u}) = \mathbf{M} \tag{1.2.3}$$

where $\mathbf{F}$ and $\mathbf{M}$ are prescribed vectors representing the resultant force and the resultant moment about O of the tractions acting on Σ_1. Accordingly, $\mathbf{R}(\cdot)$ and $\mathbf{H}(\cdot)$ are the vector-valued linear functionals defined by

$$\mathbf{R}(\mathbf{u}) = \int_{\Sigma_1} \mathbf{s}(\mathbf{u})da, \qquad \mathbf{H}(\mathbf{u}) = \int_{\Sigma_1} \mathbf{x} \times \mathbf{s}(\mathbf{u})da \tag{1.2.4}$$

Saint-Venant's problem consists in the determination of an equilibrium displacement field $\mathbf{u}$ on B subject to the conditions 1.2.3.

If $\varepsilon_{\alpha\beta}$ is the two-dimensional alternator, Equations 1.2.4 appear as

$$\begin{gathered} R_i(\mathbf{u}) = -\int_{\Sigma_1} t_{3i}(\mathbf{u})da \\ H_\alpha(\mathbf{u}) = -\int_{\Sigma_1} \varepsilon_{\alpha\beta} x_\beta t_{33}(\mathbf{u})da, \qquad H_3 = -\int_{\Sigma_1} \varepsilon_{\alpha\beta} x_\alpha t_{3\beta}(\mathbf{u})da \end{gathered} \tag{1.2.5}$$

The necessary conditions 1.2.2 for the existence of a solution to Saint-Venant's problem lead to the following relations, which are needed subsequently

$$\begin{gathered} \int_{\Sigma_2} t_{3i}(\mathbf{u})da = -R_i(\mathbf{u}), \qquad \int_{\Sigma_2} \varepsilon_{\alpha\beta} x_\alpha t_{3\beta}(\mathbf{u})da = -H_3(\mathbf{u}) \\ \int_{\Sigma_2} x_\alpha t_{33}(\mathbf{u})da = -hR_\alpha(\mathbf{u}) + \varepsilon_{\alpha\beta} H_\beta(\mathbf{u}) \end{gathered} \tag{1.2.6}$$

It is obvious that the relaxed statement of the problem fails to characterize the solution uniquely.

By a solution of Saint-Venant's problem, we mean any equilibrium displacement field that satisfies Equations 1.2.3.

Saint-Venant's formulation leads to the four basic problems of extension, bending, torsion, and flexure, characterized by

1. Extension : $F_\alpha = 0, \ M_i = 0$
2. Bending : $F_i = 0, \ M_3 = 0$
3. Torsion : $F_i = 0, \ M_\alpha = 0$
4. Flexure : $F_3 = 0, \ M_i = 0$

In the next section, we shall study the problems listed above by using the Saint-Venant's semi-inverse method of solution. This consists in making certain assumptions about the components of stress or displacement and leaving enough freedom to satisfy the basic equations and boundary conditions. Saint-Venant's results were established within the equilibrium theory of homogeneous and isotropic cylinders. In Section 1.7, we shall present a rational method of deriving Saint-Venant's solutions.

1.3 Saint-Venant's Solutions

Let B be occupied by an isotropic and homogeneous material. In this case, Saint-Venant's problem consists in the determination of a solution of the Equations 1.1.1, 1.1.4, and 1.1.8 on B which satisfies the boundary conditions 1.2.3. For convenience, in what follows, unless otherwise specified, we shall write e_{ij} for $e_{ij}(\mathbf{u})$ and t_{ij} for $t_{ij}(\mathbf{u})$. It follows from Equations 1.1.11 and 1.2.3 that the conditions on the lateral boundary can be written in the form

$$t_{\alpha i} n_\alpha = 0 \text{ on } \Pi \tag{1.3.1}$$

1.3.1 Extension

In this case, the conditions on the ends reduce to

$$\int_{\Sigma_1} t_{3\alpha} da = 0, \qquad \int_{\Sigma_1} \varepsilon_{\alpha\beta} x_\alpha t_{3\beta} da = 0 \tag{1.3.2}$$

$$\int_{\Sigma_1} t_{33} da = -F_3 \tag{1.3.3}$$

$$\int_{\Sigma_1} x_\alpha t_{33} da = 0 \tag{1.3.4}$$

The extension problem consists in the determination of the functions $u_i \in C^2(B) \cap C^1(\overline{B})$ that satisfy the Equations 1.1.1, 1.1.4, and 1.1.8 on B and the boundary conditions 1.3.1, 1.3.2, 1.3.3, and 1.3.4, where F_3 is a given constant.

Let us suppose that the rectangular cartesian coordinate frame is chosen in such a way that the origin O coincides with the centroid of Σ_1. Thus, we have

$$\int_{\Sigma_1} x_\alpha da = 0 \tag{1.3.5}$$

Following Saint-Venant, we try to solve the extension problem assuming that

$$t_{\alpha\beta} = 0, \qquad t_{33} = C, \qquad t_{\alpha 3} = 0 \tag{1.3.6}$$

where C is an unknown constant. Clearly, the equilibrium equations 1.1.8 are satisfied. From the constitutive equations 1.1.6, we find that

$$e_{\alpha\beta} = -\frac{\nu}{E} C \delta_{\alpha\beta}, \qquad e_{33} = \frac{1}{E} C, \qquad e_{3\alpha} = 0 \tag{1.3.7}$$

The equations of compatibility are identically satisfied. From Equations 1.1.1 and 1.3.7, we obtain

$$u_{\alpha,\beta} + u_{\beta,\alpha} = -\frac{2\nu}{E} C \delta_{\alpha\beta}, \qquad u_{3,\alpha} + u_{\alpha,3} = 0, \qquad u_{3,3} = \frac{1}{E} C$$

A simple calculation gives

$$u_\alpha = -\frac{\nu}{E} C x_\alpha, \qquad u_3 = \frac{1}{E} C x_3, \qquad (x_1, x_2, x_3) \in B \tag{1.3.8}$$

modulo an infinitesimal rigid displacement. We eliminate the rigid displacement by assuming that $\mathbf{u}$ and $\operatorname{curl}\mathbf{u}$ vanish at origin.

The conditions on the lateral boundary 1.3.1 and the conditions 1.3.2 are satisfied on the basis of the relations 1.3.6. It follows from Equations 1.3.5 that the conditions 1.3.4 are identically satisfied. By Equations 1.3.3 and 1.3.6 we conclude that

$$C = -\frac{1}{A} F_3 \tag{1.3.9}$$

where A is the area of the cross section.

Thus, the solution of the extension problem is given by the relations 1.3.8, where C is determined by Equation 1.3.9.

Let x_i be the coordinates of the point P_0 in the reference configuration, and let y_i be the coordinates of the corresponding point P in the deformed configuration. Then we have $y_i = x_i + u_i$. From Equations 1.3.8 and 1.3.9, we get

$$y_\alpha = \left(1 + \frac{\nu}{EA} F_3\right) x_\alpha, \qquad y_3 = \left(1 - \frac{1}{EA} F_3\right) x_3$$

Let $F_3 = -p$, $p > 0$. In this case the resultant force of the tractions acting on the end located at $x_3 = h$ is $p\mathbf{e}_3$ and the point O is fixed. The point N_0 which, prior to deformation, had the coordinates $(0, 0, h)$ goes into point N with the coordinates $(0, 0, h')$, where

$$h' = \left(1 + \frac{1}{EA} p\right) h$$

A tensile test on an elastic specimen could be utilized to obtain the material constants.

1.3.2 Bending by Terminal Couples

We assume that $\mathbf{F} = \mathbf{0}$ and $\mathbf{M} = M_1 \mathbf{e}_1$. The conditions on Σ_1 become

$$\int_{\Sigma_1} t_{3\alpha} da = 0, \qquad \int_{\Sigma_1} \varepsilon_{\alpha\beta} x_\alpha t_{3\beta} da = 0 \tag{1.3.10}$$

$$\int_{\Sigma_1} x_2 t_{33} da = -M_1 \tag{1.3.11}$$

$$\int_{\Sigma_1} t_{33} da = 0, \qquad \int_{\Sigma_1} x_1 t_{33} da = 0 \tag{1.3.12}$$

The bending problem consists in the determination of a solution of the Equations 1.1.1, 1.1.4, and 1.1.8 on B which satisfies the conditions 1.3.1, 1.3.10, 1.3.11, and 1.3.12.

We choose the cartesian coordinate frame in such a way that the x_α-axes are principal centroidal axes of the cross section Σ_1, that is,

$$\int_{\Sigma_1} x_\alpha da = 0, \qquad \int_{\Sigma_1} x_1 x_2 da = 0 \tag{1.3.13}$$

We seek the solution of the bending problem assuming that

$$t_{\alpha\beta} = 0, \qquad t_{33} = C_1 x_2, \qquad t_{3\alpha} = 0 \tag{1.3.14}$$

where C_1 is an unknown constant. It is obvious that the equations of equilibrium 1.1.8 are satisfied. The conditions 1.3.1 and 1.3.10 are satisfied on the basis of the assumptions 1.3.14. It follows from Equations 1.3.13 and 1.3.14 that the conditions 1.3.12 are also satisfied. By Equations 1.3.11 and 1.3.14 we obtain

$$C_1 = -\frac{1}{I} M_1 \tag{1.3.15}$$

where I is the moment of inertia of the cross section about the x_1-axis,

$$I = \int_{\Sigma_1} x_2^2 da$$

From Equations 1.1.6 and 1.3.14, we get

$$e_{\alpha\beta} = -\frac{\nu}{E} C_1 x_2 \delta_{\alpha\beta}, \qquad e_{33} = \frac{1}{E} C_1 x_2, \qquad e_{3\alpha} = 0 \tag{1.3.16}$$

Thus, in view of Equations 1.1.1, we obtain the following equations for the functions u_i

$$\begin{aligned} u_{\alpha,\beta} + u_{\beta,\alpha} &= -\frac{2\nu}{E} C_1 x_2 \delta_{\alpha\beta} \\ u_{3,\alpha} + u_{\alpha,3} &= 0, \qquad u_{3,3} = \frac{1}{E} C_1 x_2 \end{aligned} \tag{1.3.17}$$

The equations of compatibility are satisfied. We assume that there is no rigid displacement at the origin. The integration of Equations 1.3.17 yields

$$\begin{aligned} u_1 &= \frac{M_1 \nu}{EI} x_1 x_2, \qquad u_2 = \frac{M_1}{2EI} \left[x_3^2 + \nu (x_2^2 - x_1^2) \right] \\ u_3 &= -\frac{M_1}{EI} x_2 x_3, \qquad (x_1, x_2, x_3) \in B \end{aligned} \tag{1.3.18}$$

The coordinates of a generic point in the deformed configuration are

$$\begin{aligned} y_1 &= \left(1 + \frac{\nu}{EI} M_1 x_2 \right) x_1 \\ y_2 &= x_2 + \frac{1}{2EI} M_1 \left[x_3^2 + \nu (x_2^2 - x_1^2) \right] \\ y_3 &= \left(1 - \frac{1}{EI} M_1 x_2 \right) x_3, \qquad (x_1, x_2, x_3) \in B \end{aligned} \tag{1.3.19}$$

Since the displacements are infinitesimal we can see that $M_1/(EI)$ is infinitesimal. Then we can write

$$y_3 = \left(1 - \frac{1}{EI} M_1 y_2\right) x_3$$

It follows that the points located at the plane $x_3 =$ const. remain in a plane after deformation. By the relations 1.3.19, we see that the points on the x_3-axis go into the parabola

$$y_1 = 0, \qquad y_2 = \frac{1}{2EI} M_1 x_3^2, \qquad y_3 = x_3$$

The curvature of this curve is $M_1/(EI)$. This result is known as Bernoulli–Euler law.

Similarly, we can study the case when $\mathbf{M} = M_2 \mathbf{e}_2$.

1.3.3 Torsion

We now suppose that $\mathbf{F} = \mathbf{0}$ and $\mathbf{M} = M_3 \mathbf{e}_3$. Thus, the conditions for $x_3 = 0$ reduce to

$$\int_{\Sigma_1} t_{3\alpha} da = 0 \tag{1.3.20}$$

$$\int_{\Sigma_1} t_{33} da = 0, \qquad \int_{\Sigma_1} x_\alpha t_{33} da = 0 \tag{1.3.21}$$

$$\int_{\Sigma_1} (x_1 t_{32} - x_2 t_{31}) da = -M_3 \tag{1.3.22}$$

The torsion problem consists in the determination of the vector field $\mathbf{u} \in C^2(B) \cap C^1(\overline{B})$ that satisfies the Equations 1.1.1, 1.1.4, and 1.1.8 on B and the boundary conditions 1.3.1, 1.3.20, 1.3.21, and 1.3.22.

We seek the solution of the torsion problem in the form

$$u_1 = -\tau x_2 x_3, \qquad u_2 = \tau x_1 x_3, \qquad u_3 = \tau \varphi(x_1, x_2) \tag{1.3.23}$$

where φ is an unknown function of x_1 and x_2, $\varphi \in C^2(\Sigma_1) \cap C^1(\overline{\Sigma}_1)$, and τ is an unknown constant. From Equations 1.1.1 and 1.3.23, we obtain

$$e_{\alpha\beta} = 0, \qquad e_{33} = 0, \qquad 2e_{13} = \tau(\varphi_{,1} - x_2), \qquad 2e_{23} = \tau(\varphi_{,2} + x_1)$$

so that Equation 1.1.4 implies that

$$t_{\alpha\beta} = 0, \qquad t_{33} = 0, \qquad t_{13} = \mu\tau(\varphi_{,1} - x_2), \qquad t_{23} = \mu\tau(\varphi_{,2} + x_1) \tag{1.3.24}$$

The equations of equilibrium 1.1.8 reduce to

$$t_{13,1} + t_{23,2} = 0 \tag{1.3.25}$$

It follows from Equations 1.3.24 and 1.3.25 that the equilibrium equations will be satisfied if φ satisfies the equation

$$\Delta\varphi = 0 \text{ on } \Sigma_1 \tag{1.3.26}$$

where Δ is the two-dimensional Laplacian. Since $t_{\alpha\beta} = 0$, the conditions 1.3.1 reduce to

$$t_{13}n_1 + t_{23}n_2 = 0 \text{ on } \Gamma \tag{1.3.27}$$

In view of Equations 1.3.24, the condition 1.3.27 becomes

$$\frac{\partial\varphi}{\partial n} = x_2n_1 - x_1n_2 \text{ on } \Gamma \tag{1.3.28}$$

where $\partial\varphi/\partial n = \varphi_{,\alpha}n_\alpha$. Thus, the *torsion function* φ satisfies the Neumann problem 1.3.26 and 1.3.28.

Let us consider the boundary-value problem

$$\Delta w = f \text{ on } \Sigma_1, \qquad \frac{\partial w}{\partial n} = g \text{ on } \Gamma \tag{1.3.29}$$

It is known that a necessary condition for the existence of a solution of this problem is

$$\int_{\Sigma_1} f da - \int_\Gamma g ds = 0 \tag{1.3.30}$$

If Γ is a regular curve [119, Section 5], f is continuous on $\overline{\Sigma}$, and g is piecewise continuous on Γ, then the condition 1.3.30 is sufficient [55] for the existence of a solution of the boundary-value problem 1.3.29.

We note that

$$\int_\Gamma (x_2n_1 - x_1n_2)ds = \int_\Gamma x_1dx_1 + x_2dx_2 = 0$$

Thus, in the case of the boundary-value problem 1.3.26 and 1.3.28, the condition 1.3.30 is satisfied. The function φ is determined to within a constant. This constant is nonessential since it generates a rigid body translation.

The conditions 1.3.20 are satisfied on the basis of the equilibrium equations and the conditions on the lateral boundary. Thus, with the aid of Equations 1.3.25, 1.3.27, and the divergence theorem, we have

$$\int_{\Sigma_1} t_{3\alpha}da = \int_{\Sigma_1} (t_{3\alpha} + x_\alpha t_{\beta 3,\beta})da = \int_{\Sigma_1} (x_\alpha t_{\beta 3})_{,\beta}da = \int_\Gamma x_\alpha t_{\beta 3}n_\beta ds = 0$$

Since $t_{33} = 0$, it follows that the conditions 1.3.21 are satisfied. By Equations 1.3.22 and 1.3.24, we obtain

$$\tau D = -M_3 \tag{1.3.31}$$

where the constant D is defined by

$$D = \mu \int_{\Sigma_1} (x_1^2 + x_2^2 + x_1\varphi_{,1} - x_2\varphi_{,1})da \tag{1.3.32}$$

Let us show that D is different from zero. If we take into account Equations 1.3.26, 1.3.28, and the divergence theorem, then we get

$$\int_{\Sigma_1} (x_1\varphi_{,2} - x_2\varphi_{,1})da = \int_{\Sigma_1} [(x_1\varphi)_{,2} - (x_2\varphi)_{,1}]da = \int_\Gamma \varphi(x_1n_2 - x_2n_1)ds$$
$$= -\int_\Gamma \varphi\frac{\partial\varphi}{\partial n}ds = -\int_{\Sigma_1} \varphi_{,\alpha}\varphi_{,\alpha}da$$

Thus, we have

$$\int_{\Sigma_1} (x_1\varphi_{,2} - x_2\varphi_{,1} + \varphi_{,\alpha}\varphi_{,\alpha})da = 0 \tag{1.3.33}$$

It follows from Equations 1.3.32 and 1.3.33 that

$$D = \mu\int_{\Sigma_1} [(\varphi_{,2} + x_1)^2 + (\varphi_{,1} - x_2)^2]da \tag{1.3.34}$$

If we take into account the relations 1.1.5 and the fact that φ is of class C^2, then we conclude from Equation 1.3.34 that

$$D > 0 \tag{1.3.35}$$

Thus, the constant τ is determined by Equation 1.3.31. The constant D is called the *torsional rigidity* of the cylinder.

The solution of the torsion problem is given by the relations 1.3.23, where φ satisfies the boundary-value problem 1.3.26 and 1.3.28, and τ is given by Equation 1.3.31.

Let us show that the Neumann problem 1.3.26 and 1.3.28 can be reduced to a Dirichlet problem. Since φ is harmonic, there exists an analytic function q such that φ is the real part of q,

$$q(z) = \varphi(x_1, x_2) + i\psi(x_1, x_2), \quad (x_1, x_2) \in \Sigma_1$$

where $z = x_1 + ix_2$, and ψ is related to φ by Cauchy–Riemann equations

$$\psi_{,1} = -\varphi_{,2}, \qquad \psi_{,2} = \varphi_{,1} \tag{1.3.36}$$

The function ψ satisfies the equation

$$\Delta\psi = 0 \text{ on } \Sigma_1 \tag{1.3.37}$$

We assume that the curve Γ is a piecewise smooth curve parametrized by its arc length s,

$$x_\alpha = \widehat{x}_\alpha(s), \quad s \in [0, s_*] \tag{1.3.38}$$

Then, we have

$$n_1 = \frac{dx_2}{ds}, \qquad n_2 = -\frac{dx_1}{ds} \tag{1.3.39}$$

so that the condition 1.3.28 becomes

$$\frac{d\psi}{ds} = \frac{1}{2}\frac{d}{ds}(x_1^2 + x_2^2) \text{ on } \Gamma$$

The above condition can be written in the form

$$\psi = \frac{1}{2}(x_1^2 + x_2^2) + k \text{ on } \Gamma \tag{1.3.40}$$

where k is an arbitrary constant. From Equations 1.3.36, we see that the replacement of ψ by $\psi + c$, where c is an arbitrary constant, does not change the function φ. Since the domain Σ_1 is simply-connected, we can replace the above condition by

$$\psi = \frac{1}{2}(x_1^2 + x_2^2) \text{ on } \Gamma \tag{1.3.41}$$

In the case of a multiply-connected domain, the constant k in Equation 1.3.40 may have a different value on each contour forming the boundary of Σ_1 and only on one of these contours it can be fixed arbitrarily. For the study of the torsion problem in this case, we refer to the works of Mushelishvili [241] and Solomon [315].

We note that the function ψ satisfies the Dirichlet problem 1.3.37 and 1.3.41. We introduce the *stress function* of Prandtl by

$$\Psi = \psi(x_1, x_2) - \frac{1}{2}(x_1^2 + x_2^2), \quad (x_1, x_2) \in \Sigma_1 \tag{1.3.42}$$

It follows from Equations 1.3.37, 1.3.41, and 1.3.42 that the function Ψ satisfies the equation

$$\Delta\Psi = -2 \text{ on } \Sigma_1 \tag{1.3.43}$$

and the boundary condition

$$\Psi = 0 \text{ on } \Gamma \tag{1.3.44}$$

In view of Equations 1.3.24, 1.3.36, and 1.3.42, we find that

$$t_{13} = \mu\tau\Psi_{,2}, \qquad t_{23} = -\mu\tau\Psi_{,1} \tag{1.3.45}$$

Moreover, by Equations 1.3.34, 1.3.36, 1.3.42, 1.3.44, and the divergence theorem we obtain

$$\begin{aligned} D &= -\mu\int_{\Sigma_1}(x_1\Psi_{,1} + x_2\Psi_{,2})da \\ &= -\mu\int_{\Sigma_1}[(x_1\Psi)_{,1} + (x_2\Psi)_{,2} - 2\Psi]da = 2\mu\int_{\Sigma_1}\Psi da \end{aligned} \tag{1.3.46}$$

Thus, instead of solving the boundary-value problem 1.3.37 and 1.3.41, we can solve the Dirichlet problem 1.3.43 and 1.3.44.

We denote by P the magnitude of the stress vector

$$\mathbf{t}_3 = t_{13}\mathbf{e}_1 + t_{23}\mathbf{e}_2$$

It follows from Equations 1.3.45 that

$$P^2 = \mu^2\tau^2[(\Psi_{,1})^2 + (\Psi_{,2})^2] = \mu^2\tau^2\Psi_{,\beta}\Psi_{,\beta}$$

Let f be a function of class C^2 on Σ_1 that satisfies the inequality

$$\Delta f \geq 0 \text{ on } \Sigma_1$$

Then f is either identically a constant or else it attains its maximum on the boundary of Σ_1. Clearly,

$$\begin{aligned}\Delta P^2 &= (P^2)_{,\alpha\alpha} = (2\mu^2\tau^2\Psi_{,\beta\alpha}\Psi_{,\beta})_{,\alpha} \\ &= 2\mu^2\tau^2(\Psi_{,\beta\alpha}\Psi_{,\beta\alpha} + \Psi_{,\beta\alpha\alpha}\Psi_{,\beta}) = 2\mu^2\tau^2\Psi_{,\beta\alpha}\Psi_{,\beta\alpha} \geq 0\end{aligned}$$

We conclude that in the case of torsion, the maximum of the shear stress occurs on the boundary of Σ_1.

1.3.4 Flexure

Let us suppose that $\mathbf{F} = F_1\mathbf{e}_1$ and $\mathbf{M} = \mathbf{0}$. In this case, the conditions on Σ_1 become

$$\int_{\Sigma_1} t_{31}da = -F_1, \qquad \int_{\Sigma_1} t_{32}da = 0 \tag{1.3.47}$$

$$\int_{\Sigma_1} t_{33}da = 0, \qquad \int_{\Sigma_1} x_\alpha t_{33}da = 0 \tag{1.3.48}$$

$$\int_{\Sigma_1} (x_1 t_{32} - x_2 t_{31})da = 0 \tag{1.3.49}$$

where F_1 is a given constant.

The flexure problem consists in the determination of a solution of the Equations 1.1.1, 1.1.4, and 1.1.8 on B which satisfies the conditions 1.3.1, 1.3.47, 1.3.48, and 1.3.49. We suppose that the cartesian coordinate frame is chosen in such a way that the relations 1.3.13 hold.

We try to solve the problem assuming that

$$t_{\alpha\beta} = 0 \text{ on } B \tag{1.3.50}$$

Then, the equations of equilibrium become

$$t_{31,3} = 0, \qquad t_{32,3} = 0, \qquad t_{j3,j} = 0 \tag{1.3.51}$$

By Equations 1.1.6 and 1.3.50,

$$\begin{aligned} e_{11} = e_{22} &= -\frac{\nu}{E}t_{33}, \qquad e_{33} = \frac{1}{E}t_{33} \\ e_{\alpha 3} &= \frac{1+\nu}{E}t_{\alpha 3}, \qquad e_{12} = 0 \end{aligned} \tag{1.3.52}$$

It follows from Equations 1.3.51 that $t_{\alpha 3}$ are independent of x_3 and that t_{33} is a linear function of x_3. Thus, in view of Equations 1.3.52, the equations of compatibility 1.1.18 reduce to

$$t_{33,11} = 0, \qquad t_{33,22} = 0, \qquad t_{33,12} = 0 \tag{1.3.53}$$

$$\begin{aligned} (t_{23,1} - t_{13,2})_{,1} &= \frac{\nu}{1+\nu}t_{33,23} \\ (t_{23,1} - t_{13,2})_{,2} &= -\frac{\nu}{1+\nu}t_{33,13} \end{aligned} \tag{1.3.54}$$

Since t_{33} depends on the axial coordinate at most linearly, from Equations 1.3.53, we obtain

$$t_{33} = E[(A_1x_1 + B_1x_2 + C_1)x_3 + A_2x_1 + B_2x_2 + C_2] \tag{1.3.55}$$

where A_α, B_α, and C_α are arbitrary constants. By Equations 1.3.13 and 1.3.48, we find that $A_2 = B_2 = C_2 = 0$, so that

$$t_{33} = E(A_1x_1 + B_1x_2 + C_1)x_3 \tag{1.3.56}$$

We note that on the basis of equations of equilibrium 1.1.8 and the boundary conditions 1.3.1, we can write

$$\begin{aligned} \int_{\Sigma_1} t_{3\alpha}da &= \int_{\Sigma_1} [t_{\alpha 3} + x_\alpha(t_{13,1} + t_{23,2} + t_{33,3})]da \\ &= \int_\Gamma x_\alpha t_{\beta 3}n_\beta ds + \int_{\Sigma_1} x_\alpha t_{33,3}da = \int_{\Sigma_1} x_\alpha t_{33,3}da \end{aligned} \tag{1.3.57}$$

Thus, the conditions 1.3.47 reduce to

$$\int_{\Sigma_1} x_1 t_{33,3} = -F_1, \qquad \int_{\Sigma_1} x_2 t_{33,3}da = 0 \tag{1.3.58}$$

From Equations 1.3.56 and 1.3.58, we obtain

$$A_1 = -\frac{1}{EI^*}F_1, \qquad B_1 = 0 \tag{1.3.59}$$

where

$$I^* = \int_{\Sigma_1} x_1^2 da$$

In view of Equation 1.3.56, the relations 1.3.54 become

$$(t_{23,1} - t_{13,2})_{,1} = 0, \qquad (t_{23,1} - t_{13,2})_{,2} = -\frac{\nu}{1+\nu} E A_1$$

so that

$$t_{23,1} - t_{13,2} = \frac{E}{1+\nu}(\tau - A_1 \nu x_2) \tag{1.3.60}$$

where τ is an arbitrary constant. The relation 1.3.60 can be written in the form

$$\left[t_{23} - \frac{E}{2(1+\nu)}\tau x_1\right]_{,1} = \left[t_{13} - \frac{E}{2(1+\nu)}(A_1 \nu x_2^2 - \tau x_2)\right]_{,2}$$

We conclude that there exists a function $G \in C^2(\Sigma_1) \cap C^1(\overline{\Sigma}_1)$ such that

$$\begin{aligned} t_{23} &= \frac{E}{2(1+\nu)}(G_{,2} + \tau x_1) \\ t_{13} &= \frac{E}{2(1+\nu)}(G_{,1} + A_1 \nu x_2^2 - \tau x_2), \quad (x_1, x_2) \in \Sigma_1 \end{aligned} \tag{1.3.61}$$

The stress tensor satisfies the equations of equilibrium 1.3.51 if the function G satisfies the equation

$$\Delta G = -2(1+\nu)(A_1 x_1 + C_1) \tag{1.3.62}$$

The first two conditions of Equations 1.3.1 are satisfied on the basis of the relation 1.3.50. From the last relation of Equations 1.3.1, we obtain the following condition for the function G,

$$\frac{\partial G}{\partial n} = -\nu A_1 x_2^2 n_1 + \tau(x_2 n_1 - x_1 n_2) \text{ on } \Gamma \tag{1.3.63}$$

If we take into account the relations 1.3.30 and 1.3.13, then the necessary and sufficient condition to solve the boundary-value problem 1.3.62 and 1.3.63 implies that

$$C_1 = 0 \tag{1.3.64}$$

We introduce the function Φ by

$$G = \Phi + \tau \varphi \tag{1.3.65}$$

where φ is the torsion function. It follows from Equations 1.3.26, 1.3.28, 1.3.62, 1.3.63, and 1.3.64 that the function Φ satisfies the equation

$$\Delta \Phi = -2(1+\nu) A_1 x_1 \text{ on } \Sigma_1 \tag{1.3.66}$$

and the boundary condition

$$\frac{\partial \Phi}{\partial n} = -\nu A_1 x_2^2 n_1 \text{ on } \Gamma \tag{1.3.67}$$

The necessary and sufficient condition for the existence of a solution to the boundary-value problem 1.3.66 and 1.3.67 is satisfied. In what follows we assume that the functions φ and Φ are known.

From the relations 1.3.61 and 1.3.65, we get

$$t_{23} = \mu[\Phi_{,2} + \tau(\varphi_{,2} + x_1)], \qquad t_{13} = \mu\left[\Phi_{,1} + \tau(\varphi_{,1} - x_2) + \nu A_1 x_2^2\right]$$

The condition 1.3.49 reduces to

$$\tau D = -M^* \tag{1.3.68}$$

where D is the torsional rigidity and M^* is given by

$$M^* = \mu \int_{\Sigma_1} (x_1 \Phi_{,2} - x_2 \Phi_{,1}) da \tag{1.3.69}$$

Since $D \neq 0$, the relation 1.3.68 determines the constants τ.

The equations of compatibility 1.1.18 are satisfied so that we can determine the displacement field. From Equations 1.1.1, 1.3.52, 1.3.56, 1.3.59, 1.3.61, and 1.3.64, we obtain the following system of equations

$$u_{1,1} = -\nu A_1 x_1 x_3, \qquad u_{2,2} = -\nu A_1 x_1 x_3, \qquad u_{3,3} = A_1 x_1 x_3$$
$$u_{1,2} + u_{2,1} = 0, \qquad u_{2,3} + u_{3,2} = G_{,2} + \tau x_1$$
$$u_{1,3} + u_{3,1} = G_{,1} + \nu A_1 x_2^2 - \tau x_2$$

The integration of the above equations yields

$$\begin{aligned} u_1 &= -\frac{1}{6} A_1 x_3^3 - \frac{1}{2}\nu A_1 x_3 (x_1^2 - x_2^2) - \tau x_2 x_3 \\ u_2 &= -\nu A_1 x_1 x_2 x_3 + \tau x_1 x_3 \\ u_3 &= \frac{1}{2} A_1 x_1 x_3^2 + \frac{1}{2}\nu A_1 x_1 \left(\frac{1}{3} x_1^2 + x_2^2\right) + \tau\varphi + \Phi, \quad (x_1, x_2, x_3) \in B \end{aligned} \tag{1.3.70}$$

In a similar manner we can study the case in which $\mathbf{F} = F_2 \mathbf{e}_2$ and $\mathbf{M} = \mathbf{0}$.

1.4 Unified Treatment

In Ref. 52, Clebsch proved that Saint-Venant's solution can be derived from the assumption that the stress vector on any plane normal to the cross sections of the cylinder is parallel to its generators. In this section, we present a unified treatment of Saint-Venant's problem which rests only on the hypotheses 1.3.50. The solution is established without any special choice of the cartesian coordinate frame.

We consider the general problem in which the conditions for $x_3 = 0$ are

$$\int_{\Sigma_1} t_{3\alpha} da = -F_\alpha \tag{1.4.1}$$

$$\int_{\Sigma_1} t_{33} da = -F_3 \tag{1.4.2}$$

$$\int_{\Sigma_1} x_\alpha t_{33} da = \varepsilon_{\alpha\beta} M_\beta \tag{1.4.3}$$

$$\int_{\Sigma_1} \varepsilon_{\alpha\beta} x_\alpha t_{3\beta} da = -M_3 \tag{1.4.4}$$

where F_k and M_k are prescribed constants. In this case, the problem consists in the determination of the displacement field $\mathbf{u}$ which satisfies the Equations 1.1.1, 1.1.4, and 1.1.8 on B and the boundary conditions 1.3.1 and the conditions for $x_3 = 0$. We try to solve the problem assuming that Equations 1.3.50 holds. Then, the equilibrium equations reduce to Equations 1.3.51, and the constitutive equations can be written in the form 1.3.52. The compatibility equations 1.1.18 reduce to Equations 1.3.53 and 1.3.54. We conclude, as in the preceding section, that Equation 1.3.55 holds.

Now, from Equations 1.3.54 and 1.3.55 we obtain

$$t_{23,1} - t_{13,2} = \frac{E}{1+\nu}(B_1 \nu x_1 - A_1 \nu x_2 + \tau)$$

where τ is an arbitrary constant. The above relation can be expressed as

$$\left[t_{23} - \frac{E}{2(1+\nu)}(\nu B_1 x_1 + \tau)x_1\right]_{,1} = \left[t_{13} - \frac{E}{2(1+\nu)}(\nu A_1 x_2 - \tau)x_2\right]_{,2}$$

Thus, there exists a function S of class C^2 on Σ_1 such that

$$\begin{aligned} t_{23} &= \frac{E}{2(1+\nu)}(S_{,2} + \nu B_1 x_1^2 + \tau x_1) \\ t_{13} &= \frac{E}{2(1+\nu)}(S_{,1} + \nu A_1 x_2^2 - \tau x_2) \end{aligned} \tag{1.4.5}$$

From the equations of equilibrium, we find that S satisfies the following equation

$$\Delta S = -2(1+\nu)(A_1 x_1 + B_1 x_2 + C_1) \text{ on } \Sigma_1 \tag{1.4.6}$$

The first two conditions on the lateral boundary are identically satisfied. The third condition of the relations 1.3.1 becomes

$$\frac{\partial S}{\partial n} = -\nu(A_1 x_2^2 n_1 + B_1 x_1^2 n_2) + \tau(x_2 n_1 - x_1 n_2) \text{ on } \Gamma \tag{1.4.7}$$

In view of Equation 1.3.30, the necessary and sufficient condition for the existence of a solution to the boundary-value problem 1.4.6 and 1.4.7 is

$$C_1 = -A_1 x_1^0 - A_2 x_2^0 \tag{1.4.8}$$

where x_α^0 are the coordinates of the centroid of Σ_1,

$$A x_\alpha^0 = \int_{\Sigma_1} x_\alpha da, \quad A = \int_{\Sigma_1} da \tag{1.4.9}$$

It follows from the relations 1.4.2 and 1.3.55 that

$$C_2 = -\frac{1}{EA}F_3 - A_2 x_1^0 - B_2 x_2^0 \tag{1.4.10}$$

In view of the relations 1.4.8 and 1.4.10,

$$\begin{aligned} t_{33} = E\{&[A_1(x_1 - x_1^0) + B_1(x_2 - x_2^0)]x_3 \\ &+ A_2(x_1 - x_1^0) + B_2(x_2 - x_2^0)\} - \frac{1}{A}F_3 \end{aligned} \tag{1.4.11}$$

If we use Equations 1.3.57 and 1.4.11, then the conditions 1.4.1 reduce to the following system for the constants A_1 and B_1,

$$J_{\alpha 1} A_1 + J_{\alpha 2} B_1 = -\frac{1}{E} F_\alpha \tag{1.4.12}$$

where

$$J_{\alpha\beta} = \int_{\Sigma_1} (x_\alpha - x_\alpha^0)(x_\beta - x_\beta^0) da$$

Since $J_{11}J_{22} - J_{12}^2 \neq 0$, from Equations 1.4.12, we can determine the constants A_1 and B_1. By Equations 1.4.11 and 1.4.3, we obtain the system

$$J_{\alpha 1} A_2 + J_{\alpha 2} B_2 = \frac{1}{E}(\varepsilon_{\alpha\beta} M_\beta + x_\alpha^0 F_3) \tag{1.4.13}$$

which determines the constants A_2 and B_2. In what follows we assume that A_α and B_α are known.

Let us introduce the function χ by

$$S = \chi + \tau\varphi \tag{1.4.14}$$

where φ is the solution of the boundary-value problem 1.3.26 and 1.3.28. By Equations 1.4.6, 1.4.7, 1.4.14, 1.3.26, and 1.3.28 we find that χ satisfies the equation

$$\Delta\chi = -2(1+\nu)(A_1 x_1 + B_1 x_2 + C_1) \text{ on } \Sigma_1 \tag{1.4.15}$$

and the boundary condition

$$\frac{\partial \chi}{\partial n} = -\nu(A_1 x_2^2 n_1 + B_1 x_1^2 n_2) \text{ on } \Gamma \tag{1.4.16}$$

We note that A_1 and B_1 are given by Equations 1.4.12, and that the necessary and sufficient condition for the existence of a solution to the boundary-value problem 1.4.15 and 1.4.16 is satisfied. By Equations 1.4.5 and 1.4.14, we get

$$\begin{aligned} t_{23} &= \mu\left[\chi_{,2} + \tau(\varphi_{,2} + x_1) + \nu B_1 x_1^2\right] \\ t_{13} &= \mu\left[\chi_{,1} + \tau(\varphi_{,1} - x_2) + \nu A_1 x_2^2\right] \end{aligned}$$

so that the condition 1.4.4 reduces to

$$\tau D = -M_3 - \widehat{M} \tag{1.4.17}$$

where D is given by Equation 1.3.34 and $\widehat{M}$ is defined by

$$\widehat{M} = -\mu \int_{\Sigma_1} \left[x_1\left(\chi_{,2} + \nu B_1 x_1^2\right) - x_2\left(\chi_{,1} + \nu A_1 x_2^2\right)\right] da$$

In view of the relation 1.3.35, we can determine τ by Equation 1.4.17.

Since the equations of compatibility are satisfied, we can find the displacement field. It follows from Equations 1.1.1, 1.3.52, 1.4.5, and 1.4.11 that

$$\begin{aligned} u_{1,1} &= -\nu\left\{\left[A_1\left(x_1 - x_1^0\right) + B_1\left(x_2 - x_2^0\right)\right] x_3 \right. \\ &\quad \left. + A_2(x_1 - x_1^0) + B_2\left(x_2 - x_2^0\right)\right\} + \frac{\nu}{EA} F_3 \\ u_{2,2} &= -\nu\left\{\left[A_1\left(x_1 - x_1^0\right) + B_1\left(x_2 - x_2^0\right)\right] x_3 \right. \\ &\quad \left. + A_2\left(x_1 - x_1^0\right) + B_2\left(x_2 - x_2^0\right)\right\} + \frac{\nu}{EA} F_3 \\ u_{3,3} &= \left[A_1(x_1 - x_1^0) + B_1\left(x_2 - x_2^0\right)\right] x_3 + A_2\left(x_1 - x_1^0\right) \\ &\quad + B_2\left(x_2 - x_2^0\right) - \frac{1}{EA} F_3 \\ u_{1,2} + u_{2,1} &= 0 \\ u_{2,3} + u_{3,2} &= S_{,2} + \nu B_1 x_1^2 + \tau x_1 \\ u_{1,3} + u_{3,1} &= S_{,1} + \nu A_1 x_2^2 - \tau x_2 \end{aligned} \tag{1.4.18}$$

The first three equations of 1.4.18 imply that

$$
\begin{aligned}
u_1 &= -\nu x_1 \left\{ \left[A_1 \left(\frac{1}{2} x_1 - x_1^0 \right) + B_1 (x_2 - x_2^0) \right] x_3 \right. \\
&\quad \left. + A_2 \left(\frac{1}{2} x_1 - x_1^0 \right) + B_2 (x_2 - x_2^0) \right\} + \frac{\nu}{EA} F_3 x_1 + f_1(x_2, x_3) \\
u_2 &= -\nu x_2 \left\{ \left[A_1 (x_1 - x_1^0) + B_1 \left(\frac{1}{2} x_2 - x_2^0 \right) \right] x_3 \right. \\
&\quad \left. + A_2 (x_1 - x_1^0) + B_2 \left(\frac{1}{2} x_2 - x_2^0 \right) \right\} + \frac{\nu}{EA} F_3 x_2 + f_2(x_1, x_3) \\
u_3 &= \left\{ \frac{1}{2} [A_1 (x_1 - x_1^0) + B_1 (x_2 - x_2^0)] x_3 \right. \\
&\quad \left. + A_2 (x_1 - x_1^0) + B_2 (x_2 - x_2^0) \right\} x_3 - \frac{1}{EA} F_3 x_3 + f_3(x_1, x_2)
\end{aligned}
\tag{1.4.19}
$$

where f_k are arbitrary functions. Substituting the functions 1.4.19 into the last three equations of 1.4.18, we find

$$
\begin{aligned}
f_{1,2} + f_{2,1} &= \nu x_3 (A_1 x_2 + B_1 x_1) + \nu (A_2 x_2 + B_2 x_1) \\
f_{2,3} + f_{3,2} &= S_{,2} + \nu \left[A_1 (x_1 - x_1^0) + B_1 \left(\frac{1}{2} x_2 - x_2^0 \right) \right] x_2 \\
&\quad + \nu B_1 x_1^2 + \tau x_1 - B_2 x_3 - \frac{1}{2} B_1 x_3^2 \\
f_{3,1} + f_{1,3} &= S_{,1} + \nu \left[A_1 \left(\frac{1}{2} x_1 - x_1^0 \right) + B_1 (x_2 - x_2^0) \right] x_1 \\
&\quad + \nu A_1 x_2^2 - \tau x_2 - A_2 x_3 - \frac{1}{2} A_1 x_3^2
\end{aligned}
\tag{1.4.20}
$$

It follows from Equations 1.4.20 that

$$
\begin{aligned}
&f_{1,22} = \nu (A_1 x_3 + A_2), \qquad f_{1,33} = -A_1 x_3 - A_2, \qquad f_{1,23} = \nu A_1 x_2 - \tau \\
&f_{2,11} = \nu (B_1 x_3 + B_2), \qquad f_{2,33} = -B_1 x_3 - B_2, \qquad f_{2,13} = \nu B_1 x_1 + \tau \\
&f_{3,11} = S_{,11} + \nu [A_1 (x_1 - x_1^0) + B_1 (x_2 - x_2^0)] \\
&f_{3,22} = S_{,22} + \nu [A_1 (x_1 - x_1^0) + B_1 (x_2 - x_2^0)] \\
&f_{3,12} = S_{,12} + \nu (B_1 x_1 + A_1 x_2)
\end{aligned}
\tag{1.4.21}
$$

It is easy to find the functions f_k from Equations 1.4.21. These functions must be so determined as to satisfy Equation 1.4.20. Finally, from the relations 1.4.19 we obtain

$$\begin{aligned}
u_1 = & -\frac{1}{6}A_1x_3^3 - \frac{1}{2}A_2x_3^2 \\
& - x_3\left\{\nu x_1\left[A_1\left(\frac{1}{2}x_1 - x_1^0\right) + B_1(x_2 - x_2^0)\right] - \frac{1}{2}\nu A_1x_2^2\right\} - \tau x_2x_3 \\
& - \nu x_1\left[A_2\left(\frac{1}{2}x_1 - x_1^0\right) + B_2(x_2 - x_2^0) - \frac{1}{EA}F_3\right] + \frac{1}{2}\nu A_2x_2^2 \\
u_2 = & -\frac{1}{6}B_1x_3^3 - \frac{1}{2}B_2x_3^2 \\
& - x_3\left\{\nu x_2\left[A_1(x_1 - x_1^0) + B_1(x_2 - x_2^0)\right] - \frac{1}{2}\nu B_1x_1^2\right\} + \tau x_1x_3 \\
& - \nu x_2\left[A_2(x_1 - x_1^0) + B_2\left(\frac{1}{2}x_2 - x_2^0\right) - \frac{1}{EA}F_3\right] + \frac{1}{2}\nu B_2x_1^2 \\
u_3 = & \frac{1}{2}x_3^2\left[A_1(x_1 - x_1^0) + B_1(x_2 - x_2^0)\right] + x_3\left[A_2(x_1 - x_1^0)\right. \\
& \left. + B_2(x_2 - x_2^0) - \frac{1}{EA}F_3\right] + \frac{1}{2}\nu x_1^2\left[A_1\left(\frac{1}{3}x_1 - x_1^0\right) + B_1(x_2 - x_2^0)\right] \\
& + \frac{1}{2}\nu x_2^2\left[A_1(x_1 - x_1^0) + B_1\left(\frac{1}{3}x_2 - x_2^0\right)\right] + \tau\varphi + \chi \qquad (1.4.22)
\end{aligned}$$

modulo an infinitesimal rigid displacement.

Thus, the solution of the problem is given by the relations 1.4.22, where A_α, B_α, and C_α are given by Equations 1.4.12, 1.4.13, 1.4.8, and 1.4.10, φ is the torsion function, χ is characterized by Equations 1.4.15 and 1.4.16, and τ is defined by Equation 1.4.17.

1.5 Plane Deformation

In this section, we present some results concerning the plane strain problem of homogeneous and isotropic elastic cylinders. The relationship between the plane strain problem and Saint-Venant's problem will be discussed in Section 1.7.

1.5.1 Statement of Problem

Throughout this section, we assume that the body occupying the cylinder B is a homogeneous and isotropic elastic material, and that a continuous body force $\mathbf{f}$ is prescribed on B. We consider that on the lateral boundary is prescribed the surface displacement $\widetilde{\mathbf{u}}$ or the surface force $\widetilde{\mathbf{t}}$. We suppose that the surface displacement $\widetilde{\mathbf{u}}$, the surface traction $\widetilde{\mathbf{t}}$, and the body force $\mathbf{f}$ are all independent of x_3 and parallel to the x_1, x_2-plane.

The state of plane strain, parallel to the x_1, x_2-plane, of the cylinder B is characterized by

$$u_\alpha = u_\alpha(x_1, x_2), \qquad u_3 = 0, \quad (x_1, x_2) \in \Sigma_1 \tag{1.5.1}$$

The above restrictions, in conjunction with the strain–displacement relations 1.1.1 and the stress–strain relations 1.1.4, imply that e_{ij} and t_{ij} are all independent of x_3.

The nonzero components of the strain tensor are given by

$$e_{\alpha\beta} = \frac{1}{2}(u_{\alpha,\beta} + u_{\beta,\alpha}) \tag{1.5.2}$$

The constitutive equations show that the nonzero components of the stress tensor are $t_{\alpha\beta}$ and t_{33}. Further,

$$t_{\alpha\beta} = \lambda e_{\rho\rho}\delta_{\alpha\beta} + 2\mu e_{\alpha\beta} \tag{1.5.3}$$

and $t_{33} = \lambda\mu_{\rho,\rho}$. The equilibrium equations reduce to

$$t_{\beta\alpha,\beta} + f_\alpha = 0 \text{ on } \Sigma_1 \tag{1.5.4}$$

If the displacement field is prescribed on the lateral boundary, then we have the boundary conditions

$$u_\alpha = \widetilde{u}_\alpha \text{ on } \Gamma \tag{1.5.5}$$

where $\widetilde{u}_\alpha$ are continuous functions. The associated problem is called the *first boundary-value problem* (or the *displacement problem*).

If the stress vector is prescribed on Π, then the boundary conditions reduce to

$$t_{\beta\alpha}n_\beta = \widetilde{t}_\alpha \text{ on } \Gamma \tag{1.5.6}$$

where $\widetilde{t}_\alpha$ are piecewise regular functions. In this case we refer to the resulting problem as the *second boundary-value problem* (or *the traction problem*).

In view of Equations 1.5.2, the relation 1.5.3 becomes

$$t_{\alpha\beta} = \lambda u_{\rho,\rho}\delta_{\alpha\beta} + \mu(u_{\alpha,\beta} + u_{\beta,\alpha}) \tag{1.5.7}$$

Thus, Equations 1.5.4 imply the following displacement equations of equilibrium for plane strain

$$\mu\Delta u_\alpha + (\lambda + \mu)u_{\beta,\beta\alpha} + f_\alpha = 0 \text{ on } \Sigma_1 \tag{1.5.8}$$

The first boundary-value problem consists in the determination of the functions $u_\alpha \in C^2(\Sigma_1) \cap C^0(\overline{\Sigma}_1)$ that satisfy Equations 1.5.8 on Σ_1 and the boundary conditions 1.5.5.

In view of Equations 1.5.7, the boundary conditions 1.5.6 can be expressed as

$$[\lambda u_{\rho,\rho}\delta_{\alpha\beta} + \mu(u_{\alpha,\beta} + u_{\beta,\alpha})]n_\beta = \widetilde{t}_\alpha \text{ on } \Gamma \tag{1.5.9}$$

The second boundary-value problem consists in finding of the functions $u_\alpha \in C^2(\Sigma_1) \cap C^1(\overline{\Sigma}_1)$ which satisfy Equations 1.5.8 on Σ_1 and the boundary conditions 1.5.9 on Γ.

The component t_{33} can be determined after the displacements u_α are found. Thus, we can calculate the surface tractions over the ends, which maintain the cylinder in equilibrium. If the ends of the cylinder are free, the solution can be found by superposing, on the solution of the plane strain problem, the solution of a Saint-Venant's problem.

1.5.2 Uniqueness Results

The *elastic potential associated with* **u**, *in the case of the plane strain*, is defined by

$$W_*(\mathbf{u}) = \frac{1}{2}\lambda e_{\rho\rho}(\mathbf{u})e_{\gamma\gamma}(\mathbf{u}) + \mu e_{\alpha\beta}(\mathbf{u})e_{\alpha\beta}(\mathbf{u}) \tag{1.5.10}$$

To avoid repeated regularity assumptions, we suppose that

(*i*) f_α are continuous on $\overline{\Sigma}_1$

(*ii*) $\widetilde{u}_\alpha$ are continuous on Γ

(*iii*) $\widetilde{t}_\alpha$ are piecewise regular on Γ

(*iv*) Γ is a piecewise smooth curve

Theorem 1.5.1 *Assume that the elastic potential W_* is a positive definite quadratic form. Then*

(α) *the first boundary-value problem has at most one solution;*

(β) *any two solutions of the second boundary-value problem are equal, modulo a plane rigid displacement.*

Proof. It follows from Equations 1.5.3 and 1.5.10 that

$$t_{\alpha\beta}e_{\alpha\beta} = 2W_* \tag{1.5.11}$$

On the other hand, by Equations 1.5.2, 1.5.3, and 1.5.4, we find

$$t_{\alpha\beta}e_{\alpha\beta} = t_{\alpha\beta}u_{\alpha,\beta} = (u_\alpha t_{\beta\alpha})_{,\beta} + f_\alpha u_\alpha \tag{1.5.12}$$

From the relations 1.5.11 and 1.5.12, we get

$$2W_* = (u_\alpha t_{\beta\alpha})_{,\beta} + f_\alpha u_\alpha$$

If we integrate this relation over Σ_1, we conclude, with the aid of divergence theorem, that

$$2\int_{\Sigma_1} W_* da = \int_\Gamma u_\alpha t_{\beta\alpha} n_\beta ds + \int_{\Sigma_1} f_\alpha u_\alpha da \tag{1.5.13}$$

Suppose that there are two solutions of a boundary-value problem. Then their difference $\mathbf{u}^0$ is a solution of a plane strain problem corresponding to null external data. From Equation 1.5.13, we obtain

$$\int_{\Sigma_1} W_*(\mathbf{u}^0)da = 0 \tag{1.5.14}$$

Since the elastic potential is positive definite, from Equation 1.5.14, we find $e_{\alpha\beta}(\mathbf{u}^0) = 0$ and therefore

$$u_1^0 = \alpha_1 - \beta_3 x_2, \qquad u_2^0 = \alpha_2 + \beta_3 x_1 \tag{1.5.15}$$

where α_ρ and β_3 are arbitrary constants. In the case of the first boundary-value problem, we get $\alpha_\rho = 0, \beta_3 = 0$. □

The functions u_α^0 given by Equations 1.5.15 are the components of a *plane rigid displacement.*

Let us note that W_* is positive definite if and only if

$$\mu > 0, \qquad \lambda + \mu > 0 \tag{1.5.16}$$

We record the following existence results [194,241].

Theorem 1.5.2 *Assume that the hypotheses (i)–(iv) hold and that W_* is positive definite. Then*

(α_1) *the first boundary-value problem has solution;*

(β_1) *the second boundary-value problem has solution if and only if* $\mathbf{f}$ *and* $\widetilde{\mathbf{t}}$ *satisfy the conditions*

$$\int_{\Sigma_1} f_\alpha da + \int_\Gamma \widetilde{t}_\alpha ds = 0, \qquad \int_{\Sigma_1} \varepsilon_{\alpha\beta} x_\alpha f_\beta da + \int_\Gamma \varepsilon_{\alpha\beta} x_\alpha \widetilde{t}_\beta ds = 0 \tag{1.5.17}$$

The conditions 1.5.17 demand that the external forces be in equilibrium.

1.5.3 Airy Function

In what follows we assume that Equation 1.5.16 hold. Let us suppose that the body forces vanish. Then, the equilibrium equations become

$$t_{\beta\alpha,\beta} = 0 \tag{1.5.18}$$

Let χ be a scalar field of class C^4 on Σ_1, and let

$$t_{\alpha\beta} = \delta_{\alpha\beta}\Delta\chi - \chi_{,\alpha\beta} \tag{1.5.19}$$

Then, the stresses $t_{\alpha\beta}$ given by the relations 1.5.19 satisfy Equations 1.5.18. The representation 1.5.19 is due to G. Airy (1863). Since

$$t_{\rho\rho} = 2(\lambda + \mu)e_{\alpha\alpha}$$

from Equations 1.5.3 and 1.5.19, we get

$$2\mu e_{\alpha\beta} = t_{\alpha\beta} - \nu t_{\rho\rho}\delta_{\alpha\beta} = (1-\nu)\delta_{\alpha\beta}\Delta\chi - \chi_{,\alpha\beta} \tag{1.5.20}$$

where ν is defined by Equations 1.1.7. In the case of a plane strain, the compatibility equations 1.1.18 reduce to

$$e_{11,22} + e_{22,11} = 2e_{12,12} \tag{1.5.21}$$

It follows from Equations 1.5.20 and 1.5.21 that the function χ satisfies the equation

$$\Delta\Delta\chi = 0 \text{ on } \Sigma_1 \tag{1.5.22}$$

The relations 1.5.19 can be written in the form

$$t_{\alpha\beta} = \varepsilon_{\alpha\lambda}\varepsilon_{\beta\tau}\chi_{,\lambda\tau} \tag{1.5.23}$$

The function χ is called the Airy function. We note that any two Airy functions χ and $\widetilde{\chi}$ generating the same stresses differ by a linear function,

$$\chi(x_1, x_2) = \widetilde{\chi}(x_1, x_2) + C_\alpha x_\alpha + C_0 \tag{1.5.24}$$

where C_α and C_0 are arbitrary constants.

In the case of the second boundary-value problem, the conditions 1.5.6 become

$$\varepsilon_{\alpha\rho}\varepsilon_{\beta\gamma}\chi_{,\rho\gamma}n_\beta = \widetilde{t}_\alpha \text{ on } \Gamma \tag{1.5.25}$$

Thus, if the body forces are absent, then the second boundary-value problem reduces to finding a biharmonic function χ that satisfies the boundary conditions 1.5.25.

The boundary conditions 1.5.25 can be presented in another form. Thus, in view of Equations 1.3.39 and 1.5.19, we find that

$$t_{\beta 1}n_\beta = \frac{d}{ds}(\chi_{,2}), \qquad t_{\beta 2}n_\beta = -\frac{d}{ds}(\chi_{,1}) \tag{1.5.26}$$

on Γ. It follows from Equations 1.5.6 and 1.5.26 that

$$\chi_{,\alpha} = g_\alpha + c_\alpha \text{ on } \Gamma \tag{1.5.27}$$

where c_α are constants of integration, and g_α are given by

$$g_\alpha(s) = -\int_0^s \varepsilon_{\alpha\beta}\widetilde{t}_\beta(\sigma)d\sigma, \quad s \in [0, s_*] \tag{1.5.28}$$

By Equation 1.5.27, we get

$$\chi = G_1 + c_\alpha x_\alpha + c_0, \qquad \frac{\partial\chi}{\partial n} = G_2 + c_1\frac{dx_2}{ds} - c_2\frac{dx_1}{ds} \text{ on } \Gamma \tag{1.5.29}$$

where c_0 is an arbitrary constant, and G_α are defined by

$$G_1(s) = \int_0^s g_\alpha(\sigma)\frac{dx_\alpha}{ds}(\sigma)d\sigma$$
$$G_2(s) = \int_0^s \left[g_1(\sigma)\frac{dx_2}{ds}(\sigma) - g_2(\sigma)\frac{dx_1}{ds}(\sigma)\right] d\sigma, \quad s \in [0, s_*] \tag{1.5.30}$$

In view of Equation 1.5.24, we can choose the constants C_α and C_0 such that χ satisfies the boundary conditions

$$\chi = G_1, \qquad \frac{\partial \chi}{\partial n} = G_2 \text{ on } \Gamma \tag{1.5.31}$$

Thus, the second boundary-value problem reduces to finding a biharmonic function χ that satisfies the boundary conditions 1.5.31. If Σ_1 is multiply-connected, then its boundary is the union of a finite number of closed curves $\Gamma_1, \Gamma_2, \ldots, \Gamma_M$. In this case the relations 1.5.29 will hold on each Γ_k, the constants of integration will, in general, be different on each curve forming the boundary of Σ_1,

$$\chi = G_1 + c_\alpha^{(k)} x_\alpha + c_0^{(k)}$$
$$\frac{\partial \chi}{\partial n} = G_2 + c_1^{(k)}\frac{dx_2}{ds} - c_2^{(k)}\frac{dx_1}{ds} \text{ on } \Gamma_k \quad (k = 1, 2, \ldots, M) \tag{1.5.32}$$

The constants $c_\alpha^{(k)}$ and $c_0^{(k)}$ can be set equal to zero on one of the curves $\Gamma_k, (k = 1, 2, \ldots, M)$, while the other constants can be determined from the conditions that the displacements be single-valued (see, e.g., [113,119,241]).

Remark. From Equations 1.5.22, 1.5.23, and 1.5.31, we conclude that the stresses corresponding to a solution of the second boundary-value problem for a simply-connected domain Σ_1 are independent of the elastic constants. This result is due to M. Lévi (1898).

1.5.4 Complex Potentials

For the remainder of this section we continue to assume that the body forces are zero and that the relations 1.5.16 hold. We now establish a representation of the displacements in terms of a pair of complex analytic functions of the complex variable $z = x_1 + ix_2$. The boundary-value problems can be reduced to the determination of these functions from prescribed values of certain combinations of these functions on the boundary of Σ_1. We introduce the complex coordinates z and $\overline{z}$ on Σ_1 by

$$z = x_1 + ix_2, \qquad \overline{z} = x_1 - ix_2 \tag{1.5.33}$$

We define the *complex displacement* w by

$$w = u_1 + iu_2 \tag{1.5.34}$$

The constitutive equations 1.5.7 can be expressed in the form

$$\begin{aligned} t_{11} + t_{22} &= 2(\lambda + \mu)u_{\rho,\rho} \\ t_{11} - t_{22} + 2it_{12} &= 2\mu[u_{1,1} - u_{2,2} + i(u_{1,2} + u_{2,1})] \end{aligned} \tag{1.5.35}$$

We note that

$$\begin{aligned} \frac{\partial w}{\partial z} &= \frac{1}{2}[u_{1,1} + u_{2,2} + i(u_{2,1} - u_{1,2})] \\ \frac{\partial \overline{w}}{\partial \overline{z}} &= \frac{1}{2}[u_{1,1} + u_{2,2} - i(u_{2,1} - u_{1,2})] \\ \frac{\partial w}{\partial \overline{z}} &= \frac{1}{2}[u_{1,1} - u_{2,2} + i(u_{1,2} + u_{2,1})] \end{aligned} \tag{1.5.36}$$

where a bar over a letter designates the complex conjugate. Thus, the constitutive equations 1.5.35 can be written in the form

$$\begin{aligned} t_{11} + t_{22} &= 2(\lambda + \mu)\left(\frac{\partial w}{\partial z} + \frac{\partial \overline{w}}{\partial \overline{z}}\right) \\ t_{11} - t_{22} + 2it_{12} &= 4\mu\frac{\partial w}{\partial \overline{z}} \end{aligned} \tag{1.5.37}$$

We note that

$$u_{\beta,\beta} = \frac{\partial w}{\partial z} + \frac{\partial \overline{w}}{\partial \overline{z}}, \qquad \Delta = 4\frac{\partial^2}{\partial z \partial \overline{z}} \tag{1.5.38}$$

In complex coordinates, the system of equations of equilibrium 1.5.8, with zero body forces, can be expressed in the form

$$2\mu\frac{\partial^2 w}{\partial z \partial \overline{z}} + (\lambda + \mu)\frac{\partial}{\partial \overline{z}}\left(\frac{\partial w}{\partial z} + \frac{\partial \overline{w}}{\partial \overline{z}}\right) = 0 \tag{1.5.39}$$

Equation 1.5.39 may be integrated to give the result

$$2\mu\frac{\partial w}{\partial z} + (\lambda + \mu)\left(\frac{\partial w}{\partial z} + \frac{\partial \overline{w}}{\partial \overline{z}}\right) = \frac{2(\lambda + 2\mu)}{\lambda + \mu}\Omega'(z) \tag{1.5.40}$$

where Ω is an arbitrary analytic complex function on z, and $\Omega'(z) = d\Omega(z)/dz$. The conjugate of this relation is

$$2\mu\frac{\partial \overline{w}}{\partial \overline{z}} + (\lambda + \mu)\left(\frac{\partial w}{\partial z} + \frac{\partial \overline{w}}{\partial \overline{z}}\right) = \frac{2(\lambda + \mu)}{\lambda + \mu}\overline{\Omega}'(\overline{z}) \tag{1.5.41}$$

It follows from Equations 1.5.40 and 1.5.41 that

$$\frac{\partial w}{\partial z} + \frac{\partial \overline{w}}{\partial \overline{z}} = \frac{1}{\lambda + \mu}[\Omega'(z) + \overline{\Omega}'(\overline{z})] \tag{1.5.42}$$

In view of Equation 1.5.42, Equation 1.5.40 becomes

$$2\mu\frac{\partial w}{\partial z} = \kappa\Omega'(z) - \overline{\Omega}'(\overline{z}) \tag{1.5.43}$$

where

$$\kappa = 3 - 4\nu \tag{1.5.44}$$

Equation 1.5.43 may be integrated to give

$$2\mu w = \kappa\Omega(z) - z\overline{\Omega}'(\overline{z}) - \overline{\omega}(\overline{z}) \tag{1.5.45}$$

where ω is an arbitrary analytic complex function on z. The relation 1.5.45 gives a representation of complex displacement in terms of the complex analytic functions Ω and ω.

A simple calculation shows that the constitutive equations 1.5.37 may be written as

$$\begin{aligned} t_{11} + t_{22} &= 2[\Omega'(z) + \overline{\Omega}'(\overline{z})] \\ t_{11} - t_{22} + 2it_{12} &= -2[z\overline{\Omega}''(\overline{z}) + \overline{\omega}'(\overline{z})] \end{aligned} \tag{1.5.46}$$

The functions Ω and ω are called the *complex potentials.* The representations 1.5.45 and 1.5.46 were deduced by Kolosov [186] (see also Refs. 113, 119, 313, 315, 324).

It follows from Equations 1.1.11 that

$$s_1 + is_2 = (t_{\beta 1} + it_{\beta 2})n_\beta \tag{1.5.47}$$

In view of the relations 1.3.39 and 1.5.33, we obtain

$$n_1 = -\frac{1}{2}i\left(\frac{dz}{ds} - \frac{d\overline{z}}{ds}\right), \qquad n_2 = -\frac{1}{2}\left(\frac{dz}{ds} + \frac{d\overline{z}}{ds}\right) \tag{1.5.48}$$

By Equations 1.5.47 and 1.5.48,

$$2(s_1 + is_2) = -i(t_{11} + t_{22})\frac{dz}{ds} + i(t_{11} - t_{22} + 2it_{12})\frac{d\overline{z}}{ds} \tag{1.5.49}$$

From Equations 1.5.46 and 1.5.49, we get

$$s_1 + is_2 = -i\frac{d}{ds}[\Omega(z) + z\overline{\Omega}'(\overline{z}) + \overline{\omega}(\overline{z})] \tag{1.5.50}$$

Let $\mathcal{R}_\alpha$ be the components of the resultant vector associated to the contour $\mathcal{C}$. It follows from Equation 1.5.50 that

$$\mathcal{R}_1 + i\mathcal{R}_2 = \int_{\mathcal{C}} (s_1 + is_2)ds = -i\{\Omega(z) + z\Omega'(z) + \overline{\omega}(\overline{z})\}_P^P \tag{1.5.51}$$

where $\{g\}_P^P$ denotes the change in value of the function g on passing once round the contour $\mathcal{C}$ in the conventional sense.

Let us investigate the arbitrariness and the structure of complex potentials for several domains of interest. First, we investigate what is the difference in the forms of two sets of potentials (Ω, ω) and (Ω_*, ω_*) that correspond to the same stresses. The relations 1.5.46 demand that

$$\Re e\,[\Omega'(z)] = \Re e\,[\Omega'_*(z)], \qquad \overline{z}\Omega''(z) + \omega'(z) = \overline{z}\Omega''_*(z) + \omega'_*(z) \tag{1.5.52}$$

where $\Re e\,[f]$ denotes the real part of f. From Equation 1.5.52, we conclude that

$$\Omega(z) = \Omega_*(z) + icz + \alpha, \qquad \omega(z) = \omega_*(z) + \beta \tag{1.5.53}$$

where c is a real constant, and α and β are complex constants. If the origin O is taken within Σ_1, the functions Ω and ω will be determined uniquely if c, α, and β are chosen so that

$$\Omega(0) = 0, \qquad \Im m[\Omega'(0)] = 0, \qquad \omega(0) = 0 \tag{1.5.54}$$

Here, $\Im m\,[f]$ denotes the imaginary part of f.

Consider now the situation in which the two sets of potentials correspond to the same displacements. In this case the extent of arbitrariness in choosing the potentials cannot be greater than that indicated in Equation 1.5.53. From Equation 1.5.45, the equality of displacements requires that

$$c = 0, \qquad \kappa\alpha = \overline{\beta} \tag{1.5.55}$$

In this case we can choose α so that

$$\Omega(0) = 0 \tag{1.5.56}$$

We note that in a bounded simply-connected region, Ω and ω are single-valued analytic functions. Let us consider the case when the domain Σ_1 is multiply-connected and bounded.

We assume that the boundary of Σ_1 consists of $m+1$ simple closed contours Γ_k such that the exterior contour Γ_{m+1} contains within it the contours $\Gamma_k, (k = 1, 2, \ldots, m)$ (Figure 1.2).

In what follows we assume that the functions u_α and $t_{\alpha\beta}$ are single-valued. From Equation $1.5.46_1$ we see that the real part of Ω' is single-valued, but, in describing once each interior contour Γ_k, the imaginary part of Ω' acquires a constant increment denoted by $2\pi A_k$. Since the function Ω' acquires the

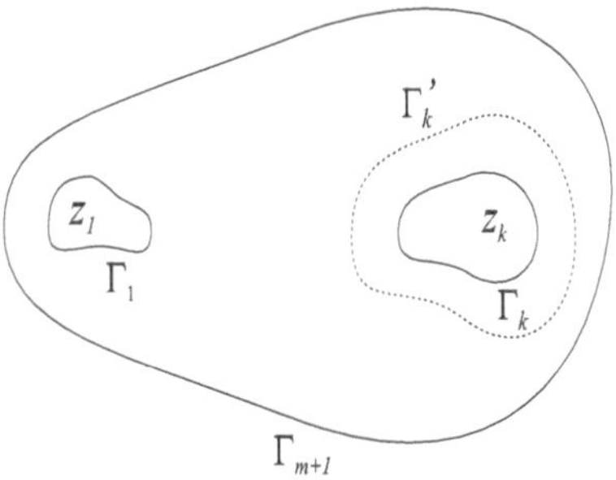

FIGURE 1.2 A multiply-connected domain.

increment $2\pi i A_k$, then the function

$$G(z) = \Omega'(z) - \sum_{k=1}^{m} A_k \log(z - z_k) \tag{1.5.57}$$

where z_k is a point in the simply-connected region S_k, bounded by Γ_k, is single-valued and analytic in Σ_1. By integration of Equation 1.5.57 we get

$$\Omega(z) = z \sum_{k=1}^{m} A_k \log(z - z_k) + \sum_{k=1}^{m} \gamma_k \log(z - z_k) + \Omega_0(z) \tag{1.5.58}$$

where γ_k are complex constants, and Ω_0 is an analytic and single-valued function on Σ_1. Since Ω'' is a single-valued function and the left-hand members of Equations 1.5.46 are single-valued, it follows that ω' is also single-valued on Σ_1. Thus, we have

$$\omega(z) = \sum_{k=1}^{m} C_k \log(z - z_k) + \omega_0(z) \tag{1.5.59}$$

where C_k are complex constants, and ω_0 is analytic and single-valued on Σ_1.

If we assume that u_α are single-valued functions, then from Equations 1.5.45, 1.5.58, and 1.5.59, we find that

$$2\pi i[(1+\kappa)A_k z + \kappa\gamma_k + \overline{C}_k] = 0$$

so that

$$A_k = 0, \qquad \kappa\gamma_k + \overline{C}_k = 0, \quad (k = 1, 2, \ldots, m) \tag{1.5.60}$$

In the case of the second boundary-value problem we denote by (X_k, Y_k) the resultant vector of external forces applied to the contour Γ_k,

$$X_k + iY_k = \int_{L_k} (\tilde{t}_1 + i\tilde{t}_2)ds, \quad (k = 1, 2, \ldots, m) \tag{1.5.61}$$

It follows from Equations 1.5.51, 1.5.58, 1.5.59, and 1.5.60 that

$$X_k + iY_k = -2\pi(\gamma_k - \overline{C}_k) \tag{1.5.62}$$

By Equations 1.5.60 and 1.5.62, we find

$$\gamma_k = -\frac{1}{2\pi(1+\kappa)}(X_k + iY_k), \qquad C_k = \frac{\kappa}{2\pi(1+\kappa)}(X_k - iY_k) \tag{1.5.63}$$

Thus, in this case the complex potentials have the forms

$$\begin{aligned} \Omega(z) &= -\frac{1}{2\pi(1+\kappa)} \sum_{k=1}^{m} (X_k + iY_k) \log(z - z_k) + \Omega_0(z) \\ \omega(z) &= \frac{\kappa}{2\pi(1+\kappa)} \sum_{k=1}^{m} (X_k - iY_k) \log(z - z_k) + \omega_0(z) \end{aligned} \tag{1.5.64}$$

We suppose now that the domain Σ_1 is unbounded, with certain contours $\Gamma_1, \Gamma_2, \ldots, \Gamma_m$ as internal boundaries. We assume that the origin of coordinates is taken outside Σ_1, and that the stresses are bounded in the neighborhood of the point at infinity. We consider the circle $\mathcal{C}_R$ of equation $|z| = R$, and suppose that R is so large that $\mathcal{C}_R$ contains within it the contours Γ_k, $(k = 1, 2, \ldots, m)$. Then, for any z such that $|z| > R$, we have $|z| > |z_k|$, so that

$$\log(z - z_k) = \log z - \frac{z_k}{z} - \frac{1}{2}\left(\frac{z_k}{z}\right)^2 - \cdots = \log z + h(z)$$

where h is a single-valued analytic function in the region $|z| > R$. It follows from Equations 1.5.64 that

$$\begin{aligned} \Omega(z) &= -\frac{1}{2\pi(1+\kappa)}(X + iY)\log z + \Omega^*(z) \\ \omega(z) &= \frac{\kappa}{2\pi(1+\kappa)}(X - iY)\log z + \omega^*(z) \end{aligned} \tag{1.5.65}$$

where

$$X = \sum_{k=1}^{m} X_k, \qquad Y = \sum_{k=1}^{m} Y_k \tag{1.5.66}$$

and Ω^* and ω^* are single-valued analytic functions for $|z| > R$. For sufficiently large $|z|$, the functions Ω^* and ω^* can be represented in the forms

$$\Omega^*(z) = \sum_{-\infty}^{\infty} a_n z^n, \qquad \omega^*(z) = \sum_{-\infty}^{\infty} b_n z^n$$

Since the stresses are bounded at infinity, then

$$\Re e\,\Omega'(z) \quad \text{and} \quad \overline{z}\Omega''(z) + \omega'(z)$$

must be bounded at infinity. It follows that

$$a_n = \overline{a}_n = 0, \qquad b_n = 0 \quad \text{for } n \geq 2$$

Thus, we find

$$\begin{aligned} \Omega(z) &= -\frac{1}{2\pi(1+\kappa)}(X + iY)\log z + (B + iC) + \widetilde{\Omega}(z) \\ \omega(z) &= \frac{\kappa}{2\pi(1+\kappa)}(X - iY)\log z + (B_1 + iC_1)z + \widetilde{\omega}(z) \end{aligned} \tag{1.5.67}$$

where $\widetilde{\Omega}$ and $\widetilde{\omega}$ are single-valued analytic functions on Σ_1 including the point at infinity,

$$\widetilde{\Omega}(z) = \sum_{n=0}^{\infty} \frac{c_n}{z^n}, \qquad \widetilde{\omega}(z) = \sum_{n=0}^{\infty} \frac{b_n}{z^n} \tag{1.5.68}$$

Let $g^{(\infty)}$ be the limiting value of $g(P)$ as the point P tends to infinity. It follows from Equations 1.5.46 that

$$2B - B_1 = t_{11}^{(\infty)}, \qquad 2B + B_1 = t_{22}^{(\infty)}, \qquad C_1 = t_{12}^{(\infty)} \tag{1.5.69}$$

The constant C is related to the rigid rotation at infinity. We introduce the notation

$$\epsilon = \frac{1}{2}(u_{2,1} - u_{1,2}) \tag{1.5.70}$$

It follows from Equations 1.5.36 and 1.5.45 that

$$\epsilon = \Im m\left(\frac{\partial D}{\partial z}\right) = \frac{1+\kappa}{4\pi i}[\Omega'(z) - \overline{\Omega}'(\overline{z})] \tag{1.5.71}$$

By Equations 1.5.67, 1.5.68, and 1.5.71, we get

$$\epsilon^{(\infty)} = \frac{1+\kappa}{2\mu}C \tag{1.5.72}$$

In view of the relations 1.5.67 and 1.5.68, from Equation 1.5.45 we find that

$$\begin{aligned} 2\mu w = {} & -\frac{\kappa}{2\pi(1+\kappa)}(X+iY)\log(z\overline{z}) \\ & + [(\kappa-1)B + i(1+\kappa)C]z - (B_1 - iC_1)\overline{z} + g(z) \end{aligned}$$

where g is bounded at infinity. If the displacements are to be bounded at infinity, then

$$X = Y = 0, \qquad B = C = B_1 = C_1 = 0$$

We note that the requirement for the displacements to be bounded at infinity imply that the stresses vanish at infinity.

Let us show that the boundary-value problems can be reduced to the determination of the functions Ω and ω from prescribed values of certain combinations of these functions on Γ. We consider a generic point $P \in \Gamma$, and denote by $\widehat{x}_\alpha(s)$ the cartesian coordinates of P. Let

$$\sigma = \widehat{x}_1(s) + i\widehat{x}_2(s), \quad s \in [0, s_*] \tag{1.5.73}$$

In the case of the second boundary-value problem, the boundary conditions 1.5.6 can be written in the form

$$s_1 + is_2 = \widetilde{t}_1 + i\widetilde{t}_2 \text{ on } \Gamma$$

In view of Equation 1.5.50, these conditions reduce to

$$\Omega(\sigma) + \sigma\overline{\Omega}'(\overline{\sigma}) + \overline{\omega}(\overline{\sigma}) = T(\sigma) + d \text{ on } \Gamma \tag{1.5.74}$$

where

$$T(\sigma) = T_1(\sigma) + iT_2(\sigma) = i\int_0^s [\widetilde{t}_1(s') + i\widetilde{t}_2(s')]ds', \quad s \in [0, s_*] \tag{1.5.75}$$

and d is an arbitrary complex constant. We saw that the replacement of Ω by $\Omega + icz + \alpha$ and of ω by $\omega + \beta$ does not change the state of stress. The relation 1.5.74 becomes

$$\Omega(\sigma) + \sigma\overline{\Omega}'(\overline{\sigma}) + \overline{\omega}(\overline{\sigma}) + \alpha + \overline{\beta} = T(\sigma) + d \text{ on } \Gamma$$

We can choose α and β so that $\alpha + \overline{\beta} = d$. With this choice we can impose only two conditions: $\Im m\,\{\Omega'(0)\} = 0$ and one of the conditions $\Omega(0) = 0$, $\omega(0) = 0$. If the domain Σ_1 is multiply-connected, the constant d can be set equal to zero on one of the curves forming the boundary of Σ_1. On the remaining curves, the integration constants can be evaluated using the requirement that the displacement be single-valued [113,119,241].

In the case of the first boundary-value problem, from Equations 1.5.5 and 1.5.45, we obtain the following form of the boundary conditions

$$\kappa\Omega(\sigma) - \sigma\overline{\Omega}'(\overline{\sigma}) - \overline{\omega}(\overline{\sigma}) = 2\mu(\widetilde{u}_1 + i\widetilde{u}_2) \text{ on } \Gamma \tag{1.5.76}$$

Thus, the first boundary-value problem is reduced to the finding of the complex analytic functions Ω and ω on Σ_1 which satisfy the boundary condition 1.5.76.

The boundary conditions 1.5.74 and 1.5.76 can be used to obtain Fredholm integral equations for determination of the complex potentials. The existence of the functions Ω and ω which satisfy the above boundary conditions has been investigated in many studies (see, e.g., [241]). Existence theorems for the boundary-value problems of the plane strain problem follow directly from the results presented in Section 4.9.

We now investigate how the relations 1.5.74 and 1.5.76 transform under conformal representation. We suppose that Σ_1 is simply-connected. Let

$$z = \vartheta(\zeta) \tag{1.5.77}$$

be the function that maps Σ_1 on the unit circle $|\zeta| \le 1$. Clearly, $d\vartheta(\zeta)/d\zeta \neq 0$. We introduce the notations

$$\Omega_1(\zeta) = \Omega[\vartheta(\zeta)], \qquad \omega_1(\zeta) = \omega[\vartheta(\zeta)] \tag{1.5.78}$$

Since

$$\Omega'(z) = \frac{1}{\vartheta'(\zeta)}\Omega_1'(\zeta)$$

the relations 1.5.45 and 1.5.50 become

$$2\mu w = \kappa\Omega_1(\zeta) - \frac{\vartheta(\zeta)}{\overline{\vartheta}'(\overline{\zeta})}\overline{\Omega}_1'(\overline{\zeta}) - \overline{\omega}_1(\overline{\zeta})$$

$$s_1 + is_2 = -i\frac{d}{ds}\left[\Omega_1(\zeta) + \frac{\vartheta(\zeta)}{\overline{\vartheta}'(\overline{\zeta})}\overline{\Omega}_1'(\overline{\zeta}) + \overline{\omega}_1(\overline{\zeta})\right], \quad |\zeta| \le 1$$

The conditions 1.5.74 and 1.5.76 become

$$\begin{aligned}
&\Omega_1(\eta) + \frac{\vartheta(\eta)}{\overline{\vartheta}'(\overline{\eta})}\overline{\Omega}_1'(\overline{\eta}) + \overline{\omega}_1(\overline{\eta}) = N_1(\eta) \text{ on } |\eta| = 1 \\
&-\kappa\Omega_1(\eta) + \frac{\vartheta(\eta)}{\overline{\vartheta}'(\overline{\eta})}\overline{\Omega}_1'(\overline{\eta}) + \overline{\omega}_1(\overline{\eta}) = N_2(\eta) \text{ on } |\eta| = 1
\end{aligned} \tag{1.5.79}$$

respectively, where N_α are uniquely determined by the prescribed data. From the relations 1.5.75 and 1.5.77, we get

$$N_1(\eta) = T[\vartheta(\eta)]$$

If Σ_1 is a bounded simply-connected region, then Ω_1 and ω_1 have the representations

$$\Omega_1(\zeta) = \sum_{n=0}^{\infty} a_n \zeta^n, \qquad \omega_1(\zeta) = \sum_{n=0}^{\infty} b_n \zeta^n, \quad |\zeta| \leq 1 \tag{1.5.80}$$

The substitution of Ω_1 and ω_1 from Equations 1.5.80 into 1.5.79 leads to a system of equations for the coefficients a_n and b_n.

An account of the historical development of the complex variable technique as well as references to various contributions may be found in the works of Muskhelishvili [241], Green and Zerna [113], and Gurtin [119].

1.6 Properties of Solutions to Saint-Venant's Problem

In what follows we denote by (P) the Saint-Venant's problem corresponding to the resultants **F** and **M**. Let $K(\mathbf{F}, \mathbf{M})$ denote the class of solutions to the problem (P). The classification of the problem rests on various assumptions concerning the resultants **F** and **M**. Throughout this section it is convenient to use the decomposition of the problem into problems (P_1) and (P_2) characterized by

(P_1) (extension–bending–torsion): $F_\alpha = 0$
(P_2) (flexure): $F_3 = M_i = 0$

For further economy it is helpful to denote by $K_I(F_3, M_1, M_2, M_3)$ the class of solutions to the problem (P_1) and by $K_{II}(F_1, F_2)$ the class of solutions to the problem (P_2). We assume for the remainder of this chapter that the material is homogeneous and isotropic.

We denote by $\mathscr{D}$ the set of all equilibrium displacement fields **u** that satisfy the condition $\mathbf{s}(\mathbf{u}) = \mathbf{0}$ on the lateral boundary. Theorem 1.6.1 will be of future use.

Theorem 1.6.1 ([159]). *If* $\mathbf{u} \in \mathscr{D}$ *and* $\mathbf{u}_{,3} \in C^1(\overline{B})$, *then* $\mathbf{u}_{,3} \in \mathscr{D}$ *and*

$$\mathbf{R}(\mathbf{u}_{,3}) = 0, \qquad H_\alpha(\mathbf{u}_{,3}) = \varepsilon_{\alpha\beta} R_\beta(\mathbf{u}), \qquad H_3(\mathbf{u}_{,3}) = 0 \tag{1.6.1}$$

Proof. We note that the first assertion follows at once from the fact that $\mathbf{t}(\mathbf{u}_{,3}) = \partial\mathbf{t}(\mathbf{u})/\partial x_3$ and the proposition: if $\mathbf{u}$ is an elastic displacement field corresponding to null body forces, then so also is $\mathbf{u}_{,k} = \partial\mathbf{u}/\partial x_k$ (cf. [119], Section 42). Next, with the aid of the equations of equilibrium 1.1.8 we find that

$$t_{3i}(\mathbf{u}_{,3}) = (t_{3i}(\mathbf{u}))_{,3} = -(t_{\rho i}(\mathbf{u}))_{,\rho}$$

$$\varepsilon_{\alpha\beta} x_\beta t_{33}(\mathbf{u}_{,3}) = -\varepsilon_{\alpha\beta} x_\beta (t_{\rho 3}(\mathbf{u}))_{,\rho} = -\varepsilon_{\alpha\beta}[(x_\beta t_{\rho 3}(\mathbf{u}))_{,\rho} - t_{\beta 3}(\mathbf{u})]$$

$$\varepsilon_{\alpha\beta} x_\alpha t_{3\beta}(\mathbf{u}_{,3}) = -\varepsilon_{\alpha\beta} x_\alpha (t_{\rho\beta}(\mathbf{u}))_{,\rho} = -\varepsilon_{\alpha\beta}(x_\alpha t_{\rho\beta}(\mathbf{u}))_{,\rho} + \varepsilon_{\alpha\beta} t_{\alpha\beta}(\mathbf{u})$$

In view of Equations 1.2.5, the divergence theorem, and the symmetry of $\mathbf{S}$, we find

$$\mathbf{R}(\mathbf{u}_{,3}) = \int_\Gamma \mathbf{s}(\mathbf{u}) ds$$

$$H_\alpha(\mathbf{u}_{,3}) = \int_\Gamma \varepsilon_{\alpha\beta} x_\beta s_3(\mathbf{u}) ds + \varepsilon_{\alpha\beta} R_\beta(\mathbf{u}) \tag{1.6.2}$$

$$H_3(\mathbf{u}_{,3}) = \int_\Gamma \varepsilon_{\alpha\beta} x_\alpha s_\beta(\mathbf{u}) ds$$

The desired result follows from Equations 1.6.2 and hypothesis. □

Since $\mathbf{u}$ is an equilibrium displacement field, $\mathbf{u}$ is analytic (cf. [119], Section 42). Theorem 1.6.1 has the following immediate consequences.

Corollary 1.6.1 *If* $\mathbf{u} \in K_I(F_3, M_1, M_2, M_3)$ *and* $\mathbf{u}_{,3} \in C^1(\overline{B})$, *then* $\mathbf{u}_{,3} \in \mathscr{D}$ *and*

$$\mathbf{R}(\mathbf{u}_{,3}) = \mathbf{0}, \qquad \mathbf{H}(\mathbf{u}_{,3}) = \mathbf{0}$$

Corollary 1.6.2 *If* $\mathbf{u} \in K_{II}(F_1, F_2)$ *and* $\mathbf{u}_{,3} \in C^1(\overline{B})$, *then*

$$\mathbf{u}_{,3} \in K_I(0, F_2, -F_1, 0)$$

Corollary 1.6.3 *If* $\mathbf{u} \in \mathscr{D}$ *and* $\partial^n\mathbf{u}/\partial x_3^n \in C^1(\overline{B})$, *then* $\partial^n\mathbf{u}/\partial x_3^n \in \mathscr{D}$ *and*

$$\mathbf{R}\left(\frac{\partial^n \mathbf{u}}{\partial x_3^n}\right) = \mathbf{0}, \qquad \mathbf{R} = \mathbf{H}\left(\frac{\partial^n \mathbf{u}}{\partial x_3^n}\right) = \mathbf{0} \quad \text{for } n \geq 2$$

1.7 New Method of Solving Saint-Venant's Problem

In this section, we shall prove that Corollary 1.6.1 allows us to establish a simple method of deriving Saint-Venant's solution to the problem (P_1). We denote by Q the class of solutions to the Saint-Venant's problem corresponding to $\mathbf{F} = \mathbf{0}$ and $\mathbf{M} = \mathbf{0}$. We note that if $\mathbf{u} \in K_I(F_3, M_1, M_2, M_3)$ and $\mathbf{u}_{,3} \in C^1(\overline{B})$, then by Corollary 1.6.1, $\mathbf{u}_{,3} \in Q$. Let us note that a rigid displacement field belongs to Q. It is natural to enquire whether there exists a solution $\mathbf{v}$ of the problem (P_1) such that $\mathbf{v}_{,3}$ is a rigid displacement field. This question is settled in Theorem 1.7.1.

Theorem 1.7.1 *Let $\mathbf{v} \in C^1(\overline{B}) \cap C^2(B)$ be a vector field such that $\mathbf{v}_{,3}$ is a rigid displacement field. Then $\mathbf{v}$ is a solution of the problem (P_1) if and only if $\mathbf{v}$ is Saint-Venant's solution.*

Proof. We suppose that $\mathbf{v} \in C^1(\overline{B}) \cap C^2(B)$ is a vector field such that

$$\mathbf{v}_{,3} = \boldsymbol{\alpha} + \boldsymbol{\beta} \times \mathbf{x} \tag{1.7.1}$$

where $\boldsymbol{\alpha}$ and $\boldsymbol{\beta}$ are constant vectors. Then,

$$\begin{aligned} v_\alpha &= -\frac{1}{2}a_\alpha x_3^2 - a_4\varepsilon_{\alpha\beta}x_\beta x_3 + w_\alpha(x_1, x_2) \\ v_3 &= (a_1x_1 + a_2x_2 + a_3)x_3 + w_3(x_1, x_2) \end{aligned} \tag{1.7.2}$$

except for an additive rigid displacement field. In Equations 1.7.2 $\mathbf{w}$ is an arbitrary vector field independent of x_3, and we have used the notations $a_\alpha = \varepsilon_{\rho\alpha}\beta_\rho, a_3 = \alpha_3, a_4 = \beta_3$. Let us prove that the functions w_i and the constants $a_s, (s = 1, 2, 3, 4)$ can be determined so that $\mathbf{v} \in K_I(F_3, M_1, M_2, M_3)$. The stress–displacement relations imply that

$$\begin{aligned} t_{\alpha\beta}(\mathbf{v}) &= \lambda(a_\rho x_\rho + a_3)\delta_{\alpha\beta} + T_{\alpha\beta}(\mathbf{w}) \\ t_{3\alpha}(\mathbf{v}) &= \mu(w_{3,\alpha} - a_4\varepsilon_{\alpha\rho}x_\rho) \\ t_{33}(\mathbf{v}) &= (\lambda + 2\mu)(a_\rho x_\rho + a_3) + \lambda w_{\rho,\rho} \end{aligned} \tag{1.7.3}$$

where

$$T_{\alpha\beta}(\mathbf{w}) = \mu(w_{\alpha,\beta} + w_{\beta,\alpha}) + \lambda\delta_{\alpha\beta}w_{\rho,\rho} \tag{1.7.4}$$

The equations of equilibrium and the conditions on the lateral boundary reduce to

$$(T_{\alpha\beta}(\mathbf{w}))_{,\beta} + f_\alpha = 0 \text{ on } \Sigma_1, \qquad T_{\alpha\beta}(\mathbf{w})n_\beta = p_\alpha \text{ on } \Gamma \tag{1.7.5}$$

$$\Delta w_3 = 0 \text{ on } \Sigma_1, \qquad \frac{\partial w_3}{\partial n} = a_4\varepsilon_{\alpha\beta}n_\alpha x_\beta \text{ on } \Gamma \tag{1.7.6}$$

where

$$f_\alpha = \lambda a_\alpha, \qquad p_\alpha = -\lambda(a_\rho x_\rho + a_3)n_\alpha \tag{1.7.7}$$

Clearly, Equations 1.7.4, 1.7.5, and 1.7.7 constitute a two-dimensional boundary-value problem (cf. Section 1.6). The necessary and sufficient conditions to solve this problem are

$$\int_{\Sigma_1} f_\alpha da + \int_\Gamma p_\alpha ds = 0, \qquad \int_{\Sigma_1} \varepsilon_{\alpha\beta}x_\alpha f_\beta da + \int_\Gamma \varepsilon_{\alpha\beta}x_\alpha p_\beta ds = 0 \tag{1.7.8}$$

From Equations 1.7.7 and the divergence theorem, we see that the conditions 1.7.8 are satisfied. We note that the boundary-value problem 1.7.5 is satisfied if one chooses

$$T_{\alpha\beta}(\mathbf{w}) = -\lambda(a_\rho x_\rho + a_3)\delta_{\alpha\beta}$$

The above stresses satisfy the compatibility condition. It follows from Equations 1.7.4 that

$$w_{1,1} = w_{2,2} = -\frac{\lambda}{2(\lambda+\mu)}(a_\rho x_\rho + a_3), \qquad w_{1,2} + w_{2,1} = 0$$

The integration of these equations yields

$$w_\alpha = a_1 w_\alpha^{(1)} + a_2 w_\alpha^{(2)} + a_3 w_\alpha^{(3)}$$

where

$$w_\alpha^{(\beta)} = \nu\left(\frac{1}{2}x_\rho x_\rho \delta_{\alpha\beta} - x_\alpha x_\beta\right), \qquad w_\alpha^{(3)} = -\nu x_\alpha \tag{1.7.9}$$

modulo a plane rigid displacement. Here ν designates Poisson's ratio defined in Equations 1.1.7. It follows from Equations 1.7.6 that $w_3 = a_4\varphi$, where φ is the torsion function, characterized by

$$\Delta\varphi = 0 \text{ on } \Sigma_1, \qquad \frac{\partial\varphi}{\partial n} = \varepsilon_{\alpha\beta} n_\alpha x_\beta \text{ on } \Gamma \tag{1.7.10}$$

The vector field $\mathbf{v}$ can be written in the form

$$\mathbf{v} = \sum_{j=1}^{4} a_j \mathbf{v}^{(j)} \tag{1.7.11}$$

where the vectors $\mathbf{v}^{(j)}, (j = 1, 2, 3, 4)$, are defined by

$$\begin{aligned} &v_\alpha^{(\beta)} = -\frac{1}{2}x_3^2\delta_{\alpha\beta} + w_\alpha^{(\beta)}, \qquad v_3^{(\beta)} = x_\beta x_3, \quad (\beta = 1, 2) \\ &v_\alpha^{(3)} = w_\alpha^{(3)}, \qquad v_3^{(3)} = x_3, v_\alpha^{(4)} = \varepsilon_{\beta\alpha} x_\beta x_3, \qquad v_3^{(4)} = \varphi \end{aligned} \tag{1.7.12}$$

It is easy to see that $\mathbf{v}^{(j)} \in \mathscr{D}, (j = 1, 2, 3, 4)$. The conditions on the end Σ_1 furnish the following system for the unknown constants

$$\begin{aligned} &E(I_{\alpha\beta} a_\beta + A x_\alpha^0 a_3) = \varepsilon_{\alpha\beta} M_\beta \\ &EA(a_1 x_1^0 + a_2 x_2^0 + a_3) = -F_3, \qquad D a_4 = -M_3 \end{aligned} \tag{1.7.13}$$

where A is the area of the cross section, x_α^0 are the coordinates of the centroid of Σ_1, E designates Young's modulus, D is the torsional rigidity defined by Equation 1.3.32, and

$$I_{\alpha\beta} = \int_{\Sigma_1} x_\alpha x_\beta da \tag{1.7.14}$$

If the rectangular cartesian coordinate frame is chosen in such a way that the x_α-axes are principal centroidal axes of the cross section Σ_1, then Equations 1.7.11 and 1.7.13 lead to the Saint-Venant's solutions presented in Section 1.3. □

We present Saint-Venant's solution which are needed subsequently.

1. *Saint-Venant's extension solution:*

$$\begin{array}{lll} \mathbf{v} = a_3\mathbf{v}^{(3)}, & v_\alpha^{(3)} = -\nu x_\alpha, & v_3^{(3)} = x_3 \\ t_{\alpha\beta}(\mathbf{v}) = 0, & t_{3\alpha}(\mathbf{v}) = 0, & t_{33}(\mathbf{v}) = Ea_3 \end{array} \tag{1.7.15}$$

where

$$F_3 = -EAa_3 \tag{1.7.16}$$

The relation 1.7.16 is known as *Saint-Venant's formula for extension.*

2. *Saint-Venant's bending solution:*

$$\begin{gathered} \mathbf{v} = a_1\mathbf{v}^{(1)}, \qquad v_1^{(1)} = \frac{1}{2}\left(\nu x_2^2 - \nu x_1^2 - x_3^2\right) \\ v_2^{(1)} = -\nu x_1 x_2, \qquad v_3^{(1)} = x_1 x_3 \\ t_{\alpha\beta}(\mathbf{v}) = 0, \qquad t_{3\alpha}(\mathbf{v}) = 0, \qquad t_{33}(\mathbf{v}) = Ea_1 x_1 \end{gathered} \tag{1.7.17}$$

where

$$M_2 = EI_{11}a_1 \tag{1.7.18}$$

The relation 1.7.18 is called *Saint-Venant's formula for bending.*

3. *Saint-Venant's torsion solution:*

$$\begin{array}{lll} \mathbf{v} = a_4\mathbf{v}^{(4)}, & v_\alpha^{(4)} = \varepsilon_{\beta\alpha}x_\beta x_3, & v_3^{(4)} = \varphi \\ t_{\alpha\beta}(\mathbf{v}) = 0, & t_{33}(\mathbf{v}) = 0, & t_{3\alpha}(\mathbf{v}) = \mu a_4(\varphi_{,\alpha} - \varepsilon_{\alpha\rho}x_\rho) \end{array} \tag{1.7.19}$$

where

$$M_3 = -Da_4 \tag{1.7.20}$$

The relation 1.7.20 is known as *Saint-Venant's formula for torsion.*

We note that the vectors $\mathbf{v}^{(j)}, (j = 1, 2, 3, 4)$, defined by the relations 1.7.12 depend only on the cross section and the elasticity field. Let $\widehat{a}$ be the four-dimensional vector (a_1, a_2, a_3, a_4). We will write $\mathbf{v}\{\widehat{a}\}$ for the displacement vector $\mathbf{v}$ defined by Equations 1.7.11 and 1.7.12, indicating thus its dependence on the constants $a_s, (s = 1, 2, 3, 4)$.

On the basis of Corollaries 1.6.1 and 1.6.2 and Theorem 1.7.1, it is natural to seek a solution of the problem (P_2) in the form

$$\mathbf{u}^0 = \int_0^{x_3} \mathbf{v}\{\widehat{b}\}dx_3 + \mathbf{v}\{\widehat{c}\} + \mathbf{w}^0 \tag{1.7.21}$$

where $\widehat{b}=(b_1,b_2,b_3,b_4)$ and $\widehat{c}=(c_1,c_2,c_3,c_4)$ are two constant four-dimensional vectors, and $\mathbf{w}^0$ is a vector field independent of x_3 such that $\mathbf{w}^0 \in C^1(\overline{\Sigma}_1) \cap C^2(\Sigma_1)$.

Theorem 1.7.2 *The vector field* $\mathbf{u}^0$ *defined by* 1.7.21 *is a solution of the problem* (P_2) *if and only if* $\mathbf{u}^0$ *is Saint-Venant's solution.*

Proof. We have to prove that the vector field $\mathbf{w}^0$ and the constants $b_s, c_s, (s = 1,2,3,4)$, can be determined so that $\mathbf{u}^0 \in K_{II}(F_1,F_2)$. It is interesting to note that the determination of $\widehat{b}$ from the condition $\mathbf{u}^0 \in K_{II}(F_1,F_2)$ can be made in a simple way. Thus, if $\mathbf{u}^0 \in K_{II}(F_1,F_2)$, then by Corollary 1.6.2 and Equation 1.7.21

$$\mathbf{v}\{\widehat{b}\} \in K_I(0,F_2,-F_1,0) \tag{1.7.22}$$

With the help of Equations 1.7.13 and 1.7.22, we get

$$E(I_{\alpha\beta}b_\beta + ax_\alpha^0 b_3) = -F_\alpha, \qquad b_\rho x_\rho^0 + b_3 = 0, \qquad b_4 = 0 \tag{1.7.23}$$

From Equations 1.7.11, 1.7.12, 1.7.21, and 1.7.23, we obtain

$$\begin{aligned} u_\alpha^0 &= -\frac{1}{6}b_\alpha x_3^3 - \frac{1}{2}c_\alpha x_3^2 - c_4\varepsilon_{\alpha\beta}x_\beta x_3 + \sum_{j=1}^{3}(c_j + x_3 b_j)w_\alpha^{(j)} + w_\alpha^0 \\ u_3^0 &= \frac{1}{2}(b_\rho x_\rho + b_3)x_3^2 + (c_\rho x_\rho + c_3)x_3 + c_4\varphi + \psi \end{aligned}$$

where we have used the notation $w_3^0 = \psi$. The stress–displacement relations imply that

$$\begin{aligned} t_{\alpha\beta}(\mathbf{u}^0) &= T_{\alpha\beta}(\mathbf{w}^0) \\ t_{\alpha 3}(\mathbf{u}^0) &= \mu\left[c_4(\varphi_{,\alpha} - \varepsilon_{\alpha\beta}x_\beta) - \nu x_\alpha(b_\rho x_\rho + b_3) + \frac{1}{2}b_\alpha \nu x_\rho x_\rho + \psi_{,\alpha}\right] \\ t_{33}(\mathbf{u}^0) &= E[(b_\rho x_\rho + b_3)x_3 + c_\rho x_\rho + c_3] + \lambda w_{\rho,\rho}^0 \end{aligned}$$

where

$$T_{\alpha\beta}(\mathbf{w}^0) = \mu(w_{\alpha,\beta}^0 + w_{\beta,\alpha}^0) + \lambda\delta_{\alpha\beta}w_{\rho,\rho}^0 \tag{1.7.24}$$

The equations of equilibrium and the conditions on the lateral boundary reduce to

$$(T_{\alpha\beta}(\mathbf{w}^0))_{,\beta} = 0 \text{ on } \Sigma_1, \qquad T_{\alpha\beta}(\mathbf{w}^0)n_\beta = 0 \text{ on } \Gamma \tag{1.7.25}$$

$$\begin{aligned} \Delta\psi &= -2(b_\rho x_\rho + b_3) \text{ on } \Sigma_1 \\ \frac{\partial\psi}{\partial n} &= b_\alpha \nu x_\rho\left(x_\alpha n_\rho - \frac{1}{2}n_\alpha x_\rho\right) + b_3\nu x_\alpha n_\alpha \text{ on } \Gamma \end{aligned} \tag{1.7.26}$$

We see from Equations 1.7.24 and 1.7.25 that w_α^0 and $T_{\alpha\beta}(\mathbf{w}^0)$ characterize a plane elastic state corresponding to zero body forces and null boundary data.

We conclude that $w_\alpha^0 = 0$ (modulo a plane rigid displacement). Thus, the equations of equilibrium and the conditions on the lateral boundary are satisfied if and only if the function ψ is characterized by Equations 1.7.26 and $w_\alpha^0 = 0$. The necessary and sufficient condition to solve the boundary-value problem 1.7.26 is satisfied on the basis of the second relation of Equation 1.7.23.

The conditions $R_3(\mathbf{u}^0) = 0$ and $\mathbf{H}(\mathbf{u}^0) = \mathbf{0}$ are satisfied if

$$\begin{gathered} I_{\alpha\beta}c_\beta + Ax_\alpha^0 c_3 = 0, \qquad c_\rho x_\rho^0 + c_3 = 0 \\ Dc_4 = -\mu\int_{\Sigma_1} \varepsilon_{\alpha\beta}x_\alpha\left(\psi_{,\beta} + \frac{1}{2}b_\beta\nu x_\rho x_\rho\right)da \end{gathered} \tag{1.7.27}$$

Since $H_\alpha(\mathbf{u}^0, 3) = \varepsilon_{\alpha\beta}R_\beta(\mathbf{u}^0)$ and $\mathbf{u}_{,3}^0 = \mathbf{v}\{\widehat{b}\} \in K_I(0, F_2, -F_1, 0)$, it follows that $R_\alpha(\mathbf{u}^0) = F_\alpha$. We conclude that $\widehat{b}$ is determined by Equations 1.7.23, ψ is characterized by Equations 1.7.26, $c_i = 0$, and c_4 is given by Equations 1.7.27. If the rectangular cartesian coordinate frame is chosen in such a way that x_α-axes are principal centroidal axes of the cross section Σ_1, then $\mathbf{u}^0$ reduces to Saint-Venant's solution. □

We have established Equations 1.7.27 from the conditions $R_3(\mathbf{u}^0) = 0$ and $\mathbf{H}(\mathbf{u}^0) = \mathbf{0}$. If we replace these conditions by $R_3(\mathbf{u}^0) = F_3$ and $\mathbf{H}(\mathbf{u}^0) = \mathbf{M}$, then we arrive at

$$\begin{gathered} E(I_{\alpha\beta}c_\beta + Ax_\alpha^0 c_3) = \varepsilon_{\alpha\beta}M_\beta, \qquad AE(c_\rho x_\rho + c_3) = -F_3 \\ Dc_4 = -M_3 - \mu\int_{\Sigma_1} \varepsilon_{\alpha\beta}x_\alpha\left(\psi_{,\beta} + \frac{1}{2}b_\beta\nu x_\rho x_\rho\right)da \end{gathered} \tag{1.7.28}$$

If $\widehat{b}$ is given by Equations 1.7.23, ψ is characterized by Equations 1.7.26, and $\widehat{c}$ is determined by Equations 1.7.28, then $\mathbf{u}^0 \in K(\mathbf{F}, \mathbf{M})$. Thus, we have the following result.

Theorem 1.7.3 *The vector field* $\mathbf{u}^0$ *defined by Equation* 1.7.21 *is a solution of the problem* (P) *if and only if* $\mathbf{u}^0$ *is Saint-Venant's solution.*

Theorem 1.7.4 presents a property of solutions of the problem of flexure.

Theorem 1.7.4 *Let* $\mathbf{u}$ *be a solution of the problem* (P_2). *Then* $\mathbf{u}$ *admits the decomposition*

$$\mathbf{u} = \mathbf{u}' + \mathbf{u}'' \tag{1.7.29}$$

where $\mathbf{u}' \in \mathscr{D}$, $\mathbf{u}'_{,3} \in K_I(0, F_2, -F_1, 0)$ *and*

$$\mathbf{u}'' \in K_I(-R_3(\mathbf{u}'), -H_1(\mathbf{u}'), -H_2(\mathbf{u}'), -H_3(\mathbf{u}'))$$

Proof. We suppose that $\mathbf{u}' \in \mathscr{D}, \mathbf{u}'_{,3} \in K_I(0, F_2, -F_1, 0)$. In view of Theorem 1.6.1, we find

$$R_\alpha(\mathbf{u}') = \varepsilon_{\beta\alpha}H_\beta(\mathbf{u}'_{,3}) = F_\alpha$$

We consider $\mathbf{u} \in K_{II}(F_1, F_2)$. If we define $\mathbf{u}''$ by $\mathbf{u}'' = \mathbf{u} - \mathbf{u}'$, then $\mathbf{u}'' \in \mathscr{D}$ and

$$R_\alpha(\mathbf{u}'') = R_\alpha(\mathbf{u}) - R_\alpha(\mathbf{u}') = 0$$
$$R_3(\mathbf{u}'') = -R_3(\mathbf{u}'), \quad \mathbf{H}(\mathbf{u}'') = -\mathbf{H}(\mathbf{u}')$$

We conclude that the decomposition 1.7.29 holds. □

We assume for the remainder of this chapter that the x_α-axes are principal centroidal axes of Σ_1. In this case, from $\mathbf{u}^0 \in K_{II}(F, 0)$ and Equations 1.7.23, it follows that $b_1 = b, b_2 = b_3 = b_4 = 0$, where b is given by

$$F = -EI_{11}b \tag{1.7.30}$$

This is *Saint-Venant's formula for flexure.*

This method of deriving Saint-Venant's solutions has been established in Ref. 159.

1.8 Minimum Energy Characterizations of Solutions

In Ref. 322, Sternberg and Knowles have characterized Saint-Venant's solutions in terms of certain associated minimum strain-energy properties. Thus, the extension and bending solutions are uniquely determined by the fact that they render the total strain energy an absolute minimum over that subset of the solutions to the respective relaxed problem which results from holding the resultant load or bending couple fixed and from requiring the shearing tractions to vanish pointwise on the ends of the cylinder. Similarly, among all solutions of the torsion problem that correspond to a fixed torque and to vanishing normal tractions on the ends of the cylinder, Saint-Venant's solution is uniquely distinguished by the fact that it furnishes the absolute minimum of the total strain energy. Other results concerning the status of Saint-Venant's solutions as minimizers of energy have been established by Maisonneuve [213] and Ericksen [80]. In this section, we present the result of Sternberg and Knowles [322] concerning the minimum strain-energy characterizations of Saint-Venant's extension, bending, and torsion solutions.

Let Y_E denote the set of all equilibrium displacement fields $\mathbf{u}$ that satisfy the conditions

$$\mathbf{s}(\mathbf{u}) = \mathbf{0} \text{ on } \Pi, \qquad t_{3\beta}(\mathbf{u}) = 0 \text{ on } \Sigma_\alpha, \qquad R_3(\mathbf{u}) = F \tag{1.8.1}$$

Theorem 1.8.1 *Let* $\mathbf{v}$ *be Saint-Venant's extension solution corresponding to a scalar load* F. *Then*

$$U(\mathbf{v}) \leq U(\mathbf{u})$$

for every $\mathbf{u} \in Y_E$, *and equality holds only if* $\mathbf{u} = \mathbf{v}$ *modulo a rigid displacement.*

Proof. We consider $\mathbf{u} \in Y_E$ and define

$$\mathbf{u}' = \mathbf{u} - \mathbf{v} \tag{1.8.2}$$

Then $\mathbf{u}'$ is an equilibrium displacement field that satisfies

$$\mathbf{s}(\mathbf{u}') = \mathbf{0} \text{ on } \Pi, \qquad t_{3\beta}(\mathbf{u}') = 0 \text{ on } \Sigma_\alpha, \qquad R_3(\mathbf{u}') = 0 \tag{1.8.3}$$

From Equations 1.1.12, 1.1.14, and 1.8.2, we get

$$U(\mathbf{u}) = U(\mathbf{u}') + U(\mathbf{v}) + \langle \mathbf{u}', \mathbf{v} \rangle$$

It follows from Equations 1.1.16, 1.1.17, 1.2.6, 1.7.15, and 1.8.3 that

$$\langle \mathbf{u}', \mathbf{v} \rangle = \int_{\Sigma_2} t_{3i}(\mathbf{u}') v_i da - \int_{\Sigma_1} t_{3i}(\mathbf{u}') v_i da = -a_3 h R_3(\mathbf{u}') = 0$$

Thus $U(\mathbf{u}) \geq U(\mathbf{v})$ and $U(\mathbf{u}) = U(\mathbf{v})$ only if $\mathbf{u}'$ is a rigid displacement. □

We denote by Y_B the set of all equilibrium displacement fields $\mathbf{u}$ that satisfy the conditions

$$\mathbf{s}(\mathbf{u}) = \mathbf{0} \text{ on } \Pi, \qquad t_{3\beta}(\mathbf{u}) = 0 \text{ on } \Sigma_\alpha, \qquad H_2(\mathbf{u}) = M_2 \tag{1.8.4}$$

Theorem 1.8.2 *Let* $\mathbf{v}$ *be Saint-Venant's bending solution corresponding to a couple of scalar moment* M_2*. Then*

$$U(\mathbf{v}) \leq U(\mathbf{u})$$

for every $\mathbf{u} \in Y_B$*, and equality holds only if* $\mathbf{u} = \mathbf{v}$ *modulo a rigid displacement.*

Proof. We consider $\mathbf{u} \in Y_B$. Since $\mathbf{v} \in Y_B$ it follows that the field

$$\mathbf{u}' = \mathbf{u} - \mathbf{v}$$

is an equilibrium displacement field that satisfies

$$\mathbf{s}(\mathbf{u}') = \mathbf{0} \text{ on } \Pi, \qquad t_{3\beta}(\mathbf{u}') = 0 \text{ on } \Sigma_\alpha, \qquad H_2(\mathbf{u}') = 0 \tag{1.8.5}$$

With the help of Equations 1.1.16, 1.1.17, 1.2.6, 1.7.17, and 1.8.5, we find

$$\langle \mathbf{u}', \mathbf{v} \rangle = \int_{\Sigma_2} t_{33}(\mathbf{u}') v_3 da - \int_{\Sigma_1} t_{33}(\mathbf{u}') v_3 da = h a_1 H_2(\mathbf{u}') = 0$$

Thus,

$$U(\mathbf{u}) = U(\mathbf{u}') + U(\mathbf{v})$$

The conclusion is immediate. □

It is a simple matter to verify that the above minimum strain-energy characterizations also hold if the conditions

$$t_{3\beta}(\mathbf{u}) = 0 \text{ on } \Sigma_\alpha$$

which appear in Equations 1.8.1 and 1.8.4 are replaced by

$$R_\alpha(\mathbf{u}) = 0, \qquad [t_{3\beta}(\mathbf{u})](x_1, x_2, h) = [t_{3\beta}(\mathbf{u})](x_1, x_2, 0), \qquad (x_1, x_2) \in \Sigma_1$$

We denote by Y_T the set of all equilibrium displacement fields $\mathbf{u}$ that satisfy the conditions

$$\mathbf{s}(\mathbf{u}) = \mathbf{0} \text{ on } \Pi, \qquad t_{33}(\mathbf{u}) = 0 \text{ on } \Sigma_\alpha, \qquad H_3(\mathbf{u}) = M_3 \tag{1.8.6}$$

Theorem 1.8.3 *Let* $\mathbf{v}$ *be Saint-Venant's torsion solution corresponding to the scalar torque* M_3. *Then*

$$U(\mathbf{v}) \leq U(\mathbf{u})$$

for every $\mathbf{u} \in Y_T$, *and equality holds only if* $\mathbf{u} = \mathbf{v}$ *modulo a rigid displacement.*

Proof. Clearly, $\mathbf{v} \in Y_T$. We consider $\mathbf{u} \in Y_T$, and define $\mathbf{u}'$ by $\mathbf{u}' = \mathbf{u} - \mathbf{v}$. Then $\mathbf{u}'$ is an equilibrium displacement field such that

$$\mathbf{s}(\mathbf{u}') = \mathbf{0} \text{ on } \Pi, \qquad t_{33}(\mathbf{u}') = 0 \text{ on } \Sigma_\alpha, \qquad H_3(\mathbf{u}') = 0 \tag{1.8.7}$$

If we apply Equations 1.1.16 and 1.1.17, we conclude, with the aid of Equations 1.8.7 and 1.7.19, that

$$\langle \mathbf{u}', \mathbf{v} \rangle = -a_4 h \int_{\Sigma_2} \varepsilon_{\alpha\beta} x_\beta t_{3\alpha}(\mathbf{u}') da = -a_4 h H_3(\mathbf{u}') = 0$$

Thus,

$$U(\mathbf{u} - \mathbf{v}) = U(\mathbf{u}) - U(\mathbf{v}) \tag{1.8.8}$$

The proof follows from Equation 1.8.8. □

If we replace in Equations 1.8.6 the conditions

$$t_{33}(\mathbf{u}) = 0 \text{ on } \Sigma_\alpha$$

by

$$[t_{33}(\mathbf{u})](x_1, x_2, h) = [t_{33}(\mathbf{u})](x_1, x_2, 0), \quad (x_1, x_2) \in \Sigma_1$$

the above theorem also remains valid.

We denote by Y_F the set of all equilibrium displacement fields $\mathbf{u}$ that satisfy the conditions

$$\begin{gathered} \mathbf{u}_{,3} \in C^1(\overline{B}), \qquad \mathbf{s}(\mathbf{u}) = \mathbf{0} \text{ on } \Pi, \qquad R_\alpha(\mathbf{u}) = F_\alpha \\ [t_{3\beta}(\mathbf{u}_{,3})](x_1, x_2, h) = [t_{3\beta}(\mathbf{u}_{,3})](x_1, x_2, 0), \quad (x_1, x_2) \in \Sigma_1 \end{gathered} \tag{1.8.9}$$

Theorem 1.8.4 *Let $\mathbf{u}^0$ be Saint-Venant's flexure solution corresponding to the scalar loads F_1 and F_2. Then*

$$U(\mathbf{u}^0_{,3}) \leq U(\mathbf{u}_{,3})$$

for every $\mathbf{u} \in Y_F$, and equality holds only if $\mathbf{u}_{,3} = \mathbf{u}^0_{,3}$.

Proof. We consider $\mathbf{u} \in Y_F$ and define $\mathbf{u}' = \mathbf{u} - \mathbf{u}^0$. Then $\mathbf{u}'$ is an equilibrium displacement field that satisfies

$$\begin{gathered} \mathbf{u}'_{,3} \in C^1(\overline{B}), \qquad \mathbf{s}(\mathbf{u}') = \mathbf{0} \text{ on } \Pi, \qquad R_\alpha(\mathbf{u}') = 0 \\ [t_{3\beta}(\mathbf{u}'_{,3})](x_1, x_2, h) = [t_{3\beta}(\mathbf{u}'_{,3})](x_1, x_2, 0), \quad (x_1, x_2) \in \Sigma_1 \end{gathered} \tag{1.8.10}$$

With the help of Equations 1.1.12 and 1.7.21 and Theorem 1.7.1, we find

$$U(\mathbf{u}_{,3}) = U(\mathbf{u}'_{,3} + \mathbf{u}^0_{,3}) = U(\mathbf{u}'_{,3} + \mathbf{v}\{\widehat{b}\}) = U(\mathbf{u}'_{,3}) + U(\mathbf{u}^0_{,3}) + \langle \mathbf{u}'_{,3}, \mathbf{v}\{\widehat{b}\}\rangle.$$

On the basis of Theorem 1.6.1 and Equations 1.2.6 and 1.8.10, we get

$$\langle \mathbf{u}'_{,3}, \mathbf{v}\{\widehat{b}\}\rangle = -\frac{1}{2} b_\alpha h^2 R_\alpha(\mathbf{u}'_{,3}) + h[b_1 H_2(\mathbf{u}'_{,3}) - b_2 H_1(\mathbf{u}'_{,3})] = 0$$

Thus,

$$U(\mathbf{u}_{,3}) - U(\mathbf{u}^0_{,3}) = U(\mathbf{u}_{,3} - \mathbf{v}\{\widehat{b}\})$$

The desired conclusion is immediate. □

The above results concerning the minimum strain-energy characterizations of Saint-Venant's solutions are based on a comparison with a subset rather than with the complete class of solutions to the corresponding problem. It is natural to seek also those members of the class of solutions to each of the four problems that minimize the strain energy over the complete class of solutions to the corresponding problem.

1.9 Truesdell's Problem

It is well-known that in the Saint-Venant's solution of the torsion problem, corresponding to a couple of scalar moment M_3, the specific angle of twist a_4 is given by Equation 1.7.20. We denote by K_T the set of all displacement fields that correspond to the solutions of the foregoing torsion problem. In Refs. 331, 334, and 336, Truesdell proposed the following problem: to define the functional $\tau(\cdot)$ on K_T such that

$$M_3 = -D\tau(\mathbf{u}), \quad \text{for each } \mathbf{u} \in K_T$$

Following Day [62], $\tau(\mathbf{u})$ is called the generalized twist at $\mathbf{u}$. In Ref. 62, Day established a solution of Truesdell's problem. A study of Truesdell's problem rephrased for extension and bending is presented in Ref. 271. Solution of Truesdell's problem for flexure has been established in Ref. 159. In this section we present these results.

We denote by Q_T the set of all equilibrium displacement fields $\mathbf{u}$ that satisfy the conditions

$$\mathbf{s}(\mathbf{u}) = \mathbf{0} \text{ on } \Pi, \qquad t_{33}(\mathbf{u}) = 0 \text{ on } \Sigma_\alpha, \qquad R_\alpha(\mathbf{u}) = 0, \qquad H_3(\mathbf{u}) = M_3 \tag{1.9.1}$$

If $\mathbf{u} \in Q_T$, then $R_3(\mathbf{u}) = 0, H_\alpha(\mathbf{u}) = 0$, so that $\mathbf{u} \in K_T$. Day [62] considered the real function

$$\alpha \to \|\mathbf{u} - \alpha \mathbf{v}^{(4)}\|_e^2 \tag{1.9.2}$$

where $\mathbf{u} \in Q_T$ and $\mathbf{v}^{(4)}$ is the displacement field given by the relations 1.7.19. The field $\alpha \mathbf{v}^{(4)}$ is called the torsion field with twist α.

The function 1.9.2 attains its minimum at

$$\gamma(\mathbf{u}) = \frac{\langle \mathbf{u}, \mathbf{v}^{(4)} \rangle}{\|\mathbf{v}^{(4)}\|_e^2} \tag{1.9.3}$$

Thus, $\gamma(\mathbf{u})$ is the twist of that torsion field which approximates $\mathbf{u}$ most closely. Let us prove that

$$\gamma(\mathbf{u}) = \tau(\mathbf{u}), \quad \text{for every } \mathbf{u} \in Q_T$$

With the help of Equations 1.1.16, 1.1.17, 1.2.6, 1.7.19, and 1.9.1, we find

$$\begin{aligned}
\langle \mathbf{u}, \mathbf{v}^{(4)} \rangle &= \int_{\partial B} \mathbf{s}(\mathbf{u}) \cdot \mathbf{v}^{(4)} da = \int_{\Sigma_2} [h\varepsilon_{\beta\alpha} x_\beta t_{3\alpha}(\mathbf{u}) + \varphi t_{33}(\mathbf{u})] da \\
&= h \int_{\Sigma_2} \varepsilon_{\beta\alpha} x_\beta t_{3\alpha}(\mathbf{u}) da = -hH_3(\mathbf{u}) \\
\|\mathbf{v}^{(4)}\|_e^2 &= h \int_{\Sigma_2} \varepsilon_{\beta\alpha} x_\beta t_{3\alpha}(\mathbf{v}^{(4)}) da = hD
\end{aligned} \tag{1.9.4}$$

From Equations 1.9.3 and 1.9.4, we get

$$H_3(\mathbf{u}) = -D\gamma(\mathbf{u})$$

for any $\mathbf{u} \in Q_T$. Clearly, $\gamma(\mathbf{u}) = \tau(\mathbf{u})$ for each $\mathbf{u} \in Q_T$. Thus, Saint-Venant's formula 1.7.20 applies to the displacement fields $\mathbf{u}$ which belong to Q_T.

By Equations 1.1.16 and 1.7.19, we find

$$\langle \mathbf{u}, \mathbf{v}^{(4)} \rangle = \mu \left[\int_{\Sigma_2} u_\alpha (\varphi_{,\alpha} - \varepsilon_{\alpha\rho} x_\rho) da - \int_{\Sigma_1} u_\alpha (\varphi_{,\alpha} - \varepsilon_{\alpha\rho} x_\rho) da \right] \tag{1.9.5}$$

We conclude from Equations 1.9.3, 1.9.4, and 1.9.5 that the generalized twist $\tau(\mathbf{u})$ associated with any $\mathbf{u} \in Q_T$ is given by

$$\tau(\mathbf{u}) = \frac{\mu}{hD} \left[\int_{\Sigma_2} u_\alpha (\varphi_{,\alpha} - \varepsilon_{\alpha\rho} x_\rho) da - \int_{\Sigma_1} u_\alpha (\varphi_{,\alpha} - \varepsilon_{\alpha\rho} x_\rho) da \right]$$

Since $\operatorname{div} \mathbf{v}^{(4)} = 0$, it follows that

$$\langle \mathbf{u}, \mathbf{u}^{(4)} \rangle = \mu \int_B e_{ij}(\mathbf{u}) e_{ij}(\mathbf{v}^{(4)}) dv$$

Thus, the energy norm which appears in the relation 1.9.2 can be replaced by the strain norm.

We consider now Saint-Venant's formula 1.7.16. Truesdell's problem can be set also for the extension problem. Let Q_E denote the set of all equilibrium displacement fields $\mathbf{u}$ that satisfy the conditions

$$\mathbf{s}(\mathbf{u}) = \mathbf{0} \text{ on } \Pi, \qquad t_{3\beta}(\mathbf{u}) = 0 \text{ on } \Sigma_\alpha, \qquad H_\alpha(\mathbf{u}) = 0, \qquad R_3(\mathbf{u}) = F_3 \tag{1.9.6}$$

Clearly, if $\mathbf{u} \in Q_E$, then $R_\alpha(\mathbf{u}) = 0$ and $H_3(\mathbf{u}) = 0$, so that $\mathbf{u} \in K_I(F_3, 0, 0, 0)$. Following Ref. 62, we consider the function

$$\beta \rightarrow \|\mathbf{u} - \beta \mathbf{v}^{(3)}\|_e^2 \tag{1.9.7}$$

where $\mathbf{u} \in Q_E$ and $\mathbf{v}^{(3)}$ is the displacement field given by Equations 1.7.15. The field $\beta \mathbf{v}^{(3)}$ is called the extension field with axial strain β. The function 1.9.7 attains its minimum at

$$\varepsilon(\mathbf{u}) = \frac{\langle \mathbf{u}, \mathbf{v}^{(3)} \rangle}{\|\mathbf{v}^{(3)}\|_e^2} \tag{1.9.8}$$

From Equations 1.1.16, 1.1.17, 1.2.6, 1.7.15, and 1.9.6, we get

$$\begin{aligned} \langle \mathbf{u}, \mathbf{v}^{(3)} \rangle &= \int_{\partial B} \mathbf{s}(\mathbf{u}) \cdot \mathbf{v}^{(3)} da = h \int_{\Sigma_2} t_{33}(\mathbf{u}) da = -h R_3(\mathbf{u}) \\ \|\mathbf{v}^{(3)}\|_e^2 &= hEA \end{aligned} \tag{1.9.9}$$

In view of the relations 1.9.8 and 1.9.9,

$$R_3(\mathbf{u}) = -EA\varepsilon(\mathbf{u})$$

for each $\mathbf{u} \in Q_E$. Thus, Saint-Venant's formula 1.7.16 applies to any displacement field $\mathbf{u} \in Q_E$. We call $\varepsilon(\mathbf{u})$ the generalized axial strain associated with the displacement field $\mathbf{u}$. From the relations 1.1.16 and 1.7.15, we have

$$\langle \mathbf{u}, \mathbf{v}^{(3)} \rangle = \int_{\partial B} \mathbf{s}(\mathbf{v}^{(3)}) \cdot \mathbf{u} da = E\left(\int_{\Sigma_2} u_3 da - \int_{\Sigma_1} u_3 da \right) \tag{1.9.10}$$

In view of Equations 1.9.8 and 1.9.9, we conclude that the generalized axial strain $\varepsilon(\mathbf{u})$ associated with any $\mathbf{u} \in Q_E$ is given by

$$\varepsilon(\mathbf{u}) = \frac{1}{hA}\left(\int_{\Sigma_2} u_3 da - \int_{\Sigma_1} u_3 da \right)$$

Let us consider the bending problem. We denote by Q_B the set of all equilibrium displacement fields $\mathbf{u}$ that satisfy the conditions

$$\mathbf{s}(\mathbf{u}) = \mathbf{0} \text{ on } \Pi, \qquad t_{3\beta}(\mathbf{u}) = 0 \text{ on } \Sigma_\alpha$$
$$R_3(\mathbf{u}) = 0, \qquad H_1(\mathbf{u}) = 0, \qquad H_2(\mathbf{u}) = M_2$$

If $\mathbf{u} \in Q_B$, then $\mathbf{u} \in K_I(0, 0, M_2, 0)$. In the same manner, we are led to generalized axial curvature $\kappa(\mathbf{u})$, associated with any $\mathbf{u} \in Q_B$,

$$\kappa(\mathbf{u}) = \frac{1}{hI_{11}}\left(\int_{\Sigma_2} x_1 u_3 da - \int_{\Sigma_1} x_1 u_3 da\right)$$

Moreover, the formula of Saint-Venant's type

$$H_2(\mathbf{u}) = EI_{11}\kappa(\mathbf{u})$$

applies for each $\mathbf{u} \in Q_B$.

Next, we study Truesdell's problem for flexure. Let Q_F denote the set of all equilibrium displacement fields $\mathbf{u}$ that satisfy the conditions

$$\begin{gathered}\mathbf{u}_{,3} \in C^1(\overline{B}), \qquad t_{3\beta}(\mathbf{u}_{,3}) = 0 \text{ on } \Sigma_\alpha \\ \mathbf{s}(\mathbf{u}) = \mathbf{0} \text{ on } \Pi, \quad R_1(\mathbf{u}) = F, \quad R_2(\mathbf{u}) = 0, \quad R_3(\mathbf{u}) = 0, \quad \mathbf{H}(\mathbf{u}) = \mathbf{0}\end{gathered} \tag{1.9.11}$$

Clearly, if $\mathbf{u} \in Q_F$, then $\mathbf{u} \in K_{II}(F, 0)$. Moreover, if $\mathbf{u} \in Q_F$, then by Corollary 1.6.2, $\mathbf{u}_{,3} \in K_I(0, 0, -F, 0)$ and $t_{3\rho}(\mathbf{u}_{,3}) = 0$ on Σ_α. In view of Theorem 1.8.4 we are led to consider the function

$$\xi \to \|\mathbf{u}_{,3} - \xi\mathbf{v}^{(1)}\|_e^2 \tag{1.9.12}$$

where $\mathbf{u} \in Q_F$ and $\mathbf{v}^{(1)}$ is the displacement field defined by Equations 1.7.17. The function 1.9.12 attains its minimum at

$$\eta(\mathbf{u}) = \frac{\langle \mathbf{u}_{,3}, \mathbf{v}^{(1)} \rangle}{\|\mathbf{v}^{(1)}\|_e^2} \tag{1.9.13}$$

From Equations 1.1.16, 1.1.17, 1.2.6, 1.7.17, and 1.9.11, we find that

$$\langle \mathbf{u}_{,3}, \mathbf{v}^{(1)} \rangle = \int_{\partial B} \mathbf{s}(\mathbf{u}_{,3}) \cdot \mathbf{v}^{(1)} da = hH_2(\mathbf{u}_{,3}) - h^2 R_1(\mathbf{u}_{,3})$$

With the help of Theorem 1.6.1, we get

$$\langle \mathbf{u}_{,3}, \mathbf{v}^{(1)} \rangle = -hR_1(\mathbf{u}) \tag{1.9.14}$$

A simple calculation using $t_{3i}(\mathbf{v}^{(1)}) = Ex_1\delta_{i3}$ yields

$$\|\mathbf{v}^{(1)}\|_e^2 = hEI_{11} \tag{1.9.15}$$

From the relations 1.9.13 and 1.9.14, we get

$$R_1(\mathbf{u}) = -EI_{11}\eta(\mathbf{u})$$

for every $\mathbf{u} \in Q_F$. Thus, we have obtained a formula of Saint-Venant's type applicable to any displacement field $\mathbf{u} \in Q_F$.

In view of Equation 1.1.16 we find

$$\langle \mathbf{u}_{,3}, \mathbf{v}^{(1)} \rangle = \int_{\partial B} \mathbf{s}(\mathbf{v}^{(1)}) \cdot \mathbf{u}_{,3} da = E\left(\int_{\Sigma_2} x_1 u_{3,3} da - \int_{\Sigma_1} x_1 u_{3,3} da \right) \tag{1.9.16}$$

We conclude from Equations 1.9.13, 1.9.15, and 1.9.16 that

$$\eta(\mathbf{u}) = \frac{1}{hI_{11}} \left(\int_{\Sigma_2} x_1 u_{3,3} da - \int_{\Sigma_1} x_1 u_{3,3} da \right)$$

and interpret the right-hand side as the global measure of strain appropriate to flexure, associated with the displacement field $\mathbf{u} \in Q_F$.

1.10 Saint-Venant's Principle

In this section we present a study of Saint-Venant's principle. The broader significance of Saint-Venant's solutions to the problem for load distributions that are statically equivalent to, but distinct from those implied by Saint-Venant's results, depends on the validity of the principle bearing his name. Saint-Venant's principle was originally enunciated in order to justify the use of Saint-Venant's solutions. This principle is usually taken to mean that a system of loads having zero resultant force and moment at each end produces a strain field that is negligible away from the ends. The first general statement of Saint-Venant's principle was given by Boussinesq [29]. Mises [232] pointed out that the formulation presented in Ref. 29 is ambiguous, and suggested an amended version of the principle.

The first precise general treatment of any version of Saint-Venant's principle was that of Sternberg [321], who formulated and proved the version suggested by Mises. Two alternative versions of Saint-Venant's principle were established by Toupin [329] and Knowles [182]. These authors arrived at estimates for the strain energy U_z contained in that portion of the body which lies beyond a distance z from the load region. The idea of using U_z in the formulation of Saint-Venant's principle is due to Zanaboni [343,344]. Knowles' results are confined to the case of the two-dimensional problems. Toupin considered the problem of an anisotropic elastic cylinder of arbitrary length subject to self-equilibrated surface tractions on one of its ends, and free of surface tractions on the remainder of its boundary. In Ref. 90, Fichera extended Toupin's result to the case of an elastic cylinder subject to self-equilibrated surface tractions on each of its ends, and free of surface traction on the lateral boundary. This is the case involved by Saint-Venant's conjecture.

Various authors have studied a nonlinear version of Saint-Venant's principle. We mention the works by Roseman [283], Breuer and Roseman [31], Muncaster [236], Horgan and Knowles [128], and Knops and Payne [180]. For the history of the problem and the detailed analysis of various results on Saint-Venant's principle, we refer to the works of Gurtin [119], Djanelidze [68], Fichera [89], Horgan and Knowles [129], and Horgan [130].

In what follows, we present the results due to Toupin [329] and Fichera [90], which provide the mathematical formulation and proof of Saint-Venant's principle in the context for which it was originally intended.

Let $\mathbf{u}'$ be Saint-Venant's solution of the relaxed Saint-Venant's problem, and let $\mathbf{u}''$ be the solution of Saint-Venant's problem with the pointwise assignment of the terminal tractions. We define the displacement field $\mathbf{u}$ on B by $\mathbf{u} = \mathbf{u}'' - \mathbf{u}'$. Then, $\mathbf{u}$ is an equilibrium displacement field that satisfies the conditions

$$\mathbf{s}(\mathbf{u}) = \mathbf{0} \text{ on } \Pi$$
$$\int_{\Sigma_\alpha} \mathbf{s}(\mathbf{u})da = \mathbf{0}, \qquad \int_{\Sigma_\alpha} \mathbf{x} \times \mathbf{s}(\mathbf{u})da = \mathbf{0}, \quad (\alpha = 1, 2) \tag{1.10.1}$$

We conclude that $\mathbf{u}$ is a displacement field corresponding to null body forces and to surface tractions which vanish on the lateral boundary and are self-equilibrated at each end.

We denote by B_z the cylinder defined by

$$B_z = \{\mathbf{x} : (x_1, x_2) \in \Sigma_1,\ z < x_3 < h - z\}, \qquad \left(0 \leq z < \frac{h}{2}\right) \tag{1.10.2}$$

We denote by $U_z(\mathbf{u})$ the strain energy corresponding to the equilibrium displacement field $\mathbf{u}$ on B_z,

$$U_z(\mathbf{u}) = \frac{1}{2} \int_{B_z} C_{ijrs} e_{ij}(\mathbf{u}) e_{rs}(\mathbf{u}) dv \tag{1.10.3}$$

The positive-definiteness of $\mathbf{C}$ implies that $U_z(\mathbf{u})$ is a nonincreasing function of z.

Theorem 1.10.1 *Assume that B is homogeneous and anisotropic, and assume that the elasticity tensor is symmetric and positive definite. Let $\mathbf{u}$ be an equilibrium displacement field that satisfies the conditions* 1.10.1. *Then the strain energy $U_z(\mathbf{u})$ satisfies the inequality*

$$U_z(\mathbf{u}) \leq U_0(\mathbf{u}) e^{-(z-\ell)/k(\ell)}, \quad (z \geq \ell) \tag{1.10.4}$$

for any $\ell > 0$, *where*

$$k(\ell) = (\mu_M/\lambda(\ell))^{1/2}$$

μ_M *is the maximum elastic modulus, and* $\lambda(\ell)$ *is the lowest nonzero characteristic value of free vibration for a slice of the cylinder, of thickness* ℓ, *taken normal to its generators and that has its boundary traction-free.*

Proof. From Equations 1.1.12, 1.1.14, 1.1.16, and 1.10.1, we get

$$U_z(\mathbf{u}) = \frac{1}{2}\int_{\partial B_z} \mathbf{s}(\mathbf{u})\cdot\mathbf{u}da = \frac{1}{2}\left\{\int_{S_{h-z}} u_i t_{3i}(\mathbf{u})da - \int_{S_z} u_i t_{3i}(\mathbf{u})da\right\} \tag{1.10.5}$$

Here S_z denotes the cross section located at $x_3 = z$.

The resultant force and resultant moment on every part of the cylinder must vanish in equilibrium. We denote by $B(t_1, t_2)$, $(0 \le t_1 < t_2 \le h)$, the cylinder

$$B(t_1,t_2) = \{x : (x_1,x_2) \in \Sigma_1, t_1 < x_3 < t_2\}$$

The conditions of equilibrium for the parts $B(0,z)$ and $B(h-z,h)$ of the cylinder, and the conditions 1.10.1 imply that

$$\begin{aligned} &\int_{S_z} \mathbf{s}(\mathbf{u})da = \mathbf{0}, \qquad \int_{S_z} \mathbf{x}\times\mathbf{s}(\mathbf{u})da = \mathbf{0} \\ &\int_{S_{h-z}} \mathbf{s}(\mathbf{u})da = \mathbf{0}, \qquad \int_{S_{h-z}} \mathbf{x}\times\mathbf{s}(\mathbf{u})da = \mathbf{0} \end{aligned} \tag{1.10.6}$$

We introduce the vector fields $\mathbf{u}^{(\alpha)}$, $(\alpha = 1,2)$, defined by

$$\mathbf{u}^{(\alpha)} = \mathbf{u} + \mathbf{a}^{(\alpha)} + \mathbf{b}^{(\alpha)} \times \mathbf{x}, \quad (\alpha = 1,2) \tag{1.10.7}$$

where $\mathbf{a}^{(\alpha)}$ and $\mathbf{b}^{(\alpha)}$ are arbitrary constant vectors. Clearly, the vector fields $\mathbf{u}^{(\alpha)}$ differ from $\mathbf{u}$ by a rigid displacement. In view of Equations 1.10.6, the displacement field $\mathbf{u}$ which appears in the integrands of Equation 1.10.5 may be replaced by $\mathbf{u}^{(\alpha)}$ such that

$$U_z(\mathbf{u}) = \frac{1}{2}\left\{\int_{S_{h-z}} u_i^{(1)} t_{3i}(\mathbf{u})da - \int_{S_z} u_i^{(2)} t_{3i}(\mathbf{u})da\right\} \tag{1.10.8}$$

If we apply the Schwartz inequality, we find that

$$\begin{aligned} U_z(\mathbf{u}) \le \frac{1}{2}\Bigg\{&\left(\int_{S_{h-z}} |\mathbf{u}^{(1)}|^2 da \int_{S_{h-z}} |\mathbf{t}(\mathbf{u})|^2 da\right)^{1/2} \\ &+ \left(\int_{S_z} |\mathbf{u}^{(2)}|^2 da \int_{S_z} |\mathbf{t}(\mathbf{u})|^2 da\right)^{1/2}\Bigg\} \end{aligned} \tag{1.10.9}$$

where $|\mathbf{t}| = (\mathbf{t}\cdot\mathbf{t})^{1/2}$.

We shall use the geometric–arithmetic mean inequality

$$\sqrt{ab} \leq \frac{1}{2}\left(\alpha a + \frac{b}{\alpha}\right)$$

for all nonnegative scalars a, b, and α, with $\alpha > 0$. If we apply this inequality to Equation 1.10.9, we obtain

$$U_z(\mathbf{u}) \leq \frac{1}{4}\left\{\alpha \int_{S_{h-z}} |\mathbf{t}(\mathbf{u})|^2 da + \frac{1}{\alpha}\int_{S_{h-z}} |\mathbf{u}^{(1)}|^2 da + \alpha \int_{S_z} |\mathbf{t}(\mathbf{u})|^2 da \right. \tag{1.10.10}$$
$$\left. + \frac{1}{\alpha}\int_{S_z} |\mathbf{u}^{(2)}|^2 da\right\}$$

where α is an arbitrary positive constant.

Since $\mathbf{C}$ is symmetric and positive definite, the characteristic values of $\mathbf{C}$ (considered as a linear transformation on the six-dimensional space of all symmetric tensors) are all strictly positive. Following Toupin [329], we call the largest characteristic value the maximum elastic modulus, the smallest the minimum elastic modulus. We denote the maximum and minimum elastic moduli by μ_M and μ_m, respectively. It follows that

$$\mu_m |\mathbf{A}|^2 \leq \mathbf{A} \cdot \mathbf{C}[\mathbf{A}] \leq \mu_M |\mathbf{A}|^2 \tag{1.10.11}$$

for any symmetric tensor $\mathbf{A}$. The elastic potential associated with the displacement field $\mathbf{u}$ is defined by

$$W(\mathbf{u}) = \frac{1}{2} C_{ijrs} e_{ij}(\mathbf{u}) e_{rs}(\mathbf{u}) \tag{1.10.12}$$

It follows from Equations 1.1.11 and 1.1.3 that

$$W(\mathbf{u}) = \frac{1}{2} \nabla\mathbf{u} \cdot \mathbf{C}[\nabla\mathbf{u}]$$

where $\nabla\mathbf{u}$ denotes the displacement gradient.

Since the characteristic values of $\mathbf{C}^2$ are the square of the characteristic values of $\mathbf{C}$, we have

$$|\mathbf{t}(\mathbf{u})|^2 = \mathbf{C}[\nabla\mathbf{u}] \cdot \mathbf{C}[\nabla\mathbf{u}] = \nabla\mathbf{u} \cdot \mathbf{C}^2[\nabla\mathbf{u}] < \mu_M \nabla\mathbf{u} \cdot \mathbf{C}[\nabla\mathbf{u}] = 2\mu_M W(\mathbf{u}) \tag{1.10.13}$$

From the relations 1.10.5 and 1.10.13, we get

$$U_z(\mathbf{u}) \leq \frac{1}{4}\left\{2\alpha\mu_M \int_{S_{h-z}} W(\mathbf{u}) da + \frac{1}{\alpha}\int_{S_{h-z}} |\mathbf{u}^{(1)}|^2 da \right.$$
$$\left. + 2\alpha\mu_M \int_{S_z} W(\mathbf{u}) da + \frac{1}{\alpha}\int_{S_z} |\mathbf{u}^{(2)}|^2 da\right\} \tag{1.10.14}$$

We choose ℓ such that $0 < \ell < h/2$ and let

$$Q(z,\ell) = \frac{1}{\ell}\int_z^{z+\ell} U_t(\mathbf{u})dt \tag{1.10.15}$$

If we integrate the inequality 1.10.14 between the limits z and $z+\ell$, then we find that

$$\ell Q(z,\ell) \leq \frac{1}{4}\left\{2\alpha\mu_M \int_{V_1} W(\mathbf{u})dv + \frac{1}{\alpha}\int_{V_1} |\mathbf{u}^{(1)}|^2 dv \right.$$
$$\left. + 2\alpha\mu_M \int_{V_2} W(\mathbf{u})dv + \frac{1}{\alpha}\int_{V_1} |\mathbf{u}^{(2)}|^2 dv\right\} \tag{1.10.16}$$

where

$$V_1 = B(h-z-\ell, h-z), \qquad V_2 = B(z, z+\ell)$$

We denote by $\lambda(\ell)$ the lowest nonzero characteristic value of free vibration for the portion $V = B(0,\ell)$ of B. According to the minimum principle from the theory of free vibrations (cf. [119], Section 76),

$$\lambda(\ell)\int_V \mathbf{v}^2 dv \leq 2\int_V W(\mathbf{v})dv$$

for every $\mathbf{v} \in C^1(\overline{V}) \cap C^2(V)$ that satisfies

$$\int_V \mathbf{v}^2 dv \neq 0, \qquad \int_V \mathbf{v}dv = \mathbf{0}, \qquad \int_V \mathbf{x}\times\mathbf{v}dv = \mathbf{0}$$

The constant vectors $\mathbf{a}^{(\alpha)}$ and $\mathbf{b}^{(\alpha)}, (\alpha = 1, 2)$, can be chosen so that

$$\int_{V_1} \mathbf{u}^{(1)}dv = \mathbf{0}, \qquad \int_{V_1} \mathbf{x}\times\mathbf{u}^{(1)}dv = \mathbf{0}$$
$$\int_{V_2} \mathbf{u}^{(2)}dv = \mathbf{0}, \qquad \int_{V_2} \mathbf{x}\times\mathbf{u}^{(2)}dv = \mathbf{0}$$

We can write

$$\int_{V_1} |\mathbf{u}^{(1)}|^2 dv \leq \frac{2}{\lambda(\ell)}\int_{V_1} W(\mathbf{u})dv, \qquad \int_{V_2} |\mathbf{u}^{(2)}|^2 dv \leq \frac{2}{\lambda(\ell)}\int_{V_2} W(\mathbf{u})dv \tag{1.10.17}$$

By using the relations 1.10.16 and 1.10.17, we find

$$\ell Q(z,\ell) \leq \frac{1}{2}\left(\alpha\mu_M + \frac{1}{\alpha\lambda(\ell)}\right)\left[\int_{V_1} W(\mathbf{u})da + \int_{V_2} W(\mathbf{u})dv\right] \tag{1.10.18}$$

With the help of the relations 1.10.3, 1.10.12, and 1.10.15, we obtain

$$\ell\frac{\partial}{\partial z}Q(z,\ell) = U_{z+\ell}(\mathbf{u}) - U_z(\mathbf{u}) = -\int_{V_1} W(\mathbf{u})dv - \int_{V_2} W(\mathbf{u})dv \tag{1.10.19}$$

From inequality 1.10.18 and 1.10.19, we get

$$g(\alpha,\ell)\frac{\partial}{\partial z}Q(z,\ell)+Q(z,\ell)\leq 0 \tag{1.10.20}$$

where

$$g(\alpha,\ell)=\frac{1}{2}\left(\alpha\mu_M+\frac{1}{\alpha\lambda(\ell)}\right)$$

From the geometric–arithmetic mean inequality, we have

$$g(\alpha,\ell)\geq[\mu_M/\lambda(\ell)]^{1/2}=k(\ell)$$

for any $\alpha>0$. The inequality 1.10.20 implies that

$$k(\ell)\frac{\partial}{\partial z}Q(z,\ell)+Q(z,\ell)\leq 0$$

Therefore, one has

$$\frac{\partial}{\partial z}(e^{z/k(\ell)}Q(z,\ell))\leq 0 \tag{1.10.21}$$

The relation 1.10.21 implies that

$$Q(t_2,\ell)\leq e^{-(t_2-t_1)/k(\ell)}Q(t_1,\ell) \tag{1.10.22}$$

for $t_2\geq t_1$. Since $U_z(\mathbf{u})$ is a nonincreasing function of z, and $Q(z,\ell)$ is the mean value of $U_z(\mathbf{u})$ in the interval $[z,z+\ell]$, we have

$$U_{z+\ell}(\mathbf{u})\leq Q(z,\ell)\leq U_z(\mathbf{u}) \tag{1.10.23}$$

From the inequalities 1.10.22 and 1.10.23, we obtain

$$U_{t_2+\ell}(\mathbf{u})\leq e^{-(t_2-t_1)/k(\ell)}U_{t_1}(\mathbf{u}) \tag{1.10.24}$$

The inequality 1.10.24 implies the relation 1.10.4. □

According to Toupin, the parameter $\ell>0$ is to be chosen in a manner which will provide a small value for $k(\ell)$. From the inequality 1.10.4, we see that, given $\varepsilon>0$, we have

$$\frac{U_z(\mathbf{u})}{U_0(\mathbf{u})}<\varepsilon$$

provided

$$z>\ell+k(\ell)\ln\frac{1}{\varepsilon}$$

In Ref. 329, Toupin employs a mean value theorem due to Diaz and Payne [67], to obtain a pointwise estimate for the magnitude of the strain tensor at interior points of the cylinder. A similar estimate was established by Fichera [90] for an isotropic cylinder. We present here the estimate obtained in Ref. 90. Let D_0 be a bounded regular region.

Lemma 1.10.1 *Let f be a biharmonic scalar field on D_0, and suppose that $f \in L^2(D_0)$. Let d be the distance of the point $\mathbf{x}$ of D_0 from ∂D_0. Then, the following estimate holds*

$$|f(\mathbf{x})| \leq 1.9144 d^{-3/2} \left(\int_{D_0} |f|^2 dv \right)^{1/2} \tag{1.10.25}$$

Proof. We denote by Ω the ball of center $\mathbf{x}$ and unit radius. For each $\mathbf{y} \in D_0$ we set $\mathbf{y} = \mathbf{x} + r\boldsymbol{\zeta}$, where $\boldsymbol{\zeta} \in \partial\Omega$. With the help of spherical coordinates, the Laplacian operator

$$\Delta = \frac{\partial^2}{\partial y_i \partial y_i}$$

appears as

$$\Delta = \Delta_0 + \Delta_*, \qquad \Delta_0 = r^{-2} \frac{\partial}{\partial r} \left(r^2 \frac{\partial}{\partial r} \right)$$

$$\Delta_* = \frac{1}{\sin\theta} \frac{\partial}{\partial\theta} \left(\sin\theta \frac{\partial}{\partial\theta} \right) + \frac{1}{\sin^2\theta} \frac{\partial^2}{\partial\varphi^2}$$

On the basis of the relation

$$\int_{\partial\Omega} \Delta_* f(\mathbf{y}) da = 0$$

the equation

$$\Delta\Delta f = 0$$

yields

$$\Delta_0 \Delta_0 \int_{\partial\Omega} f \, da = 0$$

Thus we obtain,

$$\int_{\partial\Omega} f(\mathbf{y}) da = c_1 r^{-1} + c_2 + c_3 r + c_4 r^2$$

where c_1, c_2, c_3, and c_4 are real constants. Since

$$\lim_{r \to 0} f(\mathbf{y}) = f(\mathbf{x}), \qquad \lim_{r \to 0} r \Delta_0 f(\mathbf{y}) = 0$$

uniformly with respect to $\boldsymbol{\zeta}$, we obtain $c_1 = c_3 = 0$, and $c_2 = 4\pi f(\mathbf{x})$. Thus, we find

$$\int_{\partial\Omega} f(\mathbf{y}) da = 4\pi f(\mathbf{x}) + r^2 c_4$$

If $\mathbf{y}^* = \mathbf{x} + \alpha r \boldsymbol{\zeta}$, with $0 < \alpha < 1$, then

$$\int_{\partial\Omega} f(\mathbf{y}^*) da = 4\pi f(\mathbf{x}) + \alpha^2 r^2 c_4$$

Thus, we obtain

$$f(\mathbf{x}) = \frac{1}{4\pi(1-\alpha^2)}\left(\int_{\partial\Omega} f(\mathbf{y}^*)da - \alpha^2 \int_{\partial\Omega} f(\mathbf{y})da\right)$$

Multiplying the last equality by r^2 and integrating with respect to r from $r = 0$ to $r = d$, we find

$$\frac{1}{3}d^3 f(\mathbf{x}) = \frac{1}{4\pi(1-\alpha^2)}\left(\frac{1}{\alpha^3}\int_{S(\alpha d)} f dv - \alpha^2 \int_{S(d)} f dv\right)$$

where $S(\rho)$ is the ball with radius ρ and center at $\mathbf{x}$. If we apply the Schwartz inequality, we obtain

$$|f(\mathbf{x})| \leq g(\alpha) d^{-3/2}\left(\int_{D_0} |f|^2 dv\right)^{1/2}$$

where

$$g(\alpha) = \frac{\sqrt{3}}{2\sqrt{\pi}}\frac{1+\alpha^{7/2}}{(1-\alpha^2)\alpha^{3/2}}, \quad 0 < \alpha < 1$$

The function g attains an absolute minimum which is less that 1.9144. The last inequality implies the estimate 1.10.25. □

The proof of the inequality 1.10.25 can be obtained by the mean value theorem of Nicolesco [250]. The derivation used here follows that in Ref. 90.

If $\mathbf{u}$ is an equilibrium displacement field for a homogeneous and isotropic body, then the strain tensor $\mathbf{e}(\mathbf{u})$ is biharmonic (cf. [119], Section 42).

If we use Lemma 1.10.1 for the function $\mathbf{e}(\mathbf{u})$ on B_z, we obtain

$$|\{\mathbf{e}(\mathbf{u})\}(\mathbf{x})| \leq 1.9144 d^{-3/2}\left[\int_{B_z} |\mathbf{e}(\mathbf{u})|^2 dv\right]^{1/2} \tag{1.10.26}$$

From the relations 1.10.11 and 1.10.26, we get

$$|\{\mathbf{e}(\mathbf{u})\}(\mathbf{x})| \leq 1.9144\left[\frac{2}{\mu_m d^3} U_z(\mathbf{u})\right]^{1/2}$$

When combined with energy inequality 1.10.4, the above inequality yields pointwise exponential decay for the magnitude of the strain tensor at interior points of the cylinder.

Pointwise estimates near the boundary have been obtained by Roseman [282] and Fichera [90]. Roseman established a pointwise estimate for the stress in a homogeneous and isotropic cylinder. When combined with Toupin's energy inequality, this gives pointwise exponential decay for the stress throughout the cylinder.

1.11 Exercises

1.11.1 Study the torsion of a homogeneous and isotropic elastic cylinder which occupies the domain

$$B = \left\{ x : \frac{x_1^2}{a^2} + \frac{x_2^2}{b^2} < 1, \ 0 < x_3 < h \right\}, \quad (a > 0, b > 0)$$

1.11.2 Investigate the torsion problem for a homogeneous and isotropic right cylinder whose cross section Σ_1 is bounded by the circles C_1 and C_2 defined by

$$(C_1): \ x_1^2 + x_2^2 - 2ax_2 = 0$$
$$(C_2): \ x_1^2 + x_2^2 = b^2, \quad (0 < b < 2a)$$

1.11.3 Study the flexure of an elliptical right cylinder made of a homogeneous and isotropic material.

1.11.4 A homogeneous and isotropic elastic material occupies a right hollow cylinder B with the cross section $\Sigma_1 = \{x : R_1^2 < x_1^2 + x_2^2 < R_2^2, x_3 = 0\}$. The body is in equilibrium in the absence of body forces. Investigate the plane strain of the cylinder when the lateral boundaries are subjected to constant pressures.

1.11.5 Show that

$$\chi = \frac{1}{40a^3} q x_2^5 - \frac{1}{8a^3} q x_1^2 x_2^3 - \frac{1}{4a^3}\left(m + \frac{1}{5}qa^2 - \frac{1}{2}qh^2\right) x_2^3 + \frac{3}{8a} q x_1^2 x_2 + \frac{1}{4} q x_1^2$$

where a, q, h, and m are constants, is suitable for use as an Airy stress function and investigate the stress state in the plane domain $\Sigma_1 = \{(x_1, x_2) : -h < x_1 < h, -a < x_2 < a\}$.

1.11.6 Determine, with the use of the complex potentials, the solution of the plane strain traction problem for a circular region in the absence of the body forces.

1.11.7 Investigate, with the use of the complex potentials, the solution of the plane strain traction problem for a circular ring.

1.11.8 A homogeneous and isotropic elastic right cylinder with the cross section $\Sigma_1 = \{x : x_1^2/a^2 + x_2^2/b^2 < 1, x_3 = 0\}$, $(a > 0, b > 0)$, is subjected to extension and bending by terminal couples. Determine the displacement field and the stress tensor.

1.11.9 Investigate the torsion of a homogeneous and isotropic right cylinder whose cross section is an equilateral triangle.

1.11.10 Study the flexure of a homogeneous and isotropic right cylinder with rectangular cross section.

1.11.11 A homogeneous and isotropic elastic material occupies the domain $B = \{x : a_2^2 < x_1^2 + x_2^2 < a_1^2, 0 < x_3 < h\}$, $(a_1 > a_2 > 0)$. Investigate the torsion of the tube.

1.11.12 Investigate the flexure of a homogeneous and isotropic right cylinder whose cross section is bounded by two confocal ellipses.

Chapter 2

Theory of Loaded Cylinders

2.1 Problems of Almansi and Michell

This chapter is concerned with the generalization of Saint-Venant's problem to the case when the cylinder is subjected to body forces and surface tractions on the lateral boundary. This problem was initiated by Almansi [6] and Michell [221] and was developed in various later works [28,163,175,313].

We assume that a continuous body force field $\mathbf{f}$ is prescribed on B. By an equilibrium displacement field on B, corresponding to the body force field $\mathbf{f}$, we mean a vector field $\mathbf{u} \in C^2(B) \cap C^1(\overline{B})$ that satisfies the displacement equations of equilibrium

$$t_{ji}(\mathbf{u})_{,j} + f_i = 0 \tag{2.1.1}$$

on B. We assume that the boundary conditions 1.2.3 are replaced by

$$\mathbf{s}(\mathbf{u}) = \widetilde{\mathbf{t}} \text{ on } \Pi, \qquad \mathbf{R}(\mathbf{u}) = \mathbf{F}, \qquad \mathbf{H}(\mathbf{u}) = \mathbf{M} \tag{2.1.2}$$

where $\widetilde{\mathbf{t}}$ is a vector-valued function preassigned on Π, and $\mathbf{F}$ and $\mathbf{M}$ are prescribed vectors. Suppose that $\widetilde{\mathbf{t}}$ is piecewise regular on Π.

When $\mathbf{f}$ and $\widetilde{\mathbf{t}}$ are independent of the axial coordinate, the problem was first considered by Almansi [6] and Michell [221]. This particular case defines what is nowadays known in the literature as the *Almansi–Michell problem.*

In Ref. 6, Almansi also studied the case when the prescribed forces are polynomials in the axial coordinate. This problem is known as the *Almansi problem.*

We assume that the body is homogeneous and isotropic. Let us suppose that

$$\begin{aligned} f_i &= \sum_{k=1}^{r} F_{ik}(x_1, x_2) x_3^k, \quad (x_1, x_2, x_3) \in B \\ \widetilde{t}_i &= \sum_{k=1}^{r} p_{ik}(x_1, x_2) x_3^k, \quad (x_1, x_2, x_3) \in \Pi \end{aligned} \tag{2.1.3}$$

where F_{ik} and p_{ik} are prescribed functions.

Almansi problem consists in finding an equilibrium displacement filed on B that corresponds to the body force field $\mathbf{f}$ and satisfies the boundary conditions 2.1.2, when $\mathbf{f}$ and $\widetilde{\mathbf{t}}$ have the form 2.1.3.

Let (A) denote the problem of determination of the functions $u_k \in C^2(B) \cap C^1(\overline{B})$ that satisfy the Equations 1.1.1, 1.1.4, and 2.1.1 on B and the boundary conditions 2.1.2, when $\mathbf{f}$ and $\widetilde{\mathbf{t}}$ have the form

$$\begin{aligned} f_i &= F_{in}(x_1, x_2)x_3^n, \quad (x_1, x_2, x_3) \in B \\ \widetilde{t}_i &= p_{in}(x_1, x_2)x_3^n, \quad (x_1, x_2, x_3) \in \Pi \end{aligned} \tag{2.1.4}$$

where n is a positive integer or zero, and F_{in} and p_{in} are prescribed functions. Obviously, if we know the solution of the problem (A) for any n, then, according to the linearity of the theory, we can determine the solution of Almansi problem.

We denote by (A_0) the problem (A) for $n = 0$, and by $(B^{(s)})$ the problem (A) when $n = s$, $(s = 1, 2, \ldots, r)$, and $\mathbf{F} = \mathbf{0}, \mathbf{M} = \mathbf{0}$. Let $\mathbf{U}^{(0)}$ be a solution of the problem (A_0), and let $\mathbf{U}^{(s)}$ be a solution of the problem $(B^{(s)})$, $(s = 1, 2, \ldots, r)$. Then, the solution $\mathbf{u}$ of Almansi's problem is given by

$$\mathbf{u} = \sum_{m=0}^{r} \mathbf{U}^{(m)}$$

To solve Almansi problem, we use the method of induction. In Section 2.2, we shall solve the Almansi–Michell problem (A_0). Then, in Section 2.3, we shall establish the solution of the problem $(B^{(n+1)})$ once a solution of the problem $(B^{(n)})$ is known. Throughout this chapter, we assume that the cylinder is occupied by an isotropic and homogeneous material. Moreover, we suppose that the elastic potential is a positive definite quadratic form in the components of the strain tensor.

The researches devoted to the theory of loaded cylinders are based on the semi-inverse method. Generally, the authors used various assumptions regarding the structure of the stress field. In Ref. 145, the solution was presented, for the first time, in terms of displacement vector field. In Section 2.4, we shall present a generalization of the results from Section 1.7 to provide a rational tool to solve Almansi problem. The method offers a systematic approach which avoids artificial *a priori* assumptions.

2.2 Almansi–Michell Problem

We assume that

$$f_i = G_i(x_1, x_2), \quad (x_1, x_2) \in \Sigma_1, \qquad \widetilde{t}_i = p_i(x_1, x_2), \quad (x_1, x_2) \in \Gamma \tag{2.2.1}$$

The Almansi–Michell problem consists in the determination of the vector field $\mathbf{u} \in C^2(B) \cap C^1(\overline{B})$ that satisfies Equations 1.1.1, 1.1.4, and 2.1.1 on B and the boundary conditions 2.1.2, when $\mathbf{f}$ and $\widetilde{\mathbf{t}}$ have the form 2.2.1.

Following Ref. 161, we seek the solution of Almansi–Michell problem in the form

$$u_\alpha = -\frac{1}{2}a_\alpha x_3^2 - \frac{1}{6}b_\alpha x_3^3 - \frac{1}{24}c_\alpha x_3^4 + \varepsilon_{\beta\alpha}\left(\tau_1 x_3 + \frac{1}{2}\tau_2 x_3^2\right)x_\beta$$
$$+\sum_{k=1}^{3}\left(a_k + b_k x_3 + \frac{1}{2}c_k x_3^2\right)w_\alpha^{(k)} + v_\alpha(x_1, x_2) \tag{2.2.2}$$
$$u_3 = (a_\rho x_\rho + a_3)x_3 + \frac{1}{2}(b_\rho x_\rho + b_3)x_3^2 + \frac{1}{6}(c_\rho x_\rho + c_3)x_3^3$$
$$+ (\tau_1 + \tau_2 x_3)\varphi(x_1, x_2) + \psi(x_1, x_2) + x_3\chi(x_1, x_2), \quad (x_1, x_2, x_3) \in B$$

where $w_\alpha^{(k)}$ are defined by the relations 1.7.9, φ is the torsion function defined by Equations 1.3.26 and 1.3.28, a_k, b_k, c_k, and τ_ρ are unknown constants, and $v_\alpha, \psi, \chi \in C^2(\Sigma_1) \cap C^1(\overline{\Sigma}_1)$ are unknown functions. Justification of the form 2.2.2 of solution is presented in Section 2.4.

It follows from Equations 2.2.2, 1.1.1, 1.1.4, and 1.7.9 that

$$t_{\alpha\beta} = \lambda(\chi + \tau_2\varphi)\delta_{\alpha\beta} + s_{\alpha\beta}$$
$$t_{33} = E\left[a_\rho x_\rho + a_3 + (b_\rho x_\rho + b_3)x_3 + \frac{1}{2}(c_\rho x_\rho + c_3)x_3^2\right]$$
$$+ (\lambda + 2\mu)(\chi + \tau_2\varphi) + \lambda\gamma_{\rho\rho} \tag{2.2.3}$$
$$t_{\alpha 3} = \mu[(\tau_1 + \tau_2 x_3)(\varphi_{,\alpha} + \varepsilon_{\beta\alpha}x_\beta) + x_3\chi_{,\alpha} + \psi_{,\alpha}]$$
$$+ \mu\sum_{s=1}^{3}(b_s + c_s x_3)w_\alpha^{(s)}$$

where

$$s_{\alpha\beta} = \lambda\gamma_{\rho\rho}\delta_{\alpha\beta} + 2\mu\gamma_{\alpha\beta}$$
$$\gamma_{\alpha\beta} = \frac{1}{2}(v_{\alpha,\beta} + v_{\beta,\alpha}) \tag{2.2.4}$$

The first two equations of equilibrium 2.1.1 and the first two conditions on the lateral boundary become

$$s_{\alpha\beta,\beta} + g_\alpha = 0 \text{ on } \Sigma_1$$
$$s_{\beta\alpha}n_\beta = q_\alpha \text{ on } \Gamma \tag{2.2.5}$$

where

$$g_\alpha = G_\alpha + \lambda(\chi + \tau_2\varphi)_{,\alpha} + \mu[\tau_2(\varphi_{,\alpha} + \varepsilon_{\beta\alpha}x_\beta) + \chi_{,\alpha}] + \mu\sum_{s=1}^{3}c_s w_\alpha^{(s)} \tag{2.2.6}$$
$$q_\alpha = p_\alpha - \lambda(\chi + \tau_2\varphi)n_\alpha$$

Thus, from Equations 2.2.4 and 2.2.5, we conclude that v_α are the displacements in a plane strain problem corresponding to the body forces g_α and to the surface forces q_α. The necessary and sufficient conditions to solve the boundary-value problem 2.2.4 and 2.2.5 are

$$\int_{\Sigma_1} g_\alpha da + \int_\Gamma q_\alpha ds = 0, \qquad \int_{\Sigma_1} \varepsilon_{\alpha\beta} x_\alpha g_\beta da + \int_\Gamma \varepsilon_{\alpha\beta} x_\alpha q_\beta ds = 0 \tag{2.2.7}$$

By the divergence theorem, and the relations 2.2.3 and 2.2.6, we get

$$\begin{aligned} &\int_{\Sigma_1} g_\alpha da + \int_\Gamma q_\alpha ds = \int_{\Sigma_1} G_\alpha da + \int_\Gamma p_\alpha ds + \int_{\Sigma_1} t_{\alpha 3,3} da \\ &\int_{\Sigma_1} \varepsilon_{\alpha\beta} x_\alpha g_\beta da + \int_\Gamma \varepsilon_{\alpha\beta} x_\alpha q_\beta ds \\ &\quad = \int_{\Sigma_1} \varepsilon_{\alpha\beta} x_\alpha G_\beta da + \int_\Gamma \varepsilon_{\alpha\beta} x_\alpha p_\beta ds + \int_{\Sigma_1} \varepsilon_{\alpha\beta} x_\alpha t_{3\beta,3} da \end{aligned} \tag{2.2.8}$$

In view of equations of equilibrium, we have

$$t_{\alpha 3,3} = [t_{\alpha 3} + x_\alpha(t_{j3,j} + G_3)]_{,3} = (x_\alpha t_{\beta 3})_{,\beta 3} + x_\alpha t_{33,33} \tag{2.2.9}$$

Thus, by using the divergence theorem and the conditions on the lateral boundary, we obtain

$$\int_{\Sigma_1} t_{\alpha 3,3} da = \int_{\Sigma_1} x_\alpha t_{33,33} da \tag{2.2.10}$$

In view of the relations 2.2.3 and 2.2.10, the first two conditions from 2.2.7 reduce to

$$E(I_{\alpha\beta} c_\beta + A x_\alpha^0 c_3) = -\int_{\Sigma_1} G_\alpha da - \int_\Gamma p_\alpha ds \tag{2.2.11}$$

where $I_{\alpha\beta}$ are defined in the relations 1.7.14, and x_α^0 and A are given by Equation 1.4.9. It follows from the relations 2.2.3, 2.2.8, and 1.7.9 that the third condition from Equations 2.2.7 can be written in the form

$$\begin{aligned} D\tau_2 = &-\int_{\Sigma_1} \varepsilon_{\alpha\beta} x_\alpha G_\beta da - \int_\Gamma \varepsilon_{\alpha\beta} x_\alpha p_\beta da \\ &- \mu \int_{\Sigma_1} \varepsilon_{\alpha\beta} x_\alpha \left(\chi_{,\beta} + \frac{1}{2} \nu c_\beta x_\rho x_\rho \right) da \end{aligned} \tag{2.2.12}$$

The third equation of equilibrium implies that ψ and χ must satisfy the following equations

$$\mu \Delta \psi = -G_3 - 2\mu(b_\rho x_\rho + b_3) \text{ on } \Sigma_1 \tag{2.2.13}$$

$$\Delta \chi = -2(c_\rho x_\rho + c_3) \text{ on } \Sigma_1 \tag{2.2.14}$$

In view of Equations 2.2.3 and 1.3.28, the last condition on the lateral boundary implies that

$$\mu \frac{\partial \psi}{\partial n} = p_3 + \mu\nu n_\alpha x_\alpha (b_\rho x_\rho + b_3) - \frac{1}{2} b_\alpha \mu\nu n_\alpha x_\rho x_\rho \text{ on } \Gamma \tag{2.2.15}$$

and

$$\frac{\partial\chi}{\partial n} = c_\alpha \nu x_\rho \left(x_\alpha n_\rho - \frac{1}{2} n_\alpha x_\rho \right) + c_3 \nu x_\alpha n_\alpha \text{ on } \Gamma \tag{2.2.16}$$

The necessary and sufficient condition to solve the boundary-value problem 2.2.13 and 2.2.15 is

$$AE(b_\rho x_\rho^0 + b_3) = -\int_{\Sigma_1} G_3 da - \int_\Gamma p_3 ds \tag{2.2.17}$$

Let us consider now the boundary-value problem 2.2.14 and 2.2.16. The necessary and sufficient condition to solve this boundary-value problem reduces to

$$c_\rho x_\rho^0 + c_3 = 0 \tag{2.2.18}$$

The system 2.2.11 and 2.2.18 can always be solved for c_1, c_2, and c_3. Thus, from Equations 2.2.14 and 2.2.16, we can determine the function χ. Then, the relation 2.2.12 determines the constant τ_2.

We consider now the conditions 1.4.1. Let us note that

$$\begin{aligned} \int_{\Sigma_1} t_{13} da &= \int_\Gamma [t_{13} + x_1(t_{j3,j} + G_3)] da \\ &= \int_{\Sigma_1} [(x_1 t_{\alpha 3})_{,\alpha} + x_1 t_{33,3} + x_1 G_3] da \\ &= \int_{\Sigma_1} x_1 G_3 da + \int_\Gamma x_1 p_3 ds + \int_{\Sigma_1} x_1 t_{33,3} da \\ \int_{\Sigma_1} t_{23} da &= \int_{\Sigma_1} x_2 G_3 da + \int_\Gamma x_2 p_3 ds + \int_{\Sigma_1} x_2 t_{33,3} da \end{aligned} \tag{2.2.19}$$

In view of Equations 2.2.19 and 2.2.3, the conditions 1.4.1 reduce to

$$E\big(I_{\alpha\beta} b_\beta + A x_\alpha^0 b_3\big) = -F_\alpha - \int_{\Sigma_1} x_\alpha G_3 da - \int_\Gamma x_\alpha p_3 ds \tag{2.2.20}$$

In what follows, we consider that the constants b_k are determined by the system 2.2.17 and 2.2.20. We can assume that the functions ψ and v_α are known.

In view of relations 2.2.3, the conditions 1.4.2 and 1.4.3 become

$$\begin{aligned} EA\big(a_\rho x_\rho^0 + a_3\big) &= -F_3 - \int_{\Sigma_1} [(\lambda + 2\mu)(\chi + \tau_2 \varphi) + \lambda \gamma_{\rho\rho}] da \\ E\big(I_{\alpha\beta} a_\beta + A x_\alpha^0 a_3\big) &= \varepsilon_{\alpha\beta} M_\beta - \int_{\Sigma_1} x_\alpha [(\lambda + 2\mu)(\chi + \tau_2 \varphi) + \lambda \gamma_{\rho\rho}] da \end{aligned} \tag{2.2.21}$$

The system 2.2.21 determines the constants a_j. From Equations 2.2.3 and 1.4.4, we obtain

$$D\tau_1 = -M_3 - \mu \int_{\Sigma_1} \varepsilon_{\alpha\beta} x_\alpha \left(\chi_{,\beta} + \sum_{s=1}^{3} b_s w_\beta^{(s)} \right) da \tag{2.2.22}$$

The relation 2.2.22 determines the constant τ_1.

Thus, the displacement field 2.2.2 is a solution of Almansi–Michell problem if the constants a_j, b_j, c_j, and τ_α are given by Equations 2.2.21, 2.2.17, 2.2.20, 2.2.11, 2.2.18, 2.2.22, and 2.2.12, the functions ψ and χ are characterized by the boundary-value problems 2.2.13, 2.2.15, 2.2.14, and 2.2.10, and the functions v_α are the displacements in the plane strain problem defined by Equations 2.2.4 and 2.2.5.

We call the displacement vector field 2.2.2 the *Almansi–Michell solution.*

2.3 Almansi Problem

In the case of Almansi's problem, the body forces f_i and the tractions $\widetilde{t}_i$ have the form 2.1.3. In the previous section, we obtained a solution of the problem (A_0). Our task is to establish a solution of the problem $(B^{(n+1)})$ once a solution of the problem $(B^{(n)})$ is known.

By induction hypothesis, we know to derive a solution of the problem in which

$$f_i = F_{i(n+1)}(x_1, x_2)x_3^n, \qquad \widetilde{t}_i = p_{i(n+1)}(x_1, x_2)x_3^n, \quad F_i = 0, \quad M_i = 0$$

Thus, the problem can be presented as follows: to find the functions $u_k \in C^2(B) \cap C^1(\overline{B})$ which satisfy the equations

$$\begin{gathered} 2e_{ij}(\mathbf{u}) = u_{i,j} + u_{j,i}, \qquad t_{ij}(\mathbf{u}) = \lambda e_{rr}(\mathbf{u})\delta_{ij} + 2\mu e_{ij}(\mathbf{u}) \\ (t_{ji}(\mathbf{u}))_{,j} + \Lambda_i(x_1, x_2)x_3^{n+1} = 0 \text{ on } B \end{gathered} \tag{2.3.1}$$

and the boundary conditions

$$\int_{\Sigma_1} t_{3i}(\mathbf{u})da = 0, \qquad \int_{\Sigma_1} \varepsilon_{ijk}x_j t_{3k}(\mathbf{u})da = 0 \tag{2.3.2}$$

$$t_{\alpha i}(\mathbf{u})n_\alpha = \sigma_i(x_1, x_2)x_3^{n+1}, \quad (x_1, x_2, x_3) \in \Pi \tag{2.3.3}$$

when we know the solution of the equations

$$\begin{gathered} 2e_{ij}(\mathbf{u}^*) = u^*_{i,j} + u^*_{j,i}, \qquad t_{ij}(\mathbf{u}^*) = \lambda e_{rr}(\mathbf{u}^*)\delta_{ij} + 2\mu e_{ij}(\mathbf{u}^*) \\ (t_{ji}(\mathbf{u}^*))_{,j} + \Lambda_i(x_1, x_2)x_3^n = 0 \end{gathered} \tag{2.3.4}$$

on B, with the boundary conditions

$$\int_{\Sigma_1} t_{3i}(\mathbf{u}^*)da = 0, \qquad \int_{\Sigma_1} \varepsilon_{ijk}x_j t_{3k}(\mathbf{u}^*)da = 0 \tag{2.3.5}$$

$$t_{\alpha i}(\mathbf{u}^*)n_\alpha = \sigma_i(x_1, x_2)x_3^n, \quad (x_1, x_2, x_3) \in \Pi \tag{2.3.6}$$

In the above relations, Λ_i and σ_i are prescribed functions. We assume that Λ_i are continuous on Σ_1 and that σ_i are piecewise regular on Γ.

Following Ref. 6, we seek the solution in the form

$$u_i = (n+1)\left[\int_0^{x_3} u_i^* dx_3 + \overline{w}_i\right] \tag{2.3.7}$$

where $\overline{w}_i \in C^2(B) \cap C^1(\overline{B})$ are unknown functions. It follows from Equations 2.3.7 and 2.3.1 that

$$t_{ij}(\mathbf{u}) = (n+1)\left[\int_0^{x_3} t_{ij}(\mathbf{u}^*) dx_3 + \tau_{ij}(\overline{\mathbf{w}}) + k_{ij}\right] \tag{2.3.8}$$

where

$$\tau_{ij}(\overline{\mathbf{w}}) = \lambda\gamma_{rr}(\overline{\mathbf{w}})\delta_{ij} + 2\mu\gamma_{ij}(\overline{\mathbf{w}}), \qquad \gamma_{ij}(\overline{\mathbf{w}}) = \frac{1}{2}(\overline{w}_{i,j} + \overline{w}_{j,i}) \tag{2.3.9}$$

and

$$\begin{gathered} k_{\alpha\beta} = \lambda\delta_{\alpha\beta}u_3^*(x_1, x_2, 0), \qquad k_{33} = (\lambda + 2\mu)u_3^*(x_1, x_2, 0) \\ k_{\alpha 3} = k_{3\alpha} = \mu u_\alpha^*(x_1, x_2, 0), \quad (x_1, x_2) \in \Sigma_1 \end{gathered} \tag{2.3.10}$$

With the help of Equations 2.3.4 and 2.3.8, the equations of equilibrium reduce to

$$\tau_{ji}(\overline{\mathbf{w}})_{,j} + P_i = 0 \text{ on } B \tag{2.3.11}$$

where

$$P_i = k_{\alpha i,\alpha} + [t_{3i}(\mathbf{u}^*)](x_1, x_2, 0) \tag{2.3.12}$$

We note that the functions P_i are independent of x_3. In view of Equations 2.3.6 and 2.3.8, the conditions 2.3.3 become

$$\tau_{\beta i}(\overline{\mathbf{w}})n_\beta = \eta_i \text{ on } \Pi \tag{2.3.13}$$

where

$$\eta_i = -k_{\alpha i}n_\alpha$$

The functions η_i are independent of the axial coordinate. By Equations 2.3.8 and 2.3.5, the conditions 2.3.2 reduce to

$$\int_{\Sigma_1} \tau_{3i}(\overline{\mathbf{w}})da = -\mathcal{F}_i, \qquad \int_{\Sigma_1} \varepsilon_{ijk}x_j\tau_{3k}(\overline{\mathbf{w}})da = -\mathcal{M}_i \tag{2.3.14}$$

where

$$\mathfrak{F}_i = \int_{\Sigma_1} k_{3i}da, \qquad \mathcal{M}_i = \int_{\Sigma_1} \varepsilon_{irs}x_r k_{3s}da$$

We conclude that the functions $\overline{w}_i$ satisfy Equations 2.3.9 and 2.3.11 on B and the boundary conditions 2.3.13 and 2.3.14. Thus, $\overline{w}_i$ satisfy an Almansi–Michell problem. The solution of this problem was studied in the previous section. The justification of the form 2.3.7 of the solution is presented in the next section.

2.4 Characterization of Solutions

Since the Almansi problem fails to characterize the solution uniquely, it is natural to ask for intrinsic criteria that distinguish the above solutions among all solutions of the problem. In the first part of this section, we present a relation between the solutions of the Saint-Venant's problem and the solution of the Almansi–Michell problem. Then, a characterization of the Almansi–Michell solution is established. A justification of the solution 2.3.7 is also presented. The results we give here have been established in Ref. 161.

Let $K_{III}(\mathbf{F}, \mathbf{M}, \mathbf{G}, \mathbf{p})$ denote the class of solutions to the Almansi–Michell problem (A_0). Theorem 2.4.1 will be of future use.

Theorem 2.4.1 *If* $\mathbf{u} \in C^2(B) \cap C^1(\overline{B})$, *then*

$$\begin{aligned}
R_i(\mathbf{u}_{,3}) &= \int_\Gamma s_i(\mathbf{u})ds - \int_{\Sigma_1} [t_{ji}(\mathbf{u})]_{,j} da \\
H_\alpha(\mathbf{u}_{,3}) &= \int_\Gamma \varepsilon_{\alpha\beta} x_\beta s_3(\mathbf{u})ds - \int_{\Sigma_1} \varepsilon_{\alpha\beta} x_\beta [t_{j3}(\mathbf{u})]_{,j} da + \varepsilon_{\alpha\beta} R_\beta(\mathbf{u}) \\
H_3(\mathbf{u}_{,3}) &= \int_\Gamma \varepsilon_{\alpha\beta} x_\alpha s_\beta(\mathbf{u})ds - \int_{\Sigma_1} \varepsilon_{\alpha\beta} x_\alpha [t_{j\beta}(\mathbf{u})]_{,j} da
\end{aligned} \tag{2.4.1}$$

Proof. Let us note that $t_{ji}(\mathbf{u}_{,3}) = [t_{ij}(\mathbf{u})]_{,3}$. Thus, we have

$$\begin{aligned}
t_{3i}(\mathbf{u}_{,3}) &= [t_{ji}(\mathbf{u})]_{,j} - [t_{\alpha i}(\mathbf{u})]_{,\alpha} \\
\varepsilon_{\alpha\beta} x_\beta t_{33}(\mathbf{u}_{,3}) &= \varepsilon_{\alpha\beta} x_\beta [t_{j3}(\mathbf{u})]_{,j} - \varepsilon_{\alpha\beta} [x_\beta t_{\rho 3}(\mathbf{u})]_{,\rho} + \varepsilon_{\alpha\beta} t_{\beta 3}(\mathbf{u}) \\
\varepsilon_{\alpha\beta} x_\alpha t_{3\beta}(\mathbf{u}_{,3}) &= \varepsilon_{\alpha\beta} x_\alpha [t_{j\beta}(\mathbf{u})]_{,j} - \varepsilon_{\alpha\beta} [x_\alpha t_{\rho\beta}(\mathbf{u})]_{,\rho}
\end{aligned} \tag{2.4.2}$$

By Equations 1.2.5 and 2.4.2, the divergence theorem, and Equation 1.1.11, we obtain the desired result. □

Recall that $K(\mathbf{F}, \mathbf{M})$ denotes the class of solutions to the Saint-Venant's problem corresponding to the resultants $\mathbf{F}$ and $\mathbf{M}$.

Theorem 2.4.2 *If* $\mathbf{u} \in K_{III}(\mathbf{F}, \mathbf{M}, \mathbf{G}, \mathbf{p})$ *and* $\mathbf{u}_{,3} \in C^2(B) \cap C^1(\overline{B})$, *then* $u_{,3} \in K(\mathbf{P}, \mathbf{Q})$, *where*

$$\begin{aligned}
\mathbf{P} &= \int_{\Sigma_1} \mathbf{G} da + \int_\Gamma \mathbf{p} ds \\
Q_\alpha &= \int_{\Sigma_1} \varepsilon_{\alpha\beta} x_\beta G_3 da + \int_\Gamma \varepsilon_{\alpha\beta} x_\beta p_3 ds + \varepsilon_{\alpha\beta} F_\beta \\
Q_3 &= \int_{\Sigma_1} \varepsilon_{\alpha\beta} x_\alpha G_\beta da + \int_\Gamma \varepsilon_{\alpha\beta} x_\alpha p_\beta ds
\end{aligned} \tag{2.4.3}$$

Theorem 2.4.2 is a direct consequence of Theorem 2.4.1.

The preceding theorem allows us to establish a simple method of deriving Almansi–Michell's solution. Let $\mathbf{u}^0 \in K(\mathbf{P}, \mathbf{Q})$ be Saint-Venant's solution corresponding to the resultant force $\mathbf{P}$ and the resultant moment $\mathbf{Q}$ of the tractions acting on Σ_1. Theorem 2.4.2 asserts that the partial derivative with respect to x_3 of any solution $\mathbf{u} \in C^3(B) \cap C^2(\overline{B})$ of the problem (A_0) belongs to $K(\mathbf{P}, \mathbf{Q})$. It is natural to enquire whether there exists a solution $\mathbf{w}$ of the problem (A_0) such that $\mathbf{w}_{,3}$ and $\mathbf{u}^0$ are equal modulo a rigid displacement. This question is settled in Theorem 2.4.3. We assume that the material is homogeneous and isotropic.

Theorem 2.4.3 *Let $\mathbf{u}^0 \in K(\mathbf{P}, \mathbf{Q})$ be Saint-Venant's solution. Let $\mathbf{w} \in C^3(B) \cap C^2(\overline{B})$ be a displacement vector field such that $\mathbf{w}_{,3}$ and $\mathbf{u}^0$ are equal modulo a rigid displacement. Then $\mathbf{w} \in K_{III}(\mathbf{F}, \mathbf{M}, \mathbf{G}, \mathbf{p})$ if and only if $\mathbf{w}$ is the Almansi–Michell solution.*

Proof. Let $\mathbf{u}^0$ be Saint-Venant's solution in the class $K(\mathbf{P}, \mathbf{Q})$. Following Equation 1.7.21, the vector $\mathbf{u}^0$ has the form

$$\mathbf{u}^0 = \int_0^{x_3} \mathbf{v}\{\widehat{c}\} dx_3 + \mathbf{v}\{\widehat{b}\} + \mathbf{w}^* \tag{2.4.4}$$

where $\widehat{b} = (b_1, b_2, b_3, b_4)$ and $\widehat{c} = (c_1, c_2, c_3, c_4)$ are two constant four-dimensional vectors, and $\mathbf{w}^*$ is a vector field independent of x_3 such that $\mathbf{w}^* \in C^2(\Sigma_1) \cap C^1(\overline{\Sigma}_1)$.

In view of Theorem 1.7.3, Equations 1.7.23, 1.7.27, and 2.4.3, we conclude that

$$\begin{gathered} E(I_{\alpha\beta} c_\beta + A x_\alpha^0 c_3) = -\int_{\Sigma_1} G_\alpha da - \int_\Gamma p_\alpha ds \\ c_\alpha x_\alpha^0 + c_3 = 0, \qquad c_4 = 0 \end{gathered} \tag{2.4.5}$$

and $w_\alpha^* = 0$, $w_3^* = \chi$, where χ is characterized by Equations 2.2.14 and 2.2.16. Moreover, from Equations 1.7.28 and 2.4.3, we obtain Equations 2.2.17, 2.2.20, and

$$Db_4 = -\int_{\Sigma_1} \varepsilon_{\alpha\beta} x_\beta \left[G_\beta + \mu\left(\chi_{,\beta} + \frac{1}{2} \nu c_\beta x_\rho x_\rho \right) \right] da - \int_\Gamma \varepsilon_{\alpha\beta} x_\alpha p_\beta ds$$

Let $\mathbf{w}$ be a vector field such that

$$\mathbf{w}_{,3} = \mathbf{u}^0 + \boldsymbol{\alpha} + \boldsymbol{\beta} \times \mathbf{x} \tag{2.4.6}$$

where $\mathbf{u}^0$ is defined by Equation 2.4.4, and $\boldsymbol{\alpha}$ and $\boldsymbol{\beta}$ are constant vectors. Then, it follows from Equations 1.7.11, 1.7.12, 2.4.5, and 2.4.6 that

$$
\begin{aligned}
w_\alpha &= -\frac{1}{2}a_\alpha x_3^2 - \frac{1}{6}b_\alpha x_3^3 - \frac{1}{24}c_\alpha x_3^4 + \frac{1}{2}b_4\varepsilon_{\beta\alpha}x_3^2 x_\beta \\
&\quad - \tau_1\varepsilon_{\alpha\beta}x_\beta x_3 + \sum_{k=1}^{3}\left(a_k + b_k x_3 + \frac{1}{2}c_k x_3^2\right)w_\alpha^{(k)} + v_\alpha(x_1,x_2) \\
w_3 &= (a_\rho x_\rho + a_3)x_3 + \frac{1}{2}(b_\rho x_\rho + b_3)x_3^2 + \frac{1}{6}(c_\rho x_\rho + c_3)x_3^3 \\
&\quad + b_4 x_3\varphi + \tau_1\varphi + x_3\chi + \psi(x_1,x_2)
\end{aligned}
$$

where v_α and ψ are arbitrary functions independent of x_3, and we have used the notations $a_\alpha = \varepsilon_{\alpha\rho}\beta_\rho$, $a_3 = \alpha_3, \tau_1 = \beta_3$. For convenience, on the basis of the arbitrariness of the functions v_α and ψ, we have introduced the terms $\sum_{k=1}^{3} a_k w_\alpha^{(k)}$ and $\tau_1\varphi$. In Section 2.2, we have shown that we can determine the functions v_α, ψ and the constants a_k and τ_1, such that $\mathbf{w} \in K_{III}(\mathbf{F}, \mathbf{M}, \mathbf{G}, \mathbf{p})$. The proof is complete. □

Let us present now a justification of the form 2.3.7 of the solution.

Let $Q_n\{\mathbf{F}_n x_3^n, \mathbf{p}_n x_3^n\}$, $(n = 1, 2, \ldots, r)$, be the class of solutions to the problem $(B^{(n)})$. By induction hypothesis, we know to derive a solution $\hat{\mathbf{u}} \in Q_n\{\mathbf{F}_n x_3^n, \mathbf{p}_n x_3^n\}$. It follows that we also know a solution $\mathbf{u}^* \in Q_n\{\mathbf{F}_{(n+1)} x_3^n, \mathbf{p}_{(n+1)} x_3^n\}$. Thus we are led to the following problem: to find a vector field $\mathbf{u}'' \in Q_{n+1}\{\mathbf{F}_{(n+1)} x_3^{n+1}, \mathbf{p}_{(n+1)} x_3^{n+1}\}$ when $\mathbf{u}^* \in Q_n\{\mathbf{F}_{(n+1)} x_3^n, p_{(n+1)} x_3^n\}$ is given. We refer to this problem as the problem $(\mathcal{K})$. To solve this problem, we need the following result.

Lemma 2.4.1 *If* $\mathbf{u} \in Q_{n+1}\{\mathbf{F}_{(n+1)} x_3^{n+1}, \mathbf{p}_{(n+1)} x_3^{n+1}\}$ *and* $\mathbf{u}_{,3} \in C^2(B) \cap C^1(\overline{B})$, *then*

$$
(n+1)^{-1}\mathbf{u}_{,3} \in Q_n\{\mathbf{F}_{(n+1)} x_3^n, \mathbf{p}_{(n+1)} x_3^n\}
$$

Proof. Let $\mathbf{u} \in Q_{n+1}\{\mathbf{F}_{(n+1)} x_3^{n+1}, \mathbf{p}_{(n+1)} x_3^{n+1}\}$ such that $\mathbf{u}_{,3} \in C^2(B) \cap C^1(\overline{B})$. It follows from Equations 2.1.1 and 2.1.2 that

$$
\begin{aligned}
t_{ji}(\mathbf{u}_{,3})_{,j} + (n+1)F_{i(n+1)} x_3^n &= 0 \text{ on } B \\
\mathbf{s}(\mathbf{u}_{,3}) &= (n+1)\mathbf{p}_{(n+1)} x_3^n \text{ on } \Pi
\end{aligned}
$$

Since the theory under consideration is linear, the vector field $\mathbf{u}' = (n+1)^{-1}\mathbf{u}_{,3}$ is an equilibrium displacement field on B that corresponds to the body force field $\mathbf{F}_{(n+1)} x_3^n$ and satisfies the condition $\mathbf{s}(\mathbf{u}') = \mathbf{p}_{(n+1)} x_3^n$ on Π. In view of Theorem 2.4.1, we find $\mathbf{R}(\mathbf{u}') = \mathbf{0}$, $\mathbf{H}(\mathbf{u}') = \mathbf{0}$. This completes the proof of the lemma. □

Lemma 2.4.1 allows us to solve the problem $(\mathcal{K})$. Thus, in view of this lemma, we are led to seek the vector field $\mathbf{u}''$ such that $(n+1)^{-1}\mathbf{u}''_{,3} = \mathbf{u}^*$

modulo a rigid displacement, that is,

$$(n+1)^{-1}\mathbf{u}''_{,3} = \mathbf{u}^* + \boldsymbol{\alpha} + \boldsymbol{\beta} \times \mathbf{x} \tag{2.4.7}$$

where $\boldsymbol{\alpha}$ and $\boldsymbol{\beta}$ are constant vectors. Then it follows that

$$\begin{aligned} u''_\alpha &= (n+1)\left[\int_0^{x_3} u^*_\alpha dx_3 - \frac{1}{2}a_\alpha x_3^2 - a_4 \varepsilon_{\alpha\beta} x_\beta x_3 + w^*_\alpha(x_1, x_2)\right] \\ u''_3 &= (n+1)\left[\int_0^{x_3} u^*_3 dx_3 + (a_\rho x_\rho + a_3)x_3 + w^*_3(x_1, x_2)\right] \end{aligned} \tag{2.4.8}$$

except for an additive rigid displacement. Here $\mathbf{w}^*$ is an arbitrary vector field independent of x_3, and we have used the notations $a_\alpha = \varepsilon_{\rho\alpha}\beta_\rho$, $a_3 = \alpha_3$, $\alpha_4 = \beta_3$.

Theorem 2.4.4 *Let $\mathbf{u}^* \in Q_n\{\mathbf{F}_{(n+1)}x_3^n, \mathbf{p}_{(n+1)}x_3^n\}$, and let Y be the set of all vector fields of the form 2.4.8. Then there exists a vector field $\mathbf{u}'' \in Y$ such that $\mathbf{u}'' \in Q_{(n+1)}\{\mathbf{F}_{(n+1)}x_3^{n+1}, \mathbf{p}_{(n+1)}x_3^{n+1}\}$.*

Proof. Let us prove that the functions w_i^* and the constants a_s, $(s = 1, 2, 3, 4)$, can be determined so that $\mathbf{u}'' \in Q_{n+1}\{\mathbf{F}_{(n+1)}x_3^{n+1}, \mathbf{p}_{(n+1)}x_3^{n+1}\}$.

We introduce the vector field $\mathbf{w}'$ by

$$w^*_\alpha = \sum_{i=1}^{3} a_i \mathbf{w}_\alpha^{(i)} + w'_\alpha, \qquad w^*_3 = a_4\varphi + w'_3$$

where the functions $w_\alpha^{(i)}$ have the form 1.7.9, and the function φ is the torsion function.

From Equations 2.4.8, we obtain

$$\begin{aligned} u''_1 &= (n+1)\bigg[\int_0^{x_3} u^*_1 dx_3 - \frac{1}{2}a_1 x_3^2 - a_4 x_2 x_3 - \frac{1}{2}a_1\nu(x_1^2 - x_2^2) \\ &\quad - a_2\nu x_1 x_2 - a_3 \nu x_1 + w'_1\bigg] \\ u''_2 &= (n+1)\bigg[\int_0^{x_3} u^*_2 dx_3 - \frac{1}{2}a_2 x_3^2 + a_4 x_1 x_3 - a_1 \nu x_1 x_2 \\ &\quad - \frac{1}{2}a_2\nu(x_2^2 - x_1^2) - a_3\nu x_2 + w'_2\bigg] \\ u''_3 &= (n+1)\left[\int_0^{x_3} u^*_3 dx_3 + (a_\rho x_\rho + a_3)x_3 + a_4\varphi + w'_3\right] \end{aligned} \tag{2.4.9}$$

The stress–displacement relations imply

$$t_{\alpha\beta}(\mathbf{u}'') = (n+1)\left[\int_0^{x_3} t_{\alpha\beta}(\mathbf{u}^*)dx_3 + T_{\alpha\beta}(\mathbf{w}) + \lambda\delta_{\alpha\beta}u_3^*(x_1,x_2,0)\right]$$

$$t_{33}(\mathbf{u}'') = (n+1)\left[\int_0^{x_3} t_{33}(\mathbf{u}^*)dx_3 + E(a_\rho x_\rho + a_3) + \lambda w'_{\rho,\rho} \right.$$

$$\left. + (\lambda+2\mu)u_3^*(x_1,x_2,0)\right] \tag{2.4.10}$$

$$t_{\alpha 3}(\mathbf{u}'') = (n+1)\left[\int_0^{x_3} t_{\alpha 3}(\mathbf{u}^*)dx_3 + \mu a_4(\varphi_{,\alpha} - \varepsilon_{\alpha\beta}x_\beta) \right.$$

$$\left. + w'_{3,\alpha} + \mu u_\alpha^*(x_1,x_2,0)\right]$$

We have

$$\begin{aligned} (t_{\alpha i}(\mathbf{u}''))_{,i} &= (n+1)\left[\int_0^{x_3} (t_{\alpha i}(\mathbf{u}^*))_{,i}dx_3 + (T_{\alpha\beta}(\mathbf{w}'))_{,\beta} + g_\alpha\right] \\ (t_{3i}(\mathbf{u}''))_{,i} &= (n+1)\left[\int_0^{x_3} (t_{3i}(\mathbf{u}^*))_{,i}dx_3 + \mu\Delta w'_3 + g\right] \end{aligned} \tag{2.4.11}$$

where

$$\begin{aligned} g_\alpha &= [t_{3\alpha}(\mathbf{u}^*)](x_1,x_2,0) + \lambda u^*_{3,\alpha}(x_1,x_2,0) \\ g &= [t_{33}(\mathbf{u}^*)](x_1,x_2,0) + \mu u^*_{\alpha,\alpha}(x_1,x_2,0) \end{aligned} \tag{2.4.12}$$

Since $\mathbf{u}^* \in Q_n\{\mathbf{F}_{(n+1)}x_3^n, \mathbf{p}_{(n+1)}x_3^n\}$, the equations of equilibrium and the conditions on the lateral boundary reduce to

$$[T_{\alpha\beta}(\mathbf{w}')]_{,\beta} + g_\alpha = 0 \text{ on } \Sigma_1, \qquad T_{\alpha\beta}(\mathbf{w}')n_\beta = q_\alpha \text{ on } \Gamma \tag{2.4.13}$$

$$\mu\Delta w'_3 + g = 0 \text{ on } \Sigma_1, \qquad \mu\frac{\partial w'_3}{\partial n} = q \text{ on } \Gamma \tag{2.4.14}$$

where

$$q_\alpha = -\lambda n_\alpha u_3^*(x_1,x_2,0), \qquad q = -\mu n_\alpha u_\alpha^*(x_1,x_2,0)$$

We conclude from Equations 2.4.13 that $\{w'_\alpha, T_{\alpha\beta}(\mathbf{w}')\}$ is a plane elastic state corresponding to the body forces g_α and to the surface forces q_α. It is a simple matter to see that the necessary and sufficient conditions to solve the plane strain problem 2.4.13 are satisfied.

The function w'_3 is characterized by the boundary-value problem 2.4.14. On the basis of Theorem 2.4.1, we find that $R_\alpha(\mathbf{u}'') = \varepsilon_{\beta\alpha}H_\beta((n+1)\mathbf{u}^*) = 0$. The conditions $R_3(\mathbf{u}'') = 0$, $\mathbf{H}(\mathbf{u}'') = \mathbf{0}$ reduce to

$$\begin{aligned} E(I_{\alpha\beta}a_\beta + Ax_\alpha^0 a_3) &= -\int_{\Sigma_1} x_\alpha[\lambda w_{\rho,\rho} + (\lambda+2\mu)u_3^*(x_1,x_2,0)]da \\ AE(a_\rho x_\rho^0 + a_3) &= -\int_{\Sigma_1} [\lambda w'_{\rho,\rho} + (\lambda+2\mu)u_3^*(x_1,x_2,0)]da \\ Da_4 &= -\mu\int_{\Sigma_1} \varepsilon_{\alpha\beta}x_\alpha[w'_{3,\beta} + u_\beta^*(x_1,x_2,0)]da \end{aligned} \tag{2.4.15}$$

The system 2.4.15 determines a_1, a_2, a_3, and a_4. □

Remark. It follows from Equations 1.7.11, 1.7.12, 2.4.8, and 2.4.9 that the solution $\mathbf{u}''$ may be written in the form

$$\mathbf{u}'' = (n+1)\left[\int_0^{x_3} \mathbf{u}^* dx_3 + \mathbf{v}\{\widehat{a}\} + \mathbf{w}'\right] \tag{2.4.16}$$

Here w'_α are the components of the displacement filed in the plane strain problem 2.4.13, w'_3 is characterized by the problem 2.4.14 and $\widehat{a}$ is determined by Equations 2.4.15.

The above results yield a rational scheme to derive a solution to the Almansi problem.

2.5 Direct Method

In this section, we present another method of solving Almansi problem. The advantage of this method is that it does not involve the method of induction and avoids the use of some auxiliary functions and constants. For convenience, we assume that the body forces and the tractions applied on the lateral surface are given in the form

$$f_i = \sum_{k=0}^{m} \frac{1}{k!} F_i^{(k)} x_3^k, \qquad \widetilde{t}_i = \sum_{k=0}^{m} \frac{1}{k!} P_i^{(k)} x_3^k \tag{2.5.1}$$

where $F_i^{(k)}$ and $P_i^{(k)}$ are prescribed functions independent of the axial coordinate.

The problem consists in the determination of a solution $\mathbf{u} \in C^2(B) \cap C^1(\overline{B})$ of Equations 1.1.1, 1.1.4, and 2.1.1 on B that satisfies the boundary conditions on Σ_1 and

$$t_{\alpha i} n_\alpha = \widetilde{t}_i \text{ on } \Pi \tag{2.5.2}$$

when the body forces and the lateral loading have the form 2.5.1.

The recurrence process presented in Section 2.3 lead us to seek the solution in the form

$$\begin{aligned} u_\alpha &= \sum_{k=0}^{m+2}\left[-\frac{1}{(k+2)!}C_\alpha^{(k)}x_3^{k+2}+\frac{1}{k!}x_3^k\sum_{j=1}^{3}C_j^{(k)}w_\alpha^{(j)}\right] \\ &+ \sum_{k=0}^{m}\frac{1}{k!}v_\alpha^{(k)}x_3^k+\varepsilon_{\beta\alpha}x_\beta\sum_{k=1}^{m+2}\frac{1}{k!}T^{(k)}x_3^k \\ u_3 &= \sum_{k=0}^{m+2}\frac{1}{(k+1)!}(C_1^{(k)}x_1+C_2^{(k)}x_2+C_3^{(k)})x_3^{k+1} \\ &+ \sum_{k=0}^{m+1}\frac{1}{k!}(T^{(k+1)}\varphi+\psi^{(k+1)})x_3^k \end{aligned} \tag{2.5.3}$$

where $C_j^{(k)}$, $(k=0,1,2,\ldots,m+2)$, and $T^{(s)}$, $(s=1,2,\ldots,m+2)$, are unknown constants, $v_\alpha^{(k)}$ $(k=0,1,2,\ldots,m)$ are unknown functions of class $C^2(\Sigma_1)\cap C^1(\overline{\Sigma}_1)$, independent of the axial coordinate, $w_\alpha^{(k)}$ are defined by relations 1.7.9, and $\psi^{(r)}$, $(r=1,2,\ldots,m+2)$, are unknown functions of class $C^2(\Sigma_1)\cap C^1(\overline{\Sigma}_1)$ which depend only on x_1 and x_2. Let us prove that the functions $v_\alpha^{(k)}$ and $\psi^{(r)}$ and the constants $C_j^{(k)}$ and $T^{(s)}$ can be determined so that **u** be a solution of Almansi problem.

We introduce the notations

$$\gamma_{\alpha\beta}^{(k)}=\frac{1}{2}\left(v_{\alpha,\beta}^{(k)}+v_{\beta,\alpha}^{(k)}\right),\quad (k=0,1,2,\ldots,m) \tag{2.5.4}$$

It follows from Equations 1.1.1, 1.1.4, 1.7.9, and 2.5.3 that

$$\begin{aligned} t_{\alpha\beta} &= \sum_{k=0}^{m}\frac{1}{k!}[\lambda(T^{(k+2)}\varphi+\psi^{(k+2)})\delta_{\alpha\beta}+s_{\alpha\beta}^{(k)}]x_3^k \\ t_{33} &= \lambda\sum_{k=0}^{m}\frac{1}{k!}\gamma_{\rho\rho}^{(k)}x_3^k+E\sum_{k=0}^{m+2}\frac{1}{k!}(C_1^{(k)}x_1+C_2^{(k)}x_2+C_3^{(k)})x_3^k \\ &+ (\lambda+2\mu)\sum_{k=0}^{m}\frac{1}{k!}(T^{(k+2)}\varphi+\psi^{(k+2)})x_3^k \\ t_{\alpha3} &= \mu\sum_{k=0}^{m+1}\frac{1}{k!}\left[T^{(k+1)}(\varphi_{,\alpha}+\varepsilon_{\beta\alpha}x_\beta)+\psi_{,\alpha}^{(k+1)}\right. \\ &\left.+\sum_{j=1}^{3}C_j^{(k+1)}w_\alpha^{(j)}\right]x_3^k+\mu\sum_{k=0}^{m-1}\frac{1}{k!}v_\alpha^{(k+1)}x_3^k \end{aligned} \tag{2.5.5}$$

where

$$s_{\alpha\beta}^{(k)}=\lambda\gamma_{\rho\rho}^{(k)}\delta_{\alpha\beta}+2\mu\gamma_{\alpha\beta}^{(k)},\quad (k=0,1,2,\ldots,m) \tag{2.5.6}$$

It follows from Equation 2.5.5 that the first two equations of equilibrium 2.1.1 and the first two conditions on the lateral surface 2.5.2 reduce to

$$s^{(k)}_{\beta\alpha,\beta} + G^{(k)}_\alpha = 0 \text{ on } \Sigma_1, \qquad s^{(k)}_{\beta\alpha} n_\beta = q^{(k)}_\alpha \text{ on } \Gamma, \quad (k = 0, 1, 2, \ldots, m) \quad (2.5.7)$$

where

$$\begin{aligned} G^{(k)}_\alpha &= F^{(k)}_\alpha + \lambda\big(T^{(k+2)}\varphi_{,\alpha} + \psi^{(k+2)}_{,\alpha}\big) + \Gamma^{(k)}_\alpha \\ \Gamma^{(k)}_\alpha &= \mu\Bigg[\psi^{(k+2)}_{,\alpha} + v^{(k+2)}_\alpha + T^{(k+2)}(\varphi_{,\alpha} + \varepsilon_{\beta\alpha}x_\beta) \\ &\quad + \sum_{j=1}^{3} C^{(k+2)}_j w^{(j)}_\alpha\Bigg] \\ q^{(k)}_\alpha &= p^{(k)}_\alpha - \lambda\big(T^{(k+2)}\varphi + \psi^{(k+2)}\big)n_\alpha \\ &k = 0, 1, 2, \ldots, m, \;\; v^{(m+\rho)}_\alpha \equiv 0, \;\; \rho = 1, 2 \end{aligned} \quad (2.5.8)$$

The last equation of equilibrium and the last condition on the lateral boundary become

$$\mu\Delta\psi^{(k)} = g^{(k)} \text{ on } \Sigma_1, \qquad \mu\frac{\partial\psi^{(k)}}{\partial n} = \Lambda^{(k)} \text{ on } \Gamma, \quad (k = 1, 2, \ldots, m+2) \quad (2.5.9)$$

where

$$\begin{aligned} g^{(k)} &= -F^{(k-1)}_3 - 2\mu\big(C^{(k)}_1 x_1 + C^{(k)}_2 x_2 + C^{(k)}_3\big) \\ &\quad - (\lambda + 2\mu)\big(T^{(k+2)}\varphi + \psi^{(k+2)}\big) - (\lambda + \mu)v^{(k)}_{\rho,\rho} \\ \Lambda^{(k)} &= p^{(k-1)}_3 - \mu v^{(k)}_\alpha n_\alpha - \mu n_\alpha \sum_{j=1}^{3} C^{(k)}_j w^{(j)}_\alpha, \quad (k = 1, 2, \ldots, m+2) \\ F^{(m+1)}_3 &\equiv 0, \qquad \psi^{(m+2+\rho)} \equiv 0, \qquad T^{(m+2+\rho)} \equiv 0, \quad \rho = 1, 2 \end{aligned} \quad (2.5.10)$$

The necessary and sufficient conditions to solve the boundary-value problems 2.5.9 are

$$\int_{\Sigma_1} g^{(k)}da = \int_\Gamma \Lambda^{(k)}ds, \quad (k = 1, 2, \ldots, m+2) \quad (2.5.11)$$

By using Equations 2.5.10, 1.7.9, and the divergence theorem, the conditions 2.5.11 reduce to

$$\begin{aligned} EA\big(C^{(k)}_1 x^0_1 + C^{(k)}_2 x^0_2 + C^{(k)}_3\big) &= -\int_{\Sigma_1} F^{(k-1)}_3 da - \int_\Gamma p^{(k-1)}_3 ds \\ &\quad - \int_{\Sigma_1}\big[\lambda\gamma^{(k)}_{\rho\rho} + (\lambda + 2\mu)\big(T^{(k+2)}\varphi + \psi^{(k+2)}\big)\big]da, \quad (k = 1, 2, \ldots, m+2) \end{aligned} \quad (2.5.12)$$

From Equations 2.5.4, 2.5.6, and 2.5.7, we conclude that the functions $v_\alpha^{(k)}$, $(k = 0, 1, 2, \dots, m)$, are the displacements in the plane strain problems corresponding to the body forces $G_\alpha^{(k)}$ and the tractions $q_\alpha^{(k)}$. The necessary and sufficient conditions for the existence of the functions $v_\alpha^{(k)}$ are

$$\begin{aligned} &\int_{\Sigma_1} G_\alpha^{(k)} da + \int_\Gamma q_\alpha^{(k)} ds = 0 \\ &\int_{\Sigma_1} \varepsilon_{\alpha\beta} x_\alpha G_\beta^{(k)} da + \int_\Gamma \varepsilon_{\alpha\beta} x_\alpha q_\beta^{(k)} ds = 0, \quad (k = 0, 1, 2, \dots, m) \end{aligned} \tag{2.5.13}$$

It follows from Equations 2.5.8 and the divergence theorem that

$$\int_{\Sigma_1} G_\alpha^{(k)} da + \int_\Gamma q_\alpha^{(k)} ds = \int_{\Sigma_1} F_\alpha^{(k)} da + \int_\Gamma p_\alpha^{(k)} ds + \int_{\Sigma_1} \Gamma_\alpha^{(k)} da \tag{2.5.14}$$

By use of the same procedure as that used to prove Equations 2.2.9, we obtain

$$t_{\alpha 3,3} = (x_\alpha t_{\beta 3,3})_{,\beta} + x_\alpha (t_{33,33} + f_{3,3})$$

Thus, by using the divergence theorem and Equations 2.5.2, we find that

$$\int_{\Sigma_1} t_{\alpha 3,3} da = \frac{d}{dx_3} \int_\Gamma x_\alpha \tilde{t}_3 ds + \int_{\Sigma_1} x_\alpha (t_{33,33} + f_{3,3}) da \tag{2.5.15}$$

In view of the relations 2.5.5 and 2.5.8, from Equations 2.5.14, we get

$$\begin{aligned} \int_{\Sigma_1} \Gamma_\alpha^{(k)} da = &\int_{\Sigma_1} x_\alpha F_3^{(k+1)} da + \int_\Gamma x_\alpha p_3^{(k+1)} ds \\ &+ E\big(I_{\alpha 1} C_1^{(k+2)} + I_{\alpha 2} C_2^{(k+2)} + A x_\alpha^0 C_3^{(k+2)}\big) \\ &+ \int_{\Sigma_1} x_\alpha \big[\lambda \gamma_{\rho\rho}^{(k+2)} + (\lambda + 2\mu)\big(T^{(k+4)} \varphi + \psi^{(k+4)}\big)\big] da \end{aligned}$$

$$k = 0, 1, 2, \dots, m, \quad T^{(m+\rho)} \equiv 0, \quad \psi^{(m+\rho)} \equiv 0, \quad \rho = 3, 4 \tag{2.5.16}$$

By Equations 2.5.14 and 2.5.16, the first two conditions from Equations 2.5.13 reduce to

$$\begin{aligned} &E\big(I_{\alpha 1} C_1^{(k+2)} + I_{\alpha 2} C_2^{(k+2)} + A x_\alpha^0 C_3^{(k+2)}\big) = -\int_{\Sigma_1} \{F_\alpha^{(k)} \\ &+ x_\alpha F_3^{(k+1)} + x_\alpha \big[\lambda \gamma_{\rho\rho}^{(k+2)} + (\lambda + 2\mu)\big(T^{(k+4)} \varphi + \psi^{(k+4)}\big)\big]\} da \\ &- \int_\Gamma \big[p_\alpha^{(k)} + x_\alpha p_3^{(k+1)}\big] ds, \quad (k = 0, 1, 2, \dots, m) \end{aligned} \tag{2.5.17}$$

In view of the relations 2.5.8, the remaining condition from Equations 2.5.13 reduces to

$$DT^{(k+2)} = -\int_{\Sigma_1} \varepsilon_{\alpha\beta} x_\alpha F_\beta^{(k)} da - \int_{\Gamma} \varepsilon_{\alpha\beta} x_\alpha p_\beta^{(k)} ds - \mu \int_{\Sigma_1} \varepsilon_{\alpha\beta} x_\alpha \left[\psi_{,\beta}^{(k+2)} + v_\beta^{(k+2)} + \sum_{j=1}^{3} C_j^{(k+2)} w_\beta^{(j)} \right] da, \quad (k = 0, 1, 2, \ldots, m) \quad (2.5.18)$$

where D is the torsional rigidity. Let us study now the conditions on Σ_1. We note that in the presence of body forces the relations 1.3.57 become

$$\int_{\Sigma} t_{3\alpha} da = \int_{\partial\Sigma} x_\alpha t_{\beta 3} n_\beta ds + \int_{\Sigma} x_\alpha (t_{33,3} + f_3) da$$

Thus, for $x_3 = 0$, we obtain

$$\begin{aligned} \int_{\Sigma_1} t_{3\alpha} da = & \int_{\Gamma} x_\alpha p_3^{(0)} ds + \int_{\Sigma_1} x_\alpha F_3^{(0)} da \\ & + \int_{\Sigma_1} x_\alpha \left[\lambda \gamma_{\rho\rho}^{(1)} + (\lambda + 2\mu)(T^{(3)} \varphi + \psi^{(3)}) \right] da \\ & + E\left(I_{\alpha 1} C_1^{(1)} + I_{\alpha 2} C_2^{(1)} + A x_\alpha^0 C_3^{(1)} \right) \end{aligned} \quad (2.5.19)$$

In view of Equation 2.5.19, the conditions 1.4.1 reduce to

$$\begin{aligned} E\left(I_{\alpha 1} C_1^{(1)} + I_{\alpha 2} C_2^{(1)} + A x_\alpha^0 C_3^{(1)} \right) = -F_\alpha - \int_{\Sigma_1} x_\alpha F_3^{(0)} da \\ - \int_{\Gamma} x_\alpha p_3^{(0)} ds - \int_{\Sigma_1} x_\alpha \left[\lambda \gamma_{\rho\rho}^{(1)} + (\lambda + 2\mu)(T^{(3)} \varphi + \psi^{(3)}) \right] da \end{aligned} \quad (2.5.20)$$

From the condition 1.4.2 and 2.5.5, we obtain

$$\begin{aligned} & EA\left(C_1^{(0)} x_1^0 + C_2^{(0)} x_2^0 + C_3^{(0)} \right) \\ & \quad = -F_3 - \int_{\Sigma_1} \left[\lambda \gamma_{\rho\rho}^{(0)} + (\lambda + 2\mu)(T^{(2)} \varphi + \psi^{(2)}) \right] da \end{aligned} \quad (2.5.21)$$

The conditions 1.4.3 reduce to

$$\begin{aligned} E\left(I_{\alpha 1} C_1^{(0)} + I_{\alpha 2} C_2^{(0)} + A x_\alpha^0 C_3^{(0)} \right) = \varepsilon_{\alpha\beta} M_\beta \\ - \int_{\Sigma_1} x_\alpha \left[\lambda \gamma_{\rho\rho}^{(0)} + (\lambda + 2\mu)(T^{(2)} \varphi + \psi^{(2)}) \right] da \end{aligned} \quad (2.5.22)$$

It follows from the condition 1.4.4 and 2.5.5 that

$$DT^{(1)} = -M_3 - \mu \int_{\Sigma_1} \varepsilon_{\alpha\beta} x_\alpha \left[\psi^{(1)}_{,\beta} + v^{(1)}_\beta + \sum_{j=1}^{3} C^{(1)}_j w^{(j)}_\beta \right] da \tag{2.5.23}$$

First, we determine the torsion function φ and calculate the torsional rigidity D from Equation 1.3.32. Since $\gamma^{(m+2)}_{\rho\rho} = 0, T^{(m+4)} = 0, \psi^{(m+4)} = 0$, we see that from Equations 2.5.12, with $k = m+2$, and 2.5.17, with $k = m$, we can determine the constants $C^{(m+2)}_j$ in terms of the body forces and lateral tractions. Then, from Equations 2.5.9, for $k = m+2$, we determine the function $\psi^{(m+2)}$. We note that in the relations 2.5.10, we have $T^{(m+4)} = 0, \psi^{(m+4)} = 0$, $v^{(m+2)}_\alpha = 0$. Next, from Equation 2.5.18, we can determine $T^{(m+2)}$. Since $C^{(m+2)}_j$, $T^{(m+2)}$ and $\psi^{(m+2)}$ are known, from the plane strain problem 2.5.4, 2.5.6, and 2.5.7, for $k = m$, we can obtain the functions $v^{(m)}_\alpha$. Then, from Equations 2.5.12, with $k = m+1$, and 2.5.17, with $k = m-1$, we determine the constants $C^{(m+1)}_i$. From Equations 2.5.9, we determine the function $\psi^{(m+1)}$. The constant $T^{(m+1)}$ is given by Equation 2.5.18. The plane strain problem 2.5.4, 2.5.6, and 2.5.7, with $k = m-1$, determines the functions $v^{(m-1)}_\alpha$, and so on. The constants $C^{(1)}_j$ are determined by Equations 2.5.12 and 2.5.20. The function $\psi^{(1)}$ is given by Equations 2.5.9 and the constant $T^{(1)}$ can be found from Equations 2.5.23. Finally, from Equations 2.5.11 and 2.5.22, we obtain the constants $C^{(0)}_j$. Thus, we conclude that the displacement vector field defined by Equations 2.5.3 is a solution of Almansi problem.

In the case of uniformly loaded cylinders, we have

$$f_i = F^{(0)}_i, \qquad \tilde{t}_i = p^{(0)}_i$$

From Equations 2.5.3, for $m = 0$, we obtain the following solution of Almansi–Michell problem

$$\begin{aligned} u_\alpha &= \sum_{k=0}^{2} \left[-\frac{1}{(k+2)!} C^{(k)}_\alpha x_3^{k+2} + \frac{1}{k!} x_3^k \sum_{j=1}^{3} C^{(k)}_j w^{(j)}_\alpha \right] \\ &\quad + \varepsilon_{\alpha\beta} x_\beta \left(T^{(1)} x_3 + \frac{1}{2} T^{(2)} x_3^2 \right) + v^{(0)}_\alpha \\ u_3 &= \sum_{k=0}^{2} \frac{1}{(k+1)!} \left(C^{(k)}_1 x_1 + C^{(k)}_2 x_2 + C^{(k)}_3 \right) x_3^{k+1} \\ &\quad + T^{(1)} \varphi + \psi^{(1)} + x_3 \left(T^{(2)} \varphi + \psi^{(2)} \right) \end{aligned} \tag{2.5.24}$$

From Equations 2.5.12 and 2.5.17, we find the following system for the constants $C^{(2)}_j$

$$\begin{aligned} &C^{(2)}_1 x^0_1 + C^{(2)}_2 x^0_2 + C^{(2)}_3 = 0 \\ &E\left(I_{\alpha 1} C^{(2)}_1 + I_{\alpha 2} C^{(2)}_2 + A x^0_\alpha C^{(2)}_3 \right) = -\int_\Gamma p^{(0)}_\alpha ds - \int_{\Sigma_1} F^{(0)}_\alpha da \end{aligned} \tag{2.5.25}$$

The function $\psi^{(2)}$ is characterized by

$$\begin{aligned}\Delta\psi^{(2)} &= -2\big(C_1^{(2)}x_1 + C_2^{(2)}x_2 + C_3^{(2)}\big) \text{ on } \Sigma_1\\ \frac{\partial\psi^{(2)}}{\partial n} &= -n_\alpha \sum_{j=1}^{3} C_j^{(2)} w_\alpha^{(j)} \text{ on } \Gamma\end{aligned} \tag{2.5.26}$$

It follows from Equation 2.5.18 that the constant $T^{(2)}$ is given by

$$\begin{aligned}DT^{(2)} = &-\int_\Gamma \varepsilon_{\alpha\beta}x_\alpha p_\beta^{(0)}\,ds - \int_{\Sigma_1} \varepsilon_{\alpha\beta}x_\alpha F_\beta^{(0)}\,da\\ &- \mu\int_{\Sigma_1} \varepsilon_{\alpha\beta}x_\alpha\left[\sum_{j=1}^{3} C_j^{(2)} w_\beta^{(j)} + \psi_{,\beta}^{(2)}\right]da\end{aligned} \tag{2.5.27}$$

The constants $C_j^{(1)}$ are determined from Equations 2.5.12 and 2.5.20, so that

$$\begin{aligned}EA\big(C_1^{(1)}x_1^0 + C_2^{(1)}x_2^0 + C_3^{(1)}\big) &= -\int_\Gamma p_3^{(0)}\,ds - \int_{\Sigma_1} F_3^{(0)}\,da\\ E\big(I_{\alpha1}C_1^{(1)} + I_{\alpha2}C_2^{(1)} + Ax_\alpha^0 C_3^{(1)}\big) &= -F_\alpha - \int_\Gamma x_\alpha p_3^{(0)}\,ds - \int_{\Sigma_1} x_\alpha F_3^{(0)}\,da\end{aligned} \tag{2.5.28}$$

From Equations 1.7.9 and 2.5.9, we obtain the following boundary-value problem for $\psi^{(1)}$

$$\begin{aligned}\mu\Delta\psi^{(1)} &= -F_3^{(0)} - 2\mu\big(C_1^{(1)}x_1 + C_2^{(1)}x_2 + C_3^{(1)}\big) \text{ on } \Sigma_1\\ \mu\frac{\partial\psi^{(1)}}{\partial n} &= p_3^{(0)} - \mu n_\alpha \sum_{j=1}^{3} C_j^{(1)} w_\alpha^{(j)} \text{ on } \Gamma\end{aligned} \tag{2.5.29}$$

In this case, the constant $T^{(1)}$ is given by

$$DT^{(1)} = -M_3 - \mu\int_{\Sigma_1} \varepsilon_{\alpha\beta}x_\alpha\left[\sum_{j=1}^{3} C_j^{(1)} w_\beta^{(j)} + \psi_{,\beta}^{(1)}\right]da \tag{2.5.30}$$

It follows from Equations 2.5.4, 2.5.6, and 2.5.7 that $v_\alpha^{(0)}$ are the components of the displacement vector in the plane strain problem characterized by the geometrical equations

$$\gamma_{\alpha\beta}^{(0)} = \frac{1}{2}\big(v_{\alpha,\beta}^{(0)} + v_{\beta,\alpha}^{(0)}\big) \tag{2.5.31}$$

the constitutive equations

$$s_{\alpha\beta}^{(0)} = \lambda\gamma_{\rho\rho}^{(0)}\delta_{\alpha\beta} + 2\mu\gamma_{\alpha\beta}^{(0)} \tag{2.5.32}$$

the equilibrium equations

$$s^{(0)}_{\alpha\beta,\beta} + F^{(0)}_\alpha + (\lambda+\mu)\big(T^{(2)}\varphi_{,\alpha} + \psi^{(2)}_{,\alpha}\big) + \mu\left(T^{(2)}\varepsilon_{\beta\alpha}x_\beta + \sum_{j=1}^{3} C^{(2)}_j w^{(j)}_\alpha\right) = 0 \tag{2.5.33}$$

on Σ_1 and the boundary conditions

$$s^{(0)}_{\alpha\beta}n_\beta = p^{(0)}_\alpha - \lambda\big(T^{(2)}\varphi + \psi^{(2)}\big)n_\alpha \text{ on } \Gamma \tag{2.5.34}$$

The constants $C^{(0)}_j$ are determined from the system 2.5.21 and 2.5.22.

If $f_i = 0$ and $\tilde{t}_i = 0$, then the solution 2.5.24 reduces to Saint-Venant's solution to the relaxed Saint-Venant's problem.

2.6 Applications

2.6.1 Deformation of a Circular Cylinder Subject to Uniform Load

Let us study the deformation of a homogeneous and isotropic circular cylinder that occupies the region $B = \{x : x_1^2 + x_2^2 < a^2, 0 < x_3 < h\}$, $(a > 0)$, and is subjected to a uniform load. We assume that the lateral surface is subjected to a constant pressure and that the body force is axial. Thus, we have

$$\tilde{t}_\alpha = Pn_\alpha, \qquad \tilde{t}_3 = 0, \qquad f_\alpha = 0, \qquad f_3 = Q \tag{2.6.1}$$

where P and Q are given constants. Clearly, $n_\alpha = x_\alpha/a$ so that we can take $\varphi = 0$. From Equation 1.3.32, we obtain

$$D = \frac{1}{2}\pi a^4 \mu \tag{2.6.2}$$

By using Equations 1.7.14 and 2.6.1, and taking into account that $p^{(0)}_\alpha = Pn_\alpha$, $p^{(0)}_3 = 0$, $F^{(0)}_\alpha = 0$, and $F^{(0)}_3 = Q$, we have

$$I_{\alpha\beta} = I\delta_{\alpha\beta}, \qquad I = \frac{1}{4}\pi a^4, \qquad x^0_\alpha = 0, \qquad \int_\Gamma p^{(0)}_\alpha ds = 0$$

so that the system 2.5.25 implies that

$$C^{(2)}_j = 0 \tag{2.6.3}$$

From Equation 2.5.26, we conclude that $\psi^{(2)}=0$ on Σ_1. Clearly, Equations 2.5.27 and 2.6.2 imply that $T^{(2)}=0$. It follows from Equation 2.5.28 that

$$C^{(1)}_\alpha = -\frac{1}{EI}F_\alpha, \qquad C^{(1)}_3 = -\frac{1}{E}Q \tag{2.6.4}$$

The boundary-value problem 2.5.29 reduces to

$$\Delta\psi^{(1)} = -\frac{1}{\mu}Q - 2\big(C_1^{(1)}x_1 + C_2^{(1)}x_2 + C_3^{(1)}\big) \text{ on } \Sigma_1$$

$$\frac{\partial\psi^{(1)}}{\partial n} = \frac{1}{2}\nu a\big(C_1^{(1)}x_1 + C_2^{(1)}x_2 + 2C_3^{(1)}\big) \text{ on } \Gamma$$

The solution of this problem is given by

$$\begin{aligned}\psi^{(1)} = &-\frac{1}{4EI}\left\{F_1\left[a^2(3+2\nu)\,x_1 - \left(x_1^3 + x_1x_2^2\right)\right]\right.\\ &\left.+ F_2\left[a^2(3+2\nu)\,x_2 - \left(x_2^3 + x_1^2x_2\right)\right]\right\}\\ &-\frac{\nu}{2E}Q\left(x_1^2 + x_2^2\right), \quad (x_1, x_2) \in \Sigma_1 \end{aligned} \tag{2.6.5}$$

In view of the relations 2.6.1, the equilibrium equations 2.5.33 reduce to

$$s_{\alpha\beta,\beta}^{(0)} = 0 \text{ on } \Sigma_1 \tag{2.6.6}$$

The boundary conditions 2.5.34 become

$$s_{\alpha\beta}^{(0)}n_\beta = Pn_\alpha \text{ on } \Gamma \tag{2.6.7}$$

The solution of the boundary-value problem 2.5.31, 2.5.32, 2.6.6, and 2.6.7 is given by

$$v_\alpha^{(0)} = \frac{1}{2\,(\lambda+\mu)}Px_\alpha \text{ on } \Sigma_1 \tag{2.6.8}$$

In view of Equations 1.7.9 and 2.6.5, from Equation 2.5.30, we obtain

$$DT^{(1)} = -M_3 \tag{2.6.9}$$

It follows from Equations 2.5.21, 2.5.22, and 2.6.8 that the constants $C_j^{(0)}$ are given by

$$C_\alpha^{(0)} = \frac{1}{EI}\,\varepsilon_{\alpha\beta}M_\beta, \qquad C_3^{(0)} = -\frac{1}{EA}F_3 - \frac{2\nu}{E}P \tag{2.6.10}$$

We conclude that the solution of the problem has the form

$$u_\alpha = \sum_{k=0}^{1}\left[-\frac{1}{(k+2)!}C_\alpha^{(k)}x_3^{k+2} + \frac{1}{k!}x_3^k\sum_{j=1}^{3}C_j^{(k)}w_\alpha^{(j)}\right] + \varepsilon_{\alpha\beta}T^{(1)}x_\beta x_3 + v_\alpha^{(0)}$$

$$u_3 = \sum_{k=0}^{1}\frac{1}{(k+1)!}\big(C_1^{(k)}x_1 + C_2^{(k)}x_2 + C_3^{(k)}\big)x_3^{k+1} + \psi^{(1)}$$

Here, the constants $C_j^{(k)}$, $(k = 0, 1)$, are given by Equations 2.6.4 and 2.6.10, $T^{(1)}$ and D are given by Equations 2.6.2 and 2.6.9, the function $\psi^{(1)}$ is defined in Equations 2.6.5, $v_\alpha^{(0)}$ are given by Equation 2.6.8, and the functions $w_\alpha^{(j)}$ have the expressions 1.7.9. If $Q = 0$ and $P = 0$, then we obtain the solution of Saint-Venant's problem.

2.6.2 Thermoelastic Deformation of Cylinders

Let us use the results presented in Sections 2.2 and 2.3 to study the problem of thermal stresses in homogeneous and isotropic cylinders within the linear theory of thermoelastostatics.

Let T be the absolute temperature measured from the constant absolute temperature in the reference configuration. In the equilibrium theory of linear thermoelasticity, the temperature field T can be found by solving the heat boundary-value problem associated with the heat conduction and energy equations. In this section, we shall treat the temperature field T as having already been so determined.

As is usual in thermoelastostatics, we assume that the mechanical loads are absent. Thus, the principal attention is devoted to the deformation due to the temperature field.

We consider a formulation of the problem in which the detailed assignment of the terminal tractions is abandoned in favor of prescribing merely the appropriate stress resultants.

According to the body force analogy (cf. [38], Section 11), the thermoelastic problem reduces to the problem of finding an equilibrium displacement field $\mathbf{u}$ on B that corresponds to the body force field $\mathbf{f} = -\beta \operatorname{grad} T$ and satisfies the conditions

$$\begin{gathered} \mathbf{s}(\mathbf{u}) = \mathbf{p} \text{ on } \Pi, \qquad R_\alpha(\mathbf{u}) = 0, \qquad R_3(\mathbf{u}) = -\int_{\Sigma_1} \beta T da \\ H_\alpha(\mathbf{u}) = -\int_{\Sigma_1} \beta \varepsilon_{\alpha\rho} x_\rho T da, \qquad H_3(\mathbf{u}) = 0 \end{gathered} \tag{2.6.11}$$

where $\mathbf{p} = \beta T \mathbf{n}$. Here β is the stress-temperature modulus. We refer to the foregoing problem as the problem (Z).

2.6.3 Plane Temperature Field

We now consider the case when the temperature field is independent of the axial coordinate, that is,

$$T = T_0(x_1, x_2), \quad (x_1, x_2) \in \Sigma_1$$

where $T_0 \in C^2(\Sigma_1) \cap C^1(\overline{\Sigma}_1)$ is a prescribed field.

Clearly, in this case, the problem (Z) reduces to the Almansi–Michell problem which consists in finding a vector field $\mathbf{u} \in K_{III}(\mathbf{F}, \mathbf{M}, \mathbf{f}, \mathbf{p})$ where

$$\begin{gathered} F_\alpha = 0, \qquad F_3 = -\int_{\Sigma_1} \beta T_0 da, \qquad M_\alpha = -\int_{\Sigma_1} \beta \varepsilon_{\alpha\rho} x_\rho T_0 da \\ M_3 = 0, \qquad f_\alpha = -\beta T_{0,\alpha}, \qquad f_3 = 0, \qquad p_\alpha = \beta T_0 n_\alpha, \qquad p_3 = 0 \end{gathered} \tag{2.6.12}$$

A solution of this problem is given by Equations 2.2.2. In view of the relations 2.6.12,

$$\int_{\Sigma_1} f_\alpha da + \int_\Gamma p_\alpha ds = 0, \qquad \int_{\Sigma_1} \varepsilon_{\alpha\beta} x_\alpha f_\beta da + \int_\Gamma \varepsilon_{\alpha\beta} x_\alpha p_\beta ds = 0 \tag{2.6.13}$$

It follows from Equations 2.2.11, 2.2.18, and 2.6.13 that $c_i = 0$. Clearly, $\chi = 0$ is a solution of the boundary-value problem 2.2.14 and 2.2.16. Then it follows from Equations 2.2.17, 2.2.20, and 2.6.13 that $b_s = 0$. Now we can see that $\psi = 0$ is a solution of the boundary-value problem 2.2.13 and 2.2.15. The functions v_α are characterized by the plane strain problem

$$(s_{\alpha\beta}(\mathbf{v}))_{,\beta} + f_\alpha = 0 \text{ on } \Sigma_1, \qquad s_{\alpha\beta}(\mathbf{v}) n_\beta = p_\alpha \text{ on } \Gamma \tag{2.6.14}$$

where f_α and p_α are given by the relations 2.6.12. The system 2.2.21 reduces to

$$\begin{aligned} E\big(I_{\alpha\beta} a_\beta + A x_\alpha^0 a_3\big) &= \int_{\Sigma_1} \beta x_\alpha T_0 da - \lambda \int_{\Sigma_1} x_\alpha v_{\rho,\rho} da \\ AE\big(a_\alpha x_\alpha^0 + a_3\big) &= \int_{\Sigma_1} \beta T_0 da - \lambda \int_{\Sigma_1} v_{\rho,\rho} da \end{aligned} \tag{2.6.15}$$

Thus we conclude that a solution of the problem is given by

$$\begin{aligned} u_1 &= -\frac{1}{2} a_1 x_3^2 - \frac{1}{2} a_1 \nu \big(x_1^2 - x_2^2\big) - a_2 \nu x_1 x_2 - a_3 \nu x_1 + v_1 \\ u_2 &= -\frac{1}{2} a_2 x_3^2 - a_1 \nu x_1 x_2 - \frac{1}{2} a_2 \nu \big(x_1^2 - x_2^2\big) - a_3 \nu x_2 + v_2 \\ u_3 &= (a_1 x_1 + a_2 x_2 + a_3) x_3 \end{aligned} \tag{2.6.16}$$

If $T = T^*$, where T^* is a given constant, then

$$v_\alpha = \frac{\beta}{2(\lambda + \mu)} T^* x_\alpha \tag{2.6.17}$$

Let us suppose that the coordinate frame is chosen in such a way that the origin O coincides with the centroid of Σ_1. Then, it follows from Equations 2.6.15 and 2.6.17 that

$$a_\alpha = 0, \qquad a_3 = \beta T^* / (3\lambda + 2\mu)$$

2.7 Exercises

2.7.1 Study the deformation of an isotropic and homogeneous elastic cylinder which is subjected to a temperature field that is a polynomial in the axial coordinate.

2.7.2 Study the deformation of a homogeneous and isotropic circular cylinder which is subjected to the gravity force.

2.7.3 A homogeneous and isotropic material occupies the domain $B = \{x : a_2^2 < x_1^2 + x_2^2 < a_1^2, 0 < x_3 < h\}, (a_1 > 0, a_2 > 0)$. Investigate the extension and bending of the cylinder if the lateral surfaces are subjected to constant pressures.

2.7.4 Investigate the deformation of a circular cylinder when the lateral boundary is subjected to a pressure which is linear in the axial coordinate.

2.7.5 A homogeneous and isotropic elliptical cylinder is subjected to the loads $f_j = 0, \widetilde{t}_\alpha = Pn_\alpha, \widetilde{t}_3 = 0, F_3 = Q$, $F_\alpha = 0, M_j = 0$, where P and Q are prescribed constants. Study the deformation of the body.

2.7.6 Investigate the deformation of the right circular cylinder $B = \{x : x_1^2 + x_2^2 < a^2, 0 < x_3 < h\}$, $(a > 0)$ which is subjected on the lateral surface to the tractions $\widetilde{t}_1 = -px_2x_3$, $\widetilde{t}_2 = px_1x_3, \widetilde{t}_3 = 0$, where p is a constant.

2.7.7 Investigate the deformation of an isotropic and homogeneous elliptical cylinder which is subjected to a temperature field that is a polynomial in the axial coordinate, with constant coefficients.

2.7.8 Study the deformation of an isotropic and homogeneous circular cylinder subjected to the external loading $\{f_\alpha = 0, f_3 = G, \widetilde{t}_\alpha = (P_0 + P_1x_3 + P_2x_3^2)n_\alpha, \widetilde{t}_3 = H, F_1 = R, F_2 = F_3 = 0, M_j = 0\}$, where G, P_0, P_1, P_2, H, and R are prescribed constants.

2.7.9 An elliptical right cylinder is made of a homogeneous and isotropic elastic material. Let $(\tau_1, \tau_2, 0)$ designate the tangent unit vector along the boundary of the generic cross section. Study the deformation of the body which is subjected on the lateral boundary to the tractions $\widetilde{t}_1 = P\tau_1, \widetilde{t}_2 = P\tau_2$, $\widetilde{t}_3 = 0$, where P is a given constant.

Chapter 3

Deformation of Nonhomogeneous Cylinders

3.1 Preliminaries

This chapter is devoted to the study of the deformation of nonhomogeneous and isotropic cylinders. Most of the works concerned with Saint-Venant's problem are restricted to homogeneous cylinders. However, some investigations are devoted to Saint-Venant's problem for nonhomogeneous cylinders where the elastic coefficients are independent of the axial coordinate, they being prescribed functions of the remaining coordinates. This theory is of interest from both the mathematical and technical points of view [3,75,88,130]. According to Toupin [329], the proof of Saint-Venant's principle presented in Section 1.10 also remains valid for this kind of nonhomogeneous elastic bodies. The study of Saint-Venant's problem for nonhomogeneous cylinders was initiated by Nowinski and Turski [256] and was developed in various later works [150,279,303,318]. An account of the historical developments of the theory of nonhomogeneous elastic bodies as well as references to various contributions may be found in Refs. 175, 209, 219, and 290. Many works concerned with Saint-Venant's problem for nonhomogeneous cylinders are restricted to the case when the Poisson's ratio is constant. A method to solve the problem, which avoids this restriction, was presented in Ref. 149.

The equilibrium problem for heterogeneous elastic bodies was studied in various works [88,196,241]. Fichera [88] was the first to consider the case of the bodies compounded of different nonhomogeneous and anisotropic elastic materials. The deformation of cylinders compounded of different homogeneous and isotropic materials was first studied by Muskhelishvili [241] and his treatment was extended in various works [28,175,204,313]. Most of the works dealing with this problem are restricted to piecewise homogeneous cylinders. In Refs. 151 and 152, we established a solution of Saint-Venant's problem for a cylinder composed of two different nonhomogeneous elastic materials, where the elastic coefficients are independent of the axial coordinate. The mathematical formulation of the problems of extension, bending, torsion, and flexure of compound cylinders differs from that for homogeneous cylinders only in added boundary conditions on the interfaces of the media with different elastic properties. We shall assume that Σ_1 is a C^1-smooth domain ([88], p. 369). Let Γ_1 and Γ_2 be complementary subsets of Γ and let Γ_0 be a curve contained in Σ_1 with the property that $\overline{\Gamma}_0 \cup \overline{\Gamma}_\rho$, $(\rho = 1, 2)$, is the boundary of a regular domain A_ρ contained in Σ_1 such that $A_1 \cap A_2 = \emptyset$. We denote by B_ρ the cylinder that is defined by $B_\rho = \{x : (x_1, x_2) \in A_\rho, 0 < x_3 < h\}$, $(\rho = 1, 2)$. We assume that B_ρ is occupied by an elastic material with the elasticity field $\mathbf{C}^{(\rho)}, (\rho = 1, 2)$, and that $\mathbf{C}^{(\rho)}$ is symmetric, positive definite, and smooth on $\overline{B}_\rho$. Let Π_0 denote the surface of separation of the two materials. Clearly, $\Pi_0 = \{x : (x_1, x_2) \in \Gamma_0, 0 \leq x_3 \leq h\}$. We can consider that the cylinder B is composed of two materials which are welded together along Π_0. Let Π_1 and Π_2 be the complementary subsets of Π defined by $\Pi_\rho = \{x : (x_1, x_2) \in \Gamma_\rho, 0 \leq x_3 \leq h\}$. Assume that in the course of deformation, there is no separation of material along Π_0. The displacement vector field and the stress vector field are continuous in passing from one medium to another. Accordingly, we have the conditions

$$[u_i]_1 = [u_i]_2, \qquad [t_{i\beta}(\mathbf{u})]_1 n^0_\beta = [t_{i\beta}(\mathbf{u})]_2 n^0_\beta \text{ on } \Pi_0 \tag{3.1.1}$$

where we have indicated that the expressions in brackets are calculated for the material corresponding to the regions B_1 and B_2, respectively, and $(n^0_1, n^0_2, 0)$ are the components of the unit normal $\mathbf{n}^0$ of Π_0, outward to B_1.

In the first part of this chapter, we study the deformation of nonhomogeneous and isotropic cylinders when the elastic coefficients are independent of the axial coordinate. Then, the case of elastic cylinders composed of different nonhomogeneous and isotropic materials is investigated. This chapter points out the importance of the plane strain problem in the treatment of Saint-Venant's problem.

3.2 Plane Strain Problem: Auxiliary Plane Strain Problems

3.2.1 Basic Equations

In Section 1.5, we have studied the plane strain problem for homogeneous and isotropic elastic cylinders. In this section, we suppose that the cylinder B is made of a nonhomogeneous and isotropic elastic material for which the constitutive coefficients are independent of the axial coordinate, that is,

$$\lambda = \lambda(x_1, x_2), \qquad \mu = \mu(x_1, x_2), \quad (x_1, x_2) \in \Sigma_1 \tag{3.2.1}$$

We suppose that the domain Σ_1 is C^∞-smooth [88], and that the functions λ and μ belong to C^∞ and satisfy the conditions 1.5.16. We restrict our attention to the second boundary-value problem and assume that f_α and $\widetilde{t}_\alpha$ are independent of x_3 and are prescribed functions of class C^∞. We consider only a C^∞-*theory* but it is possible to get a classical solution of the problem for more general domains and more general assumptions of regularity for the above functions [88]. We have chosen these hypotheses to best emphasize the method for the solving of Saint-Venant's problem.

The second boundary-value problem consists in finding of the functions u_α of class $C^2(\Sigma_1)\cap C^1(\overline{\Sigma}_1)$ that satisfy Equations 1.5.2, 1.5.3, and 1.5.4 and the boundary conditions 1.5.6, when λ and μ are prescribed functions of the form 3.2.1. The first boundary-value problem can be introduced as in Section 1.5.

Under the above assumptions of regularity for the domain Σ_1 and the prescribed functions, Fichera [88] established the following result.

Theorem 3.2.1 *The second boundary-value problem has solution belonging to $C^\infty(\overline{\Sigma}_1)$ if and only if the conditions* 1.5.17 *hold.*

We note that Theorem 1.5.1 remains valid for the nonhomogeneous bodies considered in this section.

From the basic equations, we obtain the equations of equilibrium expressed in terms of the displacement vector field,

$$\mu\Delta u_1 + (\lambda+\mu)\frac{\partial\vartheta}{\partial x_1} + \vartheta\frac{\partial\lambda}{\partial x_1} + 2\frac{\partial\mu}{\partial x_1}\frac{\partial u_1}{\partial x_1} + \frac{\partial\mu}{\partial x_2}\left(\frac{\partial u_1}{\partial x_2} + \frac{\partial u_2}{\partial x_1}\right) + f_1 = 0$$

$$\mu\Delta u_2 + (\lambda+\mu)\frac{\partial\vartheta}{\partial x_2} + \vartheta\frac{\partial\lambda}{\partial x_2} + 2\frac{\partial\mu}{\partial x_2}\frac{\partial u_2}{\partial x_2} + \frac{\partial\mu}{\partial x_1}\left(\frac{\partial u_1}{\partial x_2} + \frac{\partial u_2}{\partial x_1}\right) + f_2 = 0 \text{ on } \Sigma_1 \tag{3.2.2}$$

where we have used the notation $\vartheta = u_{\rho,\rho}$. Thus, we have an alternative formulation of the second boundary-value problem: to find the functions u_α of class $C^\infty(\overline{\Sigma}_1)$ that satisfy Equations 3.2.2 on Σ_1 and the boundary conditions 1.5.9 on Γ. It follows from Equations 1.1.7 that

$$\lambda = \frac{2\nu\mu}{1-2\nu} \tag{3.2.3}$$

Let us assume that the Poisson's ratio is constant. Then, in view of Equation 3.2.3, Equations 3.2.2 reduce to

$$\Delta u_1 + \eta\frac{\partial\vartheta}{\partial x_1} + 2\frac{\partial \ln\mu}{\partial x_1}\left(\frac{\partial u_1}{\partial x_1} + \nu\eta\vartheta\right) + \frac{\partial \ln\mu}{\partial x_2}\left(\frac{\partial u_1}{\partial x_2} + \frac{\partial u_2}{\partial x_1}\right) + \frac{1}{\mu}f_1 = 0$$

$$\Delta u_2 + \eta\frac{\partial\vartheta}{\partial x_2} + 2\frac{\partial \ln\mu}{\partial x_2}\left(\frac{\partial u_2}{\partial x_2} + \nu\eta\vartheta\right) + \frac{\partial \ln\mu}{\partial x_1}\left(\frac{\partial u_1}{\partial x_2} + \frac{\partial u_2}{\partial x_1}\right) + \frac{1}{\mu}f_2 = 0 \text{ on } \Sigma_1 \tag{3.2.4}$$

where η is defined by

$$\eta = \frac{1}{1-2\nu}$$

3.2.2 Airy Function

If the body forces vanish, then the equations of equilibrium 1.5.4 reduce to Equations 1.5.18. These equations are satisfied if the stresses $t_{\alpha\beta}$ are expressed by Equation 1.5.19 in terms of the Airy function χ. Let us impose the compatibility equation 1.5.21. As in Ref. 209, we express the constitutive equations 1.5.3 in the form

$$e_{\alpha\beta} = (\gamma - q)t_{\rho\rho}\delta_{\alpha\beta} + qt_{\alpha\beta} \tag{3.2.5}$$

where

$$\gamma = \frac{1-\nu^2}{E}, \qquad q = \frac{1+\nu}{E}$$

It follows from Equations 1.5.19 and 3.2.5 that

$$e_{\alpha\beta} = \delta_{\alpha\beta}\gamma\Delta\chi - q\chi_{,\alpha\beta} \tag{3.2.6}$$

In view of Equation 3.2.6, the compatibility equation 1.5.21 reduces to the following equation [209]

$$\Delta(\gamma\Delta\chi) = q_{,22}\chi_{,11} + q_{,11}\chi_{,22} - 2q_{,12}\chi_{,12} \text{ on } \Sigma_1 \tag{3.2.7}$$

When the body is homogeneous, Equation 3.2.7 takes the form 1.5.22.

We assume that Σ_1 is a simply-connected domain. Then, in the case of nonhomogeneous bodies, the second boundary-value problem reduces to finding of the Airy function χ that satisfies Equation 3.2.7 on Σ_1 and the boundary conditions 1.5.25 on Γ.

In contrast with the case of homogeneous bodies, the stresses $t_{\alpha\beta}$ depend on the constitutive coefficients. Other results concerning the plane strain problem and the solutions of particular problems may be found in the work of Lomakin [209].

3.2.3 Auxiliary Plane Strain Problems

We will have occasion to use three special problems $\mathcal{D}^{(k)}$, $(k = 1, 2, 3)$, of plane strain. The problem $\mathcal{D}^{(1)}$ is characterized by the body forces

$$f_\alpha = (\lambda x_1)_{,\alpha} \text{ on } \Sigma_1$$

and the following tractions

$$\widetilde{t}_\alpha = -\lambda x_1 n_\alpha \text{ on } \Gamma$$

In the problem $\mathcal{D}^{(2)}$, the body forces are given by

$$f_\alpha = (\lambda x_2)_{,\alpha} \text{ on } \Sigma_1$$

and the tractions are

$$\widetilde{t}_\alpha = -\lambda x_2 n_\alpha \text{ on } \Gamma$$

The problem $\mathcal{D}^{(3)}$ is characterized by the body forces

$$f_\alpha = \lambda_{,\alpha} \text{ on } \Sigma_1$$

and the tractions

$$\widetilde{t}_\alpha = -\lambda n_\alpha \text{ on } \Gamma$$

In what follows, we denote by $u_\alpha^{(k)}, e_{\alpha\beta}^{(k)}$, and $t_{\alpha\beta}^{(k)}$, respectively, the components of the displacement vector, the components of the strain tensor, and the components of the stress tensor from the problems $\mathcal{D}^{(k)}$. The problems $\mathcal{D}^{(k)}$ are characterized by the strain–displacement relations

$$e_{\alpha\beta}^{(k)} = \frac{1}{2}(u_{\alpha,\beta}^{(k)} + u_{\beta,\alpha}^{(k)}) \tag{3.2.8}$$

the constitutive equations

$$t_{\alpha\beta}^{(k)} = \lambda e_{\rho\rho}^{(k)}\delta_{\alpha\beta} + 2\mu e_{\alpha\beta}^{(k)} \tag{3.2.9}$$

the equations of equilibrium

$$\begin{gathered} t_{\beta\alpha,\beta}^{(1)} + (\lambda x_1)_{,\alpha} = 0, \qquad t_{\beta\alpha,\beta}^{(2)} + (\lambda x_2)_{,\alpha} = 0 \\ t_{\beta\alpha,\beta}^{(3)} + \lambda_{,\alpha} = 0 \text{ on } \Sigma_1 \end{gathered} \tag{3.2.10}$$

and the following boundary conditions

$$t_{\beta\alpha}^{(1)} n_\beta = -\lambda x_1 n_\alpha, \qquad t_{\beta\alpha}^{(2)} n_\beta = -\lambda x_2 n_\alpha, \qquad t_{\beta\alpha}^{(3)} n_\beta = -\lambda n_\alpha \text{ on } \Gamma \tag{3.2.11}$$

It is easy to prove that the necessary and sufficient conditions 1.5.17 for the existence of the solution are satisfied for each boundary-value problem $\mathcal{D}^{(k)}$.

We note that the solutions of the problem $\mathcal{D}^{(k)}$ depend only on the domain Σ_1 and the elastic coefficients.

It is easy to see that for homogeneous and isotropic bodies, the solutions of the problems $\mathcal{D}^{(k)}$ are

$$\begin{aligned} u_1^{(1)} &= -\frac{\lambda}{4(\lambda+\mu)}(x_1^2 - x_2^2), \qquad u_2^{(1)} = -\frac{\lambda}{2(\lambda+\mu)}x_1x_2 \\ u_1^{(2)} &= -\frac{\lambda}{2(\lambda+\mu)}x_1x_2, \qquad u_2^{(2)} = \frac{\lambda}{4(\lambda+\mu)}(x_1^2 - x_2^2) \\ u_\alpha^{(3)} &= -\frac{\lambda}{2(\lambda+\mu)}x_\alpha, \qquad (x_1, x_2) \in \Sigma_1 \end{aligned} \tag{3.2.12}$$

Remark. In the case of homogeneous and isotropic elastic bodies, the solutions $u_\alpha^{(k)}$ of the problems $\mathcal{D}^{(k)}$ are identical with the functions $w_\alpha^{(k)}$ defined in Equations 1.7.9.

3.3 Extension and Bending of Nonhomogeneous Cylinders

Let the loading applied on Σ_1 be statically equivalent to a force $\mathbf{F} = F_3\mathbf{e}_3$ and a moment $\mathbf{M} = M_\alpha \mathbf{e}_\alpha$. Thus, the conditions on Σ_1 reduce to

$$\int_{\Sigma_1} t_{3\alpha} da = 0, \qquad \int_{\Sigma_1} \varepsilon_{\alpha\beta} x_\alpha t_{3\beta} da = 0 \tag{3.3.1}$$

$$\int_{\Sigma_1} t_{33} da = -F_3, \qquad \int_{\Sigma_1} x_\alpha t_{33} da = \varepsilon_{\alpha\beta} M_\beta \tag{3.3.2}$$

The problem consists in the finding of a displacement vector field that satisfies Equations 1.1.1, 1.1.4, and 1.1.8 on B and the boundary conditions 1.3.1, 3.3.1, and 3.3.2, when λ and μ have the form 3.2.1.

The results presented in this section have been established in Ref. 149. We seek the solution in the form

$$u_\alpha = -\frac{1}{2} a_\alpha x_3^2 + \sum_{k=1}^{3} a_k u_\alpha^{(k)}, \qquad u_3 = (a_1 x_1 + a_2 x_2 + a_3) x_3 \tag{3.3.3}$$

where $u_\alpha^{(k)}$ are the components of displacement vector field from the problem $\mathcal{D}^{(k)}$, $(k = 1, 2, 3)$, and a_k are unknown constants. From Equations 1.1.1 and 3.3.3, we obtain

$$e_{\alpha\beta} = \sum_{k=1}^{3} a_k e_{\alpha\beta}^{(k)}, \qquad e_{\alpha 3} = 0, \qquad e_{33} = a_1 x_1 + a_2 x_2 + a_3 \tag{3.3.4}$$

where $e_{\alpha\beta}^{(k)}$ are given by Equation 3.2.8. By Equations 1.1.4 and 3.3.4, we get

$$\begin{aligned} t_{\alpha\beta} &= \lambda(a_1 x_1 + a_2 x_2 + a_3)\delta_{\alpha\beta} + \sum_{k=1}^{3} a_k t_{\alpha\beta}^{(k)}, \qquad t_{\alpha 3} = 0 \\ t_{33} &= (\lambda + 2\mu)(a_1 x_1 + a_2 x_2 + a_3) + \lambda \sum_{k=1}^{3} a_k e_{\alpha\alpha}^{(k)} \end{aligned} \tag{3.3.5}$$

where $t_{\alpha\beta}^{(k)}$ are the stresses from the plane strain problem $\mathcal{D}^{(k)}$. The equations of equilibrium 1.1.8 and the boundary conditions 1.3.1 are satisfied on the basis of Equations 3.2.10 and 3.2.11.

The conditions 3.3.1 are identically satisfied on the basis of the relations 3.3.5. By Equations 3.3.2 and 3.3.5, we obtain the following system for the unknown constants a_1, a_2, and a_3

$$D_{ij} a_j = C_i \tag{3.3.6}$$

where

$$\begin{aligned} D_{\alpha\beta} &= \int_{\Sigma_1} x_\alpha\big[(\lambda+2\mu)x_\beta + \lambda e_{\rho\rho}^{(\beta)}\big]\,da \\ D_{\alpha 3} &= \int_{\Sigma_1} x_\alpha\big[\lambda+2\mu + \lambda e_{\rho\rho}^{(3)}\big]\,da \\ D_{3\alpha} &= \int_{\Sigma_1} \big[(\lambda+2\mu)x_\alpha + \lambda e_{\rho\rho}^{(\alpha)}\big]\,da \\ D_{33} &= \int_{\Sigma_1} \big[\lambda+2\mu + \lambda e_{\rho\rho}^{(3)}\big]\,da \end{aligned} \tag{3.3.7}$$

and

$$C_\alpha = \varepsilon_{\alpha\beta} M_\beta, \qquad C_3 = -F_3 \tag{3.3.8}$$

Clearly, the constants D_{ij} can be calculated after the displacement $u_\alpha^{(k)}$ are determined. Let us prove that the system 3.3.6 can always be solved for a_1, a_2, and a_3. The relations 3.3.3 and 3.3.5 can be written in the form

$$u_i = \sum_{k=1}^{3} a_k \omega_i^{(k)}, \qquad t_{ij} = \sum_{k=1}^{3} a_k \tau_{ij}^{(k)} \tag{3.3.9}$$

where

$$\begin{aligned} &\omega_\alpha^{(b)} = -\frac{1}{2}x_3^2\delta_{\alpha\beta} + u_\alpha^{(\beta)}, \qquad \omega_\alpha^{(3)} = u_\alpha^{(3)}, \qquad \omega_3^{(\alpha)} = x_\alpha x_3 \\ &\omega_3^{(3)} = x_3, \qquad \tau_{\alpha\beta}^{(\rho)} = \lambda x_\rho \delta_{\alpha\beta} + t_{\alpha\beta}^{(\rho)}, \qquad \tau_{\alpha\beta}^{(3)} = \lambda\delta_{\alpha\beta} + t_{\alpha\beta}^{(3)} \\ &\tau_{\alpha 3}^{(k)} = 0, \qquad \tau_{33}^{(\rho)} = (\lambda+2\mu)x_\rho + \lambda e_{\alpha\alpha}^{(\rho)}, \qquad \tau_{33}^{(3)} = \lambda+2\mu+\lambda e_{\alpha\alpha}^{(3)}, \quad (\rho=1,2) \end{aligned} \tag{3.3.10}$$

In view of Equations 1.1.12, 1.1.13, and 3.3.9,

$$U(\mathbf{u}) = \sum_{i,j=1}^{3} U(\boldsymbol{\omega}^{(i)}, \boldsymbol{\omega}^{(j)}) a_i a_j \tag{3.3.11}$$

By Equations 1.1.14, 1.1.16, and 1.1.17, we get

$$U(\boldsymbol{\omega}^{(i)}, \boldsymbol{\omega}^{(j)}) = U(\boldsymbol{\omega}^{(j)}, \boldsymbol{\omega}^{(i)}) \tag{3.3.12}$$

$$2U(\boldsymbol{\omega}^{(i)}, \boldsymbol{\omega}^{(j)}) = \int_{\partial B} \omega_k^{(i)} \tau_{pk}^{(j)} n_p\, da \tag{3.3.13}$$

Since the elastic potential is positive definite, we have

$$\det(U(\boldsymbol{\omega}^{(i)}, \boldsymbol{\omega}^{(j)})) \neq 0 \tag{3.3.14}$$

Let us apply the relations 3.3.13 for $\boldsymbol{\omega}^{(1)}$ and $\boldsymbol{\omega}^{(2)}$. We note that on Σ_2 we have

$$\omega_\alpha^{(\beta)} = -\frac{1}{2}h^2\delta_{\alpha\beta} + u_\alpha^{(\beta)}, \qquad \omega_3^{(\alpha)} = x_\alpha h$$
$$\tau_{p\alpha}^{(\beta)}n_p = \tau_{3\alpha}^{(\beta)} = 0, \qquad \tau_{p3}^{(\beta)}n_p = \tau_{33}^{(\beta)} = (\lambda + 2\mu)x_\beta + \lambda e_{\alpha\alpha}^{(\beta)} \tag{3.3.15}$$

Similarly, on Σ_1 we get

$$\tau_{p\alpha}^{(\beta)}n_p = 0, \qquad \omega_\alpha^{(\beta)} = u_\alpha^{(\beta)}, \qquad \omega_3^{(\beta)} = 0 \tag{3.3.16}$$

It follows from Equations 1.3.1, 3.3.7, 3.3.13, 3.3.15, and 3.3.16 that

$$\begin{aligned} 2U(\boldsymbol{\omega}^{(1)}, \boldsymbol{\omega}^{(2)}) &= \int_{\partial B} \omega_k^{(1)}\tau_{pk}^{(1)}n_p da \\ &= h\int_{\Sigma_1} x_1[(\lambda + 2\mu)x_1 + \lambda e_{\rho\rho}^{(1)}]da = hD_{11} \\ 2U(\boldsymbol{\omega}^{(1)}, \boldsymbol{\omega}^{(2)}) &= hD_{12} \end{aligned}$$

In a similar way, we find

$$2U(\boldsymbol{\omega}^{(i)}, \boldsymbol{\omega}^{(j)}) = hD_{ij} \tag{3.3.17}$$

We note that Equations 3.3.12 and 3.3.17 imply that $D_{ij} = D_{ji}$. By Equations 3.3.14 and 3.3.17,

$$\det(D_{ij}) \neq 0 \tag{3.3.18}$$

so that the system 3.3.6 uniquely determines the constants a_k. Thus, we conclude that the constants a_k can be determined so that the functions 3.3.3 be a solution of the problem of extension and bending.

If the material is homogeneous and isotropic, then Equations 3.2.12 and 3.3.7 imply

$$D_{\alpha\beta} = EI_{\alpha\beta}, \qquad D_{\alpha 3} = D_{3\alpha} = EAx_\alpha^0, \qquad D_{33} = EA \tag{3.3.19}$$

where $I_{\alpha\beta}, x_\alpha^0$, and A are defined by Equations 1.4.9 and 1.7.14. It is easy to see that in this case we rediscover the Saint-Venant's solution of the problem.

Remark. The form 3.3.3 of the solution is justified by Theorem 1.7.1, which holds also when λ and μ are functions of the variables x_α.

3.4 Torsion

The torsion problem consists in the determination of a displacement vector field that satisfies Equations 1.1.1, 1.1.4, and 1.1.8 on B and the boundary conditions 1.3.1 and the condition for $x_3 = 0$, when the elastic coefficients are functions independent of x_3.

We seek a solution of the torsion problem in the form 1.3.23, where φ is an unknown function of x_1 and x_2, and τ is an unknown constant. It follows from Equations 1.1.1, 1.1.4, and 1.3.23 that the components of the stress tensor are given by Equations 1.3.24. The equations of equilibrium 1.1.8 are satisfied if the function φ satisfies the equation

$$(\mu\varphi_{,\alpha})_{,\alpha} = \varepsilon_{\rho\beta}(\mu x_\beta)_{,\rho} \text{ on } \Sigma_1 \tag{3.4.1}$$

In view of Equations 1.3.24, the conditions 1.3.1 on the lateral boundary reduce to

$$\frac{\partial\varphi}{\partial n} = \varepsilon_{\alpha\beta}x_\beta n_\alpha \text{ on } \Gamma \tag{3.4.2}$$

Let us consider the boundary-value problem

$$(\mu u_{,\alpha})_{,\alpha} = F \text{ on } \Sigma_1, \qquad \mu u_{,\alpha}n_\alpha = G \text{ on } \Gamma \tag{3.4.3}$$

where μ, F, and G are prescribed functions of class C^∞. Necessary and sufficient condition to solve the boundary-value problem 3.4.3 is (cf. [55,88])

$$\int_{\Sigma_1} F da = \int_\Gamma G ds \tag{3.4.4}$$

It is easy to see that in the case of the boundary-value problem 3.4.1 and 3.4.2, the condition 3.4.4 is satisfied. In what follows, we shall assume that the function φ is known.

In view of Equations 1.3.24 and 1.3.57, the conditions 1.3.20 and 1.3.21 are identically satisfied. The condition 1.3.22 reduces to

$$D_*\tau = -M_3 \tag{3.4.5}$$

where

$$D_* = \int_{\Sigma_1} \mu(\varepsilon_{\alpha\beta}x_\alpha\varphi_{,\beta} + x_\rho x_\rho)da \tag{3.4.6}$$

By using Equations 3.4.1, 3.4.2, and divergence theorem,

$$\begin{aligned}\int_{\Sigma_1} \mu\varepsilon_{\alpha\beta}x_\alpha\varphi_{,\beta}da &= \int_\Gamma \mu\varepsilon_{\alpha\beta}x_\alpha\varphi n_\beta ds - \int_{\Sigma_1}\varphi(\mu\varepsilon_{\alpha\beta}x_\alpha)_{,\beta}da\\ &= -\int_\Gamma \mu\varphi\varphi_{,\alpha}n_\alpha ds + \int_{\Sigma_1}\varphi(\mu\varphi_{,\alpha})_{,\alpha}da\\ &= -\int_{\Sigma_1}\mu\varphi_{,\alpha}\varphi_{,\alpha}da\end{aligned}$$

Thus, we can express the constant D_* in the form

$$\begin{aligned}D_* &= \int_{\Sigma_1}\mu(\varphi_{,\alpha}\varphi_{,\alpha} + 2\varepsilon_{\alpha\beta}x_\alpha\varphi_{,\beta} + x_\rho x_\rho)\\ &= \int_{\Sigma_1}\mu(\varphi_{,\alpha} + \varepsilon_{\beta\alpha}x_\beta)(\varphi_{,\alpha} + \varepsilon_{\rho\alpha}x_\rho)da\end{aligned} \tag{3.4.7}$$

It follows from the relations 1.1.5 and 3.4.7 that $D_* > 0$. The relation 3.4.5 determines the constant τ. We conclude that the displacement vector field 1.3.23, where φ is the solution of the boundary-value problem 3.4.1, 3.4.2, and τ is given by Equation 3.4.5, is a solution of the torsion problem. Clearly, if the material is homogeneous, then we rediscover Saint-Venant's solution.

We note that Equation 3.4.1 can be written as follows

$$\Delta\varphi + (\ln\mu)_{,\alpha}(\varphi_{,\alpha} - \varepsilon_{\alpha\beta}x_\beta) = 0 \text{ on } \Sigma_1 \tag{3.4.8}$$

A form for μ that is commonly used [209] is

$$\mu = \mu_0 \exp(\alpha x_1 + \beta x_2) \tag{3.4.9}$$

where μ_0, α, and β are prescribed constants. For the law 3.4.9, Equation 3.4.8 becomes

$$\Delta\varphi + \alpha(\varphi_{,1} - x_2) + \beta(\varphi_{,2} + x_1) = 0 \text{ on } \Sigma_1 \tag{3.4.10}$$

The torsion problem can be formulated in terms of the stress function χ defined by

$$\mu(\varphi_{,1} - x_2) = \chi_{,2}, \qquad \mu(\varphi_{,2} + x_2) = -\chi_{,1} \tag{3.4.11}$$

It follows from Equation 3.4.11 that χ satisfies the following equation

$$\left(\frac{1}{\mu}\chi_{,\alpha}\right)_{,\alpha} = -2 \text{ on } \Sigma_1 \tag{3.4.12}$$

In view of Equations 1.3.39 and 3.4.11, the function χ satisfies the following condition on the boundary of the simply-connected domain Σ_1

$$\chi = 0 \text{ on } \Gamma \tag{3.4.13}$$

By Equations 1.3.24 and 3.4.11, we find that

$$t_{13} = \tau\chi_{,2}, \qquad t_{23} = -\tau\chi_{,1} \tag{3.4.14}$$

Using Equations 3.4.6, 3.4.11, and 3.4.13, we can express D_* as follows

$$D_* = 2\int_{\Sigma_1} \chi da \tag{3.4.15}$$

Consider the family of curves in Σ_1 defined by

$$\chi(x_1, x_2) = 0 \tag{3.4.16}$$

By Equations 1.3.39, for any curve of this family we have

$$\chi_{,1}n_2 - \chi_{,2}n_1 = 0$$

In view of Equation 3.4.14, the last relation implies that the stress vector $\mathbf{T} = t_{\alpha 3}\mathbf{e}_\alpha$ is directed along the tangent to the curve. The curves 3.4.16 are called the lines of shearing stress. The magnitude of the tangential stress $\mathbf{T}$ is

$$|\mathbf{T}| = \mu(\chi_{,\alpha}\chi_{,\alpha})^{1/2}$$

We note that, instead of solving the Neumann problem 3.4.1 and 3.4.2, we can equally well solve the Dirichlet problem 3.4.12 and 3.4.13.

3.5 Flexure

We assume that the loading applied on Σ_1 is statically equivalent to the force $\mathbf{F} = F_\alpha \mathbf{e}_\alpha$ and the moment $\mathbf{M} = \mathbf{0}$. The conditions on Σ_1 are given by Equations 1.3.48, 1.3.49, and 1.4.1. The flexure problem consists in the finding of a displacement vector field that satisfies the Equations 1.1.1, 1.1.4, and 1.1.8 on B and the boundary conditions 1.3.1, 1.3.48, 1.3.49, and 1.4.1, when λ and μ have the form 3.2.1.

In view of Theorem 1.7.2, we seek the solution of the flexure problem in the form

$$\begin{aligned} u_\alpha &= -\frac{1}{6} b_\alpha x_3^3 + x_3 \sum_{k=1}^{3} b_k u_\alpha^{(k)} - \tau \varepsilon_{\alpha\beta} x_\beta x_3 \\ u_3 &= \frac{1}{2}(b_1 x_2 + b_2 x_2 + b_3) x_3^2 + \tau\varphi + G(x_1, x_2) \end{aligned} \tag{3.5.1}$$

where $u_\alpha^{(k)}$ are the components of displacement vector from the problem $\mathcal{D}^{(k)}$, $(k = 1, 2, 3)$, φ is the solution of the boundary-value problem 3.4.1 and 3.4.2, G is an unknown function of x_1 and x_2, and b_j and τ are unknown constants.

By Equations 1.1.1, 1.1.4, and 3.5.1, we get

$$\begin{aligned} t_{\alpha\beta} &= \lambda(b_1 x_1 + b_2 x_2 + b_3) x_3 \delta_{\alpha\beta} + x_3 \sum_{k=1}^{3} b_k t_{\alpha\beta}^{(k)} \\ t_{\alpha 3} &= \mu\tau(\varphi_{,\alpha} - \varepsilon_{\alpha\beta} x_\beta) + \mu\left[G_{,\alpha} + \sum_{k=1}^{3} b_k u_\alpha^{(k)}\right] \\ t_{33} &= (\lambda + 2\mu)(b_1 x_1 + b_2 x_2 + b_3) x_3 + \lambda x_3 \sum_{k=1}^{3} b_k e_{\alpha\alpha}^{(k)} \end{aligned} \tag{3.5.2}$$

where $t_{\alpha\beta}^{(k)}$ are given by Equations 3.2.8 and 3.2.9.

The first two equations of equilibrium 1.1.8 and the first two conditions 1.3.1 are satisfied on the basis of the relations 3.2.10 and 3.2.11. In view of Equations 3.4.1 and 3.5.2, the third equation of equilibrium 1.1.8 reduces to

$$(\mu G_{,\alpha})_{,\alpha} = p \text{ on } \Sigma_1 \tag{3.5.3}$$

where

$$p = -(\lambda + 2\mu)(b_1 x_1 + b_2 x_2 + b_3) - \sum_{k=1}^{3} b_k \left[\left(\mu u_\beta^{(k)}\right)_{,\beta} + \lambda e_{\rho\rho}^{(k)}\right] \tag{3.5.4}$$

By Equations 3.4.2 and 3.5.2, we see that the last of conditions 1.3.1 on the lateral boundary becomes

$$\mu G_{,\alpha} n_\alpha = q \text{ on } \Gamma \tag{3.5.5}$$

where

$$q = -\mu n_\alpha \sum_{k=1}^{3} b_k u_\alpha^{(k)} \tag{3.5.6}$$

Thus, the function G is solution of the boundary-value problem 3.5.3 and 3.5.5. It follows from Equations 3.3.7, 3.5.4, and 3.5.6 that

$$\int_{\Sigma_1} p da - \int_\Gamma q ds = -\int_{\Sigma_1} \left[(\lambda + 2\mu)(b_1 x_1 + b_2 x_2 + b_3) + \lambda \sum_{k=1}^{3} b_k e_{\rho\rho}^{(k)} \right] da$$
$$= -D_{3j} b_j$$

so that the necessary and sufficient condition to solve the boundary-value problem 3.5.3 and 3.5.5 is

$$D_{3j} b_j = 0 \tag{3.5.7}$$

In view of Equations 1.3.57, 3.3.7, and 3.5.2, we find that the conditions 1.4.1 reduce to

$$D_{\alpha j} b_j = -F_\alpha \tag{3.5.8}$$

It follows from the relation 3.3.18 that the systems 3.5.7 and 3.5.8 determine the constants b_1, b_2, and b_3. We consider that in the functions 3.5.4 and 3.5.6, the constants b_k are given by Equations 3.5.7 and 3.5.8. In what follows, we suppose that G is known.

If we use Equations 3.4.6 and 3.5.2, we find that the condition 1.3.49 reduces to

$$D_* \tau = -\mathfrak{M} \tag{3.5.9}$$

where

$$\mathfrak{M} = \int_{\Sigma_1} \mu \varepsilon_{\alpha\beta} x_\alpha \left[G_{,\beta} + \sum_{k=1}^{3} b_k u_\beta^{(k)} \right] da \tag{3.5.10}$$

We conclude that the constant τ is determined by Equation 3.5.9. The conditions 1.3.48 are identically satisfied. Thus, the flexure problem has a solution of the form 3.5.1.

3.6 Elastic Cylinders Composed of Different Nonhomogeneous and Isotropic Materials

In this section, we study the deformation of composed cylinders introduced in Section 3.1. We suppose that B_ρ is occupied by an isotropic material with the Lamé moduli $\lambda^{(\rho)}$ and $\mu^{(\rho)}$, and that

$$\lambda^{(\rho)} = \lambda^{(\rho)}(x_1, x_2), \qquad \mu^{(\rho)} = \mu^{(\rho)}(x_1, x_2), \quad (x_1, x_2) \in A_\rho \tag{3.6.1}$$

We can consider B as being occupied by an elastic medium which, in general, has elastic coefficients discontinuous along Π_0. We assume that $\lambda^{(\rho)}$ and $\mu^{(\rho)}$ belongs to $C^{(\infty)}$ and that the elastic potential corresponding to the material which occupies B_ρ is positive definite.

Saint-Venant's problem for heterogeneous cylinders consists in finding of a displacement vector field $\mathbf{u} \in C^2(B_1) \cap C^2(B_2) \cap C^1(\overline{B}_1) \cap C^1(\overline{B}_2) \cap C^0(B)$ that satisfies Equations 1.1.1, 1.1.4, and 1.1.8 on B_ρ, the conditions 3.1.1 on the surface of separation Π_0, the conditions for $x_3 = 0$ and the boundary conditions 1.3.1.

3.6.1 Auxiliary Plane Strain Problems

Let us consider the state of plane strain of composed cylinders. The displacement field has the form 1.5.1. Given elastic coefficients $\lambda^{(\rho)}$ and $\mu^{(\rho)}$, body forces $\mathbf{f}^{(\rho)}$ on B_ρ, surface tractions $\tilde{\mathbf{t}}^{(\rho)}$ on Π_ρ, with $\mathbf{f}^{(\rho)}$ and $\tilde{\mathbf{t}}^{(\rho)}$ independent of x_3 and parallel to the x_1, x_2-plane, the second boundary-value problem consists in finding an elastic state on B that satisfies the strain–displacement, the stress–strain relations, the equations of equilibrium, the conditions on the surface of separation, and the tractions condition. The first boundary-value problem can be defined as in Section 1.5. In what follows, we restrict our attention to the second boundary-value problem. The basic equations of the plane strain problem consist of the strain–displacement relations

$$e_{\alpha\beta} = \frac{1}{2}(u_{\alpha,\beta} + u_{\beta,\alpha}) \tag{3.6.2}$$

the stress–strain relations

$$t_{\alpha\beta} = \lambda^{(\rho)} e_{\eta\eta}\delta_{\alpha\beta} + 2\mu^{(\rho)} e_{\alpha\beta} \tag{3.6.3}$$

and the equations of equilibrium

$$t_{\beta\alpha,\beta} + f_\alpha^{(\rho)} = 0 \tag{3.6.4}$$

on A_ρ. The conditions on the surface of separation Π_0 reduce to

$$[u_\alpha]_1 = [u_\alpha]_2, \qquad [t_{\beta\alpha}]_1\, n_\beta^0 = [t_{\beta\alpha}]_2\, n_\beta^0 \text{ on } \Gamma_0 \tag{3.6.5}$$

The conditions on the lateral boundary become

$$[t_{\beta\alpha} n_\beta]_\rho = \tilde{t}_\alpha^{(\rho)} \text{ on } \Gamma_\rho \tag{3.6.6}$$

We assume that the functions $f_\alpha^{(\rho)}$ and $\tilde{t}_\alpha^{(\rho)}$ belong to C^∞. From the general theory developed by Fichera ([88], Section 13), it follows that under suitable smoothness hypotheses on the arcs Γ_ρ and Γ_0, a solution $u_\alpha \in C^{(\infty)}(\overline{A}_1) \cap C^\infty(\overline{A}_2) \cap C^\infty(\Sigma_1)$ of the second boundary-value problem exists if and only if

$$\sum_{\rho=1}^{2}\left[\int_{A_\rho} f_\alpha^{(\rho)}da + \int_{\Gamma_\rho} \widetilde{t}_\alpha^{(\rho)}ds\right] = 0$$
$$\sum_{\rho=1}^{2}\left[\int_{A_\rho} \varepsilon_{\alpha\beta}x_\alpha f_\beta^{(\rho)}da + \int_{\Gamma_\rho} \varepsilon_{\alpha\beta}x_\alpha \widetilde{t}_\beta^{(\rho)}ds\right] = 0 \tag{3.6.7}$$

In what follows, we assume that the requirements which insure this result are fulfilled. It can be shown that if the conditions 3.6.5 are replaced by

$$[u_\alpha]_1 = [u_\alpha]_2, \qquad [t_{\beta\alpha}]_1 n_\beta^0 = [t_{\beta\alpha}]_2 n_\beta^0 + g_\alpha \text{ on } \Gamma_0 \tag{3.6.8}$$

where g_α are C^∞ functions, then the conditions 3.6.7 are replaced by

$$\sum_{\rho=1}^{2}\left[\int_{A_\rho} f_\alpha^{(\rho)}da + \int_{\Gamma_\rho} \widetilde{t}_\alpha^{(\rho)}da\right] + \int_{\Gamma_0} g_\alpha ds = 0$$
$$\sum_{\rho=1}^{2}\left[\int_{A_\rho} \varepsilon_{\alpha\beta}x_\alpha f_\beta^{(\rho)}da + \int_{\Gamma_\rho} \varepsilon_{\alpha\beta}x_\alpha \widetilde{t}_\beta^{(\rho)}ds\right] + \int_{\Gamma_0} \varepsilon_{\alpha\beta}x_\alpha g_\beta ds = 0 \tag{3.6.9}$$

These conditions have been established by Sherman [308] and Muskhelishvili [241] within the theory of piecewise homogeneous cylinders.

We will have occasion to use three special problems $\mathcal{P}^{(\kappa)}$, $(k = 1, 2, 3)$, of plane strain for the composed cylinder B. In what follows we denote by $v_\alpha^{(k)}, \gamma_{\alpha\beta}^{(k)}$, and $\sigma_{\alpha\beta}^{(k)}$ the components of displacement vector, the components of the strain tensor, and the components of the stress tensor for the problem $\mathcal{P}^{(k)}$, respectively. The problems $\mathcal{P}^{(k)}$ are characterized by the equations

$$\gamma_{\alpha\beta}^{(k)} = \frac{1}{2}\left(v_{\alpha,\beta}^{(k)} + v_{\beta,\alpha}^{(k)}\right) \tag{3.6.10}$$

$$\sigma_{\alpha\beta}^{(k)} = \lambda^{(\rho)}\gamma_{\eta\eta}^{(k)}\delta_{\alpha\beta} + 2\mu^{(\rho)}\gamma_{\alpha\beta}^{(k)} \tag{3.6.11}$$

$$\sigma_{\beta\alpha,\beta}^{(\kappa)} + \left(\lambda^{(\rho)}x_\kappa\right)_{,\alpha} = 0, \qquad \sigma_{\beta\alpha,\beta}^{(3)} + \lambda_{,\alpha}^{(\rho)} = 0 \text{ on } A_\rho, \quad (\kappa = 1, 2) \tag{3.6.12}$$

and the conditions

$$[v_\alpha^{(k)}]_1 = [v_\alpha^{(k)}]_2, \qquad [\sigma_{\beta\alpha}^{(k)}]_1 n_\beta^0 = [\sigma_{\beta\alpha}^{(k)}]_2 n_\beta^0 + g_\alpha^{(k)} \text{ on } \Gamma_0 \tag{3.6.13}$$

$$[\sigma_{\beta\alpha}^{(\kappa)}n_\beta]_\rho = -\lambda^{(\rho)}x_\kappa n_\alpha, \qquad [\sigma_{\beta\alpha}^{(3)}n_\beta]_\rho = -\lambda^{(\rho)}n_\alpha \text{ on } \Gamma_\rho \tag{3.6.14}$$

where

$$g_\alpha^{(k)} = (\lambda^{(2)} - \lambda^{(1)})x_\kappa n_\alpha^0, \qquad g_\alpha^{(3)} = (\lambda^{(2)} - \lambda^{(1)})n_\alpha^0, \quad (\kappa = 1, 2) \tag{3.6.15}$$

It is easy to prove that the necessary and sufficient conditions 3.6.9 for the existence of the solution are satisfied for each boundary-value problem $\mathcal{P}^{(k)}$. In what follows, we shall consider that the functions $v_\alpha^{(k)}$, $\gamma_{\alpha\beta}^{(k)}$, and $\sigma_{\alpha\beta}^{(k)}$ are known.

3.6.2 Extension and Bending

The problem of extension and bending for the composed cylinder B consists in the determination of a displacement vector field that satisfies Equations 1.1.1, 1.1.4, and 1.1.8 on B_ρ, the conditions 3.1.1 on Π_0, the conditions 3.3.1 and 3.3.2 on Σ_1, and the conditions 1.3.1 on Π. On the basis of Theorem 1.7.1, we try to solve the problem assuming that the displacement vector field has the form

$$u_\alpha = -\frac{1}{2}d_\alpha x_3^2 + \sum_{k=1}^{3} d_k v_\alpha^{(k)}, \qquad u_3 = (d_1x_1 + d_2x_2 + d_3)x_3 \tag{3.6.16}$$

where $v_\alpha^{(k)}$ are the solutions of the problems $\mathcal{P}^{(k)}$, $(k = 1, 2, 3)$, and the d_k are unknown constants. It follows from Equations 1.1.1, 1.1.4, and 3.6.16 that

$$\begin{aligned} t_{\alpha\beta} &= \lambda^{(\rho)}(d_1x_1 + d_2x_2 + d_3)\delta_{\alpha\beta} + \sum_{k=1}^{3} d_k\sigma_{\alpha\beta}^{(k)}, \qquad t_{\alpha3} = 0 \\ t_{33} &= (\lambda^{(\rho)} + 2\mu^{(\rho)})(d_1x_1 + d_2x_2 + d_3) + \lambda^{(\rho)}\sum_{k=1}^{3} d_k\gamma_{\alpha\alpha}^{(k)} \text{ on } A_\rho \end{aligned} \tag{3.6.17}$$

where $\gamma_{\alpha\beta}^{(k)}$ and $\sigma_{\alpha\beta}^{(k)}$ are given by Equations 3.6.10 and 3.6.11, respectively. It is easy to verify that the equations of equilibrium 1.1.8 and the boundary conditions 1.3.1 on Π are satisfied on the basis of the relations 3.6.12 and 3.6.14. The conditions 3.1.1 on the surface of separation Π_0 are satisfied in view of the relations 3.6.13 and 3.6.15.

If we take into account Equation 3.6.11, we see that the conditions 3.3.1 are satisfied. It follows from Equations 3.3.2 and 3.6.11 that the constants d_k satisfy the following equations

$$L_{\alpha j}d_j = \varepsilon_{\alpha\beta}M_\beta, \qquad L_{3j}d_j = -F_3 \tag{3.6.18}$$

where

$$\begin{aligned} L_{\alpha\beta} &= \sum_{\rho=1}^{2}\int_{A_\rho} x_\alpha\left[(\lambda^{(\rho)} + 2\mu^{(\rho)})x_\beta + \lambda^{(\rho)}\gamma_{\eta\eta}^{(\beta)}\right]da \\ L_{\alpha3} &= \sum_{\rho=1}^{2}\int_{A_\rho} x_\alpha\left[\lambda^{(\rho)} + 2\mu^{(\rho)} + \lambda^{(\rho)}\gamma_{\beta\beta}^{(3)}\right]da \\ L_{3\alpha} &= \sum_{\rho=1}^{2}\int_{A_\rho}\left[(\lambda^{(\rho)} + 2\mu^{(\rho)})x_\alpha + \lambda^{(\rho)}\gamma_{\beta\beta}^{(\alpha)}\right]da \\ L_{33} &= \sum_{\rho=1}^{2}\int_{A_\rho}(\lambda^{(\rho)} + 2\mu^{(\rho)} + \lambda^{(\rho)}\gamma_{\alpha\alpha}^{(3)})da \end{aligned} \tag{3.6.19}$$

As in Section 3.3, we can show that the system 3.6.18 uniquely determines the constants d_k. Let $W^{(\rho)}(\mathbf{u})$ be the elastic potential associated with $\mathbf{u}$ on B_ρ. Clearly,

$$W^{(\rho)}(\mathbf{u}) = \frac{1}{2}\lambda^{(\rho)} e_{rr}(\mathbf{u}) e_{ss}(\mathbf{u}) + \mu^{(\rho)} e_{ij}(\mathbf{u}) e_{ij}(\mathbf{u}) = \frac{1}{2}\left[t_{ij}(\mathbf{u}) e_{ij}(\mathbf{u})\right]_\rho \quad (3.6.20)$$

We continue to assume that $W^{(\rho)}$ is a positive definite quadratic form in the variables $e_{rs}(\mathbf{u})$. Let us consider two displacement vector fields $\mathbf{u}'$ and $\mathbf{u}''$ that satisfy Equations 1.1.1, 1.1.4, and 1.1.8 on B_ρ and the conditions 3.1.1 on Π_0. We denote

$$W^{(\rho)}(\mathbf{u}', \mathbf{u}'') = \frac{1}{2}\lambda^{(\rho)} e_{rr}(\mathbf{u}') e_{ss}(\mathbf{u}'') + \mu^{(\rho)} e_{ij}(\mathbf{u}') e_{ij}(\mathbf{u}'') = \frac{1}{2}[t_{ij}(\mathbf{u}') e_{ij}(\mathbf{u}'')]_\rho \quad (3.6.21)$$

Clearly,

$$W^{(\rho)}(\mathbf{u}', \mathbf{u}'') = W^{(\rho)}(\mathbf{u}'', \mathbf{u}'), \qquad W^{(\rho)}(\mathbf{u}, \mathbf{u}) = W^{(\rho)}(\mathbf{u}) \quad (3.6.22)$$

In view of Equations 1.1.1, 1.1.8, 3.1.1, and the divergence theorem, we find that

$$2\sum_{\rho=1}^{2} \int_{B_\rho} W^{(\rho)}(\mathbf{u}', \mathbf{u}'') dv = \int_{\partial B} t_{ji}(\mathbf{u}') n_j u_i'' da = \int_{\partial B} t_{ji}(\mathbf{u}'') n_j u_i' da \quad (3.6.23)$$

The strain energy $U(\mathbf{u})$ corresponding to a displacement vector field $\mathbf{u}$ on $B_1 \cup B_2$ is given by

$$U(\mathbf{u}) = \sum_{\rho=1}^{2} \int_{B_\rho} W^{(\rho)}(\mathbf{u}) dv \quad (3.6.24)$$

By Equations 3.6.22 and 3.6.23,

$$U(\mathbf{u}) = \frac{1}{2} \int_{\partial B} t_{ji}(\mathbf{u}) n_j u_i da \quad (3.6.25)$$

It follows from Equations 3.6.16 and 3.6.17 that

$$\mathbf{u} = \sum_{j=1}^{3} d_j \widehat{\mathbf{u}}^{(j)}, \qquad t_{ij} = \sum_{k=1}^{3} d_k s_{ij}^{(k)} \quad (3.6.26)$$

Clearly, $\widehat{\mathbf{u}}^{(j)}$ satisfy Equations 1.1.1, 1.1.4, and 1.1.8 on B_ρ and the conditions 1.3.1 and 3.1.1. By Equations 3.6.25 and 3.6.26,

$$U(\mathbf{u}) = U_{ij} d_i d_j \quad (3.6.27)$$

where

$$U_{ij} = \sum_{\rho=1}^{2} \int_{B_\rho} W^{(\rho)}(\widehat{\mathbf{u}}^{(i)}, \widehat{\mathbf{u}}^{(j)}) dv \quad (3.6.28)$$

In view of Equations 3.6.16, 3.6.17, and 3.6.26, we find that

$$\widehat{u}_\alpha^{(\beta)} = -\frac{1}{2}h^2\delta_{\alpha\beta} + v_\alpha^{(\beta)}, \qquad \widehat{u}_3^{(\alpha)} = hx_\alpha, \qquad \widehat{u}_\alpha^{(3)} = v_\alpha^{(3)}, \qquad \widehat{u}_3^{(3)} = h$$

$$\begin{aligned} & s_{j\alpha}^{(k)} n_j = s_{3\alpha}^{(k)} = 0, \qquad s_{j3}^{(\alpha)} n_j = (\lambda^{(\rho)} + 2\mu^{(\rho)})x_\alpha + \lambda^{(\rho)}\gamma_{\beta\beta}^{(\alpha)} \\ & s_{j3}^{(3)} n_j = \lambda^{(\rho)} + 2\mu^{(\rho)} + \lambda^{(\rho)}\gamma_{\alpha\alpha}^{(3)} \text{ on } \Sigma_2 \end{aligned} \tag{3.6.29}$$

$$s_{j\alpha}^{(k)} n_j = 0, \qquad \widehat{u}_3^{(k)} = 0, \qquad \widehat{u}_\alpha^{(k)} = v_\alpha^{(k)} \text{ on } \Sigma_1$$

Using the relations 1.3.1, 3.6.22, 3.6.23, 3.6.28, and 3.6.29, we find that

$$2U_{11} = \int_{\Sigma_1\cup\Sigma_2\cup\Pi} s_{ji}^{(1)} n_j \widehat{u}_i^{(1)} da = \int_{\Sigma_2} s_{33}^{(1)}\widehat{u}_3^{(1)} da = hL_{11}$$

Similarly,

$$2U_{ij} = hL_{ij}$$

It follows from Equations 3.6.24, 3.6.27, and 3.6.28 that $L_{ij} = L_{ji}$ and

$$\det(L_{ij}) \neq 0 \tag{3.6.30}$$

so that the system 3.6.18 can always be solved for d_1, d_2, and d_3. Thus, the solution of the problem has the form 3.6.16, where $v_\alpha^{(k)}$ are characterized by the problem $\mathcal{P}^{(k)}$ and d_j are given by Equation 3.6.18.

3.6.3 Torsion and Flexure

Let us suppose that $\mathbf{F} = F_\alpha \mathbf{e}_\alpha$ and $\mathbf{M} = M_3\mathbf{e}_3$. Then, the conditions on Σ_1 are given by Equations 1.3.21, 1.3.22, and 1.4.1. The problem of torsion and flexure consists in the finding of a displacement vector field that satisfies Equations 1.1.1, 1.1.4, and 1.1.8 on B_ρ, the conditions 3.1.1 on the surface of separation Π_0, the conditions 1.3.21, 1.3.22, and 1.4.1 on Σ_1, and the conditions 1.3.1 on the lateral boundary of the cylinder B. Following Ref. 151, we seek a solution of the problem in the form

$$\begin{aligned} u_\alpha &= -\frac{1}{6}b_\alpha x_3^3 - \tau\varepsilon_{\alpha\beta}x_\beta x_3 + x_3\sum_{j=1}^{3} b_j v_\alpha^{(j)} \\ u_3 &= \frac{1}{2}(b_1x_1 + b_2x_2 + b_3)x_3^2 + \Phi(x_1, x_2) \end{aligned} \tag{3.6.31}$$

where $v_\alpha^{(j)}$ are the components of the displacement vector in the auxiliary plane strain problem $\mathcal{P}^{(j)}, \Phi \in C^2(A_1)\cap C^2(A_2)\cap C^1(\overline{A}_1)\cap C^2(\overline{A}_2)\cap C^0(\Sigma_1)$ is an unknown function, and b_k and τ are unknown constants. In view of

Equations 1.1.1, 1.1.4, and 3.6.31, we get

$$t_{\alpha\beta} = \lambda^{(\rho)}(b_1x_1 + b_2x_2 + b_3)x_3\delta_{\alpha\beta} + x_3\sum_{j=1}^{3} b_j\sigma_{\alpha\beta}^{(j)}$$

$$t_{\alpha 3} = \mu^{(\rho)}\left(\Phi_{,\alpha} - \tau\varepsilon_{\alpha\beta}x_\beta + \sum_{j=1}^{3} b_j v_\alpha^{(j)}\right) \tag{3.6.32}$$

$$t_{33} = (\lambda^{(\rho)} + 2\mu^{(\rho)})(b_1x_1 + b_2x_2 + b_3)x_3 + \lambda^{(\rho)}x_3\sum_{j=1}^{3} b_j\gamma_{\alpha\alpha}^{(j)} \text{ on } B_\rho$$

where $\gamma_{\alpha\beta}^{(j)}$ and $\sigma_{\alpha\beta}^{(j)}$ are defined by Equations 3.6.10 and 3.6.11.

Clearly, the conditions 1.3.21 are satisfied on the basis of Equations 3.6.32. It follows from the equations which characterize the auxiliary plane strain problems and 3.6.32 that the equations of equilibrium and the conditions 1.3.1 and 3.1.1 are satisfied if the function Φ satisfies the equation

$$(\mu^{(\rho)}\Phi_{,\alpha})_{,\alpha} = -p^{(\rho)} \text{ on } A_\rho \tag{3.6.33}$$

and the conditions

$$[\Phi]_1 = [\Phi]_2, \qquad \mu^{(1)}\left[\frac{\partial\Phi}{\partial n^0}\right]_1 = \mu^{(2)}\left[\frac{\partial\Phi}{\partial n^0}\right]_2 + q \text{ on } \Gamma_0 \tag{3.6.34}$$

$$\mu^{(\rho)}\left[\frac{\partial\Phi}{\partial n}\right]_\rho = m^{(\rho)} \text{ on } \Gamma_\rho \tag{3.6.35}$$

where

$$\begin{aligned} p^{(\rho)} &= (\lambda^{(\rho)} + 2\mu^{(\rho)})(b_1x_1 + b_2x_2 + b_3) + \lambda^{(\rho)}\sum_{j=1}^{3} b_j\gamma_{\alpha\alpha}^{(j)} \\ &\quad - \left[\mu^{(\rho)}\left(\tau\varepsilon_{\alpha\beta}x_\beta - \sum_{j=1}^{3} b_j v_\alpha^{(j)}\right)\right]_{,\alpha} \\ q &= (\mu^{(1)} - \mu^{(2)})\left(\tau\varepsilon_{\alpha\beta}x_\beta - \sum_{j=1}^{3} b_j v_\alpha^{(j)}\right)n_\alpha^0 \\ m^{(\rho)} &= \mu^{(\rho)}\left(\tau\varepsilon_{\alpha\beta}x_\beta - \sum_{j=1}^{3} b_j v_\alpha^{(j)}\right)n_\alpha \end{aligned} \tag{3.6.36}$$

Let us consider the boundary-value problem

$$\begin{aligned} &(\mu^{(\rho)}\chi_{,\alpha})_{,\alpha} = -f^{(\rho)} \text{ on } A_\rho, \qquad \mu^{(\rho)}\left[\frac{\partial\chi}{\partial n}\right]_\rho = \xi^{(\rho)} \text{ on } \Gamma_\rho \\ &[\chi]_1 = [\chi]_2, \qquad \mu^{(1)}\left[\frac{\partial\chi}{\partial n^0}\right]_1 = \mu^{(2)}\left[\frac{\partial\chi}{\partial n^0}\right]_2 + \zeta \text{ on } \Gamma \end{aligned} \tag{3.6.37}$$

where $f^{(\rho)}, \xi^{(\rho)}$, and ζ are C^∞ functions. Necessary and sufficient condition to solve the boundary-value problem 3.6.37 is (cf. [88,241,308])

$$\sum_{\rho=1}^{2}\left(\int_{A_\rho} f^{(\rho)} da + \int_{\Gamma_\rho} \xi^{(\rho)} ds\right) + \int_{\Gamma_0} \zeta da = 0 \tag{3.6.38}$$

By Equations 3.6.19 and 3.6.36, we obtain

$$\sum_{\rho=1}^{2}\left(\int_{A_\rho} p^{(\rho)} da + \int_{\Gamma_\rho} m^{(\rho)} ds\right) + \int_{\Gamma_0} q ds = L_{3j} b_j$$

Thus, the necessary and sufficient condition for the existence of a solution to the boundary-value problem 3.6.33, 3.6.34, and 3.6.35 reduces to

$$L_{3j} b_j = 0 \tag{3.6.39}$$

It is easy to verify that the relations 1.3.57 are valid in the present circumstances.

By Equations 1.3.57, 3.6.19, and 3.6.32, we conclude that the conditions 1.4.1 reduce to

$$L_{\alpha j} b_j = -F_\alpha \tag{3.6.40}$$

In view of Equation 3.6.30, the system 3.6.39 and 3.6.40 determines the constants b_1, b_2, and b_3. We introduce the function $\varphi \in C^2(A_1) \cap C^2(A_2) \cap C^1(\overline{A}_1) \cap C^1(\overline{A}_2) \cap C^0(\Sigma_1)$ which satisfies equation

$$(\mu^{(\rho)} \varphi_{,\beta})_{,\beta} = \varepsilon_{\alpha\beta}(\mu^{(\rho)} x_\beta)_{,\alpha} \text{ on } A_\rho \tag{3.6.41}$$

and the conditions

$$\begin{aligned} &[\varphi]_1 = [\varphi]_2 \\ &\mu^{(1)}\left[\frac{\partial \varphi}{\partial n^0}\right]_1 = \mu^{(2)}\left[\frac{\partial \varphi}{\partial n^0}\right]_2 + (\mu^{(1)} - \mu^{(2)})\varepsilon_{\alpha\beta} x_\beta n_\alpha^0 \text{ on } \Gamma_0 \\ &\left[\frac{\partial \varphi}{\partial n}\right]_\rho = \varepsilon_{\alpha\beta} x_\beta n_\alpha \text{ on } \Gamma_\rho \end{aligned} \tag{3.6.42}$$

It is easy to show that the necessary and sufficient condition 3.6.38 for the existence of a solution to the boundary-value problem 3.6.41 and 3.6.42 is satisfied. We introduce the function ψ by

$$\Phi = \tau\varphi + \psi \tag{3.6.43}$$

It follows from the above equations that the function ψ satisfies the equation

$$\begin{aligned} (\mu^{(\rho)} \psi_{,\alpha})_{,\alpha} = &-(\lambda^{(\rho)} + 2\mu^{(\rho)})(b_1 x_1 + b_2 x_2 + b_3) \\ &- \lambda^{(\rho)} \sum_{j=1}^{3} b_j \gamma_{\alpha\alpha}^{(j)} - \left(\mu^{(\rho)} \sum_{j=1}^{3} b_j v_\alpha^{(j)}\right)_{,\alpha} \text{ on } A_\rho \end{aligned} \tag{3.6.44}$$

and the conditions

$$[\psi]_1 = [\psi]_2$$
$$\mu^{(1)} \left[\frac{\partial \psi}{\partial n^0}\right]_1 = \mu^{(2)} \left[\frac{\partial \psi}{\partial n^0}\right]_2 - (\mu^{(1)} - \mu^{(2)}) \sum_{j=1}^{3} b_j v_\alpha^{(j)} n_\alpha^0 \text{ on } \Gamma_0 \tag{3.6.45}$$
$$\left[\frac{\partial \psi}{\partial n}\right]_\rho = -\sum_{j=1}^{3} b_j v_\alpha^{(j)} n_\alpha \text{ on } \Gamma_\rho$$

In what follows, we shall treat φ and ψ as known functions. By Equations 3.6.32 and 3.6.43, we obtain

$$t_{\alpha 3} = \tau \mu^{(\rho)} (\varphi_{,\alpha} - \varepsilon_{\alpha\beta} x_\beta) + \mu^{(\rho)} \left(\psi_{,\alpha} + \sum_{j=1}^{3} b_j v_\alpha^{(j)} \right) \text{ on } B_\rho \tag{3.6.46}$$

In view of Equation 3.6.46, the condition 1.3.22 reduces to

$$D_0 \tau = -M_3 - M^* \tag{3.6.47}$$

where D_0 is the torsional rigidity defined by

$$D_0 = \sum_{\rho=1}^{2} \int_{A_\rho} \mu^{(\rho)} \varepsilon_{\alpha\beta} x_\alpha (\varphi_{,\beta} - \varepsilon_{\beta\eta} x_\eta) da \tag{3.6.48}$$

and

$$M^* = \sum_{\rho=1}^{3} \int_{A_\rho} \varepsilon_{\alpha\beta} \mu^{(\rho)} x_\alpha \left(\psi_{,\beta} + \sum_{j=1}^{3} b_j v_\beta^{(j)} \right) da \tag{3.6.49}$$

As in Section 3.4, we can prove that $D_0 > 0$. From Equation 3.6.47, we can determine the constant τ. Thus, the problem of torsion and flexure is solved.

3.6.4 Uniformly Loaded Cylinders

We shall now consider the Almansi–Michell problem for heterogeneous cylinders. We assume that the body forces have the form

$$f_i = G_i^{(\rho)}(x_1, x_2), \quad (x_1, x_2) \in A_\rho \tag{3.6.50}$$

Let us consider the following conditions on the lateral boundary

$$[t_{ji} n_j]_\rho = p_i^{(\rho)} \text{ on } \Pi_\rho \tag{3.6.51}$$

We suppose that $G_i^{(\rho)}$ and $p_i^{(\rho)}$ are C^∞ functions which are independent of the axial coordinate.

The Almansi–Michell problem consists in finding of a displacement vector field $\mathbf{u} \in C^2(B_1) \cap C^2(B_2) \cap C^1(\overline{B}_1) \cap C^1(\overline{B}_2) \cap C^0(B)$ that satisfies Equations 1.1.1, 1.1.4, and 2.1.1 on B_ρ, the conditions for $x_3 = 0$, the conditions 3.1.1 on the surface of separation Π_0, and the boundary conditions 3.6.51, when the body forces and the surface tractions are independent of x_3. Following the results of Section 2.4, we seek the solution of Almansi–Michell problem in the form

$$\begin{aligned} u_\alpha &= -\frac{1}{2}a_\alpha x_3^2 - \frac{1}{6}b_\alpha x_3^3 - \frac{1}{24}c_\alpha x_3^4 + \varepsilon_{\beta\alpha}\left(\tau_1 x_3 + \frac{1}{2}\tau_2 x_3^2\right)x_\beta \\ &\quad + \sum_{k=1}^{3}\left(a_k + b_k x_3 + \frac{1}{2}c_k x_3^2\right)v_\alpha^{(k)} + w_\alpha(x_1,x_2) \\ u_3 &= (a_\eta x_\eta + a_3)x_3 + \frac{1}{2}(b_\eta x_\eta + b_3)x_3^2 + \frac{1}{6}(c_\eta x_\eta + c_3)x_3^3 \\ &\quad + (\tau_1 + \tau_2 x_3)\varphi + \Psi(x_1,x_2) + x_3\Lambda(x_1,x_2), \quad (x_1,x_2,x_3) \in B \end{aligned} \tag{3.6.52}$$

where $v_\alpha^{(k)}$ are the displacements from the plane strain problem $\mathcal{P}^{(k)}$, φ is torsion function characterized by Equations 3.6.41 and 3.6.42, Ψ and Λ are unknown functions, and a_k, b_k, c_k, and τ_α are unknown constants.

We introduce the notations

$$2\gamma_{\alpha\beta} = w_{\alpha,\beta} + w_{\beta,\alpha}, \qquad \pi_{\alpha\beta} = \lambda^{(\rho)}\gamma_{\nu\nu}\delta_{\alpha\beta} + 2\mu^{(\rho)}\gamma_{\alpha\beta} \text{ on } A_\rho \tag{3.6.53}$$

By Equations 1.1.1, 1.1.4, 3.6.10, 3.6.11, 3.6.52, and 3.6.53, we get

$$\begin{aligned} t_{\alpha\beta} &= \lambda^{(\rho)}\left[a_\eta x_\eta + a_3 + (b_\eta x_\eta + b_3)x_3 + \frac{1}{2}(c_\eta x_\eta + c_3)x_3^2\right]\delta_{\alpha\beta} \\ &\quad + \lambda^{(\rho)}(\Lambda + \tau_2\varphi)\delta_{\alpha\beta} + \sum_{j=1}^{3}\left(a_j + b_j x_3 + \frac{1}{2}c_j x_3^2\right)\sigma_{\alpha\beta}^{(j)} + \pi_{\alpha\beta} \\ t_{33} &= (\lambda^{(\rho)} + 2\mu^{(\rho)})\left[a_\eta x_\eta + a_3 + (b_\eta x_\eta + b_3)x_3 \right. \\ &\quad \left. + \frac{1}{2}(c_\eta x_\eta + c_3)x_3^2\right] + (\lambda^{(\rho)} + 2\mu^{(\rho)})(\Lambda + \tau_2\varphi) \\ &\quad + \lambda^{(\rho)}\sum_{j=1}^{3}\left(a_j + b_j x_3 + \frac{1}{2}c_j x_3^2\right)\gamma_{\alpha\alpha}^{(j)} + \lambda^{(\rho)}\gamma_{\alpha\alpha} \\ t_{\alpha 3} &= \mu^{(\rho)}\left[(\tau_1 + \tau_2 x_3)(\varphi_{,\alpha} + \varepsilon_{\beta\alpha}x_\beta) + \Psi_{,\alpha} + x_3\Lambda_{,\alpha} \right. \\ &\quad \left. + \sum_{j=1}^{3}(b_j + c_j x_3)v_\alpha^{(j)}\right] \end{aligned} \tag{3.6.54}$$

By using Equations 3.6.12, 3.6.41, and 3.6.54, we find that the equations of equilibrium 2.1.1 reduce to

$$\pi_{\beta\alpha,\beta} + H_\alpha^{(\rho)} = 0 \tag{3.6.55}$$

$$(\mu^{(\rho)}\Psi_{,\alpha})_{,\alpha} = g^{(\rho)} \tag{3.6.56}$$

$$(\mu^{(\rho)}\Lambda_{,\alpha})_{,\alpha} = h^{(\rho)} \tag{3.6.57}$$

on A_ρ, where

$$H_\alpha^{(\rho)} = G_\alpha^{(\rho)} + [\lambda^{(\rho)}(\Lambda + \tau_2\varphi)]_{,\alpha} + \mu^{(\rho)}[\tau_2(\varphi_{,\alpha} + \varepsilon_{\beta\alpha}x_\beta) + \Lambda_{,\alpha}] + \mu^{(\rho)}\sum_{j=1}^{3} c_j v_\alpha^{(j)}$$

$$g^{(\rho)} = -G_3^{(\rho)} - (\lambda^{(\rho)} + 2\mu^{(\rho)})(b_\eta x_\eta + b_3) - \sum_{j=1}^{3} b_j\left[\left(\mu^{(\rho)}v_\alpha^{(j)}\right)_{,\alpha} + \lambda^{(\rho)}\gamma_{\alpha\alpha}^{(j)}\right]$$

$$h^{(\rho)} = -(\lambda^{(\rho)} + 2\mu^{(\rho)})(c_\eta x_\eta + c_3) - \sum_{j=1}^{3} c_j\left[\left(\mu^{(\rho)}v_\alpha^{(j)}\right)_{,\alpha} + \lambda^{(\rho)}\gamma_{\alpha\alpha}^{(j)}\right] \tag{3.6.58}$$

In view of Equations 3.6.13, 3.6.42, and 3.6.54, the conditions 3.1.1 on the surface of separation become

$$[w_\alpha]_1 = [w_\alpha]_2, \qquad [\pi_{\beta\alpha}]_1 n_\beta^0 = [\pi_{\beta\alpha}]_2 n_\beta^0 + (\lambda^{(2)} - \lambda^{(1)})(\Lambda + \tau_2\varphi)n_\alpha^0 \tag{3.6.59}$$

$$[\Psi]_1 = [\Psi]_2, \qquad \mu^{(1)}\left[\frac{\partial\Psi}{\partial n^0}\right]_1 = \mu^{(2)}\left[\frac{\partial\Psi}{\partial n^0}\right]_2 + n_\alpha^0(\mu^{(2)} - \mu^{(1)})\sum_{j=1}^{3} b_j v_\alpha^{(j)} \tag{3.6.60}$$

$$[\Lambda]_1 = [\Lambda]_2, \qquad \mu^{(1)}\left[\frac{\partial\Lambda}{\partial n^0}\right]_1 = \mu^{(2)}\left[\frac{\partial\Lambda}{\partial n^0}\right]_2 + n_\alpha^0(\mu^{(2)} - \mu^{(1)})\sum_{j=1}^{3} c_j v_\alpha^{(j)} \tag{3.6.61}$$

on Γ_0. Using Equations 3.6.14, 3.6.42, and 3.6.54, we find that the conditions (3.6.51) reduce to

$$[\pi_{\beta\alpha}n_\beta]_\rho = P_\alpha^{(\rho)} \tag{3.6.62}$$

$$\mu^{(\rho)}\left[\frac{\partial\Psi}{\partial n}\right]_\rho = Q^{(\rho)} \tag{3.6.63}$$

$$\mu^{(\rho)}\left[\frac{\partial\Lambda}{\partial n}\right]_\rho = K^{(\rho)} \tag{3.6.64}$$

on Γ_ρ, where

$$P_\alpha^{(\rho)} = p_\alpha^{(\rho)} - \lambda^{(\rho)}(\Lambda + \tau_2\varphi)n_\alpha$$

$$Q^{(\rho)} = p_3^{(\rho)} - \mu^{(\rho)}n_\alpha\sum_{j=1}^{3} b_j v_\alpha^{(j)}, \qquad K^{(\rho)} = -\mu^{(\rho)}n_\alpha\sum_{j=1}^{3} c_j v_\alpha^{(j)} \tag{3.6.65}$$

Thus, from Equations 3.6.53, 3.6.59, and 3.6.62, we conclude that w_α are the displacements in a plane strain problem. By Equations 3.6.58, 3.6.59, and 3.6.62, we find that the necessary and sufficient conditions 3.6.9 to solve this problem become

$$\sum_{\rho=1}^{2}\left\{\int_{A_\rho} G_\alpha^{(\rho)} da + \int_{\Gamma_\rho} p_\alpha^{(\rho)} ds\right\} + \int_{\Sigma_1} t_{3\alpha,3} da = 0$$
$$\sum_{\rho=1}^{3}\left\{\int_{A_\rho} \varepsilon_{\alpha\beta} x_\alpha G_\beta^{(\rho)} da + \int_{\Gamma_\rho} \varepsilon_{\alpha\beta} x_\alpha p_\beta^{(\rho)} ds\right\} + \int_{\Sigma_1} \varepsilon_{\alpha\beta} x_\alpha t_{3\beta,3} da = 0 \tag{3.6.66}$$

It follows from Equations 2.2.10, 3.6.19, and 3.6.54 that

$$\int_{\Sigma_1} t_{3\alpha,3} da = L_{\alpha j} c_j$$

Thus, the first two conditions from Equation 3.6.66 reduce to

$$L_{\alpha j} c_j = -\sum_{\rho=1}^{2}\left[\int_{A_\rho} G_\alpha^{(\rho)} da + \int_{\Gamma_\rho} p_\alpha^{(\rho)} ds\right] \tag{3.6.67}$$

Let us consider now the boundary-value problem 3.6.57, 3.6.61, and 3.6.64. The necessary and sufficient condition to solve this problem becomes

$$L_{3j} c_j = 0 \tag{3.6.68}$$

where L_{3j} are given by Equation 3.6.19. Thus, in view of Equation 3.6.31, we conclude that the system 3.6.67 and 3.6.68 uniquely determines the constants c_1, c_2, and c_3. In what follows we shall consider Λ as a known function. By Equations 3.6.54, the last condition of Equations 3.6.66 reduces to

$$D_0 \tau_2 = -\sum_{\rho=1}^{2}\left[\int_{A_\rho} \varepsilon_{\alpha\beta} x_\alpha G_\beta^{(\rho)} da + \int_{\Gamma_\rho} \varepsilon_{\alpha\beta} x_\alpha p_\beta^{(\rho)} ds + \int_{A_\rho} \varepsilon_{\alpha\beta} x_\alpha \mu^{(\rho)}\left(\Lambda_{,\beta} + \sum_{j=1}^{3} c_j v_\beta^{(j)}\right) da\right] \tag{3.6.69}$$

where D_0 is given by Equation 3.6.48. The constant τ_2 is determined by Equation 3.6.69.

Let us consider now the boundary-value problem 3.6.56, 3.6.60, and 3.6.63. The necessary and sufficient conditions to solve this problem can be expressed in the form

$$L_{3j} b_j = -\sum_{j=1}^{3}\left[\int_{A_\rho} G_3^{(\rho)} da + \int_{\Gamma_\rho} p_3^{(\rho)} ds\right] \tag{3.6.70}$$

As in Section 2.2, we can prove that

$$\int_{\Sigma_1} t_{3\alpha} da = \int_{\Sigma_1} x_\alpha t_{33,3} da + \sum_{\rho=1}^{2} \left[\int_{A_\rho} x_\alpha G_3^{(\rho)} da + \int_{\Gamma_\rho} x_\alpha p_3^{(\rho)} ds \right] \tag{3.6.71}$$

In view of Equations 3.6.54 and 3.6.71, the conditions 1.4.1 reduce to

$$L_{\alpha j} b_j = -F_\alpha - \sum_{\rho=1}^{2} \left[\int_{A_\rho} x_\alpha G_3^{(\rho)} da + \int_{\Gamma_\rho} x_\alpha p_3^{(\rho)} ds \right] \tag{3.6.72}$$

The system 3.6.70 and 3.6.71 determines the constants b_1, b_2, and b_3. By Equations 3.6.54, the conditions 1.4.2 and 1.4.3 become

$$L_{ik} a_k = N_i \tag{3.6.73}$$

where

$$\begin{aligned} N_\alpha &= \varepsilon_{\alpha\beta} M_\beta - \sum_{\rho=1}^{2} \int_{A_\rho} x_\alpha \left[(\lambda^{(\rho)} + 2\mu^{(\rho)})(\Lambda + \tau_2 \varphi) + \lambda^{(\rho)} \gamma_{\alpha\alpha} \right] da \\ N_3 &= -F_3 - \sum_{\rho=1}^{2} \int_{A_\rho} \left[(\lambda^{(\rho)} + 2\mu^{(\rho)})(\Lambda + \tau_2 \varphi) + \lambda^{(\rho)} \gamma_{\alpha\alpha} \right] da \end{aligned} \tag{3.6.74}$$

In view of the relation 3.6.30, the system 3.6.73 can always be solved for a_1, a_2, and a_3. The condition 1.4.4 reduces to

$$D_0 \tau_1 = -M_3 - \sum_{\rho=1}^{2} \int_{A_\rho} \varepsilon_{\alpha\beta} x_\alpha \mu^{(\rho)} \left(\Psi_{,\beta} + \sum_{j=1}^{3} b_j v_\beta^{(j)} \right) da \tag{3.6.75}$$

where D_0 is defined by Equation 3.6.48. The relation 3.6.75 determines the constant τ_1. Thus, the Almansi–Michell problem is solved.

3.6.5 Almansi Problem

We now suppose that the body forces and the tractions on the lateral surface of the cylinder B have the form

$$\begin{aligned} f_i &= \sum_{k=1}^{r} F_{ik}^{(\rho)}(x_1, x_2) x_3^k, \quad (x_1, x_2, x_3) \in B_\rho \\ \widetilde{t}_i &= \sum_{k=1}^{r} p_{ik}^{(\rho)}(x_1, x_2) x_3^k, \quad (x_1, x_2, x_3) \in \Pi_\rho \end{aligned} \tag{3.6.76}$$

where $F_{ik}^{(\rho)}$ and $p_{ik}^{(\rho)}$ are prescribed functions. The Almansi problem for heterogeneous cylinders consists in determination of a displacement vector field $\mathbf{u} \in$

$C^2(B_1) \cap C^2(B_2) \cap C^1(\overline{B}_1) \cap C^1(\overline{B}_2) \cap C^0(B)$ that satisfies Equations 1.1.1, 1.1.4, and 2.1.1 on B_ρ, the conditions for $x_3 = 0$, the conditions 3.1.1 on the surface of separation Π_0, and the boundary conditions 2.5.2 on the lateral boundary, when the body forces and the surface tractions are given by Equation 3.6.76. As in Section 2.3, the Almansi problem reduces to the following: to find the functions u_i which satisfy the equations

$$\begin{gathered} t_{ji,j} + \Lambda_i^{(\rho)}(x_1, x_2) x_3^{n+1} = 0 \\ t_{ij} = \lambda^{(\rho)} e_{rr} \delta_{ij} + 2\mu^{(\rho)} e_{ij}, \qquad 2e_{ij} = u_{i,j} + u_{j,i} \text{ on } B_\rho \end{gathered} \tag{3.6.77}$$

and the conditions

$$[u_i]_1 = [u_i]_2, \qquad [t_{\alpha i}]_1 n_\alpha^0 = [t_{\alpha i}]_2 n_\alpha^0 \text{ on } \Pi_0 \tag{3.6.78}$$

$$[t_{\alpha i} n_\alpha]_\rho = \sigma_i^{(\rho)}(x_1, x_2) x_3^{n+1} \text{ on } \Pi_\rho \tag{3.6.79}$$

$$\int_{\Sigma_1} t_{3i} da = 0, \qquad \int_{\Sigma_1} \varepsilon_{ijk} x_j t_{3k} da = 0 \tag{3.6.80}$$

when the solution of the equations

$$\begin{gathered} t_{ji,j}^* + \Lambda_i^{(\rho)}(x_1, x_2) x_3^n = 0 \\ t_{ij}^* = \lambda^{(\rho)} e_{rr}^* \delta_{ij} + 2\mu^{(\rho)} e_{ij}^*, \qquad 2e_{ij}^* = u_{i,j}^* + u_{j,i}^* \text{ on } B_\rho \end{gathered} \tag{3.6.81}$$

with the conditions

$$[u_i^*]_1 = [u_i^*]_2, \qquad [t_{\alpha i}^*]_1 n_\alpha^0 = [t_{\alpha i}^*]_2 n_\alpha^0 \text{ on } \Pi_0 \tag{3.6.82}$$

$$[t_{\alpha i}^* n_\alpha]_\rho = \sigma_i^{(\rho)}(x_1, x_2) x_3^n \text{ on } \Pi_\rho \tag{3.6.83}$$

$$\int_{\Sigma_1} t_{3i}^* da = 0, \qquad \int_{\Sigma_1} \varepsilon_{ijk} x_j t_{3k}^* da = 0 \tag{3.6.84}$$

is known. In the above relations, Λ_i and σ_i are prescribed functions which belong to C^∞. As in Section 2.4, we seek the solution of the problem in the form

$$u_i = (n+1)\left[\int_0^{x_3} u_i^* dx_3 + v_i\right] \tag{3.6.85}$$

where v_i are unknown functions. By Equations 3.6.85 and 3.6.77, we get

$$t_{ij} = (n+1)\left[\int_0^{x_3} t_{ij}^* dx_3 + s_{ij} + k_{ij}^{(\rho)}\right] \tag{3.6.86}$$

where

$$s_{ij} = \lambda^{(\rho)} \eta_{rr} \delta_{ij} + 2\mu^{(\rho)} \eta_{ij}, \qquad 2\eta_{ij} = v_{i,j} + v_{j,i} \tag{3.6.87}$$

and

$$k^{(\rho)}_{\alpha\beta} = \lambda^{(\rho)} u^*_3(x_1, x_2, 0)\delta_{\alpha\beta}, \qquad k^{(\rho)}_{\alpha 3} = k^{(\rho)}_{3\alpha} = \mu^{(\rho)} u^*_\alpha(x_1, x_2, 0)$$
$$k^{(\rho)}_{33} = (\lambda^{(\rho)} + 2\mu^{(\rho)}) u^*_3(x_1, x_2, 0), \quad (x_1, x_2) \in A_\rho \tag{3.6.88}$$

In view of Equations 3.6.81 and 3.6.86, the equations of equilibrium reduce to

$$s_{ji,j} + \ell^{(\rho)}_i = 0 \text{ on } B_\rho \tag{3.6.89}$$

where

$$\ell^{(\rho)}_i = k^{(\rho)}_{\alpha i,\alpha} + t^*_{3i}(x_1, x_2, 0) \tag{3.6.90}$$

Clearly, the functions $\ell^{(\rho)}_i$ are independent of the axial coordinate. By Equations 3.6.82, 3.6.83, 3.6.85, and 3.6.86, we find that the conditions 3.6.78 and 3.6.79 become

$$[v_i]_1 = [v_i]_2, \qquad [s_{\alpha i}]_1 n^0_\alpha = [s_{\alpha i}]_2 n^0_\alpha + \kappa_i \text{ on } \Pi_0 \tag{3.6.91}$$

$$[s_{\alpha i} n_\alpha]_\rho = \tau^{(\rho)}_i \text{ on } \Pi_\rho \tag{3.6.92}$$

where

$$\kappa_i = (k^{(2)}_{\alpha i} - k^{(1)}_{\alpha i}) n^0_\alpha, \qquad \tau^{(\rho)}_i = -k^{(\rho)}_{\alpha i} n_\alpha \tag{3.6.93}$$

We note that the functions κ_i and $\tau^{(\rho)}_i$ are independent of x_3. In view of Equations 3.6.84 and 3.6.86, we conclude that the conditions 3.6.80 reduce to

$$\int_{\Sigma_1} s_{3i} da = -T_i, \qquad \int_{\Sigma_1} \varepsilon_{ijk} x_j s_{3k} da = -N_i \tag{3.6.94}$$

where

$$T_i = \sum_{\rho=1}^{2} \int_{A_\rho} k^{(\rho)}_{i3} da, \qquad N_i = \sum_{\rho=1}^{2} \int_{A_\rho} \varepsilon_{irs} x_r k^{(s)}_{3s} da$$

Thus, the functions v_i are characterized by Equations 3.6.87 and 3.6.89 on B_ρ, the conditions 3.6.91 on the surface Π_0, the conditions 3.6.92 on the lateral boundary, and the conditions 3.6.94 on Σ_1. If κ_i were to vanish, then this problem would reduce to the Almansi–Michell problem studied in the preceding section. However, it is easy to see that for $\kappa_i \neq 0$ as well the solution of this problem has the form 3.6.52. Moreover, in this case, the solution has the form 3.6.52 with $c_i = \tau_2 = b_i = 0, \Lambda = 0$. Thus, the Almansi–Michell problem is solved. The results presented in this section have been established in Ref. 151.

Remark 1. It is easy to extend the solution to the case when B is composed of n elastic bodies with different elasticities.

Remark 2. The results presented in this section continue to hold when we consider the following distribution of the two materials. Let L be a closed curve contained in Σ_1, which is the boundary of a regular domain A^*_2 contained in Σ_1. We assume that L and Γ have no common points. We denote by A^*_1 the regular domain bounded by the curves L and Γ. Clearly,

$A_1^* \cap A_2^* = \emptyset$, $A_1^* \cup A_2^* \cup L = \Sigma_1$. We denote by B_ρ^* the cylinder defined by $B_\rho^* = \{x : (x_1, x_2) \in A_\rho^*, 0 < x_3 < h\}$, $(\rho = 1, 2)$. We assume that B_ρ^* is occupied by an isotropic elastic material with the constitutive coefficients $\lambda^{(\rho)}$ and $\mu^{(\rho)}$. We continue to denote by Π_0 the surface of separation of the two materials.

The solutions 3.6.16, 3.6.31, and 3.6.52 continue to hold in this case if we consider that $v_\alpha^{(k)}$ are the solutions of the problems $\mathcal{P}_0^{(k)}$ characterized by Equations 3.6.10, 3.6.11, and 3.6.12 on A_ρ^* and the conditions

$$[v_\alpha^{(k)}]_1 = [v_\alpha^{(k)}]_2, \qquad [\sigma_{\beta\alpha}^{(k)}]_1 n_\beta^0 = [\sigma_{\beta\alpha}^{(k)}]_2 n_\beta^0 + g_\alpha^{(k)} \text{ on } L \tag{3.6.95}$$

$$[\sigma_{\beta\alpha}^{(\kappa)} n_\beta]_1 = -\lambda^{(1)} x_\kappa n_\alpha, \qquad [\sigma_{\beta\alpha}^{(3)} n_\beta]_1 = -\lambda^{(1)} n_\alpha \text{ on } \Gamma \tag{3.6.96}$$

where $g_\alpha^{(k)}$ are defined by Equation 3.6.15. The torsion function φ is the solution of the equation

$$(\mu^{(\rho)} \varphi_{,\beta})_{,\beta} = \varepsilon_{\alpha\beta} (\mu^{(\rho)} x_\beta)_{,\alpha} \text{ on } A_\rho^* \tag{3.6.97}$$

with the conditions

$$[\varphi]_1 = [\varphi]_2, \qquad \mu^{(1)} \left[\frac{\partial \varphi}{\partial n^0}\right]_1 = \mu^{(2)} \left[\frac{\partial \varphi}{\partial n}\right]_2 + (\mu^{(1)} - \mu^{(2)}) \varepsilon_{\alpha\beta} x_\beta n_\alpha^0 \text{ on } L$$

$$\left[\frac{\partial \varphi}{\partial n}\right]_1 = \varepsilon_{\alpha\beta} x_\beta n_\alpha \text{ on } \Gamma \tag{3.6.98}$$

In this case, in the relations 3.4.67, 3.6.19, 3.6.48, 3.6.49, 3.6.69, and 3.6.75, we have to replace A_ρ by A_ρ^* and to take $\Gamma_2 = 0, \Gamma_1 = \Gamma, p_j^{(1)} = p_j$. The other boundary conditions can be modified as in the case of the boundary-value problem 3.6.97 and 3.6.98.

3.7 Piecewise Homogeneous Cylinders

Muskhelishvili [241] was the first to solve Saint-Venant's problem for cylinders composed of different homogeneous and isotropic materials. The solutions for several problems of interest from a technical point of view have been established in various works [307,313,340]. An account of the historical developments of the theory as well as references to various contributions may be found in the books by Sokolnikoff [313], Bors [28], and Khatiashvili [173,175].

In this section, we derive the results established by Muskhelishvili by using the theory developed in Section 3.6. Throughout this section, we assume that the elastic coefficients $\lambda^{(\rho)}$ and $\mu^{(\rho)}$ are constants. Thus, we consider that cylinder B is composed of two homogeneous and isotropic materials which

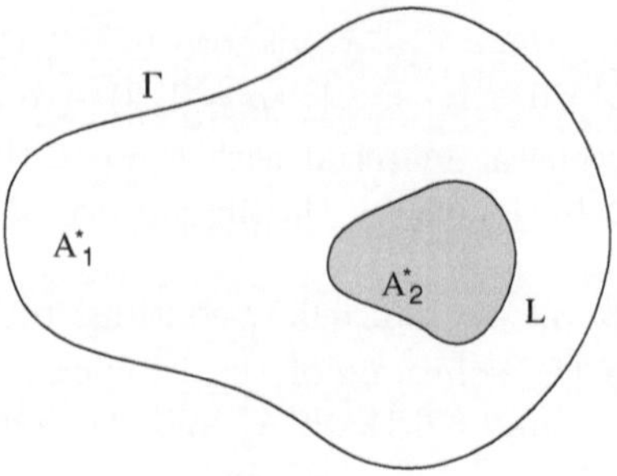

FIGURE 3.1 Cross section of a piecewise homogeneous cylinder.

are welded together along the surface Π_0. We assume that the materials are distributed as in Section 3.6, Remark 2. (Figure 3.1)

3.7.1 Alternative Form of Auxiliary Plane Strain Problems

We introduce the functions $v_\alpha^{*(k)}$ on A_ρ^* by

$$\begin{aligned} v_1^{*(1)} &= v_1^{(1)} + \frac{1}{2}\nu^{(\rho)}(x_1^2 - x_2^2), & v_2^{*(1)} &= v_2^{(1)} + \nu^{(\rho)} x_1 x_2 \\ v_1^{*(2)} &= v_1^{(2)} + \nu^{(\rho)} x_1 x_2, & v_2^{*(2)} &= v_2^{(2)} - \frac{1}{2}\nu^{(\rho)}(x_1^2 - x_2^2) \\ v_1^{*(3)} &= v_1^{(3)} + \nu^{(\rho)} x_1, & v_2^{*(3)} &= v_2^{(3)} + \nu^{(\rho)} x_2 \end{aligned} \tag{3.7.1}$$

where

$$\nu^{(\rho)} = \frac{\lambda^{(\rho)}}{2(\lambda^{(\rho)} + \mu^{(\rho)})} \text{ on } A_\rho^*, \quad (\rho = 1, 2) \tag{3.7.2}$$

We define $e_{\alpha\beta}^{*(k)}$ and $\sigma_{\alpha\beta}^{*(k)}$ by

$$\begin{aligned} \gamma_{\alpha\beta}^{*(k)} &= \frac{1}{2}\left(v_{\alpha,\beta}^{*(k)} + v_{\beta,\alpha}^{*(k)}\right) \\ \sigma_{\alpha\beta}^{*(k)} &= \lambda^{(\rho)}\gamma_{\eta\eta}^{*(k)}\delta_{\alpha\beta} + 2\mu^{(\rho)}\gamma_{\alpha\beta}^{*(k)}, \quad (k = 1, 2, 3) \end{aligned} \tag{3.7.3}$$

By Equations 3.6.10, 3.6.11, 3.7.1, and 3.7.3, we find that

$$\begin{aligned} \gamma_{\alpha\beta}^{*(\kappa)} &= \gamma_{\alpha\beta}^{(\kappa)} + \nu^{(\rho)} x_\kappa \delta_{\alpha\beta}, & \gamma_{\alpha\beta}^{*(3)} &= \gamma_{\alpha\beta}^{(3)} + \nu^{(\rho)}\delta_{\alpha\beta} \\ \sigma_{\alpha\beta}^{*(\kappa)} &= \sigma_{\alpha\beta}^{(\kappa)} + \lambda^{(\rho)} x_\kappa \delta_{\alpha\beta}, & \sigma_{\alpha\beta}^{*(3)} &= \sigma_{\alpha\beta}^{(3)} + \lambda^{(\rho)}\delta_{\alpha\beta} \end{aligned} \tag{3.7.4}$$

From Equations 3.6.12 and 3.7.4, we obtain the following form of the equations of equilibrium

$$\sigma_{\beta\alpha,\beta}^{*(k)} = 0 \text{ on } A_\rho^* \tag{3.7.5}$$

In view of Equations 3.7.1 and 3.7.4, the conditions 3.6.95 reduce to

$$\left[v_\alpha^{*(k)}\right]_1 = \left[v_\alpha^{*(k)}\right]_2 + h_\alpha^{(k)}, \qquad \left[\sigma_{\beta\alpha}^{*(k)}\right]_1 n_\beta^0 = \left[\sigma_{\beta\alpha}^{*(k)}\right]_2 n_\beta^0 \text{ on } L \tag{3.7.6}$$

where

$$\begin{aligned} &h_1^{(1)} = \frac{1}{2}\left(\nu^{(1)} - \nu^{(2)}\right)\left(x_1^2 - x_2^2\right), \qquad h_2^{(1)} = \left(\nu^{(1)} - \nu^{(2)}\right)x_1x_2 \\ &h_1^{(2)} = h_2^{(1)}, \qquad h_2^{(2)} = -h_1^{(1)}, \qquad h_\alpha^{(3)} = \left(\nu^{(1)} - \nu^{(2)}\right)x_\alpha \end{aligned} \tag{3.7.7}$$

It follows from Equations 3.7.4 that the conditions 3.6.96 become

$$[\sigma_{\beta\alpha}^{*(k)} n_\beta]_1 = 0 \text{ on } \Gamma, \quad (k = 1, 2, 3) \tag{3.7.8}$$

We denote by $\mathcal{P}_*^{(k)}$, $(k = 1, 2, 3)$, the plane strain problem characterized by Equations 3.7.3 and 3.7.5 on A_ρ^*, and the conditions 3.7.6 and 3.7.8. The problems $\mathcal{P}_*^{(k)}$ have been introduced by Muskhelishvili [241] to solve Saint-Venant's problem for composed cylinders. The existence of solutions of these problems has been established by Sherman [308].

Muskhelishvili [241] studied the plane strain problems $\mathcal{P}_*^{(k)}$ with the aid of the method of functions of a complex variable, presented in Section 1.5. Thus, in the case of the problem $\mathcal{P}_*^{(1)}$, the relation 1.5.45 implies that

$$v_1^{*(1)} + iv_2^{*(1)} = \alpha^{(\rho)}\Omega(z) - \beta^{(\rho)}z\overline{\Omega}'(\overline{z}) - \beta^{(\rho)}\overline{\omega}(\overline{z}) \text{ on } A_\rho^* \tag{3.7.9}$$

where

$$\alpha^{(\rho)} = \frac{1}{2\mu^{(\rho)}}(3 - 4\nu^{(\rho)}), \qquad \beta^{(\rho)} = \frac{1}{2\mu^{(\rho)}} \tag{3.7.10}$$

and Ω and ω are arbitrary analytic complex functions on A_ρ^*. It follows from Equations 1.5.50 and 3.7.8 that

$$\Omega(z) + z\overline{\Omega}'(\overline{z}) + \overline{\omega}(\overline{z}) = \text{const. on } \Gamma \tag{3.7.11}$$

By Equations 3.7.9 and 1.5.50, the conditions 3.7.6 imply

$$\begin{aligned} &\left[\alpha^{(1)}\Omega(z) - \beta^{(1)}z\overline{\Omega}'(\overline{z}) - \beta^{(1)}\overline{\omega}(\overline{z})\right]_1 \\ &- \left[\alpha^{(2)}\Omega(z) - \beta^{(2)}z\overline{\Omega}'(\overline{z}) - \beta^{(2)}\overline{\omega}(\overline{z})\right]_2 = f \\ &\left[\Omega(z) + z\overline{\Omega}'(\overline{z}) + \overline{\omega}(\overline{z})\right]_1 = \left[\Omega(z) + z\overline{\Omega}'(\overline{z}) + \overline{\omega}(\overline{z})\right]_2 + \text{const. on } L \end{aligned} \tag{3.7.12}$$

where, in the case of the problem $\mathcal{P}_*^{(1)}$, we have

$$f \equiv f^{(1)} = h_1^{(1)} + ih_2^{(1)} = \frac{1}{2}(\nu^{(1)} - \nu^{(2)})z^2 \tag{3.7.13}$$

Thus, the problem $\mathcal{P}_*^{(1)}$ is reduced to the finding of the complex analytic functions Ω and ω on A_ρ^* which satisfy the conditions 3.7.11 and 3.7.12. In a similar way, we can formulate the problems $\mathcal{P}_*^{(2)}$ and $\mathcal{P}_*^{(3)}$ with the aid of the complex potentials.

3.7.2 Extension and Bending of Piecewise Homogeneous Cylinders

In view of Equations 3.7.1, the solution 3.6.16 can be expressed in the form

$$
\begin{aligned}
u_1 &= -\frac{1}{2}d_1\left[x_3^2+\nu^{(\rho)}(x_1^2-x_2^2)\right]-d_2\nu^{(\rho)}x_1x_2-d_3\nu^{(\rho)}x_1+\sum_{k=1}^{3}d_k v_1^{*(k)}\\
u_2 &= -d_1\nu^{(\rho)}x_1x_2-\frac{1}{2}d_2\left[x_3^2-\nu^{(\rho)}(x_1^2-x_2^2)\right]-d_3\nu^{(\rho)}x_2+\sum_{k=1}^{3}d_k v_2^{*(k)}\\
u_3 &= (d_1x_1+d_2x_2+d_3)x_3
\end{aligned}
\tag{3.7.14}
$$

where $v_\alpha^{*(k)}$ are the displacements in the problems $\mathcal{P}_*^{(k)}$, $(k=1,2,3)$. By using Equations 3.7.4, we find that the constants L_{ij} defined by Equations 3.6.19 have the form

$$
\begin{aligned}
&L_{\alpha\beta}=L_{\beta\alpha}=\mathscr{I}_{\alpha\beta}+\mathscr{K}_{\alpha\beta}, \qquad L_{\alpha 3}=\mathscr{I}_{\alpha 3}+\mathscr{K}_{\alpha 3}=L_{3\alpha}\\
&L_{33}=\mathscr{I}_{33}+\mathscr{K}_{33}
\end{aligned}
\tag{3.7.15}
$$

where

$$
\begin{aligned}
&\mathscr{I}_{\alpha\beta}=\sum_{\rho=1}^{2}\int_{A_\rho^*}E^{(\rho)}x_\alpha x_\beta, \qquad \mathscr{K}_{\alpha\beta}=\sum_{\rho=1}^{2}\int_{A_\rho^*}\lambda^{(\rho)}x_\alpha\gamma_{\eta\eta}^{*(\beta)}da\\
&\mathscr{I}_{\alpha 3}=\mathscr{I}_{3\alpha}=\sum_{\rho=1}^{2}\int_{A_\rho^*}E^{(\rho)}x_\alpha da, \qquad \mathscr{K}_{\alpha 3}=\sum_{\rho=1}^{2}\int_{A_\rho^*}\lambda^{(\rho)}x_\alpha\gamma_{\eta\eta}^{*(3)}da\\
&\mathscr{I}_{33}=\sum_{\rho=1}^{2}\int_{A_\rho^*}E^{(\rho)}da, \qquad \mathscr{K}_{33}=\sum_{\rho=1}^{2}\int_{A_\rho^*}\lambda^{(\rho)}\gamma_{\eta\eta}^{*(3)}da
\end{aligned}
\tag{3.7.16}
$$

The constants d_j are determined by the system 3.6.18. The solution 3.7.14 has been established by Muskhelishvili ([241], Section 146).

3.7.3 Torsion and Flexure

By using the relations 3.7.1 and 3.6.43, we can write the solution 3.6.31 in the form

$$
\begin{aligned}
u_1 &= -\frac{1}{2}b_1\left[\frac{1}{3}x_3^2 + \nu^{(\rho)}\left(x_1^2 - x_2^2\right)\right]x_3 - b_2\nu^{(\rho)}x_1x_2x_3 \\
&\quad - b_3\nu^{(\rho)}x_1x_3 - \tau x_2x_3 + x_3\sum_{k=1}^{3} b_k v_1^{*(k)} \\
u_2 &= -b_1\nu^{(\rho)}x_1x_2x_3 - \frac{1}{2}b_2\left[\frac{1}{3}x_3^2 - \nu^{(\rho)}\left(x_1^2 - x_2^2\right)\right]x_3 \\
&\quad - b_3\nu^{(\rho)}x_2x_3 + \tau x_1x_3 + x_3\sum_{k=1}^{3} b_k v_2^{*(k)} \\
u_3 &= \frac{1}{2}(b_1x_1 + b_2x_2 + b_3)x_3^2 + \tau\varphi + \psi
\end{aligned}
\tag{3.7.17}
$$

In this case, the torsion function φ satisfies the equation

$$\Delta\varphi = 0 \text{ on } A_\rho^* \tag{3.7.18}$$

and the conditions

$$
\begin{aligned}
&[\varphi]_1 = [\varphi]_2 \\
&\mu^{(1)}\left[\frac{\partial\varphi}{\partial n^0}\right]_1 - \mu^{(2)}\left[\frac{\partial\varphi}{\partial n^0}\right]_2 = (\mu^{(1)} - \mu^{(2)})\varepsilon_{\alpha\beta}x_\beta n_\alpha^0 \text{ on } L \\
&\left[\frac{\partial\varphi}{\partial n}\right]_1 = \varepsilon_{\alpha\beta}x_\beta n_\alpha \text{ on } \Gamma
\end{aligned}
\tag{3.7.19}
$$

In view of Equations 3.6.44, 3.6.45, and 3.7.1, we find that the function ψ is the solution of the following boundary-value problem

$$
\begin{aligned}
&\Delta\psi = -2(b_1x_1 + b_2x_2 + b_3) - \frac{1}{\mu^{(\rho)}}(\lambda^{(\rho)} + \mu^{(\rho)})\sum_{k=1}^{3} b_k\gamma_{\alpha\alpha}^{*(k)} \text{ on } A_\rho^* \\
&[\psi]_1 = [\psi]_2, \qquad \mu^{(1)}\left[\frac{\partial\psi}{\partial n^0}\right]_1 - \mu^{(2)}\left[\frac{\partial\psi}{\partial n^0}\right]_2 = \sigma \text{ on } L \\
&\left[\frac{\partial\psi}{\partial n}\right]_1 = \eta \text{ on } \Gamma
\end{aligned}
\tag{3.7.20}
$$

where

$$
\begin{aligned}
\sigma &= (\mu^{(2)} - \mu^{(1)})\sum_{j=1}^{3} b_j v_\alpha^{*(j)} n_\alpha^0 - (\mu^{(2)}\nu^{(2)} - \mu^{(1)}\nu^{(1)})\left[\frac{1}{2}(x_1^2 - x_2^2)\right. \\
&\quad \left.(b_1n_1^0 - b_2n_2^0) + x_1x_2(b_1n_2^0 + b_2n_1^0) + b_3x_\alpha n_\alpha^0\right] \\
\eta &= -\sum_{j=1}^{3} b_j v_\alpha^{*(j)} n_\alpha + \frac{1}{2}\nu^{(1)}(b_1n_1 - b_2n_2)(x_1^2 - x_2^2) \\
&\quad + \nu^{(1)}(b_1n_2 + b_2n_1)x_1x_2 + b_3\nu^{(1)}x_\alpha n_\alpha
\end{aligned}
\tag{3.7.21}
$$

The constants b_1, b_2, and b_3 are determined by the system 3.6.39 and 3.6.40, where L_{ij} are given by Equation 3.7.15.

Readers interested in further details can find them in Ref. 241.

3.8 Applications

3.8.1 Nonhomogeneous Cylinders with Constant Poisson's Ratio

In what follows, we use the results of Sections 3.3 and 3.5 to study the deformation of nonhomogeneous and isotropic elastic cylinders when the constitutive coefficients have the form

$$E = E(x_1, x_2), \qquad \nu = \text{const.}, \quad (x_1, x_2) \in \Sigma_1 \tag{3.8.1}$$

This case has been studied in many works [209,279]. It is easy to verify that the solution of the problem $\mathcal{D}^{(1)}$, defined in Section 3.2, is

$$u_1^{(1)} = -\frac{1}{2}\nu\left(x_1^2 - x_2^2\right), \qquad u_2^{(1)} = -\nu x_1 x_2 \tag{3.8.2}$$

By Equations 3.2.8, 3.2.9, 3.8.1, and 3.8.2, we get

$$\begin{aligned} &e_{11}^{(1)} = -\nu x_1, \qquad e_{22}^{(1)} = -\nu x_1, \qquad e_{12}^{(1)} = 0 \\ &t_{11}^{(1)} = -2\nu x_1(\lambda + \mu) = -\lambda x_1, \qquad t_{22}^{(1)} = -\lambda x_1, \qquad t_{12}^{(1)} = 0 \end{aligned} \tag{3.8.3}$$

The solutions of the problems $\mathcal{D}^{(2)}$ and $\mathcal{D}^{(3)}$ are given by

$$\begin{aligned} &u_1^{(2)} = -\nu x_1 x_2, \qquad u_2^{(2)} = \frac{1}{2}\nu\left(x_1^2 - x_2^2\right) \\ &u_1^{(3)} = -\nu x_1, \qquad u_2^{(3)} = -\nu x_2 \end{aligned} \tag{3.8.4}$$

From Equation 3.8.4, we obtain

$$e_{11}^{(2)} = e_{22}^{(2)} = -\nu x_2, \qquad e_{12}^{(2)} = 0, \qquad e_{11}^{(3)} = e_{22}^{(3)} = -\nu, \qquad e_{12}^{(3)} = 0 \tag{3.8.5}$$

It follows from Equations 1.1.7, 3.8.3, and 3.8.5 that

$$\begin{aligned} &\lambda + 2\mu + \lambda e_{\rho\rho}^{(3)} = \lambda + 2\mu - 2\nu\lambda = E \\ &(\lambda + 2\mu)x_\beta + \lambda e_{\rho\rho}^{(\beta)} = E x_\beta \end{aligned} \tag{3.8.6}$$

In view of Equations 3.3.7 and 3.8.6, we find that the coefficients D_{ij} of the system 3.3.6 are given by

$$D_{\alpha\beta} = I_{\alpha\beta}^*, \qquad D_{\alpha 3} = D_{3\alpha} = \Omega\xi_\alpha^0, \qquad D_{33} = \Omega \tag{3.8.7}$$

where

$$I^*_{\alpha\beta} = \int_{\Sigma_1} x_\alpha x_\beta E(x_1, x_2) da, \qquad \xi^0_\alpha = \frac{1}{\Omega} \int_{\Sigma_1} x_\alpha E(x_1, x_2) da$$
$$\Omega = \int_{\Sigma_1} E(x_1, x_2) da \tag{3.8.8}$$

Thus, the system 3.3.6 becomes

$$I^*_{\alpha\beta} a_\beta + \Omega \xi^0_\alpha a_3 = \varepsilon_{\alpha\beta} M_\beta$$
$$a_1 \xi^0_1 + a_2 \xi^0_2 + a_3 = -\frac{1}{\Omega} F_3 \tag{3.8.9}$$

It follows from Equations 3.3.3, 3.8.2, and 3.8.4 that the solution of extension and bending problem is

$$u_1 = -\frac{1}{2} a_1 \left(x_3^2 + \nu x_1^2 - \nu x_2^2\right) - a_2 \nu x_1 x_2 - a_3 \nu x_1$$
$$u_2 = -a_1 \nu x_1 x_2 - \frac{1}{2} a_2 \left(x_3^2 - \nu x_1^2 + \nu x_2^2\right) - a_3 \nu x_2 \tag{3.8.10}$$
$$u_3 = (a_1 x_1 + a_2 x_2 + a_3) x_3$$

where the constants a_1, a_2, and a_3 are given by Equations 3.8.9.

Clearly, the solution of the torsion problem, presented in Section 3.4, cannot be simplified by the assumption that ν is constant.

By Equations 3.5.1, 3.8.2, and 3.8.4, we find that the solution of the flexure problem is given by

$$u_1 = -\frac{1}{6} b_1 x_3^3 - \frac{1}{2} b_1 \nu \left(x_1^2 - x_2^2\right) x_3 - b_2 \nu x_1 x_2 x_3 - b_3 \nu x_1 x_3 - \tau x_2 x_3$$
$$u_2 = -\frac{1}{6} b_2 x_3^3 - b_1 \nu x_1 x_2 x_3 - \frac{1}{2} b_2 \nu \left(x_2^2 - x_1^2\right) x_3 - b_3 \nu x_2 x_3 + \tau x_1 x_3 \tag{3.8.11}$$
$$u_3 = \frac{1}{2} (b_1 x_1 + b_2 x_2 + b_3) x_3^2 + \tau \varphi + G$$

From Equations 3.5.7, 3.5.8, and 3.8.7, we find that the constants b_1, b_2, and b_3 are determined by the following system

$$I^*_{\alpha\beta} b_\beta + \Omega \xi^0_\alpha b_3 = -F_\alpha$$
$$b_1 \xi^0_1 + b_2 \xi^0_2 + b_3 = 0 \tag{3.8.12}$$

It is easy to see that the function G satisfies the equation

$$(\mu G_{,\alpha})_{,\alpha} = -2\mu(b_1 x_1 + b_2 x_2 + b_3) + \mu_{,1} \nu \left[\frac{1}{2} b_1 \left(x_1^2 - x_2^2\right) + b_2 x_1 x_2 + b_3 x_1\right]$$
$$+ \mu_{,2} \nu \left[b_1 x_1 x_2 + \frac{1}{2} b_2 \left(x_2^2 - x_1^2\right) + b_3 x_2\right] \text{ on } \Sigma_1 \tag{3.8.13}$$

and the boundary condition

$$G_{,\alpha}n_\alpha = \nu\left[\frac{1}{2}b_1(x_1^2 - x_2^2) + b_2x_1x_2 + b_3x_1\right]n_1 + \nu\left[b_1x_1x_2 + \frac{1}{2}b_2(x_2^2 - x_1^2) + b_3x_2\right]n_2 \text{ on } \Gamma \quad (3.8.14)$$

The constant τ is given by Equation 3.5.9, where $\mathfrak{M}$ has the form

$$\mathfrak{M} = \int_{\Sigma_1} \mu\left[x_1G_{,2} - x_2G_{,1} - \frac{1}{2}\nu(x_1^2 + x_2^2)(b_1x_2 - b_2x_1)\right]da \quad (3.8.15)$$

The torsional rigidity D_* is given by Equation 3.4.6, where φ is the torsion function.

3.8.2 Deformation of a Nonhomogeneous Circular Cylinder

Let us study the extension, bending, and torsion of a nonhomogeneous and isotropic cylinder that occupies the domain $B = \{x : x_1^2 + x_2^2 < a^2, 0 < x_3 < h\}$, $(a > 0)$. We assume that

$$E = E(r), \qquad \nu = \text{const.} \quad (3.8.16)$$

where

$$r = \left(x_1^2 + x_2^2\right)^{1/2} \quad (3.8.17)$$

It follows from Equations 3.8.8 and 3.8.16 that

$$I_{11}^* = I_{22}^* = \pi\int_0^a r^3E(r)dr, \qquad I_{12}^* = 0$$
$$\xi_\alpha^0 = 0, \qquad \Omega = 2\pi\int_0^a rE(r)dr \quad (3.8.18)$$

By Equations 3.8.9 and 3.8.18, we obtain

$$a_1 = \frac{M_2}{I_{11}^*}, \qquad a_2 = -\frac{M_1}{I_{22}^*}, \qquad a_3 = -\frac{1}{\Omega}F_3 \quad (3.8.19)$$

The solution of extension and bending problem has the form 3.8.10 where the constants a_k are given by Equations 3.8.19. In this case, the extension is not influenced by the bending of terminal couples.

To solve the torsion problem, we consider the boundary-value problem 3.4.12 and 3.4.13. We assume that the functions φ and χ depend only on r. Let us introduce the function H by

$$\frac{1}{\mu}\chi_{,\alpha} = H_{,\alpha} \quad (3.8.20)$$

We note that the condition

$$\left(\frac{1}{\mu}\chi_{,\alpha}\right)_{,\beta} = \left(\frac{1}{\mu}\chi_{,\beta}\right)_{,\alpha}$$

is satisfied on the basis of relations

$$\left(\frac{1}{\mu}\right)_{,\beta} = \frac{x_\beta}{r}\frac{d}{dr}\left(\frac{1}{\mu}\right), \qquad \chi_{,\alpha} = \chi'\frac{x_\alpha}{r}, \qquad \chi' = \frac{d\chi}{dr}$$

Thus, the function H exists. From Equation 3.4.12, we see that H satisfies the equation

$$\Delta H = -2 \text{ on } \Sigma_1 \tag{3.8.21}$$

By Equations 3.4.14 and 3.8.20, we find that the stresses $t_{\alpha 3}$ are given by

$$t_{13} = \mu\tau H_{,2} \qquad t_{23} = -\mu\tau H_{,1} \tag{3.8.22}$$

The conditions on the lateral surface are satisfied if

$$H = 0 \text{ on } r = a \tag{3.8.23}$$

The solution of the boundary-value problem 3.8.21 and 3.8.23 is

$$H = \frac{1}{2}(a^2 - r^2) \tag{3.8.24}$$

Thus, from Equations 3.8.22 and 3.8.24, we obtain

$$t_{13} = -\mu(r)\tau x_2, \qquad t_{23} = \mu(r)\tau x_1 \tag{3.8.25}$$

By Equations 3.4.11, 3.8.20, and 3.8.24, we find that $\varphi = 0$. In view of Equation 3.4.6, we obtain the torsional rigidity,

$$D_* = 2\pi \int_0^a r^3 \mu(r) dr \tag{3.8.26}$$

The constant τ is given by Equation 3.4.5. We note that from Equations 3.8.20 and 3.8.24, we find that

$$\chi(r) = -\int_0^r t\mu(t)dt$$

3.8.3 Extension, Bending, and Torsion of Nonhomogeneous Tube

First, we study the plane strain problems $\mathcal{D}^{(k)}$ defined in Section 3.2, for a hollow cylinder. We assume that the domain Σ_1 is bounded by two concentric circles of radius R_1 and R_2, $\Sigma_1 = \{x : R_1^2 < x_1^2 + x_2^2 < R_2^2, x_3 = 0\}$. We suppose that the cylinder is occupied by an isotropic material with the following constitutive coefficients

$$\lambda = \lambda_0 r^{-m}, \qquad \mu = \mu_0 r^{-m}, \quad m > 0 \tag{3.8.27}$$

where r is given by Equation 3.8.17 and λ_0, μ_0, and m are prescribed constants. This kind of inhomogeneity has been investigated by Lekhnitskii [205] and Lomakin [209]. Let us prove that the solution of the problem $\mathcal{D}^{(1)}$ is given by

$$u_1^{(1)} = -\frac{1}{2}\nu_0(x_1^2 - x_2^2), \qquad u_2^{(1)} = -\nu_0 x_1 x_2 \tag{3.8.28}$$

where

$$\nu_0 = \frac{\lambda_0}{2(\lambda_0 + \mu_0)} \tag{3.8.29}$$

In view of Equations 3.2.8 and 3.8.2, we obtain

$$e_{11}^{(1)} = -\nu_0 x_1, \qquad e_{22}^{(1)} = -\nu_0 x_1, \qquad e_{12}^{(1)} = 0 \tag{3.8.30}$$

By the constitutive equations 3.8.9 and the relations 3.8.27, 3.8.29, and 3.8.30, we find that

$$\begin{aligned} t_{11}^{(1)} &= -2\nu_0 x_1(\lambda + \mu) = -2\nu_0 x_1(\lambda_0 + \mu_0)r^{-m} \\ &= -\lambda_0 x_1 r^{-m} = -\lambda x_1 \\ t_{22}^{(1)} &= -\lambda x_1 \quad t_{12}^{(1)} = 0 \end{aligned} \tag{3.8.31}$$

It is easy to see that the stresses 3.8.31 satisfy the equations of equilibrium 3.2.10 and the boundary conditions 3.2.11. In a similar way, we can prove that

$$u_1^{(2)} = -\nu_0 x_1 x_2, \qquad u_2^{(2)} = \frac{1}{2}\nu_0(x_1^2 - x_2^2), \qquad u_\alpha^{(3)} = -\nu_0 x_\alpha \tag{3.8.32}$$

By Equations 1.1.7 and 3.8.27, we find

$$E = E_0 r^{-m} \tag{3.8.33}$$

where

$$E_0 = \frac{\mu_0(3\lambda_0 + 2\mu_0)}{\lambda_0 + \mu_0}$$

With the aid of Equations 3.8.30, 3.8.32, and 3.8.33, we obtain

$$\begin{aligned} &\lambda + 2\mu + \lambda e_{\rho\rho}^{(3)} = \lambda + 2\mu - 2\lambda\nu_0 = (\lambda_0 + 2\mu_0 - 2\lambda_0\nu_0)r^{-m} = E \\ &(\lambda + 2\mu)x_\beta + \lambda e_{\rho\rho}^{(\beta)} = E x_\beta \end{aligned} \tag{3.8.34}$$

It follows from Equations 3.3.7, 3.8.33, and 3.8.34, that

$$\begin{aligned} &D_{11} = D_{22} = J \qquad D_{12} = D_{21} = D_{\alpha 3} = D_{3\alpha} = 0 \qquad D_{33} = J_* \\ &J = \frac{\pi}{4-m}\left[R_2^4 E(R_2) - R_1^4 E(R_1)\right] = \frac{\pi}{4-m}E_0\left(R_2^{4-m} - R_1^{4-m}\right), \ \text{for } m \neq 4 \\ &J = 2\pi E_0 \ln(R_2/R_1), \ \text{for } m = 4 \\ &J_* = \frac{2\pi}{2-m}E_0(R_2^{2-m} - R_1^{2-m}), \ \text{for } m \neq 2 \\ &J_* = 2\pi E_0 \ln(R_2/R_1), \ \text{for } m = 2 \end{aligned} \tag{3.8.35}$$

By Equations 3.3.6 and 3.8.5, we obtain

$$a_1 = \frac{M_2}{J}, \qquad a_2 = -\frac{M_1}{J}, \qquad a_3 = -\frac{F_3}{J_*} \tag{3.8.36}$$

Thus, the solution of extension and bending problem is given by

$$\begin{aligned} u_1 &= -\frac{1}{2}a_1\left(x_3^2 + \nu_0 x_1^2 - \nu_0 x_2^2\right) - a_2\nu_0 x_1 x_2 - a_3\nu_0 x_1 \\ u_2 &= -a_1\nu_0 x_1 x_2 - \frac{1}{2}a_2\left(x_3^2 - \nu_0 x_1^2 + \nu_0 x_2^2\right) - a_3\nu_0 x_2 \\ u_3 &= (a_1 x_1 + a_2 x_2 + a_3)x_3 \end{aligned} \tag{3.8.37}$$

where a_k are defined by Equations 3.8.36 and 3.8.35. In view of Equations 3.8.27, we find that

$$\varepsilon_{\rho\beta}(\mu x_\beta)_{,\rho} = \varepsilon_{\rho\beta}\left(\frac{d\mu}{dr}x_\rho x_\beta r^{-1} + \mu\delta_{\rho\beta}\right) = 0$$

so that Equation 3.4.1 for the torsion function becomes

$$(\mu\varphi_{,\alpha})_{,\alpha} = 0 \text{ on } \Sigma_1$$

Clearly, for $r = R_1$ and $r = R_2$, we have

$$\varepsilon_{\alpha\beta}x_\beta n_\alpha = 0$$

and the boundary condition 3.4.2 reduces to

$$\varphi_{,\alpha}n_\alpha = 0 \text{ on } \Gamma$$

Thus, in this case, we find that

$$\varphi = 0 \text{ on } \Sigma_1 \tag{3.8.38}$$

From Equation 3.4.6, we obtain the torsional rigidity,

$$\begin{aligned} D_* &= \frac{2\pi}{4-m}\mu_0\left(R_2^{4-m} - R_1^{4-m}\right), \quad \text{for } m \neq 4 \\ D_* &= 2\pi\mu_0 \ln\left(R_2/R_1\right), \quad \text{for } m = 4 \end{aligned} \tag{3.8.39}$$

The solution of the torsion problem is

$$u_\alpha = \tau\varepsilon_{\beta\alpha}x_3 x_\beta, \qquad u_3 = 0$$

where the constant τ is given by Equations 3.4.5 and 3.8.39.

3.8.4 Flexure of Hollow Cylinder

We now study the flexure of the hollow cylinder defined in the Section 3.8.3. We continue to assume that the elastic coefficients are given by Equations 3.8.27. We suppose that the loading applied on the end located at $x_3 = 0$ is statically equivalent to the force $\mathbf{F}=F_1\mathbf{e}_1$ and the moment $\mathbf{M}=\mathbf{0}$. The form of the solution is given by the functions 3.5.1. Since the Lamé moduli are specified by Equations 3.8.27, the solutions of the problems $\mathcal{D}^{(k)}$ are given by Equations 3.8.28 and 3.8.32. Moreover, we have seen that for the considered cylinder, the torsion function is zero. The constants b_1, b_2, and b_3 which appear in Equations 3.5.1 are determined by Equations 3.5.7 and 3.5.8. In view of Equations 3.8.35, we find that

$$b_1 = -\frac{F_1}{J}, \qquad b_2 = 0, \qquad b_3 = 0 \tag{3.8.40}$$

Thus, the boundary-value problem 3.5.3 and 3.5.5 reduces to the equation

$$(\mu G_{,\alpha})_{,\alpha} = -2\mu b_1 x_1 - b_1 \mu_{,\beta} u_\beta^{(1)} \text{ on } \Sigma_1 \tag{3.8.41}$$

and the boundary condition

$$G_{,\alpha} n_\alpha = -b_1 u_\beta^{(1)} n_\beta \text{ on } \Gamma \tag{3.8.42}$$

In view of Equations 3.8.30 and 3.8.37, we find that

$$\mu_{,\beta} u_\beta^{(1)} = \frac{1}{2}\nu_0 \mu m x_1$$

so that Equation 3.8.41 takes the form

$$\mu \Delta G + \mu_{,\alpha} G_{,\alpha} = -\mu\left(2 + \frac{1}{2}\nu_0 m\right) b_1 x_1 \text{ on } \Sigma_1 \tag{3.8.43}$$

We seek the solution of this equation in the form

$$G = x_1 \Phi(r) \tag{3.8.44}$$

where r is given by Equation 3.8.17 and Φ is an unknown function.
With the aid of relations

$$\Delta G = x_1(\Phi'' + 3r^{-1}\Phi'), \qquad \mu_{,\alpha} G_{,\alpha} = -\mu m r^{-2} x_1(\Phi + r\Phi')$$

we find that Equation 3.8.43 reduces to

$$\Phi'' + (3-m)\frac{1}{r}\Phi' - \frac{m}{r^2}\Phi = -\left(2 + \frac{1}{2} m\nu_0\right) b_1 \tag{3.8.45}$$

A particular solution of this equation is

$$\Phi_* = Br^2 \tag{3.8.46}$$

where

$$B = \frac{1}{3m-8}\left(2 + \frac{1}{2} m\nu_0\right) b_1$$

The general solution for Φ is

$$\Phi = C_1 r^{k_1} + C_2 r^{k_2} + Br^2 \tag{3.8.47}$$

where C_1 and C_2 are arbitrary constants, and

$$k_1 = \frac{1}{2}[m - 2 + (m^2 + 4)^{1/2}], \qquad k_2 = \frac{1}{2}[m - 2 - (m^2 + 4)^{1/2}] \tag{3.8.48}$$

In view of Equations 3.8.28 and 3.8.44, the condition 3.8.42 reduces to

$$\Phi(r) + r\Phi'(r) = \frac{1}{2}\nu_0 b_1 r^2 \text{ on } r = R_1 \quad \text{and} \quad r = R_2 \tag{3.8.49}$$

By Equation 3.8.47, the conditions 3.8.49 take the form

$$\begin{aligned}
(1 + k_1)R_1^{k_1} C_1 + (1 + k_2)R_1^{k_2} C_2 &= \left(\frac{1}{2}\nu_0 b_1 - 3B\right) R_1^2 \\
(1 + k_1)R_2^{k_1} C_1 + (1 + k_2)R_2^{k_2} C_2 &= \left(\frac{1}{2}\nu_0 b_1 - 3B\right) R_2^2
\end{aligned} \tag{3.8.50}$$

Since

$$(1 + k_1)(1 + k_2) \neq 0, \quad k_1 \neq k_2$$

we conclude that the system 3.8.50 can always be solved for the constants C_1 and C_2. The constant τ is given by Equation 3.5.9. It follows from Equations 3.8.28, 3.8.40, and 3.8.44 that

$$\mathfrak{M} = -\int_{\Sigma_1} x_2 \mu \left(\Phi + \frac{1}{2}\nu_0 b_1 r^2\right) da = 0$$

so that $\tau = 0$

From Equations 3.5.1, 3.8.28, 3.8.38, 3.8.40, and 3.8.44, we find the solution of the flexure of a hollow cylinder,

$$\begin{aligned}
u_1 &= -\frac{1}{2}\left(\frac{1}{3}x_3^2 + \nu_0 x_1^2 - \nu_0 x_2^2\right) b_1 x_3, \qquad u_2 = -b_1 \nu_0 x_1 x_2 x_3 \\
u_3 &= \frac{1}{2} b_1 x_1 x_3^2 + x_1 \Phi
\end{aligned}$$

where Φ has the form 3.8.47 and b_1 is given by Equation 3.8.40.

3.8.5 Plane Strain of Nonhomogeneous Tube

In this section, we investigate the plane strain traction problem for the domain $\Sigma_1 = \{x : R_1^2 < x_1^2 + x_2^2 < R_2^2, x_3 = 0\}$, $(R_2 > R_1 > 0)$, when the cylinder is in equilibrium in the absence of body forces, and the lateral boundaries are subjected to constant pressures. We assume that the tube is occupied by a nonhomogeneous and isotropic elastic material with Lamé moduli given by Equations 3.8.27. The equilibrium equations become

$$t_{\beta\alpha,\beta} = 0 \text{ on } \Sigma_1 \tag{3.8.51}$$

The boundary conditions are

$$\begin{aligned} t_{\beta\alpha} n_\beta &= -p_1 n_\alpha \text{ on } r = R_1 \\ t_{\beta\alpha} n_\beta &= -p_2 n_\alpha \text{ on } r = R_2 \end{aligned} \tag{3.8.52}$$

where p_α are prescribed constants and $r = (x_1^2 + x_2^2)^{1/2}$.

We seek the solution in the form

$$u_\alpha = x_\alpha r^{-1} G \tag{3.8.53}$$

where G is an unknown function of r. Then, we have

$$\begin{aligned} u_{\alpha,\beta} &= \delta_{\alpha\beta} r^{-1} G - x_\alpha x_\beta r^{-3} G + x_\alpha x_\beta r^{-2} G' \\ u_{\rho,\rho} &= r^{-1} G + G' \qquad G' = \frac{dG}{dr} \\ t_{\alpha\beta} &= \lambda \left(r^{-1} G + G'\right) \delta_{\alpha\beta} \\ &\quad + 2\mu(\delta_{\alpha\beta} r^{-1} G - x_\alpha x_\beta r^{-3} G + x_\alpha x_\beta r^{-2} G') \\ \Delta u_\alpha &= u_{\rho,\rho\alpha} = x_\alpha \left(r^{-1} G'' + r^{-2} G' - r^{-3} G\right) \end{aligned} \tag{3.8.54}$$

We note that

$$t_{\beta\alpha,\beta} = \mu \Delta u_\alpha + (\lambda + \mu) u_{\rho,\rho\alpha} + \lambda_{,\alpha} u_{\rho,\rho} + \mu_{,\beta}(u_{\alpha,\beta} + u_{\beta,\alpha}) \tag{3.8.55}$$

By using Equations 3.8.54, 3.8.55, and the relations

$$\lambda_{,\alpha} = \lambda' x_\alpha r^{-1}, \qquad \mu_{,\alpha} = \mu' x_\alpha r^{-1}$$

we obtain

$$t_{\beta\alpha,\beta} = x_\alpha r^{-1} \left[(\lambda + 2\mu)(G'' + r^{-1} G' - r^{-2} G) + r^{-1} \lambda' G + (\lambda' + 2\mu') G'\right] \tag{3.8.56}$$

From Equations 3.8.51 and 3.8.56, we conclude that the equilibrium equations are satisfied if the function G satisfies the equation

$$G'' + \left(\frac{1}{r} + \frac{M'}{M}\right) G' - \left(\frac{1}{r^2} - \frac{\lambda'}{rM}\right) G = 0 \tag{3.8.57}$$

where

$$M = \lambda + 2\mu \tag{3.8.58}$$

By Equations 3.8.27 and 3.8.58,

$$M' = -mr^{-1}M, \qquad \lambda' = -mr^{-1}MN_0, \qquad N_0 = \lambda_0/(\lambda_0 + 2\mu_0) \quad (3.8.59)$$

so that Equation 3.8.57 reduces to

$$G'' + \frac{1}{r}(1-m)G' - \frac{1}{r^2}(1+mN_0)G = 0 \quad (3.8.60)$$

In view of relations 1.1.5, we obtain

$$1 - N_0^2 > 0 \quad (3.8.61)$$

The general solution for G is

$$G(r) = C_1 r^{k_1} + C_2 r^{k_2} \quad (3.8.62)$$

where

$$k_1 = \frac{1}{2}\left[m + (m^2 + 4mN_0 + 4)^{1/2}\right], \qquad k_2 = \frac{1}{2}\left[m - (m^2 + 4mN_0 + 4)^{1/2}\right] \quad (3.8.63)$$

and C_1 and C_2 are arbitrary constants. It follows from inequality 3.8.61 that

$$m^2 + 4mN_0 + 4 = (m + 2N_0)^2 + 4(1 - N_0^2) > 0$$

so that k_1 and k_2 are real and distinct. On the boundary of Σ_1, we have

$$n_\alpha = -\frac{x_\alpha}{R_1} \text{ on } r = R_1, \qquad n_\alpha = \frac{x_\alpha}{R_2} \text{ on } r = R_2$$

so that the conditions 3.8.52 reduce to

$$t_{\beta\alpha}x_\beta = -p_\rho x_\alpha \text{ on } r = R_\rho \quad (3.8.64)$$

It follows from Equations 3.8.27, 3.8.54, and 3.8.62 that

$$\begin{aligned} t_{\beta\alpha}x_\beta &= x_\alpha[\lambda r^{-1}G + (\lambda + 2\mu)G'] \\ &= x_\alpha r^{-m}\{[\lambda_0 + (\lambda_0 + 2\mu_0)k_1]C_1 r^{k_1-1} + [\lambda_0 + (\lambda_0 + 2\mu_0)k_2]C_2 r^{k_2-1}\} \end{aligned} \quad (3.8.65)$$

Thus, the boundary conditions 3.8.64 become

$$\begin{aligned} &[\lambda_0 + (\lambda_0 + 2\mu_0)k_1]R_1^{k_1-1}C_1 + [\lambda_0 + (\lambda_0 + 2\mu_0)k_2]R_1^{k_2-1}C_2 = -p_1 R_1^m \\ &[\lambda_0 + (\lambda_0 + 2\mu_0)k_1]R_2^{k_1-1}C_1 + [\lambda_0 + (\lambda_0 + 2\mu_0)k_2]R_2^{k_2-1}C_2 = -p_2 R_2^m \end{aligned} \quad (3.8.66)$$

The determinant of this system is

$$(\lambda_0 + 2\mu_0)^2(N_0 + k_1)(N_0 + k_2)R_1^{k_2-1}R_2^{k_1-1}\left[\left(\frac{R_1}{R_2}\right)^{k_1-1} - \left(\frac{R_1}{R_2}\right)^{k_2-1}\right]$$

By the relations 1.1.5, 3.8.59, and 3.8.63, we get

$$\lambda_0 + 2\mu_0 > 0, \qquad k_1 \neq -N_0, \qquad k_2 \neq -N_0, \qquad k_1 \neq k_2$$

so that the system 3.8.66 uniquely determines the constants C_1 and C_2. Thus, the solution of the problem is given by Equations 3.8.53, 3.8.62, and 3.8.66.

3.8.6 Special Solutions of Plane Strain Problem

We consider the plane strain problem for nonhomogeneous bodies when the body forces vanish. In Section 3.2, we have seen that the stresses $t_{\alpha\beta}$ can be expressed in terms of the Airy function which satisfies Equation 3.2.7. First, we assume that

$$q = d_1x_1 + d_2x_2 + d_3, \qquad \gamma = d_4 \tag{3.8.67}$$

where d_k, $(k = 1, 2, 3, 4)$, are constants. In this case, Equation 3.2.7 reduces to

$$\Delta\Delta\chi = 0 \text{ on } \Sigma_1$$

We note that the boundary conditions 1.5.25 also hold for nonhomogeneous bodies. By Equations 1.5.19, 1.5.22, and 1.5.25, we see that the stresses $t_{\alpha\beta}$ in the nonhomogeneous material defined by Equations 3.8.67 are the same as the corresponding stresses in a homogeneous material, provided the material occupy cylinders of the same shape and are subject to the same surface forces. It is simple to verify that the conditions 3.8.67 correspond to the following constitutive coefficients

$$E = \frac{2(d_1x_1 + d_2x_2 + d_3) - d_4}{(d_1x_1 + d_2x_2 + d_3)^2}, \qquad \nu = 1 - \frac{d_4}{d_1x_1 + d_2x_2 + d_3}$$

Let us consider now the traction problem for a simply-connected region Σ_1 and for the following surface tractions

$$\widetilde{t}_\alpha = -pn_\alpha \tag{3.8.68}$$

where p is a given constant. In the case of homogeneous bodies, the solution of the boundary-value problem 1.5.22 and 1.5.25 is

$$\chi = -\frac{1}{2}p(x_1^2 + x_2^2), \quad (x_1, x_2) \in \Sigma_1 \tag{3.8.69}$$

The corresponding stresses are

$$t_{\alpha\beta} = -p\delta_{\alpha\beta} \text{ on } \Sigma_1 \tag{3.8.70}$$

Let us determine the class of nonhomogeneous materials which, subjected to the tractions 3.8.68, generate the stresses 3.8.70. Substituting the function 3.8.69 into Equation 3.2.7, we get

$$\Delta(2\gamma - q) = 0 \tag{3.8.71}$$

The condition 3.8.71 can be written as

$$\Delta\left[\frac{(1+\nu)(1-2\nu)}{E}\right] = 0 \tag{3.8.72}$$

If the Poisson's ratio is constant, then Equation 3.8.72 reduces to

$$\Delta(E^{-1}) = 0 \tag{3.8.73}$$

Thus, if ν is constant and the Young's modulus attains a maximum (or minimum) in interior of Σ_1, then the tractions 3.8.68 cannot produce the plane stress field 3.8.70.

For the remaining of this section, we assume that the Poisson's ratio is constant. Then, Equation 3.2.7 can be written in the form

$$\Delta(\gamma\Delta\chi) = \frac{1}{1-\nu}(\gamma_{,22}\chi_{,11} + \gamma_{,11}\chi_{,22} - 2\gamma_{,12}\chi_{,12}) \tag{3.8.74}$$

By using the relations 1.5.19, this equation can be expressed in terms of the stresses $t_{\alpha\beta}$,

$$\Delta(\gamma t_{\alpha\alpha}) = \frac{1}{1-\nu}(\gamma_{,11}t_{11} + \gamma_{,22}t_{22} + 2t_{12}\gamma_{,12}) \tag{3.8.75}$$

This equation has been established by Olszak and Rychlewski [260].

Let us assume that the loading generates the plane elastic state characterized by

$$t_{12} = T, \qquad t_{11} = t_{22} = 0 \tag{3.8.76}$$

where T is a given constant. Then, from Equation 3.8.75, we find that

$$\gamma_{,12} = 0$$

The plane stress field 3.8.76 is possible to exist if and only if

$$\gamma = h_1(x_1) + h_2(x_2)$$

where h_1 and h_2 are arbitrary functions.

We now consider the plane elastic state for which

$$t_{11} = P, \qquad t_{22} = t_{12} = 0 \tag{3.8.77}$$

where P is a given constant. In this case, Equation 3.8.75 reduces to

$$-\frac{\nu}{1-\nu}\gamma_{,11} + \gamma_{,22} = 0$$

The general solution of this equation is

$$\gamma = g_1(x_2 + \kappa x_1) + g_2(x_2 - \kappa x_1), \qquad \kappa = \left(\frac{\nu}{1-\nu}\right)^{1/2}$$

where g_1 and g_2 are arbitrary functions. Thus, for example, we can say that for nonhomogeneous bodies with

$$\gamma = x_1^3 + A_1 x_1^2 + A_2 x_1 + A_3$$

where A_k are constants, it is not possible to have the plane stress field 3.8.77. Other special plane elastic states have been discussed in Ref. 260.

3.9 Exercises

3.9.1 A continuum body occupies the domain $B = \{x : (x_1, x_2) \in \Sigma_1, 0 < x_3 < h\}$ where the cross section Σ_1 is the assembly of the regions $A_1 = \{x : -\alpha_1 < x_1 < 0, -\beta < x_2 < \beta\}$, $A_2 = \{x : 0 < x_1 < \alpha_2, -\beta < x_2 < \beta\}$, $(\alpha_1 > 0, \alpha_2 > 0, \beta > 0)$. The domains $B_1 = \{x : (x_1, x_2) \in A_1, 0 < x_3 < h\}$ and $B_2 = \{x : (x_1, x_2) \in A_2, 0 < x_3 < h\}$ are occupied by different homogeneous and isotropic elastic materials. Study the torsion of cylinder B.

3.9.2 Determine the solutions of auxiliary plane strain problems defined in Section 3.7 when L and Γ are two concentric circles.

3.9.3 Investigate the extension and bending of a piecewise homogeneous circular cylinder.

3.9.4 Study the plane strain of a circular cylinder composed of two homogeneous and isotropic elastic materials and subjected on the lateral surface to a constant pressure.

3.9.5 Investigate the deformation of a piecewise homogeneous circular cylinder which is subjected to a constant temperature variation.

3.9.6 An inhomogeneous and isotropic elastic cylinder occupies the domain $B = \{x : x_1^2 + x_2^2 < a^2, 0 < x_3 < h\}, (a > 0)$. The constitutive coefficients are given by

$$\lambda = \lambda_0 e^{-\kappa r}, \qquad \mu = \mu_0 e^{-\kappa r}, \quad \kappa > 0$$

where λ_0, μ_0, and κ are prescribed constants and $r = (x_1^2 + x_2^2)^{1/2}$. Study the deformation of the considered cylinder when it is subjected to the loads

$$f_\alpha = -U_{,\alpha}, \quad f_3 = 0, \quad \tilde{t}_\alpha = U n_\alpha, \quad \tilde{t}_3 = 0,$$
$$F_\alpha = 0, \quad F_3 = Q, \quad M_j = 0$$

where $U = U_0 e^{-\kappa r}$, and Q and U_0 are given by constants.

3.9.7 A nonhomogeneous and isotropic elastic cylinder has the constitutive coefficients independent of the axial coordinate. The body is subjected to a temperature field that is a polynomial in the axial coordinate. Study the deformation of the cylinder.

Chapter 4

Anisotropic Bodies

4.1 Preliminaries

The Saint-Venant's problem for anisotropic elastic bodies has been extensively studied [28,175,204,313]. We note that the researches devoted to Saint-Venant's problem are based on various assumptions regarding the structure of the prevailing fields of displacement or stress. It is the purpose of this chapter to extend the results derived in the previous chapters to the case of anisotropic elastic bodies with general elasticities. The procedure presented in this chapter avoids the semi-inverse method and permits a treatment of the problem even for nonhomogeneous bodies, where the elasticity tensor is independent of the axial coordinate. Saint-Venant's problem for nonhomogeneous elastic cylinders where the elastic coefficients are independent of the axial coordinate has been studied in various works [150,152,318]. According to Toupin [329], the proof of Saint-Venant's principle presented in Section 1.10 also remains valid for this kind of nonhomogeneous elastic bodies.

In the first part of the chapter, we present a solution to the Saint-Venant's problem for anisotropic elastic bodies. This solution coincides with that given in Ref. 150 and incorporates the solutions presented in Refs. 28, 175, and 204. Then, minimum energy characterizations of the solutions are established. The results of Section 1.9 are extended to study Truesdell's problem for anisotropic elastic cylinders. We also present a study of the problems of Almansi and Michell. The theory is used to study the deformation of orthotropic cylinders. Finally, the Saint-Venant's problem for elastic cylinders composed of different anisotropic materials is analyzed.

We assume for the remainder of this chapter that the elasticity field $\mathbf{C}$ is independent of the axial coordinate, that is,

$$C_{ijkl} = C_{ijkl}(x_1, x_2), \quad (x_1, x_2) \in \Sigma_1 \tag{4.1.1}$$

Moreover, we continue to assume that $\mathbf{C}$ is symmetric and positive definite.

We denote by $\mathscr{D}^*$ the set of all equilibrium displacement fields $\mathbf{u}$ that satisfy the condition $\mathbf{s}(\mathbf{u}) = \mathbf{0}$ on the lateral boundary. The following results hold true for anisotropic elastic bodies.

Theorem 4.1.1 *If* $\mathbf{u} \in \mathscr{D}^*$ *and* $\mathbf{u}_{,3} \in C^1(\overline{B}) \cap C^2(B)$, *then* $\mathbf{u}_{,3} \in \mathscr{D}^*$ *and*

$$\mathbf{R}(\mathbf{u}_{,3}) = 0, \qquad H_\alpha(\mathbf{u}_{,3}) = \varepsilon_{\alpha\beta} R_\beta(\mathbf{u}), \qquad H_3(\mathbf{u}_{,3}) = 0$$

The proof of this theorem, which we omit, is analogous to that given for Theorem 1.6.1. We continue to use notations from Section 1.6. Theorem 4.1.1 has the following consequences.

Corollary 4.1.1 *If* $\mathbf{u} \in K_I(F_3, M_1, M_2, M_3)$ *and* $\mathbf{u}_{,3} \in C^1(\overline{B}) \cap C^2(B)$, *then* $\mathbf{u}_{,3} \in \mathscr{D}^*$ *and*

$$\mathbf{R}(\mathbf{u}_{,3}) = \mathbf{0}, \qquad \mathbf{H}(\mathbf{u}_{,3}) = \mathbf{0}$$

Corollary 4.1.2 *If* $\mathbf{u} \in K_{II}(F_1, F_2)$ *and* $\mathbf{u}_{,3} \in C^1(\overline{B}) \cap C^2(B)$ *then*

$$\mathbf{u}_{,3} \in K_I(0, F_2, -F_1, 0)$$

4.2 Generalized Plane Strain Problem

The state of generalized plane strain of cylinder B is characterized by

$$\mathbf{u} = \mathbf{u}(x_1, x_2), \quad (x_1, x_2) \in \Sigma_1 \tag{4.2.1}$$

This restriction, in conjunction with the stress–displacement relation, implies that $t_{ij} = t_{ij}(x_1, x_2)$. Further,

$$t_{i\alpha}(\mathbf{u}) = C_{i\alpha k\beta} u_{k,\beta} \tag{4.2.2}$$

By an admissible displacement field, we mean a vector field with the properties

(*i*) $\mathbf{u}$ is independent of x_3 and

(*ii*) $\mathbf{u} \in C^1(\overline{\Sigma}_1) \cap C^2(\Sigma_1)$

Given body force $\mathbf{f}$ on B and surface force $\mathbf{p}$ on Π, with $\mathbf{f}$ and $\mathbf{p}$ independent of x_3, the generalized plane strain problem consists in finding an admissible displacement field $\mathbf{u}$ which satisfies the equations of equilibrium

$$(t_{i\alpha}(\mathbf{u}))_{,\alpha} + f_i = 0 \text{ on } \Sigma_1 \tag{4.2.3}$$

and the boundary conditions

$$t_{i\alpha}(\mathbf{u}) n_\alpha = p_i \text{ on } \Gamma \tag{4.2.4}$$

We note that the stress $t_{33}(\mathbf{u})$ can be determined after the displacement field $\mathbf{u}$ is found.

The generalized plane strain problem for homogeneous bodies was studied in various works (e.g., [204]).

The conditions of equilibrium for cylinder B are equivalent to

$$\int_{\Sigma_1} \mathbf{f} da + \int_{\Gamma} \mathbf{p} ds = \mathbf{0}, \qquad \int_{\Sigma_1} \varepsilon_{\alpha\beta} x_\alpha f_\beta da + \int_{\Gamma} \varepsilon_{\alpha\beta} x_\alpha p_\beta ds = 0 \qquad (4.2.5)$$

$$\int_{\Sigma_1} x_\alpha f_3 da + \int_{\Gamma} x_\alpha p_3 ds = \int_{\Sigma_1} t_{3\alpha}(\mathbf{u}) da \qquad (4.2.6)$$

From Equations 4.2.3, 4.2.4, and the divergence theorem, we get

$$\int_{\Sigma_1} t_{3\alpha}(\mathbf{u}) da = \int_{\Sigma_1} \{t_{3\alpha}(\mathbf{u}) + x_\alpha[(t_{3\rho}(\mathbf{u}))_{,\rho} + f_3]\} da$$
$$= \int_{\Sigma_1} [(x_\alpha t_{3\rho}(\mathbf{u}))_{,\rho} + x_\alpha f_3] da = \int_{\Gamma} x_\alpha p_3 ds + \int_{\Sigma_1} x_\alpha f_3 da$$

Thus, the conditions 4.2.6 are identically satisfied.

We assume for the remainder of this chapter that $\mathbf{C} \in C^\infty(\overline{\Sigma}_1)$ and that the domain Σ_1 is C^∞-smooth. Moreover, we assume that $\mathbf{f}$ and $\mathbf{p}$ belong to C^∞.

We denote by $\mathscr{P}$ the set of all admissible displacement fields. Let $\mathbf{L}$ be the operator on $\mathscr{P}$ defined by

$$L_i \mathbf{u} = -(C_{i\alpha k\beta} u_{k,\beta})_{,\alpha}$$

The equations of equilibrium 4.2.3 take the form

$$\mathbf{L}\mathbf{u} = \mathbf{f} \text{ on } \Sigma_1 \qquad (4.2.7)$$

The conditions 4.2.4 can be written as

$$\mathbf{s}(\mathbf{u}) = \mathbf{p} \text{ on } \Gamma \qquad (4.2.8)$$

We assume that $\mathbf{u}, \mathbf{v} \in \mathscr{P}$. By the divergence theorem, we find

$$\int_{\Sigma_1} (\mathbf{L}\mathbf{u}) \cdot \mathbf{v} da = 2 \int_{\Sigma_1} \widehat{W}(\mathbf{u}, \mathbf{v}) da - \int_{\Gamma} \mathbf{s}(\mathbf{u}) \cdot \mathbf{v} ds \qquad (4.2.9)$$

Here

$$2\widehat{W}(\mathbf{u}, \mathbf{v}) = C_{i\alpha k\beta} e_{i\alpha}(\mathbf{u}) e_{k\beta}(\mathbf{v})$$

is the bilinear form corresponding to the quadratic form

$$2\widehat{W}(\mathbf{u}) = C_{i\alpha k\beta} e_{i\alpha}(\mathbf{u}) e_{k\beta}(\mathbf{u})$$

Let $\mathbf{u}^*$ be a solution of the boundary-value problem 4.2.7 and 4.2.8 corresponding to $\mathbf{f} = \mathbf{0}$ and $\mathbf{p} = \mathbf{0}$. We assume that $\widehat{W}(\mathbf{u})$ is positive definite in the variables $e_{s\beta}(\mathbf{u})$. It follows from Equation 4.2.9 that

$$u_\alpha^* = a_\alpha + \varepsilon_{\alpha\beta} b x_\beta, \qquad u_3^* = a_3 \qquad (4.2.10)$$

where a_i and b are arbitrary constants. Let us consider the boundary condition

$$\mathbf{s}(\mathbf{u}) = \mathbf{0} \text{ on } \Gamma \tag{4.2.11}$$

Following Fichera [88], a C^∞ solution in Σ_1 of the boundary-value problem 4.2.7 and 4.2.11 exists if and only if

$$\int_{\Sigma_1} \mathbf{f} \cdot \mathbf{u}^* da = 0$$

for any displacement field $\mathbf{u}^*$ given by Equation 4.2.10. Thus, we derive the following result.

Theorem 4.2.1 *Let* $\mathbf{f}$ *be a vector field of class* C^∞ *on* $\overline{\Sigma}_1$. *The boundary-value problem* 4.2.7 *and* 4.2.11 *has solutions belonging to* $C^\infty(\overline{\Sigma}_1)$ *if and only if*

$$\int_{\Sigma_1} \mathbf{f} da = \mathbf{0}, \qquad \int_{\Sigma_1} \varepsilon_{\alpha\beta} x_\alpha f_\beta da = 0 \tag{4.2.12}$$

It is easy to see that in the case of the boundary-value problem 4.2.7 and 4.2.8, the conditions 4.2.12 are replaced by conditions 4.2.5.

4.3 Extension, Bending, and Torsion

We denote by $\mathcal{R}$ the set of all rigid displacement fields. In view of Corollary 4.1.1, we are led to seek a solution $\mathbf{u}^0$ of the problem of extension, bending, and torsion such that $\mathbf{u}^0_{,3} \in \mathcal{R}$.

Theorem 4.3.1 *Let* $\mathcal{I}$ *be the set of all vector fields* $\mathbf{u} \in C^1(\overline{B}) \cap C^2(B)$ *such that* $\mathbf{u}_{,3} \in \mathcal{R}$. *Then there exists a vector field* $\mathbf{u}^0 \in \mathcal{I}$ *which is solution of the problem* (P_1).

Proof. We consider $\mathbf{u}^0 \in C^1(\overline{B}) \cap C^2(B)$ such that

$$\mathbf{u}^0_{,3} = \boldsymbol{\alpha} + \boldsymbol{\beta} \times \mathbf{x}$$

where $\boldsymbol{\alpha}$ and $\boldsymbol{\beta}$ are constant vectors. Then we get

$$\begin{aligned} u^0_\alpha &= -\frac{1}{2} a_\alpha x_3^2 - a_4 \varepsilon_{\alpha\beta} x_\beta x_3 + w_\alpha \\ u^0_3 &= (a_\rho x_\rho + a_3) x_3 + w_3 \end{aligned} \tag{4.3.1}$$

modulo a rigid displacement field. Here $\mathbf{w}$ is an arbitrary vector field independent of x_3 such that $\mathbf{w} \in C^1(\overline{\Sigma}_1) \cap C^2(\Sigma_1)$, and we have used the notations $a_\alpha = \varepsilon_{\rho\alpha}\beta_\rho$, $a_3 = \alpha_3$, and $a_4 = \beta_3$. From Equation 4.3.1, we obtain

$$\begin{aligned} u^0_{k,\alpha} &= a_\alpha x_3 \delta_{k3} - a_4 \varepsilon_{\beta\alpha} x_3 \delta_{k\beta} + w_{k,\alpha} \\ u^0_{k,3} &= (a_\rho x_\rho + a_3)\delta_{k3} - \delta_{k\alpha} a_\alpha x_3 - \delta_{k\alpha} a_4 \varepsilon_{\alpha\beta} x_\beta \end{aligned}$$

The stress–displacement relations imply that

$$t_{ij}(\mathbf{u}^0) = C_{ij33}(a_\rho x_\rho + a_3) - a_4 C_{ij\alpha 3}\varepsilon_{\alpha\beta}x_\beta + T_{ij}(\mathbf{w}) \tag{4.3.2}$$

where

$$T_{ij}(\mathbf{w}) = C_{ijk\alpha}w_{k,\alpha} \tag{4.3.3}$$

The functions $T_{ij}(\mathbf{w})$ are independent of the axial coordinate.

The equations of equilibrium and the conditions on the lateral boundary reduce to

$$(T_{i\alpha}(\mathbf{w}))_{,\alpha} + g_i = 0 \text{ on } \Sigma_1, \qquad T_{i\alpha}(\mathbf{w})n_\alpha = q_i \text{ on } \Gamma \tag{4.3.4}$$

where

$$\begin{aligned} g_i &= a_\rho(C_{i\alpha 33}x_\rho)_{,\alpha} + a_3 C_{i\alpha 33,\alpha} - a_4\varepsilon_{\rho\beta}(C_{i\alpha\rho 3}x_\beta)_{,\alpha} \\ q_i &= (a_4\varepsilon_{\rho\beta}C_{i\alpha\rho 3}x_\beta - a_\rho C_{i\alpha 33}x_\rho - a_3 C_{i\alpha 33})n_\alpha \end{aligned} \tag{4.3.5}$$

We note that the relations 4.3.3, 4.3.4, and 4.3.5 constitute a generalized plane strain problem. It follows from the relations 4.3.5 and the divergence theorem that the necessary and sufficient conditions to solve this problem are satisfied for any constants a_1, a_2, a_3, and a_4. We denote by $\mathbf{w}^{(j)}$ a solution of the boundary-value problem 4.3.4 when $a_i = \delta_{ij}$, $a_4 = 0$, and by $\mathbf{w}^{(4)}$ a solution of the boundary-value problem 4.3.4 corresponding to $a_i = 0, a_4 = 1$. We can write

$$\mathbf{w} = \sum_{i=1}^{4} a_i \mathbf{w}^{(i)} \tag{4.3.6}$$

The functions $\mathbf{w}^{(s)}$ are characterized by the equations

$$\begin{aligned} &(T_{i\alpha}(\mathbf{w}^{(\beta)}))_{,\alpha} + (C_{i\alpha 33}x_\beta)_{,\alpha} = 0, \quad (\beta = 1,2) \\ &(T_{i\alpha}(\mathbf{w}^{(3)}))_{,\alpha} + C_{i\alpha 33,\alpha} = 0 \\ &(T_{i\alpha}(\mathbf{w}^{(4)}))_{,\alpha} - \varepsilon_{\rho\beta}(C_{i\alpha\rho 3}x_\beta)_{,\alpha} = 0 \text{ on } \Sigma_1 \end{aligned} \tag{4.3.7}$$

and the boundary conditions

$$\begin{aligned} &T_{i\alpha}(\mathbf{w}^{(\beta)})n_\alpha = -C_{i\alpha 33}x_\beta n_\alpha, \qquad T_{i\alpha}(\mathbf{w}^{(3)})n_\alpha = -C_{i\alpha 33}n_\alpha \\ &T_{i\alpha}(\mathbf{w}^{(4)})n_\alpha = \varepsilon_{\rho\beta}C_{i\alpha\rho 3}x_\beta n_\alpha \text{ on } \Gamma \end{aligned} \tag{4.3.8}$$

We assume that the displacement fields $\mathbf{w}^{(s)}, (s = 1, 2, 3, 4)$, are known. The vector field $\mathbf{u}^0$ can be written in the form

$$\mathbf{u}^0 = \sum_{j=1}^{4} a_j \mathbf{u}^{(j)} \tag{4.3.9}$$

where $\mathbf{u}^{(j)}$ are defined by

$$u_\alpha^{(\beta)} = -\frac{1}{2}x_3^2\delta_{\alpha\beta} + w_\alpha^{(\beta)}, \qquad u_3^{(\beta)} = x_\beta x_3 + w_3^{(\beta)}, \quad (\beta = 1, 2)$$
$$u_\alpha^{(3)} = w_\alpha^{(3)}, \qquad u_3^{(3)} = x_3 + w_3^{(3)}, \qquad u_\alpha^{(4)} = \varepsilon_{\beta\alpha}x_\beta x_3 + w_\alpha^{(4)}, \qquad u_3^{(4)} = w_3^{(4)} \tag{4.3.10}$$

From Equations 4.3.2 and 4.3.9, we get

$$t_{ij}(\mathbf{u}^0) = \sum_{k=1}^{4} a_k t_{ij}(\mathbf{u}^{(k)}) \tag{4.3.11}$$

where

$$t_{ij}(\mathbf{u}^{(\alpha)}) = C_{ij33}x_\alpha + T_{ij}(\mathbf{w}^{(\alpha)})$$
$$t_{ij}(\mathbf{u}^{(3)}) = C_{ij33} + T_{ij}(\mathbf{w}^{(3)}), \qquad t_{ij}(\mathbf{u}^{(4)}) = -C_{ij\alpha3}\varepsilon_{\alpha\beta}x_\beta + T_{ij}(\mathbf{w}^{(4)}) \tag{4.3.12}$$

By Equations 4.3.7 and 4.3.8,

$$(t_{ki}(\mathbf{u}^{(j)}))_{,k} = 0 \text{ on } B, \qquad \mathbf{s}(\mathbf{u}^{(j)}) = \mathbf{0} \text{ on } \Pi, \quad (j = 1, 2, 3, 4) \tag{4.3.13}$$

so that $\mathbf{u}^{(j)} \in \mathscr{D}^*, (j = 1, 2, 3, 4)$.

The conditions on the end Σ_1 are

$$R_\alpha(\mathbf{u}^0) = 0, \qquad R_3(\mathbf{u}^0) = F_3, \qquad \mathbf{H}(\mathbf{u}^0) = \mathbf{M} \tag{4.3.14}$$

In view of Theorem 4.1.1 and $\mathbf{u}^0_{,3} \in \mathcal{R}$, we obtain

$$R_\alpha(\mathbf{u}^0) = \varepsilon_{\beta\alpha}H_\beta(\mathbf{u}^0_{,3}) = 0$$

so that the first two conditions 4.3.14 are satisfied. The remaining conditions furnish the following system for the constants $a_s, (s = 1, 2, 3, 4)$,

$$\sum_{i=1}^{4} D^*_{\alpha i}a_i = \varepsilon_{\alpha\beta}M_\beta, \qquad \sum_{i=1}^{4} D^*_{3i}a_i = -F_3, \qquad \sum_{i=1}^{4} D^*_{4i}a_i = -M_3 \tag{4.3.15}$$

where

$$D^*_{\alpha i} = \int_{\Sigma_1} x_\alpha t_{33}(\mathbf{u}^{(i)})da, \qquad D^*_{3i} = \int_{\Sigma_1} t_{33}(\mathbf{u}^{(i)})da$$
$$D^*_{4i} = \int_{\Sigma_1} \varepsilon_{\alpha\beta}x_\alpha t_{3\beta}(\mathbf{u}^{(i)})da, \quad (i = 1, 2, 3, 4) \tag{4.3.16}$$

The constants $D^*_{rs}, (r, s = 1, 2, 3, 4)$, can be calculated after the displacement fields $\mathbf{w}^{(i)}, (i = 1, 2, 3, 4)$, are determined. Let us prove that the system 4.3.15 can always be solved for a_1, a_2, a_3, and a_4.

By Equations 1.1.2 and 4.3.9,

$$U(\mathbf{u}^0) = \frac{1}{2} \sum_{i,j=1}^{4} \langle \mathbf{u}^{(i)}, \mathbf{u}^{(j)} \rangle a_i a_j$$

Since $\mathbf{C}$ is positive definite and $\mathbf{u}^{(i)}$ is not a rigid displacement, we find that

$$\det \langle \mathbf{u}^{(i)}, \mathbf{u}^{(j)} \rangle \neq 0 \tag{4.3.17}$$

We note that $\mathbf{u}^{(i)} \in \mathscr{D}^*, (i = 1, 2, 3, 4)$. It follows from Equations 4.3.10, 4.3.12, 1.1.16, and 1.1.17 that

$$\langle \mathbf{u}^{(i)}, \mathbf{u}^{(\alpha)} \rangle = \frac{1}{2} h^2 R_\alpha(\mathbf{u}^{(i)}) + hD^*_{\alpha i}$$

$$\langle \mathbf{u}^{(i)}, \mathbf{u}^{(3)} \rangle = hD^*_{3i}, \qquad \langle \mathbf{u}^{(i)}, \mathbf{u}^{(4)} \rangle = hD^*_{4i}, \quad (i = 1, 2, 3, 4)$$

Since $\mathbf{u}^{(i)} \in \mathscr{D}^*$ and $\mathbf{u}^{(i)}_{,3} \in \mathcal{R}$, by Theorem 4.1.1, we have $R_\alpha(\mathbf{u}^{(i)}) = 0$, $(i = 1, 2, 3, 4)$. Thus, we obtain

$$\langle \mathbf{u}^{(i)}, \mathbf{u}^{(j)} \rangle = hD^*_{ji} \tag{4.3.18}$$

From relations 4.3.17 and 4.3.18, we find

$$\det (D^*_{rs}) \neq 0 \tag{4.3.19}$$

so that the system 4.3.15 uniquely determines the constants a_i, $(i = 1, 2, 3, 4)$. Thus, we have proved that the constants $a_s, (s = 1, 2, 3, 4)$, and the vector field $\mathbf{w}$ can be determined so that $\mathbf{u}^0 \in K_I(F_3, M_1, M_2, M_3)$. □

Remark. Theorem 4.3.1 offers a constructive procedure to obtain a solution of the problem (P_1) for anisotropic elastic bodies. This solution is given by Equations 4.3.9 and 4.3.10 where the vector fields $\mathbf{w}^{(j)}, (j = 1, 2, 3, 4)$, are characterized by the boundary-value problems 4.3.7 and 4.3.8, and the constants $a_s, (s = 1, 2, 3, 4)$, are determined by Equations 4.3.15.

4.4 Flexure of Anisotropic Cylinders

The flexure problem consists in finding an equilibrium displacement field $\mathbf{u}$ that satisfies the conditions

$$\mathbf{s}(\mathbf{u}) = \mathbf{0} \text{ on } \Pi, \qquad R_\alpha(\mathbf{u}) = F_\alpha, \qquad R_3(\mathbf{u}) = 0, \qquad \mathbf{H}(\mathbf{u}) = \mathbf{0}$$

We denote by $\widehat{a}$ the four-dimensional vector (a_1, a_2, a_3, a_4). We shall write $\mathbf{u}^0\{\widehat{a}\}$ for the displacement vector $\mathbf{u}^0$ defined by Equation 4.3.9, indicating thus its dependence on the constants a_1, a_2, a_3, and a_4. In view of Corollaries 4.1.1 and 4.1.2 and Theorem 4.3.1, it is natural to seek a solution of the flexure problem in the form

$$\mathbf{u} = \int_0^{x_3} \mathbf{u}^0\{\widehat{b}\}dx_3 + \mathbf{u}^0\{\widehat{c}\} + \mathbf{w}' \tag{4.4.1}$$

where $\widehat{b} = (b_1,\ b_2,\ b_3,\ b_4)$ and $\widehat{c} = (c_1,\ c_2,\ c_3,\ c_4)$ are two constant four-dimensional vectors, and $\mathbf{w}'$ is a vector field independent of x_3 such that $\mathbf{w}' \in C^1(\overline{\Sigma}_1) \cap C^2(\Sigma_1)$.

Theorem 4.4.1 *Let Y be the set of all vector fields of the form 4.4.1. Then there exists a vector field $\mathbf{u}' \in Y$ which is solution of the flexure problem.*

Proof. We have to prove that the vector field $\mathbf{w}'$ and the constants b_i, c_i, $(i = 1,\ 2,\ 3,\ 4)$, can be determined so that $\mathbf{u}' \in K_{II}(F_1,\ F_2)$. First, we determine the vector $\widehat{b}$. If $\mathbf{u}' \in K_{II}(F_1,\ F_2)$, then by Corollary 4.1.2 and Equation 4.4.1,

$$\mathbf{u}^0\{\widehat{b}\} \in K_I(0, F_2, -F_1, 0) \tag{4.4.2}$$

By Equations 4.3.15 and 4.4.2, we find that

$$\begin{gathered}\sum_{i=1}^{4} D^*_{\alpha i} b_i = -F_\alpha \\ \sum_{i=1}^{4} D^*_{3i} b_i = 0, \qquad \sum_{i=1}^{4} D^*_{4i} b_i = 0\end{gathered} \tag{4.4.3}$$

From the system 4.4.3, we can determine b_1, b_2, b_3, and b_4.

It follows from Equations 4.3.9, 4.3.10, and 4.4.1 that

$$\begin{aligned}u'_\alpha &= -\frac{1}{6} b_\alpha x_3^3 - \frac{1}{2} c_\alpha x_3^2 - \frac{1}{2} b_4 \varepsilon_{\alpha\beta} x_\beta x_3^2 - c_4 \varepsilon_{\alpha\beta} x_\beta x_3 \\ &\quad + \sum_{j=1}^{4} (c_j + x_3 b_j) w_\alpha^{(j)} + w'_\alpha \\ u'_3 &= \frac{1}{2}(b_\rho x_\rho + b_3) x_3^2 + (c_\rho x_\rho + c_3) x_3 + \sum_{j=1}^{4} (c_j + x_3 b_j) w_3^{(j)} + w'_3\end{aligned} \tag{4.4.4}$$

By Equations 1.1.2 and 4.3.12, we obtain

$$t_{rs}(\mathbf{u}') = \sum_{i=1}^{4} (c_i + x_3 b_i) t_{rs}(\mathbf{u}^{(i)}) + T_{rs}(\mathbf{w}') + k_{rs} \tag{4.4.5}$$

where

$$k_{ij} = \sum_{r=1}^{4} C_{ijk3} b_r w_k^{(r)}$$

If we substitute Equation 4.4.5 into equations of equilibrium, then we find, with the aid of Equation 4.3.13, that

$$(T_{i\alpha}(\mathbf{w}')),_{\alpha} + f_i' = 0 \text{ on } \Sigma_1 \tag{4.4.6}$$

where

$$f_i' = k_{i\alpha,\alpha} + \sum_{j=1}^{4} b_j t_{i3}(\mathbf{u}^{(j)})$$

With the help of Equations 4.3.13, the conditions on the lateral boundary reduce to

$$T_{i\alpha}(\mathbf{w}')n_\alpha = p_i' \text{ on } \Gamma \tag{4.4.7}$$

where $p_i' = -k_{i\alpha}n_\alpha$. The relations 4.4.6 and 4.4.7 constitute a generalized plane strain problem. The necessary and sufficient conditions for the existence of a solution of this problem are

$$\int_{\Sigma_1} f_i' da + \int_\Gamma p_i' ds = 0, \qquad \int_{\Sigma_1} \varepsilon_{\alpha\beta} x_\alpha f_\beta' da + \int_\Gamma \varepsilon_{\alpha\beta} x_\alpha p_\beta' ds = 0$$

We can verify that these conditions become

$$\sum_{j=1}^{4} b_j \int_{\Sigma_1} t_{i3}(\mathbf{u}^{(j)}) da = 0, \qquad \sum_{j=1}^{4} b_j \int_{\Sigma_1} \varepsilon_{\alpha\beta} x_\alpha t_{\beta 3}(\mathbf{u}^{(j)}) da = 0 \tag{4.4.8}$$

It follows from Equation 4.4.3 and $R_\alpha(\mathbf{u}^{(j)}) = 0$, $(j = 1, 2, 3, 4)$, that the conditions 4.4.8 are satisfied. In what follows, we assume that the displacement field $\mathbf{w}'$ is known.

Since $H_\alpha(\mathbf{u}'_{,3}) = \varepsilon_{\alpha\beta} R_\beta(\mathbf{u}')$ and $\mathbf{u}'_{,3} \in K_I(0,\ F_2, -F_1,\ 0)$, it follows that $R_\alpha(\mathbf{u}') = F_\alpha$. The conditions $R_3(\mathbf{u}') = 0$, $\mathbf{H}(\mathbf{u}') = \mathbf{0}$ are satisfied if and only if

$$\sum_{j=1}^{4} D_{ij}^* c_j = A_i, \quad (i = 1, 2, 3, 4) \tag{4.4.9}$$

where

$$A_\alpha = -\int_{\Sigma_1} x_\alpha [k_{33} + T_{33}(\mathbf{w}')] da, \qquad A_3 = -\int_{\Sigma_1} [k_{33} + T_{33}(\mathbf{w}')] da$$

$$A_4 = -\int_{\Sigma_1} \varepsilon_{\alpha\beta} x_\alpha [k_{\beta 3} + T_{\beta 3}(\mathbf{w})]$$

On the basis of relation 4.3.19, the system 4.4.9 can always be solved for c_1, c_2, c_3, and c_4. Thus, if $\widehat{b}$ and $\widehat{c}$ are defined by Equations 4.4.3 and 4.4.9, respectively, and the displacement field $\mathbf{w}'$ is characterized by the generalized plane strain problem 4.4.6 and 4.4.7, then the displacement field $\mathbf{u}'$ defined by Equations 4.4.4 is a solution of the flexure problem. □

We have obtained the system 4.4.9 from the conditions $R_3(\mathbf{u}') = 0$, $\mathbf{H}(\mathbf{u}') = \mathbf{0}$. If we replace these conditions by $R_3(\mathbf{u}') = F_3$, $\mathbf{H}(\mathbf{u}') = \mathbf{M}$, then we arrive at

$$\begin{aligned} &\sum_{i=1}^{4} D^*_{\alpha i} c_i = \varepsilon_{\alpha\beta} M_\beta + A_\alpha \\ &\sum_{i=1}^{4} D^*_{3i} c_i = A_3 - F_3, \qquad \sum_{i=1}^{4} D^*_{4i} c_i = A_4 - M_3 \end{aligned} \tag{4.4.10}$$

If $\widehat{b}$ is defined by Equation 4.4.3, $\widehat{c}$ is defined by Equation 4.4.10, and $\mathbf{w}'$ is characterized by the boundary-value problem 4.4.6 and 4.4.7, then $\mathbf{u}' \in K(\mathbf{F}, \mathbf{M})$.

4.5 Minimum Energy Characterizations of Solutions

In this section, we present minimum strain-energy characterizations of the solutions obtained in Sections 4.3 and 4.4. Similar results for homogeneous and isotropic bodies were given by Sternberg and Knowles [322].

We denote by Q_I the set of all equilibrium displacement fields $\mathbf{u}$ that satisfy the conditions

$$\begin{gathered} [t_{3i}(\mathbf{u})](x_1, x_2, 0) = [t_{3i}(\mathbf{u})](x_1, x_2, h), \quad (x_1, x_2) \in \Sigma_1 \\ R_\alpha(\mathbf{u}) = 0, \qquad R_3(\mathbf{u}) = F_3, \qquad \mathbf{H}(\mathbf{u}) = \mathbf{M} \\ \mathbf{s}(\mathbf{u}) = \mathbf{0} \text{ on } \Pi \end{gathered} \tag{4.5.1}$$

Theorem 4.5.1 *Let* $\mathbf{u}^0$ *be the solution* 4.3.9 *of the problem* (P_1) *corresponding to the scalar load* F_3 *and the moment* $\mathbf{M}$. *Then*

$$U(\mathbf{u}^0) \leq U(\mathbf{u})$$

for every $\mathbf{u} \in Q_I$, *and equality holds only if* $\mathbf{u} = \mathbf{u}^0$ *modulo a rigid displacement.*

Proof. Let $\mathbf{u} \in Q_I$ and define

$$\mathbf{v} = \mathbf{u} - \mathbf{u}^0$$

Then $\mathbf{v}$ is an equilibrium displacement field that satisfies

$$\begin{gathered} [t_{3i}(\mathbf{v})](x_1, x_2, 0) = [t_{3i}(\mathbf{v})](x_1, x_2, h), \quad (x_1, x_2) \in \Sigma_1 \\ \mathbf{s}(\mathbf{v}) = \mathbf{0} \text{ on } \Pi, \qquad \mathbf{R}(\mathbf{v}) = \mathbf{0}, \qquad \mathbf{H}(\mathbf{v}) = \mathbf{0} \end{gathered} \tag{4.5.2}$$

From Equations 1.1.12 and 1.1.13, we obtain

$$U(\mathbf{u}) = U(\mathbf{v}) + U(\mathbf{u}^0) + \langle \mathbf{v}, \mathbf{u}^0 \rangle$$

If we apply Equations 1.1.16 and 1.1.17, then we conclude, with the aid of Equations 4.3.9, 4.3.10, and 4.5.2, that

$$\begin{aligned}\langle \mathbf{v}, \mathbf{u}^0 \rangle = \int_{\Sigma_2} t_{3i}(\mathbf{v}) u_i^0 da - \int_{\Sigma_1} t_{3i}(\mathbf{v}) u_i^0 da = -\frac{1}{2} h^2 a_\alpha R_\alpha(\mathbf{v}) \\ + h[\varepsilon_{\alpha\beta} a_\alpha H_\beta(\mathbf{v}) - a_3 R_3(\mathbf{v}) - a_4 H_3(\mathbf{v})] = 0\end{aligned}$$

We can write

$$U(\mathbf{u}) = U(\mathbf{v}) + U(\mathbf{u}^0)$$

Thus $U(\mathbf{u}) \geq U(\mathbf{u}^0)$, and $U(\mathbf{u}) = U(\mathbf{u}^0)$ only if $\mathbf{v}$ is a rigid displacement. □

We denote by Q_{II} the set of all equilibrium displacement fields $\mathbf{u}$ that satisfy the conditions

$$\begin{aligned}&\mathbf{u}_{,3} \in C^1(\overline{B}) \cap C^2(B), \qquad \mathbf{s}(\mathbf{u}) = \mathbf{0} \text{ on } \Pi, \qquad R_\alpha(\mathbf{u}) = F_\alpha \\ &[t_{3i}(\mathbf{u}_{,3})](x_1, x_2, 0) = [t_{3i}(\mathbf{u}_{,3})](x_1, x_2, h), \quad (x_1, x_2) \in \Sigma_1\end{aligned} \tag{4.5.3}$$

Theorem 4.5.2 *Let $\mathbf{u}'$ be the solution 4.4.4 of the flexure problem corresponding to the loads F_1 and F_2. Then*

$$U(\mathbf{u}'_{,3}) \leq U(\mathbf{u}_{,3})$$

for every $\mathbf{u} \in Q_{II}$, and equality holds only if $\mathbf{u}_{,3} = \mathbf{u}'_{,3}$ (modulo a rigid displacement).

Proof. We consider $\mathbf{u} \in Q_{II}$. Since $\mathbf{u}' \in Q_{II}$ it follows that the field

$$\mathbf{v} = \mathbf{u} - \mathbf{u}'$$

is an equilibrium displacement field that satisfies

$$\begin{aligned}&\mathbf{v}_{,3} \in C^1(\overline{B}) \cap C^2(B), \qquad \mathbf{s}(\mathbf{v}) = \mathbf{0} \text{ on } \Pi, \qquad R_\alpha(\mathbf{v}) = 0 \\ &[t_{3\beta}(\mathbf{v}_{,3})](x_1, x_2, 0) = [t_{3\beta}(\mathbf{v}_{,3})](x_1, x_2, h), \quad (x_1, x_2) \in \Sigma_1\end{aligned} \tag{4.5.4}$$

With the help of Equations 1.1.12 and 4.4.1 and Theorem 4.3.1, we find

$$U(\mathbf{u}_{,3}) = U(\mathbf{v}_{,3} + \mathbf{u}'_{,3}) = U(\mathbf{v}_{,3} + \mathbf{u}^0\{\widehat{b}\}) = U(\mathbf{v}_{,3}) + U(\mathbf{u}'_{,3}) + \langle \mathbf{v}_{,3}, \mathbf{u}^0\{\widehat{b}\} \rangle$$

By Equations 4.3.9, 1.1.16, 1.1.17, and 4.5.4,

$$\begin{aligned}\langle \mathbf{v}_{,3}, \mathbf{u}^0\{\widehat{b}\} \rangle = -\frac{1}{2} b_\alpha h^2 R_\alpha(\mathbf{v}_{,3}) + h[b_1 H_2(\mathbf{v}_{,3}) \\ - b_2 H_1(\mathbf{v}_{,3}) - b_3 R_3(\mathbf{v}_{,3}) - b_4 H_3(\mathbf{v}_{,3})]\end{aligned}$$

In view of Theorem 4.1.1 and Equations 4.5.4, we conclude that $\langle \mathbf{v}_{,3}, \mathbf{u}^0\{\widehat{b}\} \rangle = 0$. We find that

$$U(\mathbf{u}_{,3}) = U(\mathbf{v}_{,3}) + U(\mathbf{u}'_{,3})$$

The desired conclusion is now immediate. □

4.6 Global Strain Measures

Truesdell's problem as formulated in Section 1.9 can be set also for anisotropic bodies. Thus we are led to the following problem: to define the functionals $\tau_i(\cdot)$, $(i = 1, 2, 3, 4)$, on $K_I(F_3, M_1, M_2, M_3)$ such that

$$\sum_{j=1}^{4} D^*_{\alpha j}\tau_j(\mathbf{u}) = \varepsilon_{\alpha\beta}M_\beta$$
$$\sum_{j=1}^{4} D^*_{3j}\tau_j(\mathbf{u}) = -F_3, \qquad \sum_{j=1}^{4} D^*_{4j}\tau_j(\mathbf{u}) = -M_3 \tag{4.6.1}$$

hold for each $\mathbf{u} \in K_I(F_3, M_1, M_2, M_3)$. We consider the set Q_I of all equilibrium displacement fields $\mathbf{u}$ that satisfy the conditions 4.5.1. Clearly, if $\mathbf{u} \in Q_I$ then $\mathbf{u} \in K_I(F_3, M_1, M_2, M_3)$. In view of Theorem 4.5.1, we are led to consider the real function f of the variables ξ_1, ξ_2, ξ_3, and ξ_4 defined by

$$f = \left\| \mathbf{u} - \sum_{j=1}^{4} \xi_j \mathbf{u}^{(j)} \right\|_e^2$$

where $\mathbf{u} \in Q_I$ and $\mathbf{u}^{(j)}$, $(j = 1, 2, 3, 4)$, are given by Equations 4.3.10. By Equation 4.3.18,

$$f = h \sum_{i,j=1}^{4} D^*_{ij}\xi_i\xi_j - 2\sum_{i=1}^{4} \xi_i \langle \mathbf{u}, \mathbf{u}^{(i)} \rangle + \|\mathbf{u}\|_e^2$$

Since the matrix (D^*_{ij}), $(i, j = 1, 2, 3, 4)$, is positive definite, f will be a minimum at $(\alpha_1(\mathbf{u}), \alpha_2(\mathbf{u}), \alpha_3(\mathbf{u}), \alpha_4(\mathbf{u}))$ if and only if $(\alpha_1(\mathbf{u}), \alpha_2(\mathbf{u}), \alpha_3(\mathbf{u}), \alpha_4(\mathbf{u}))$ is the solution of the following system of equations

$$h\sum_{j=1}^{4} D^*_{ij}\alpha_j(\mathbf{u}) = \langle \mathbf{u}, \mathbf{u}^{(i)} \rangle, \quad (i = 1, 2, 3, 4) \tag{4.6.2}$$

From Equations 4.3.18, 1.1.16, 1.1.17, and 4.5.1, we obtain

$$\begin{aligned}\langle \mathbf{u}, \mathbf{u}^{(1)} \rangle &= \int_{\partial B} \mathbf{s}(\mathbf{u}) \cdot \mathbf{u}^{(1)} da = h\int_{\Sigma_2} x_1 t_{33}(\mathbf{u}) da - \frac{1}{2}h^2 \int_{\Sigma_1} t_{31}(\mathbf{u}) da \\ &= -\frac{1}{2}h^2 R_1(\mathbf{u}) + hH_2(\mathbf{u}) = hH_2(\mathbf{u})\end{aligned}$$

Similarly,

$$\langle \mathbf{u}, \mathbf{u}^{(\alpha)} \rangle = h\varepsilon_{\alpha\beta}H_\beta(u), \qquad \langle \mathbf{u}, \mathbf{u}^{(3)} \rangle = -R_3(\mathbf{u}), \qquad \langle \mathbf{u}, \mathbf{u}^{(4)} \rangle = -H_3(\mathbf{u}) \tag{4.6.3}$$

It follows from Equations 4.6.1, 4.6.2, and 4.6.3 that $\tau_i(\mathbf{u}) = \alpha_i(\mathbf{u})$, $(i = 1, 2, 3, 4)$, for each $\mathbf{u} \in Q_I$.

On the other hand, by Equation 1.1.16 we find

$$\langle \mathbf{u}, \mathbf{u}^{(r)} \rangle = A_r(\mathbf{u}), \quad (r = 1, 2, 3, 4) \tag{4.6.4}$$

where

$$A_r(\mathbf{u}) = \int_{\Sigma_2} t_{3i}(\mathbf{u}^{(r)}) u_i da - \int_{\Sigma_1} t_{3i}(\mathbf{u}^{(r)}) u_i da \tag{4.6.5}$$

From Equations 4.6.2 and 4.6.4, we get

$$\sum_{j=1}^{4} D_{ij}^{*} \tau_j(\mathbf{u}) = \frac{1}{2} A_i(\mathbf{u}), \quad (i = 1, 2, 3, 4) \tag{4.6.6}$$

The system 4.6.6 defines $\tau_i(\mathbf{u})$, $(i = 1, 2, 3, 4)$, for every displacement field $\mathbf{u} \in Q_I$.

Truesdell's problem can be set also for the flexure of anisotropic cylinders: to define the functionals $\gamma_i(\cdot)$, $(i = 1, 2, 3, 4)$, on $K_{II}(F_1, F_2)$ such that

$$\sum_{i=1}^{4} D_{\alpha i}^{*} \gamma_i(\mathbf{u}) = -F_i, \qquad \sum_{i=1}^{4} D_{3i}^{*} \gamma_i(\mathbf{u}) = 0, \qquad \sum_{i=1}^{4} D_{4i}^{*} \gamma_i(\mathbf{u}) = 0 \tag{4.6.7}$$

hold for each $\mathbf{u} \in K_{II}(F_1, F_2)$.

We denote by $\mathcal{H}$ the set of all equilibrium displacement fields $\mathbf{u}$ that satisfy the conditions

$$\begin{gathered}
\mathbf{u}_{,3} \in C^1(\overline{B}) \cap C^2(B), \qquad \mathbf{s}(\mathbf{u}) = \mathbf{0} \text{ on } \Pi \\
R_\alpha(\mathbf{u}) = F_\alpha, \qquad R_3(\mathbf{u}) = 0, \qquad \mathbf{H}(\mathbf{u}) = \mathbf{0} \\
[t_{3i}(\mathbf{u}_{,3})](x_1, x_2, 0) = [t_{3i}(\mathbf{u}_{,3})](x_1, x_2, h), \quad (x_1, x_2) \in \Sigma_1
\end{gathered} \tag{4.6.8}$$

If $\mathbf{u} \in \mathcal{H}$ then $\mathbf{u} \in K_{II}(F_1, F_2)$. Let g be the real function of the variables ζ_i, $(i = 1, 2, 3, 4)$, defined by

$$g = \left\| \mathbf{u}_{,3} - \sum_{i=1}^{4} \zeta_i \mathbf{u}^{(i)} \right\|_e^2$$

where $\mathbf{u} \in \mathcal{H}$ and $\mathbf{u}^{(i)}$, $(i = 1, 2, 3, 4)$, are given by Equations 4.3.10. Clearly, g will be a minimum at $(\beta_1(\mathbf{u}), \beta_2(\mathbf{u}), \beta_3(\mathbf{u}), \beta_4(\mathbf{u}))$ if and only if $(\beta_1(\mathbf{u}), \beta_2(\mathbf{u}), \beta_3(\mathbf{u}), \beta_4(\mathbf{u}))$ is the solution of the following system of equations

$$h \sum_{j=1}^{4} D_{ij}^{*} \beta_j(\mathbf{u}) = \langle \mathbf{u}_{,3}, \mathbf{u}^{(i)} \rangle, \quad (i = 1, 2, 3, 4) \tag{4.6.9}$$

Let us prove that $\beta_i(\mathbf{u}) = \gamma_i(\mathbf{u})$, $(i = 1, 2, 3, 4)$, for every $\mathbf{u} \in \mathcal{H}$. By Equations 4.3.10, 1.1.16, and 4.6.8, we obtain

$$\langle \mathbf{u}_{,3}, \mathbf{u}^{(1)} \rangle = \int_{\partial B} \mathbf{s}(\mathbf{u}_{,3}) \cdot \mathbf{u}^{(1)} da = -\frac{1}{2} h^2 R_1(\mathbf{u}_{,3}) + h H_2(\mathbf{u}_{,3})$$

With the help of Theorem 4.1.1, we get

$$\langle \mathbf{u}_{,3}, \mathbf{u}^{(1)} \rangle = -hR_1(\mathbf{u})$$

Similarly,

$$\langle \mathbf{u}_{,3}, \mathbf{u}^{(\alpha)} \rangle = -hR_\alpha(\mathbf{u}), \qquad \langle \mathbf{u}_{,3}, \mathbf{u}^{(2+\alpha)} \rangle = 0, \quad (\alpha = 1, 2) \tag{4.6.10}$$

It follows from Equations 4.6.7 and 4.6.9 that $\gamma_i(\mathbf{u}) = \beta_i(\mathbf{u})$, $(i = 1, 2, 3, 4)$, for any $\mathbf{u} \in \mathcal{H}$. By Equation 1.1.16, we obtain

$$\langle \mathbf{u}_{,3}, \mathbf{u}^{(i)} \rangle = B_i(\mathbf{u}), \quad (i = 1, 2, 3, 4) \tag{4.6.11}$$

where

$$B_j(\mathbf{u}) = \int_{\Sigma_2} t_{3i}(\mathbf{u}^{(j)}) u_{i,3} da - \int_{\Sigma_1} t_{3i}(\mathbf{u}^{(j)}) u_{i,3} da$$

Thus, from Equations 4.6.9 and 4.6.11, we conclude that

$$\sum_{j=1}^{4} D^*_{ij} \gamma_j(\mathbf{u}) = \frac{1}{h} B_i(\mathbf{u}), \quad (i = 1, 2, 3, 4)$$

for each $\mathbf{u} \in \mathcal{H}$. This system defines $\gamma_i(\cdot)$ on the subclass $\mathcal{H}$ of solutions to the flexure problem.

4.7 Problem of Loaded Cylinders

In the first part of this section, we consider the Almansi–Michell problem. It is easy to verify that Theorem 2.4.1 also remains valid for anisotropic bodies where the elasticity field is independent of the axial coordinate. As in Section 2.4, we are led to consider the set V of all vector fields of the form

$$\int_0^{x_3} \int_0^{x_3} \mathbf{u}^0\{\widehat{b}\} dx_3 dx_3 + \int_0^{x_3} \mathbf{u}^0\{\widehat{c}\} dx_3 + \mathbf{u}^0\{\widehat{d}\} + x_3 \mathbf{w}' + \mathbf{w}'' \tag{4.7.1}$$

where $\widehat{b}, \widehat{c}$, and $\widehat{d}$ are four-dimensional constant vectors, and $\mathbf{w}'$ and $\mathbf{w}''$ are vector fields independent of x_3 such that $\mathbf{w}', \mathbf{w}'' \in C^1(\overline{\Sigma}_1) \cap C^2(\Sigma_1)$. Here $\mathbf{u}^0\{\widehat{a}\}$ is defined by Equation 4.3.9. We assume that the body force and surface force belong to C^∞.

Theorem 4.7.1 *Let B be anisotropic and assume that the elasticity field is independent of the axial coordinate. Then there exists a vector field $\mathbf{u}'' \in V$ such that $\mathbf{u}'' \in K_{III}(\mathbf{F}, \mathbf{M}, \mathbf{G}, \mathbf{p})$.*

Proof. If $\mathbf{u}'' \in V$ and $\mathbf{u}'' \in K_{III}(\mathbf{F}, \mathbf{M}, \mathbf{G}, \mathbf{p})$, then by Theorems 2.4.2 and 4.3.1 and Equation 4.7.1,

$$\int_0^{x_3} \mathbf{u}^0\{\widehat{b}\}dx_3 + \mathbf{u}^0\{\widehat{c}\} + \mathbf{w}' \in K(\mathbf{P}, \mathbf{Q})$$

where $\mathbf{P}$ and $\mathbf{Q}$ are defined by Equation 2.4.3. From Equations 4.4.3 and 4.4.6, we find that $\widehat{b}$ is given by

$$\begin{aligned} &\sum_{i=1}^{4} D^*_{\alpha i} b_i = -\int_{\Sigma_1} G_\alpha da - \int_\Gamma p_\alpha ds \\ &\sum_{i=1}^{4} D^*_{3i} b_i = 0, \qquad \sum_{i=1}^{4} D^*_{4i} b_i = 0 \end{aligned} \tag{4.7.2}$$

and $\mathbf{w}'$ is characterized by

$$\begin{aligned} (T_{i\alpha}(\mathbf{w}'))_{,\alpha} + \sum_{r=1}^{4} b_r[(C_{i\alpha k3} w_k^{(r)})_{,\alpha} + t_{i3}(\mathbf{u}^{(r)})] = 0 \text{ on } \Sigma_1 \\ T_{i\alpha}(\mathbf{w}')n_\alpha = -\sum_{r=1}^{4} C_{i\alpha k3} b_r w_k^{(r)} n_\alpha \text{ on } \Gamma \end{aligned} \tag{4.7.3}$$

In view of Equation 4.4.10, the vector $\widehat{c}$ is determined by

$$\sum_{j=1}^{4} D^*_{ij} c_j = C_i, \quad (i = 1, 2, 3, 4) \tag{4.7.4}$$

where

$$\begin{aligned} C_\alpha &= -\int_\Gamma x_\alpha p_3 ds - \int_{\Sigma_1} x_\alpha G_3 da - F_\alpha - \int_{\Sigma_1} x_\alpha [k_{33} + T_{33}(\mathbf{w}')] da \\ C_3 &= -\int_\Gamma p_3 ds - \int_{\Sigma_1} G_3 da - \int_{\Sigma_1} [k_{33} + T_{33}(\mathbf{w}')] da \\ C_4 &= -\int_\Gamma \varepsilon_{\alpha\beta} x_\alpha p_\beta ds - \int_{\Sigma_1} \varepsilon_{\alpha\beta} x_\alpha G_\beta da - \int_{\Sigma_1} \varepsilon_{\alpha\beta} x_{\alpha\beta} [k_{\beta 3} + T_{\beta 3}(\mathbf{w}')] da \\ k_{ij} &= \sum_{r=1}^{4} C_{ijk3} b_r w_k^{(r)} \end{aligned}$$

From Equations 4.3.9, 4.3.10, and 4.7.1, we get

$$\begin{aligned} u''_\alpha = &-\frac{1}{24} b_\alpha x_3^4 - \frac{1}{6} c_\alpha x_3^3 - \frac{1}{2} d_\alpha x_3^2 - \varepsilon_{\alpha\beta} x_\beta \left(\frac{1}{6} b_4 x_3^3 + \frac{1}{2} c_4 x_3^2 + d_4 x_3\right) \\ &+ \sum_{j=1}^{4} \left(d_j + c_j x_3 + \frac{1}{2} b_j x_3^2\right) w_\alpha^{(j)} + x_3 w'_\alpha + w''_\alpha \end{aligned}$$

$$
\begin{aligned}
u_3'' &= \frac{1}{6}(b_\rho x_\rho + b_3)x_3^3 + \frac{1}{2}(c_\rho x_\rho + c_3)x_3^2 + (d_\rho x_\rho + d_3)x_3 \\
&\quad + \sum_{j=1}^{4}\left(d_j + c_j x_3 + \frac{1}{2}b_j x_3^2\right) w_3^{(j)} + x_3 w_3' + w_3''
\end{aligned}
\tag{4.7.5}
$$

By Equations 1.1.2, 4.3.12, and 4.7.5,

$$
\begin{aligned}
t_{ij}(\mathbf{u}'') = \sum_{r=1}^{4}\left(d_r + c_r x_3 + \frac{1}{2}b_r x_3^2\right) t_{ij}(\mathbf{u}^{(r)}) + x_3 k_{ij} + k_{ij}' \\
+ T_{ij}(\mathbf{w}'') + x_3 T_{ij}(\mathbf{w}')
\end{aligned}
\tag{4.7.6}
$$

where

$$
k_{ij}' = C_{ijk3}w_k' + \sum_{s=1}^{4} c_s C_{ijk3} w_k^{(s)}
$$

The equations of equilibrium and the conditions on the lateral boundary reduce to

$$
\begin{aligned}
(T_{i\alpha}(\mathbf{w}''))_{,\alpha} + h_i &= 0 \text{ on } \Sigma_1 \\
T_{i\alpha}(\mathbf{w}'')n_\alpha &= q_i \text{ on } \Gamma
\end{aligned}
\tag{4.7.7}
$$

where

$$
\begin{aligned}
h_i &= G_i + k_{i\alpha,\alpha}' + T_{i3}(\mathbf{w}') + k_{i3} + \sum_{r=1}^{4} c_r t_{i3}(\mathbf{u}^{(r)}) \\
q_i &= p_i - k_{i\alpha}' n_\alpha
\end{aligned}
$$

Using the divergence theorem, we find that

$$
\int_{\Sigma_1} h_i da + \int_\Gamma q_i ds = \int_{\Sigma_1} G_i da + \int_\Gamma p_i ds - R_i(\mathbf{u}_{,3}'') = P_i - R_i(\mathbf{u}_{,3}'')
$$

$$
\begin{aligned}
\int_{\Sigma_1} \varepsilon_{\alpha\beta} x_\alpha h_\beta da + \int_\Gamma \varepsilon_{\alpha\beta} x_\alpha q_\beta ds = \int_{\Sigma_1} \varepsilon_{\alpha\beta} x_\alpha G_\beta da + \int_\Gamma \varepsilon_{\alpha\beta} x_\alpha p_\beta da \\
- H_3(\mathbf{u}_{,3}'') = Q_3 - H_3(\mathbf{u}_{,3}'')
\end{aligned}
$$

The conditions

$$
R_i(\mathbf{u}_{,3}'') = P_i, \qquad H_3(\mathbf{u}_{,3}'') = Q_3
$$

were used to obtain Equations 4.7.2 and 4.7.4.

We conclude that the necessary and sufficient conditions for the existence of a solution of the boundary-value problem 4.7.7 are satisfied.

From Equations 4.7.4 and 4.7.6, we obtain

$$
\begin{aligned}
H_\alpha(\mathbf{u}_{,3}'') &= \varepsilon_{\beta\alpha}\left(\sum_{i=1}^{4} D_{\beta i}^{*} c_i + \int_{\Sigma_1} x_\beta [k_{33} + T_{33}(\mathbf{w}')] da\right) \\
&= \varepsilon_{\beta\alpha}\left(\int_\Gamma x_\beta p_3 ds + \int_{\Sigma_1} x_\beta G_3 da\right) + \varepsilon_{\alpha\beta} F_\beta
\end{aligned}
\tag{4.7.8}
$$

With the help of Theorem 2.4.1, we get

$$H_\alpha(\mathbf{u}''_{,3}) = \varepsilon_{\alpha\beta}\left(\int_\Gamma x_\beta p_3 ds + \int_{\Sigma_1} x_\beta G_3 da\right) + \varepsilon_{\alpha\beta} R_\beta(\mathbf{u}'') \tag{4.7.9}$$

It follows from Equations 4.7.8 and 4.7.9 that $R_\alpha(\mathbf{u}'') = F_\alpha$. The conditions $R_3(\mathbf{u}'') = F_3$, $\mathbf{H}(\mathbf{u}'') = \mathbf{M}$ reduce to

$$\sum_{s=1}^{4} D^*_{rs} d_s = E_r, \quad (r = 1, 2, 3, 4) \tag{4.7.10}$$

where

$$E_\alpha = \varepsilon_{\alpha\beta} M_\beta - \int_{\Sigma_1} x_\alpha [k'_{33} + T_{33}(\mathbf{w}'')] da$$

$$E_3 = -F_3 - \int_{\Sigma_1} [k'_{33} + T_{33}(\mathbf{w}'')] da$$

$$E_4 = -M_3 - \int_{\Sigma_1} \varepsilon_{\alpha\beta} x_\alpha [k'_{\beta 3} + T_{\beta 3}(\mathbf{w}'')] da$$

The system 4.7.10 determines the vector $\widehat{d}$. Thus we have determined the vectors $\widehat{b}$, $\widehat{c}$, and $\widehat{d}$, and the vector fields $\mathbf{w}'$ and $\mathbf{w}''$ to have $\mathbf{u}'' \in K_{III}(\mathbf{F}, \mathbf{M}, \mathbf{G}, \mathbf{p})$. □

Let us consider the Almansi problem. Let $\mathbf{u}^*$ be an equilibrium displacement field on B corresponding to the body force $\mathbf{f} = \mathbf{g} x_3^n$ that satisfies the conditions

$$\mathbf{s}(\mathbf{u}^*) = \mathbf{q} x_3^n \text{ on } \Pi, \qquad \mathbf{R}(\mathbf{u}^*) = \mathbf{0}, \qquad \mathbf{H}(\mathbf{u}^*) = \mathbf{0} \tag{4.7.11}$$

where $\mathbf{g}$ and $\mathbf{p}$ are vector fields independent of x_3, and n is a positive integer or zero.

We denote by $\mathbf{u}$ an equilibrium displacement field on B, corresponding to the body force $\mathbf{f} = \mathbf{g} x_3^{n+1}$ that satisfies the conditions

$$\mathbf{s}(\mathbf{u}) = \mathbf{q} x_3^{n+1} \text{ on } \Pi, \qquad \mathbf{R}(\mathbf{u}) = \mathbf{0}, \qquad \mathbf{H}(\mathbf{u}) = \mathbf{0} \tag{4.7.12}$$

With the help of the results obtained in Section 2.4, the Almansi problem reduces to the problem of finding a vector field $\mathbf{u}$ once the vector field $\mathbf{u}^*$ is known. As in Section 2.4, we are led to seek the vector field $\mathbf{u}$ in the form

$$\mathbf{u} = (n+1)\left[\int_0^{x_3} \mathbf{u}^* dx_3 + \mathbf{u}^0\{\widehat{a}\} + \mathbf{w}\right] \tag{4.7.13}$$

where $\widehat{a} = (a_1, a_2, a_3, a_4)$ is an unknown vector, $\mathbf{u}^0\{\widehat{a}\}$ is given by Equation 4.3.9, and $\mathbf{w}$ is an unknown vector field independent of x_3.

By Equations 4.3.12 and 4.7.13,

$$t_{ij}(\mathbf{u}) = (n+1)\left[\int_0^{x_3} t_{ij}(\mathbf{u}^*) dx_3 + \sum_{r=1}^{4} a_r t_{ij}(\mathbf{u}^{(r)}) + T_{ij}(\mathbf{w}) + g_{ij}\right]$$

where

$$g_{ij} = C_{ijk3}u_k^*(x_1, x_2, 0)$$

The equilibrium equations and the conditions on the lateral boundary reduce to

$$\begin{aligned}(T_{i\alpha}(\mathbf{w}))_{,\alpha} + h_i' &= 0 \text{ on } \Sigma_1 \\ T_{i\alpha}(\mathbf{w})n_\alpha &= q_i' \text{ on } \Gamma\end{aligned} \tag{4.7.14}$$

where

$$h_i' = g_{i\alpha,\alpha} + [t_{i3}(\mathbf{u}^*)](x_1, x_2, 0), \qquad q_i' = -g_{i\alpha}n_\alpha$$

We can write

$$\int_{\Sigma_1} h_i' da + \int_\Gamma q_i' ds = -R_i(\mathbf{u}^*) = 0$$

$$\int_{\Sigma_1} \varepsilon_{\alpha\beta}x_\alpha h_\beta' da + \int_\Gamma \varepsilon_{\alpha\beta}x_\alpha q_\beta' ds = -Q_3(\mathbf{u}^*) = 0$$

The necessary and sufficient conditions for the existence of a solution to the boundary-value problem 4.7.14 are satisfied. We conclude that the vector field $\mathbf{w}$ is characterized by the generalized plane strain problem 4.7.14.

With the help of Theorem 2.4.1, we get $R_\alpha(\mathbf{u}) = \varepsilon_{\beta\alpha}H_\beta[(n+1)\mathbf{u}^*] = 0$. The conditions $R_3(\mathbf{u}) = 0$ and $\mathbf{H}(\mathbf{u}) = \mathbf{0}$ imply that

$$\sum_{s=1}^{4} D_{rs}^* a_s = k_r, \quad (r = 1, 2, 3, 4) \tag{4.7.15}$$

where

$$\begin{aligned}k_\alpha &= -\int_{\Sigma_1} x_\alpha[T_{33}(\mathbf{w}) + g_{33}]da \\ k_3 &= -\int_{\Sigma_1} [T_{33}(\mathbf{w}) + g_{33}]da \\ k_4 &= -\int_{\Sigma_1} \varepsilon_{\alpha\beta}x_\alpha[T_{3\beta}(\mathbf{w}) + g_{3\beta}]da\end{aligned}$$

Thus, the constant vector $\widehat{a}$ is determined by Equation 4.7.15.

The solution presented in this section coincides with the solution established in Ref. 150 using the semi-inverse method. Various applications are presented in Ref. 175.

4.8 Orthotropic Bodies

A large number of works are devoted to the deformation of anisotropic cylinders with various symmetry properties of the material. The torsion problem for an orthotropic material was first studied by Saint-Venant [291]. An account of the historical development of the subject as well as references to various contributions may be found in the works of Lekhnitskii [204], Sokolnikoff [313], Borş [28], and Khatiashvili [175].

In the first part of this section, we study Saint-Venant's problem when the material is homogeneous and orthotropic. The solution for the case when the medium is homogeneous and has a plane of elastic symmetry, normal to the axis of cylinder, can be obtained in the same manner. In the second part of this section, we present the solution of Almansi–Michell problem.

For an orthotropic material, the nonzero components of the elasticity tensor are $C_{1111}, C_{1122}, C_{1133}$, C_{2222}, $C_{2233}, C_{3333}, C_{2323}, C_{3131}$, and C_{1212}. In what follows, we use the notations

$$\begin{aligned} &A_{11} = C_{1111}, \quad A_{22} = C_{2222}, \quad A_{33} = C_{3333}, \quad A_{12} = A_{21} = C_{1122} \\ &A_{13} = A_{31} = C_{1133}, \quad A_{23} = A_{32} = C_{2233} \\ &A_{44} = C_{2323}, \quad A_{55} = C_{3131}, \quad A_{66} = C_{1212} \end{aligned} \tag{4.8.1}$$

The constitutive equations 1.1.2 reduce to

$$\begin{aligned} t_{11} &= A_{11}e_{11} + A_{12}e_{22} + A_{13}e_{33} \\ t_{22} &= A_{12}e_{11} + A_{22}e_{22} + A_{23}e_{33} \\ t_{33} &= A_{13}e_{11} + A_{23}e_{22} + A_{33}e_{33} \\ t_{23} &= 2A_{44}e_{23}, \quad t_{31} = 2A_{55}e_{31}, \quad t_{12} = 2A_{66}e_{12} \end{aligned} \tag{4.8.2}$$

where, for convenience, we have suppressed the argument $\mathbf{u}$ in the components of the stress tensor and the strain tensor. We assume that the constitutive coefficients are constant. The condition that $\mathbf{C}$ is positive definite implies

$$\begin{aligned} &A_{11} > 0, \quad A_{11}A_{22} - A_{12}^2 > 0, \quad \det(A_{ij}) > 0, \quad A_{44} > 0, \\ &A_{55} > 0, \quad A_{66} > 0 \end{aligned} \tag{4.8.3}$$

In this section, we apply the results established in the preceding sections to obtain the solution of Saint-Venant's problem for orthotropic bodies. The constitutive equations 4.2.2 for the generalized plane strain problem become

$$\begin{aligned} &t_{11} = A_{11}u_{1,1} + A_{12}u_{2,2}, \quad t_{22} = A_{12}u_{1,1} + A_{22}u_{2,2} \\ &t_{33} = A_{13}u_{1,1} + A_{23}u_{2,2}, \quad t_{23} = A_{44}u_{3,2} \\ &t_{31} = A_{55}u_{3,1}, \quad t_{12} = A_{66}(u_{1,2} + u_{2,1}) \end{aligned} \tag{4.8.4}$$

The equations of equilibrium 4.2.3 reduce to

$$\begin{aligned} &\left(A_{11}\frac{\partial^2}{\partial x_1^2} + A_{66}\frac{\partial^2}{\partial x_2^2}\right)u_1 + (A_{12} + A_{66})\frac{\partial^2 u_2}{\partial x_1 \partial x_2} + f_1 = 0 \\ &(A_{12} + A_{66})\frac{\partial^2 u_1}{\partial x_1 \partial x_2} + \left(A_{66}\frac{\partial^2}{\partial x_1^2} + A_{22}\frac{\partial^2}{\partial x_2^2}\right)u_2 + f_2 = 0 \\ &\left(A_{55}\frac{\partial^2}{\partial x_1^2} + A_{44}\frac{\partial^2}{\partial x_2^2}\right)u_3 + f_3 = 0 \text{ on } \Sigma_1 \end{aligned} \tag{4.8.5}$$

The boundary conditions 4.2.4 can be expressed in the form

$$\begin{aligned}(A_{11}u_{1,1}+A_{12}u_{2,2})n_1+A_{66}(u_{1,2}+u_{2,1})n_2&=p_1\\ A_{66}(u_{1,2}+u_{2,1})n_1+(A_{12}u_{1,1}+A_{22}u_{2,2})n_2&=p_2\\ A_{55}u_{3,1}n_1+A_{44}u_{3,2}n_2&=p_3 \text{ on } \Gamma\end{aligned} \tag{4.8.6}$$

It follows from Equations 4.8.5 and 4.8.6 that in the case of orthotropic bodies the generalized plane strain problem reduces to the solution of two boundary-value problems. The first boundary-value problem consists in the determination of the functions u_1 and u_2 which satisfy Equations $4.8.5_{1,2}$ and the boundary conditions $4.8.6_{1,2}$. This is a *plane strain problem* for homogeneous and orthotropic cylinders. The study of this problem will be presented in Section 4.9. The second boundary-value problem consists in the finding of the function w_3 which satisfies Equation $4.8.5_3$ and the boundary condition $4.8.6_3$. This is an *antiplane problem* for the considered cylinder.

4.8.1 Extension, Bending, and Torsion of Orthotropic Cylinders

We shall use the solution 4.3.1 to obtain the displacement vector field corresponding to the problem of extension, bending, and torsion of homogeneous and orthotropic cylinders. The vector field $\mathbf{w}$ from Equation 4.3.1 has the form 4.3.6, where $\mathbf{w}^{(k)}$ are the solutions of the generalized plane strain problems 4.3.7 and 4.3.8. Thus, for homogeneous and orthotropic bodies, the vector field $\mathbf{w}^{(1)}$ satisfies the equations

$$\begin{aligned}A_{11}w^{(1)}_{1,11}+A_{66}w^{(1)}_{1,22}+(A_{12}+A_{66})w^{(1)}_{2,12}+A_{13}&=0\\ (A_{12}+A_{66})w^{(1)}_{1,12}+A_{66}w^{(1)}_{2,11}+A_{22}w^{(1)}_{2,22}&=0\\ A_{55}w^{(1)}_{3,11}+A_{44}w^{(1)}_{3,22}&=0 \text{ on } \Sigma_1\end{aligned} \tag{4.8.7}$$

and the boundary conditions

$$\begin{aligned}(A_{11}w^{(1)}_{1,1}+A_{12}w^{(1)}_{2,2})n_1+A_{66}(w^{(1)}_{1,2}+w^{(1)}_{2,1})n_2&=-A_{13}x_1n_1\\ A_{66}(w^{(1)}_{1,2}+w^{(1)}_{2,1})n_1+(A_{12}w^{(1)}_{1,1}+A_{22}w^{(1)}_{2,2})n_2&=-A_{23}x_1n_2\\ A_{55}w^{(1)}_{3,1}n_1+A_{44}w^{(1)}_{3,2}n_2&=0 \text{ on } \Gamma\end{aligned} \tag{4.8.8}$$

We seek the solution of the boundary-value problem 4.8.7 and 4.8.8 in the form

$$w^{(1)}_1=-\frac{1}{2}(\nu_1x_1^2-\nu_2x_2^2),\qquad w^{(1)}_2=-\nu_2x_1x_2,\qquad w^{(1)}_3=0,\qquad (x_1,x_2)\in\Sigma_1 \tag{4.8.9}$$

where ν_1 and ν_2 are unknown constants. It is easy to see that Equations 4.8.7 and the boundary conditions 4.8.8 are satisfied if

$$A_{11}\nu_1 + A_{12}\nu_2 = A_{13}$$
$$A_{12}\nu_1 + A_{22}\nu_2 = A_{23}$$

In view of the relations 4.8.3, we can determine ν_1 and ν_2,

$$\nu_1 = \frac{1}{\delta_1}(A_{13}A_{22} - A_{23}A_{12}), \qquad \nu_2 = \frac{1}{\delta_1}(A_{23}A_{11} - A_{13}A_{12}) \tag{4.8.10}$$

where

$$\delta_1 = A_{11}A_{22} - A_{12}^2 \tag{4.8.11}$$

Similarly, we find that

$$\begin{aligned} &w_1^{(2)} = -\nu_1 x_1 x_2, \qquad w_2^{(2)} = \frac{1}{2}(\nu_1 x_1^2 - \nu_2 x_2^2), \qquad w_3^{(2)} = 0 \\ &w_1^{(3)} = -\nu_1 x_1, \qquad w_2^{(3)} = -\nu_2 x_2, \qquad w_3^{(3)} = 0, \quad (x_1, x_2) \in \Sigma_1 \end{aligned} \tag{4.8.12}$$

It follows from Equations 4.3.7 and 4.3.8 that the vector field $\mathbf{w}^{(3)}$ satisfies the equations

$$\begin{aligned} &A_{11}w_{1,11}^{(3)} + A_{66}w_{1,22}^{(3)} + (A_{12} + A_{66})w_{2,12}^{(3)} = 0 \\ &(A_{12} + A_{66})w_{1,12}^{(3)} + A_{66}w_{2,11}^{(3)} + A_{22}w_{2,22}^{(3)} = 0 \\ &A_{55}w_{3,11}^{(3)} + A_{44}w_{3,22}^{(3)} = 0 \text{ on } \Sigma_1 \end{aligned} \tag{4.8.13}$$

and the boundary conditions

$$\begin{aligned} &(A_{11}w_{1,1}^{(3)} + A_{12}w_{2,2}^{(3)})n_1 + A_{66}(w_{1,2}^{(3)} + w_{2,1}^{(3)}) = 0 \\ &A_{66}(w_{1,2}^{(3)} + w_{2,1}^{(3)})n_1 + (A_{12}w_{1,1}^{(3)} + A_{22}w_{2,2}^{(3)})n_2 = 0 \\ &A_{55}w_{3,1}^{(3)}n_1 + A_{44}w_{3,2}^{(3)}n_2 = A_{55}x_2 n_1 - A_{44}x_1 n_2 \text{ on } \Gamma \end{aligned} \tag{4.8.14}$$

The solution of the boundary-value problem 4.8.13 and 4.8.14 is given by

$$w_1^{(4)} = 0, \qquad w_2^{(4)} = 0, \qquad w_3^{(4)} = \varphi(x_1, x_2), \quad (x_1, x_2) \in \Sigma_1 \tag{4.8.15}$$

where φ satisfies the equation

$$A_{55}\varphi_{,11} + A_{44}\varphi_{,22} = 0 \text{ on } \Sigma_1 \tag{4.8.16}$$

and the boundary condition

$$A_{55}\varphi_{,1}n_1 + A_{44}\varphi_{,2}n_2 = A_{55}x_2 n_1 - A_{44}x_1 n_2 \text{ on } \Gamma \tag{4.8.17}$$

It follows from Equations 4.3.1, 4.3.6, 4.8.9, 4.8.12, and 4.8.15 that the solution of the problem of extension, bending, and torsion for orthotropic cylinders is given by

$$\begin{aligned} u_1^0 &= -\frac{1}{2}a_1(x_3^2 + \nu_1 x_1^2 - \nu_2 x_2^2) - a_2\nu_1 x_1 x_2 - a_3\nu_1 x_1 - a_4 x_2 x_3 \\ u_2^0 &= -a_1\nu_2 x_1 x_2 - \frac{1}{2}a_2(x_3^2 - \nu_1 x_1^2 + \nu_2 x_2^2) - a_3\nu_2 x_2 + a_4 x_1 x_3 \\ u_3^0 &= (a_1 x_1 + a_2 x_2 + a_3)x_3 + a_4\varphi(x_1, x_2), \quad (x_1, x_2, x_3) \in B \end{aligned} \tag{4.8.18}$$

where the constants a_k, $(k = 1, 2, 3, 4)$, are determined from Equations 4.3.15. For homogeneous and orthotropic cylinders, the system 4.3.15 has a special form. Let us study the coefficients D_{rs}^*, $(r, s = 1, 2, 3, 4)$. From Equations 4.3.10 and 4.8.9, we get

$$u_1^{(1)} = -\frac{1}{2}(x_3^2 + \nu_1 x_1^2 - \nu_2 x_2^2), \qquad u_2^{(1)} = -\nu_2 x_1 x_2, \qquad u_3^{(1)} = x_1 x_3$$

so that

$$t_{33}(\mathbf{u}^{(1)}) = -(A_{13}\nu_1 + A_{23}\nu_2 - A_{33})x_1, \qquad t_{3\beta}(\mathbf{u}^{(1)}) = 0 \tag{4.8.19}$$

By Equations 4.3.16 and 4.8.19, we obtain

$$\begin{aligned} D_{\alpha 1}^* &= (A_{33} - A_{13}\nu_1 - A_{23}\nu_2)I_{\alpha 1} \\ D_{31}^* &= (A_{33} - A_{13}\nu_1 - A_{23}\nu_2)Ax_1^0, \qquad D_{41}^* = 0 \end{aligned} \tag{4.8.20}$$

where $I_{\alpha\beta}, x_\alpha^0$, and A are defined by Equations 1.7.14 and 1.4.9. In view of Equations 4.8.10, we get

$$A_{33} - A_{13}\nu_1 - A_{23}\nu_2 = \frac{\delta_2}{\delta_1} \tag{4.8.21}$$

where $\delta_2 = \det(A_{ij})$, and δ_1 is given by Equation 4.8.11. If we introduce the notation

$$E_0 = \frac{\delta_2}{\delta_1} \tag{4.8.22}$$

then from Equations 4.8.20, we find that

$$D_{\alpha 1}^* = E_0 I_{\alpha 1}, \qquad D_{31}^* = Ax_1^0, \qquad D_{41}^* = 0$$

In the same manner, we arrive at

$$\begin{aligned} D_{\alpha\beta}^* &= E_0 I_{\alpha\beta}, \qquad D_{3\alpha}^* = E_0 A x_\alpha^0, \qquad D_{4\alpha}^* = 0 \\ D_{33}^* &= E_0 A, \qquad D_{43}^* = 0, \qquad D_{44}^* = D_0 \end{aligned} \tag{4.8.23}$$

where

$$D_0 = \int_{\Sigma_1} [A_{44}x_1(\varphi_{,2} + x_1) - A_{55}x_2(\varphi_{,1} - x_2)]da \tag{4.8.24}$$

The system 4.3.15 reduces to

$$
\begin{aligned}
&E_0(I_{\alpha\beta}a_\beta + Ax_\alpha^0 a_3) = \varepsilon_{\alpha\beta}M_\beta \\
&E_0A(a_1x_1^0 + a_2x_2^0 + a_3) = -F_3, \qquad D_0a_4 = -M_3
\end{aligned}
\tag{4.8.25}
$$

We note that the torsion problem can be treated independently of the extension and bending problems.

The solution of the torsion problem is

$$
u_\alpha = a_4\varepsilon_{\beta\alpha}x_\beta x_3, \qquad u_3 = a_4\varphi \tag{4.8.26}
$$

where φ satisfies the boundary-value problem 4.8.16 and 4.8.17, and the constant a_4 is given by Equation 4.8.25.

Remark. The finding of the torsion function for homogeneous and orthotropic cylinders can be reduced to the determination of the torsion function for certain homogeneous and isotropic cylinders. We introduce new independent variables ξ_k by

$$
x_1 = \xi_1\left(\frac{2A_{55}}{A_{44}+A_{55}}\right)^{1/2}, \qquad x_2 = \xi_2\left(\frac{2A_{44}}{A_{44}+A_{55}}\right)^{1/2}, \qquad x_3 = \xi_3 \tag{4.8.27}
$$

Let Σ_1^* be the image of Σ_1 under the mapping 4.8.27.

We assume that the curve Γ admits the representation

$$
f(x_1, x_2) = 0, \qquad x_3 = 0
$$

and denote

$$
f^*(\xi_1, \xi_2) = f[x_1(\xi_1), x_2(\xi_2)]
$$

Let Γ_* be the curve described by the equations

$$
f^*(\xi_1, \xi_2) = 0, \qquad \xi_3 = 0
$$

We introduce the function G defined by

$$
G(\xi_1, \xi_2) = \frac{A_{44}+A_{55}}{2\sqrt{A_{44}A_{55}}}\varphi[x_1(\xi_1), x_2(\xi_2)]
$$

Clearly,

$$
\begin{aligned}
&\frac{\partial G}{\partial \xi_1} = \left(\frac{A_{44}+A_{55}}{2A_{44}}\right)^{1/2}\frac{\partial \varphi}{\partial x_1}, \qquad \frac{\partial G}{\partial \xi_2} = \left(\frac{A_{44}+A_{55}}{2A_{55}}\right)^{1/2}\frac{\partial \varphi}{\partial x_2} \\
&\frac{\partial^2 G}{\partial \xi_1^2} = \left(\frac{A_{55}}{A_{44}}\right)^{1/2}\frac{\partial^2 \varphi}{\partial x_1^2}, \qquad \frac{\partial^2 G}{\partial \xi_2^2} = \left(\frac{A_{44}}{A_{55}}\right)^{1/2}\frac{\partial^2 \varphi}{\partial x_2^2} \\
&\frac{\partial f^*}{\partial \xi_1} = \left(\frac{2A_{55}}{A_{44}+A_{55}}\right)^{1/2}\frac{\partial f}{\partial x_1}, \qquad \frac{\partial f^*}{\partial \xi_2} = \left(\frac{2A_{44}}{A_{44}+A_{55}}\right)^{1/2}\frac{\partial f}{\partial x_2}
\end{aligned}
$$

It follows from Equation 4.8.16 that the function G satisfies the equations

$$\frac{\partial^2 G}{\partial \xi_1^2} + \frac{\partial^2 G}{\partial \xi_2^2} = 0 \text{ on } \Sigma_1^*$$

The boundary condition 4.8.17 reduces to

$$\frac{\partial G}{\partial \xi_1} n_1^* + \frac{\partial G}{\partial \xi_2} n_2^* = \xi_2 n_1^* - \xi_1 n_2^* \text{ on } \Gamma_*$$

where (n_1^*, n_2^*) are the components of the outward normal unit vector along Γ_*. We conclude that G is the torsion function for a homogeneous and isotropic cylinder with the cross section Σ_1^*.

4.8.2 Flexure

In the case of homogeneous and orthotropic elastic materials, the solution 4.4.4 takes a special form. First, from Equations 4.4.3 and 4.8.23, we obtain the following system for the constants b_k, $(k = 1, 2, 3, 4)$,

$$\begin{aligned} &E_0\big(I_{\alpha\beta} b_\beta + A x_\alpha^0 b_3\big) = -F_\alpha \\ &E_0 A\big(b_1 x_1^0 + b_2 x_2^0 + b_3\big) = 0, \qquad b_4 = 0 \end{aligned} \tag{4.8.28}$$

By Equations 4.3.10, 4.8.9, 4.8.12, 4.8.15, and 4.8.19, we get

$$\begin{aligned} &t_{33}(\mathbf{u}^{(\alpha)}) = E_0 x_\alpha, \ t_{3\beta}(\mathbf{u}^{(\alpha)}) = 0 \\ &t_{33}(\mathbf{u}^{(3)}) = E_0, \ t_{3\alpha}(\mathbf{u}^{(3)}) = 0, \qquad t_{33}(\mathbf{u}^{(4)}) = 0 \\ &t_{31}(\mathbf{u}^{(4)}) = A_{55}(\varphi_{,1} - x_2), \qquad t_{32}(\mathbf{u}^{(4)}) = A_{44}(\varphi_{,2} + x_1) \end{aligned} \tag{4.8.29}$$

so that the functions k_{rs} which appear in Equation 4.4.5 are given by

$$\begin{aligned} &k_{\alpha\beta} = k_{\beta\alpha} = 0, \qquad k_{33} = 0 \\ &k_{31} = k_{13} = A_{55}\big(b_1 w_1^{(1)} + b_2 w_1^{(2)} + b_3 w_1^{(3)}\big) \\ &k_{23} = k_{32} = A_{44}\big(b_1 w_2^{(1)} + b_2 w_2^{(2)} + b_3 w_2^{(3)}\big) \end{aligned} \tag{4.8.30}$$

Here, $w_\alpha^{(k)}$ are defined in Equations 4.8.9 and 4.8.12. The body forces f_i' for the generalized plane strain problem 4.4.6 and 4.4.7 reduce to

$$f_1' = 0, \qquad f_2' = 0, \qquad f_3' = (E_0 - A_{55}\nu_1 - A_{44}\nu_2)(b_1 x_1 + b_2 x_2 + b_3) \tag{4.8.31}$$

The surface tractions p_i' associated to this problem become

$$p_1' = 0, \qquad p_2' = 0, \qquad p_3' = Q \tag{4.8.32}$$

where

$$\begin{aligned} Q = {} & A_{55}\left[\frac{1}{2} b_1\big(\nu_1 x_1^2 - \nu_2 x_2^2\big) + b_2 \nu_1 x_1 x_2 + b_3 \nu_1 x_2\right] n_1 \\ & + A_{44}\left[b_1 \nu_2 x_1 x_2 + \frac{1}{2} b_2\big(\nu_2 x_2^2 - \nu_1 x_1^2\big) + b_3 \nu_2 x_2\right] n_2 \end{aligned} \tag{4.8.33}$$

By using Equations 4.8.5, 4.8.6, 4.8.31, and 4.8.32, we conclude that the solution of the generalized plane strain problem 4.4.6 and 4.4.7 is given by

$$w'_1 = 0, \qquad w'_2 = 0, \qquad w'_3 = \psi \tag{4.8.34}$$

where the function ψ satisfies the equation

$$A_{55}\psi_{,11} + A_{44}\psi_{,22} = (A_{55}\nu_1 + A_{44}\nu_2 - E_0)(b_1x_1 + b_2x_2 + b_3) \text{ on } \Sigma_1 \tag{4.8.35}$$

and the boundary condition

$$A_{55}\psi_{,1}n_1 + A_{44}\psi_{,2}n_2 = Q \text{ on } \Gamma \tag{4.8.36}$$

From Equations 4.3.3, 4.8.1, and 4.8.34, we get

$$T_{33}(\mathbf{w}') = 0, \qquad T_{23}(\mathbf{w}') = A_{44}\psi_{,2}, \qquad T_{13}(\mathbf{w}') = A_{55}\psi_{,1} \tag{4.8.37}$$

so that the system 4.4.9 reduces to

$$\begin{aligned} E_0(I_{\alpha\beta}c_\beta + Ax^0_\alpha c_3) &= 0 \\ c_1x^0_1 + c_2x^0_2 + c_3 &= 0 \end{aligned} \tag{4.8.38}$$

and

$$D_0c_4 = \int_{\Sigma_1}\left[A_{55}x_2\psi_{,1} - A_{44}x_1\psi_{,2} + \sum_{j=1}^{3}\left(A_{55}x_2b_jw_1^{(j)} - A_{44}x_1b_jw_2^{(j)}\right)\right]da \tag{4.8.39}$$

By Equation 4.8.38, we find that

$$c_i = 0, \quad (i = 1, 2, 3) \tag{4.8.40}$$

The constant c_4 is given by Equation 4.8.39.

It follows from Equations 4.4.4, 4.8.28, 4.8.34, 4.8.40, 4.8.9, 4.8.12, and 4.8.15 that the solution of the flexure problem for homogeneous and orthotropic cylinders is

$$\begin{aligned} u'_1 &= -\frac{1}{2}b_1\left(\frac{1}{3}x_3^2 + \nu_1x_1^2 - \nu_2x_2^2\right)x_3 - b_2\nu_1x_1x_2x_3 - b_3\nu_1x_1x_3 - c_4x_2x_3 \\ u'_2 &= -b_1\nu_2x_1x_2x_3 - \frac{1}{2}b_2(x_3^2 - \nu_1x_1^2 + \nu_2x_2^2)x_3 - b_3\nu_2x_2x_3 + c_4x_1x_3 \\ u'_3 &= \frac{1}{2}(b_1x_1 + b_2x_2 + b_3)x_3^2 + c_4\varphi(x_1, x_2) + \psi(x_1, x_2), \quad (x_1, x_2, x_3) \in B \end{aligned} \tag{4.8.41}$$

4.8.3 Uniformly Loaded Cylinders

The solution 4.7.5 of the Almansi–Michell problem will now be specialized to the case of homogeneous and orthotropic bodies. In view of Equations 4.8.23, the system 4.7.2 takes the form

$$\begin{aligned} E_0(I_{\alpha\beta}b_\beta + Ax_\alpha^0 b_3) &= -\int_{\Sigma_1} G_\alpha da - \int_\Gamma p_\alpha ds \\ b_1x_1^0 + b_2x_2^0 + b_3 &= 0 \\ b_4 &= 0 \end{aligned} \tag{4.8.42}$$

It follows from Equations 4.8.9, 4.8.12, 4.8.15, 4.8.29, and 4.8.30 that the solution of the boundary-value problem 4.7.3 is

$$w_1' = 0, \qquad w_2' = 0, \qquad w_3' = \psi \tag{4.8.43}$$

where ψ is characterized by the boundary-value problem 4.8.35 and 4.8.36 with b_k defined by Equations 4.8.42.

The system 4.7.4 reduces to

$$\begin{aligned} &E_0(I_{\alpha\beta}c_\beta + Ax_\alpha^0 c_3) = -\int_{\Sigma_1} x_\alpha G_3 da - \int_\Gamma x_\alpha p_3 ds - F_\alpha \\ &E_0A(c_1x_1^0 + c_2x_2^0 + c_3) = -\int_{\Sigma_1} G_3 da - \int_\Gamma p_3 ds \\ &D_0c_4 = -\int_{\Sigma_1} \varepsilon_{\alpha\beta}x_\alpha G_\beta da - \int_\Gamma \varepsilon_{\alpha\beta}x_\alpha p_\beta ds + \int_{\Sigma_1}\Big[A_{55}x_2\psi_{,1} \\ &\qquad - A_{44}x_1\psi_{,2} + \sum_{j=1}^{3}\left(A_{55}x_2b_jw_1^{(j)} - A_{44}x_1b_jw_2^{(j)}\right)\Big] da \end{aligned} \tag{4.8.44}$$

The boundary-value problem 4.7.7 reduces to the solution of two independent boundary-value problems. The first problem consists in finding of the functions w_α'' which satisfy the equations of the plane strain problem

$$\begin{aligned} A_{11}w_{1,11}'' + A_{66}w_{1,22}'' + (A_{12} + A_{66})w_{2,12}'' + h_1 &= 0 \\ (A_{12} + A_{66})w_{1,12}'' + (A_{66}w_{2,11}'' + A_{22}w_{2,22}'') + h_2 &= 0 \text{ on } \Sigma_1 \end{aligned} \tag{4.8.45}$$

and the boundary conditions

$$\begin{aligned} (A_{11}w_{1,1}'' + A_{12}w_{2,2}'')n_1 + A_{66}(w_{1,2}'' + w_{2,1}'')n_2 &= q_1 \\ A_{66}(w_{1,2}'' + w_{2,1}'')n_1 + (A_{12}w_{1,1}'' + A_{22}w_{2,2}'')n_2 &= q_2 \text{ on } \Gamma \end{aligned} \tag{4.8.46}$$

where

$$\begin{aligned} h_1 &= G_1 + (A_{13} + A_{55})(\psi_{,1} + c_4\varphi_{,1}) + A_{55}\left(\sum_{j=1}^{3} b_jw_1^{(j)} - c_4x_2\right) \\ h_2 &= G_2 + (A_{23} + A_{44})(\psi_{,2} + c_4\varphi_{,2}) + A_{44}\left(\sum_{j=1}^{3} b_jw_2^{(j)} + c_4x_1\right) \\ q_1 &= p_1 - A_{13}(\psi + c_4\varphi)n_1, \qquad q_2 = p_2 - A_{23}(\psi + c_4\varphi)n_2 \end{aligned} \tag{4.8.47}$$

We have seen in Section 4.7 that the necessary and sufficient conditions for the existence of a solution of the boundary-value problem 4.8.45 and 4.8.46 are satisfied. In what follows, we assume that w''_α are known functions.

We introduce the notation $w''_3 = \chi$. The second boundary-value problem derived from Equation 4.7.7 consists in the determination of the function χ which satisfies the equation

$$A_{55}\chi_{,11} + A_{44}\chi_{,22} = -G_3 - (E_0 - A_{55}\nu_1 - A_{44}\nu_2)(c_1x_1 + c_2x_2 + c_3) \text{ on } \Sigma_1 \tag{4.8.48}$$

and the boundary condition

$$A_{55}\chi_{,1}n_1 + A_{44}\chi_{,2}n_2 = p_3 - \sum_{j=1}^{3} c_j(A_{55}w_1^{(j)}n_1 + A_{44}w_2^{(j)}n_2) \text{ on } \Gamma \tag{4.8.49}$$

We note that $T_{33}(\mathbf{w}'') = 0, T_{31}(\mathbf{w}'') = A_{55}\chi_{,1}$ and $T_{32}(\mathbf{w}'') = A_{44}\chi_{,2}$. Thus, Equations 4.7.10 reduce to

$$\begin{aligned}
E_0\big(I_{\alpha\beta}d_\beta + Ax^0_\alpha d_3\big) &= \varepsilon_{\alpha\beta}M_\beta - \int_{\Sigma_1} x_\alpha A_{33}(\psi + c_4\varphi)da \\
E_0A\big(d_1x^0_1 + d_2x^0_2 + d_3\big) &= -F_3 - \int_{\Sigma_1} A_{33}(\psi + c_4\varphi)da \\
D_0d_4 &= -M_3 - \int_{\Sigma_1}\Bigg[A_{44}x_1\chi_{,2} - A_{55}x_2\chi_{,1} \\
&\quad + \sum_{j=1}^{3} c_j\big(A_{44}x_1w_2^{(j)} - A_{55}x_2w_1^{(j)}\big)\Bigg]da
\end{aligned} \tag{4.8.50}$$

By Equation 4.7.5, we conclude that the solution of Almansi–Michell problem for homogeneous and orthotropic cylinders is given by

$$\begin{aligned}
u''_1 &= -\frac{1}{4}b_1\left(\frac{1}{6}x_3^2 + \nu_1x_1^2 - \nu_2x_2^2\right)x_3^2 - \frac{1}{2}b_2\nu_1x_1x_2x_3^2 - \frac{1}{2}b_3\nu_1x_1x_3^2 \\
&\quad - \frac{1}{2}c_1\left(\frac{1}{3}x_3^2 + \nu_1x_1^2 - \nu_2x_2^2\right)x_3 - c_2\nu_1x_1x_2x_3 - c_3\nu_1x_1x_3 \\
&\quad - \frac{1}{2}d_1(x_3^2 + \nu_1x_1^2 - \nu_2x_2^2) - d_2\nu_1x_1x_2 - d_3\nu_1x_1 \\
&\quad - \left(\frac{1}{2}c_4x_3^2 + d_4x_3\right)x_2 + w''_1 \\
u''_2 &= -\frac{1}{2}b_1\nu_2x_1x_2x_3^2 - \frac{1}{4}b_2\left(\frac{1}{6}x_3^2 - \nu_1x_1^2 + \nu_2x_2^2\right)x_3^2 - \frac{1}{2}b_3\nu_2x_2x_3^2 \\
&\quad - c_1\nu_2x_1x_2x_3 - \frac{1}{2}c_2\left(\frac{1}{3}x_3^2 - \nu_1x_1^2 + \nu_2x_2^2\right)x_3 - c_3\nu_2x_2x_3
\end{aligned}$$

$$
\begin{aligned}
& - d_1\nu_2 x_1 x_2 - \frac{1}{2}d_2(x_3^2 - \nu_1 x_1^2 + \nu_2 x_2^2) - d_3\nu_2 x_2 \\
& + \left(\frac{1}{2}c_4 x_3^2 + d_4\right) x_1 + w_2'' \\
u_3'' &= \frac{1}{6}(b_\alpha x_\alpha + b_3)x_3^3 + \frac{1}{2}(c_\alpha x_\alpha + c_3)x_3^2 + (d_\alpha x_\alpha + d_3)x_3 \\
& + (c_4 x_3 + d_4)\varphi + x_3\psi + \chi
\end{aligned} \tag{4.8.51}
$$

The constants b_s, c_s, and d_s, $(s = 1, 2, 3, 4)$, are given by Equations 4.8.42, 4.8.44, and 4.8.50, respectively.

By using the results of Sections 2.3 and 4.7, we can also derive the solution of Almansi problem.

4.9 Plane Strain Problem of Orthotropic Bodies

In the previous section, we have seen the important role of the plane strain problem in the study of Saint-Venant's problem for orthotropic cylinders. The state of plane strain of cylinder B is defined by Equations 1.5.1. It is easy to see that the basic equations of the plane strain of orthotropic cylinders consist of the equations of equilibrium

$$t_{\beta\alpha,\beta} + f_\alpha = 0 \tag{4.9.1}$$

the constitutive equations

$$t_{11} = A_{11}e_{11} + A_{12}e_{22}, \qquad t_{22} = A_{12}e_{11} + A_{22}e_{22}, \qquad t_{12} = 2A_{66}e_{12} \tag{4.9.2}$$

and the geometrical equations

$$2e_{\alpha\beta} = u_{\alpha,\beta} + u_{\beta,\alpha} \tag{4.9.3}$$

on Σ_1. We restrict our attention to homogeneous bodies so that the constitutive coefficients $A_{\alpha\beta}$ and A_{66} are prescribed constants. We continue to assume that the elastic potential is a positive definite quadratic form. This fact implies that

$$A_{11} > 0, \qquad A_{11}A_{22} - A_{12}^2 > 0, \qquad A_{66} > 0 \tag{4.9.4}$$

The nonzero surface tractions acting at a point x on the curve Γ are given by

$$s_\alpha = t_{\beta\alpha}n_\beta \tag{4.9.5}$$

where $n_\alpha = \cos(\mathbf{n}_x, x_\alpha)$, and $\mathbf{n}_x$ is the unit vector of the outward normal to Γ at x.

In the case of the first boundary-value problem, the boundary conditions are

$$u_\alpha = \widetilde{u}_\alpha \text{ on } \Gamma \tag{4.9.6}$$

The first boundary-value problem consists in the determination of the functions $u_\alpha \in C^2(\Sigma_1) \cap C^0(\overline{\Sigma}_1)$ that satisfy Equations 4.9.1, 4.9.2, and 4.9.3 on Σ_1 and the boundary conditions 4.9.6.

The second boundary-value problem is characterized by Equations 4.9.1, 4.9.2, and 4.9.3 the following boundary conditions

$$t_{\beta\alpha} n_\beta = \widetilde{t}_\alpha \text{ on } \Gamma \tag{4.9.7}$$

The plane strain problems for homogeneous and orthotropic bodies can be studied with the aid of the method of functions of complex variables [113,204]. For isotropic bodies, this method has been presented in Section 1.5. In this section, we present the method of potentials [194,196]. This method has been applied for anisotropic bodies by various authors. Here we present some of the results established by Basheleishvili and Kupradze [10] and Basheleishvili [13]. We note that the method of potentials is a constructive one.

The equations of equilibrium can be expressed in terms of displacement vector field,

$$\begin{aligned} A_{11} u_{1,11} + A_{66} u_{1,22} + (A_{12} + A_{66}) u_{2,12} + f_1 = 0 \\ (A_{12} + A_{66}) u_{1,12} + A_{66} u_{2,11} + A_{22} u_{2,22} + f_2 = 0 \end{aligned} \tag{4.9.8}$$

on Σ_1.

4.9.1 Galerkin Representation

We present a counterpart of the Boussinesq–Somigliana–Galerkin solution in the classical elastostatics. We introduce the notation

$$\mathfrak{M} = A_{11} A_{66} \frac{\partial^4}{\partial x_1^4} + (A_{11} A_{22} - A_{12}^2 - 2A_{12} A_{66}) \frac{\partial^4}{\partial x_1^2 \partial x_2^2} + A_{22} A_{66} \frac{\partial^4}{\partial x_2^4} \tag{4.9.9}$$

Theorem 4.9.1 *Let*

$$\begin{aligned} u_1 &= \left(A_{66} \frac{\partial^2}{\partial x_1^2} + A_{22} \frac{\partial^2}{\partial x_2^2} \right) G_1 - (A_{12} + A_{66}) \frac{\partial^2 G_2}{\partial x_1 \partial x_2} \\ u_2 &= -(A_{12} + A_{66}) \frac{\partial^2 G_1}{\partial x_1 \partial x_2} + \left(A_{11} \frac{\partial^2}{\partial x_1^2} + A_{66} \frac{\partial^2}{\partial x_2^2} \right) G_2 \end{aligned} \tag{4.9.10}$$

where the fields G_α of class C^4 satisfy the equations

$$\mathfrak{M} G_1 = -f_1, \qquad \mathfrak{M} G_2 = -f_2 \tag{4.9.11}$$

Then u_1 and u_2 satisfy Equations 4.9.8.

Proof. By Equations 4.9.10,

$$
\begin{aligned}
&A_{11}u_{1,11} + A_{66}u_{1,22} + (A_{12} + A_{66})u_{2,12} \\
&= \left(A_{11}\frac{\partial^2}{\partial x_1^2} + A_{66}\frac{\partial^2}{\partial x_2^2}\right)\left[\left(A_{66}\frac{\partial^2}{\partial x_1^2} + A_{22}\frac{\partial^2}{\partial x_2^2}\right)G_1\right. \\
&\quad \left. - (A_{12} + A_{66})\frac{\partial^2 G_2}{\partial x_1 \partial x_2}\right] \\
&\quad + (A_{12} + A_{66})\frac{\partial^2}{\partial x_1 \partial x_2}\left[\left(A_{11}\frac{\partial^2}{\partial x_1^2} + A_{66}\frac{\partial^2}{\partial x_2^2}\right)G_2\right. \\
&\quad \left. - (A_{12} + A_{66})\frac{\partial^2 G_1}{\partial x_1 \partial x_2}\right] = \mathfrak{M}G_1
\end{aligned}
\tag{4.9.12}
$$

Similarly, we get

$$
(A_{12} + A_{66})u_{1,12} + A_{66}u_{2,11} + A_{22}u_{2,22} = \mathfrak{M}G_2 \tag{4.9.13}
$$

From Equations 4.9.11, we obtain the desired result. □

4.9.2 Fundamental Solutions

We now apply the representation 4.9.10 to derive the fundamental solutions of the field equations 4.9.8. First, we assume that

$$
f_1 = \delta(x - y), \qquad f_2 = 0
$$

where $\delta(\cdot)$ is the Dirac delta and $y(y_\alpha)$ is a fixed point. If we take $G_1 = g$ and $G_2 = 0$, then Equations 4.9.11 are satisfied if g satisfies the equation

$$
\mathfrak{M}g = -\delta(x - y) \tag{4.9.14}
$$

From Equations 4.9.10, we obtain the following displacements

$$
u_1^{(1)} = A_{66}g_{,11} + A_{22}g_{,22}, \qquad u_2^{(1)} = -(A_{12} + A_{66})g_{,12} \tag{4.9.15}
$$

In the case of the following body forces

$$
f_1 = 0, \qquad f_2 = \delta(x - y)
$$

Equations 4.9.11 are satisfied if $G_1 = 0$ and $G_2 = g$. By Equations 4.9.10, we find the corresponding displacements

$$
u_1^{(2)} = -(A_{12} + A_{66})g_{,12}, \qquad u_2^{(2)} = A_{11}g_{,11} + A_{66}g_{,22} \tag{4.9.16}
$$

The functions $u_\alpha^{(\beta)}$ given by Equations 4.9.15 and 4.9.16 represent the fundamental solutions of the system 4.9.8.

In view of relations 4.9.4, we conclude that the system 4.9.8 is elliptic. We consider the characteristic equation

$$A_{22}A_{66}\alpha^4 + \left(A_{11}A_{22} - A_{12}^2 - 2A_{12}A_{66}\right)\alpha^2 + A_{11}A_{66} = 0 \qquad (4.9.17)$$

The roots of Equation 4.9.17 have the form

$$\alpha_\rho = a_\rho + ib_\rho, \qquad \overline{\alpha}_\rho = a_\rho - ib_\rho, \quad b_\rho > 0, \quad \rho = 1, 2 \qquad (4.9.18)$$

We assume that $\alpha_1 \neq \alpha_2$. The equality $\alpha_1 = \alpha_2$ holds only for isotropic bodies. We introduce the notations

$$\mathcal{A} = \begin{Vmatrix} 1 & \alpha_1 & \alpha_1^2 & \alpha_1^3 \\ 1 & \overline{\alpha}_1 & \overline{\alpha}_1^2 & \overline{\alpha}_1^3 \\ 1 & \alpha_2 & \alpha_2^2 & \alpha_2^3 \\ 1 & \overline{\alpha}_2 & \overline{\alpha}_2^2 & \overline{\alpha}_2^3 \end{Vmatrix}, \qquad d = \det \mathcal{A} \qquad (4.9.19)$$

and denote by d_k the cofactor of α_k^3 divided by d. Following Levi [208] (see also Kupradze [194]), the solution of Equation 4.9.14 is

$$g = a\Im m \sum_{k=1}^{2} d_k \sigma_k^2 \ln \sigma_k \qquad (4.9.20)$$

where

$$\sigma_k = x_1 - y_1 + \alpha_k(x_2 - y_2), \qquad a = -\frac{1}{2\pi A_{22}A_{66}} \qquad (4.9.21)$$

It is easy to verify that

$$\begin{gathered} d = -4b_1b_2\left[(a_1 - a_2)^2 + (b_1 - b_2)^2\right]\left[(a_1 - a_2)^2 + (b_1 + b_2)^2\right] \\ \sum_{k=1}^{2} d_k = -\frac{1}{2}iC, \qquad \sum_{k=1}^{2} \alpha_k d_k = -\frac{1}{2}iA, \qquad \sum_{k=1}^{2} \alpha_k^2 d_k = -\frac{1}{2}iB \\ A = 2(a_2b_1 + a_1b_2)\gamma, \qquad B = 2\left[b_1\left(a_2^2 + b_2^2\right) + b_2\left(a_1^2 + b_1^2\right)\right]\gamma \\ C = 2(b_1 + b_2)\gamma, \qquad \gamma^{-1} = 2b_1b_2\left[(a_1 - a_2)^2 + (b_1 + b_2)^2\right] \end{gathered} \qquad (4.9.22)$$

Let $\Gamma(x, y)$ be the matrix of fundamental solutions of the system 4.9.8

$$\Gamma(x, y) = \|\Gamma_{\alpha\beta}(x, y)\|_{2\times 2} \qquad (4.9.23)$$

where

$$\Gamma_{\alpha\beta} = u_\alpha^{(\beta)} \qquad (4.9.24)$$

Substituting function 4.9.20 into Equations 4.9.15 and 4.9.16, we find that

$$\Gamma(x, y) = \Im m \sum_{k=1}^{2} \begin{Vmatrix} A_k & B_k \\ B_k & C_k \end{Vmatrix} \ln \sigma_k \qquad (4.9.25)$$

where

$$\begin{aligned} A_k &= 2a\left(A_{22}\alpha_k^2 + A_{66}\right)d_k \\ B_k &= -2a(A_{12} + A_{66})\alpha_k d_k, \qquad C_k = 2a\left(A_{66}\alpha_k^2 + A_{11}\right)d_k \end{aligned} \tag{4.9.26}$$

We note that

$$A_k C_k - B_k^2 = 0, \quad (k = 1, 2) \tag{4.9.27}$$

Clearly, we have

$$\Gamma(x, y) = \Gamma^*(x, y) \tag{4.9.28}$$

where M^* is the transpose of the matrix M. If $x \neq y$, each column $\Gamma^{(s)}(x, y)$, $(s = 1, 2)$, of the matrix $\Gamma(x, y)$ satisfies at x the homogeneous system 4.9.8.

We introduce the matricial differential operator

$$D\left(\frac{\partial}{\partial x}\right) = \left\| D_{\alpha\beta}\left(\frac{\partial}{\partial x}\right) \right\|_{2\times 2} \tag{4.9.29}$$

where

$$\begin{aligned} D_{11}\left(\frac{\partial}{\partial x}\right) &= A_{11}\frac{\partial^2}{\partial x_1^2} + A_{66}\frac{\partial^2}{\partial x_2^2} \\ D_{12}\left(\frac{\partial}{\partial x}\right) &= D_{21}\left(\frac{\partial}{\partial x}\right) = A_{12}\frac{\partial^2}{\partial x_1 \partial x_2} \\ D_{22}\left(\frac{\partial}{\partial x}\right) &= A_{66}\frac{\partial^2}{\partial x_1^2} + A_{22}\frac{\partial^2}{\partial x_2^2} \end{aligned} \tag{4.9.30}$$

The system 4.9.8 can be written in matricial form. Following Kupradze [195], the vector $\mathbf{v} = (v_1, v_2, \ldots, v_m)$ shall be considered as a column matrix. Thus, the product of the matrix $A = \|a_{ij}\|_{m\times m}$ and the vector $\mathbf{v} = (v_1, v_2, \ldots, v_m)$ is an m-dimensional vector. The vector $\mathbf{v}$ multiplied by the matrix A will denote the matrix product between the row matrix $\|v_1, v_2, \ldots, v_m\|$ and the matrix A. We denote

$$u = (u_1, u_2), \qquad F = (f_1, f_2) \tag{4.9.31}$$

The system 4.9.8 can be written in the form

$$D\left(\frac{\partial}{\partial x}\right)u = -F \tag{4.9.32}$$

We introduce the matricial operator

$$H\left(\frac{\partial}{\partial x}, n_x\right) = \left\| H_{\alpha\beta}\left(\frac{\partial}{\partial x}, n_x\right) \right\|_{2\times 2} \tag{4.9.33}$$

where

$$\begin{aligned}
H_{11}\left(\frac{\partial}{\partial x}, n_x\right) &= A_{11}n_1\frac{\partial}{\partial x_1} + A_{66}n_2\frac{\partial}{\partial x_2}\\
H_{12}\left(\frac{\partial}{\partial x}, n_x\right) &= A_{66}n_2\frac{\partial}{\partial x_1} + A_{12}n_1\frac{\partial}{\partial x_2}\\
H_{21}\left(\frac{\partial}{\partial x}, n_x\right) &= A_{12}n_2\frac{\partial}{\partial x_1} + A_{66}n_1\frac{\partial}{\partial x_2}\\
H_{22}\left(\frac{\partial}{\partial x}, n_x\right) &= A_{66}n_1\frac{\partial}{\partial x_1} + A_{22}n_2\frac{\partial}{\partial x_2}
\end{aligned} \tag{4.9.34}$$

If we denote

$$T = (s_1, s_2) \tag{4.9.35}$$

then the relations 4.9.5 reduce to

$$T = H\left(\frac{\partial}{\partial x}, n_x\right)u \tag{4.9.36}$$

Let $H_i(\partial/\partial x, n_x)$ be the row matrix with the elements $H_{ij}(\partial/\partial x, n_x)$, $(i, j = 1, 2)$. Clearly,

$$s_\alpha = H_\alpha\left(\frac{\partial}{\partial x}, n_x\right)u \tag{4.9.37}$$

For convenience, we denote

$$T_\alpha^{(x)}u = H_\alpha\left(\frac{\partial}{\partial x}, n_x\right)u \tag{4.9.38}$$

Let us introduce the matrix

$$\mathscr{T}_y\Gamma(x, y) = H\left(\frac{\partial}{\partial y}, n_y\right)\Gamma(x, y) \tag{4.9.39}$$

If we use the relations

$$\begin{aligned}
A_{11}A_k + A_{12}\alpha_k B_k &= -\alpha_k A_{66}(A_k\alpha_k + B_k)\\
A_{11}B_k + A_{12}\alpha_k C_k &= -\alpha_k A_{66}(B_k\alpha_k + C_k)\\
A_{66}(B_k + A_k\alpha_k) &= -\alpha_k(A_{12}A_k + A_{22}B_k\alpha_k)\\
A_{66}(B_k\alpha_k + C_k) &= -\alpha_k(A_{12}B_k + A_{22}\alpha_k C_k)\\
\frac{\partial}{\partial s_y}\ln\sigma_k &= \frac{\partial \ln\sigma_k}{\partial y_2}\cos(n_y, x_1) - \frac{\partial \ln\sigma_k}{\partial y_1}\cos(n_y, x_2)\\
&= \frac{1}{\sigma_k}[\cos(n_y, x_2) - \alpha_k\cos(n_y, x_1)]
\end{aligned}$$

then we obtain

$$
\begin{aligned}
T_1^{(y)}\Gamma^{(1)}(x,y) &= \Im m \sum_{k=1}^{2} L_k \frac{\partial}{\partial s_y} \ln \sigma_k \\
T_2^{(y)}\Gamma^{(1)}(x,y) &= \Im m \sum_{k=1}^{2} M_k \frac{\partial}{\partial s_y} \ln \sigma_k \\
T_1^{(y)}\Gamma^{(2)}(x,y) &= \Im m \sum_{k=1}^{2} N_k \frac{\partial}{\partial s_y} \ln \sigma_k \\
T_2^{(y)}\Gamma^{(2)}(x,y) &= \Im m \sum_{k=1}^{2} P_k \frac{\partial}{\partial s_y} \ln \sigma_k
\end{aligned}
\tag{4.9.40}
$$

Here we have used the notations

$$
\begin{aligned}
L_k &= \frac{1}{2\pi}[2(b_1b_2 - a_1a_2)\alpha_k d_k - (-1)^k \alpha_k/(\alpha_1 - \alpha_2)], \qquad M_k = -L_k/\alpha_k \\
N_k &= \frac{1}{2\pi}[2(b_1b_2 - a_1a_2)\alpha_k^2 d_k - (-1)^k \alpha_1\alpha_2/(\alpha_1 - \alpha_2)], \qquad P_k = -N_k/\alpha_k
\end{aligned}
\tag{4.9.41}
$$

We denote by $\Lambda(x,y)$ the matrix obtained from Equation 4.9.39 by interchanging the rows and columns,

$$
\Lambda(x,y) = \left[H\left(\frac{\partial}{\partial y}, n_y\right) \Gamma(x,y) \right]^* \tag{4.9.42}
$$

In view of Equations 4.9.40, we can write

$$
\Lambda(x,y) = \Im m \sum_{k=1}^{2} \begin{Vmatrix} L_k & M_k \\ N_k & P_k \end{Vmatrix} \frac{\partial}{\partial s_y} \ln \sigma_k \tag{4.9.43}
$$

It follows from Equations 4.9.21, 4.9.22, 4.9.26, and 4.9.41 that

$$
\begin{aligned}
&\sum_{k=1}^{2} L_k = \frac{1}{2\pi}(1 - i\kappa A), \qquad \sum_{k=1}^{2} M_k = \frac{1}{2\pi} i\kappa C, \qquad \sum_{k=1}^{2} N_k = -\frac{1}{2\pi} i\kappa B \\
&\sum_{k=1}^{2} P_k = \frac{1}{2\pi}(1 + i\kappa A), \qquad \kappa = b_1b_2 - a_1a_2
\end{aligned}
\tag{4.9.44}
$$

where A, B, and C are defined by relations 4.9.22.

It is easy to verify that for $x \neq y$, each column of the matrix $\Lambda(x,y)$ satisfies at x the homogeneous system 4.9.8.

4.9.3 Somigliana Relations

Let us consider two states of plane strain for the domain Σ, characterized by the displacements $u_\alpha^{(\kappa)}$, the components of the strain tensor $e_{\alpha\beta}^{(\kappa)}$, and the components of the stress tensor $t^{(\kappa)}_{\alpha\beta}$, $(\kappa = 1, 2)$. We assume that the state $\{u^{(\kappa)}_{\alpha}, e^{(\kappa)}_{\alpha\beta}, t^{(\kappa)}_{\alpha\beta}\}$ corresponds to the body forces $f^{(\kappa)}_{\alpha}$. Thus, we have

$$e^{(\kappa)}_{\alpha\beta} = \frac{1}{2}\left(u^{(\kappa)}_{\alpha,\beta} + u^{(\kappa)}_{\beta,\alpha}\right), \qquad t^{(\kappa)}_{\beta\alpha,\beta} + f^{(\kappa)}_{\alpha} = 0$$
$$t^{(\kappa)}_{11} = A_{11}e^{(\kappa)}_{11} + A_{12}e^{(\kappa)}_{22}, \qquad t^{(\kappa)}_{12} = 2A_{66}e^{(\kappa)}_{12}, \qquad t^{(\kappa)}_{22} = A_{12}e^{(\kappa)}_{11} + A_{22}e^{(\kappa)}_{22} \tag{4.9.45}$$

on Σ, $(\kappa = 1, 2)$. If we denote

$$W_{\rho\kappa} = t^{(\rho)}_{\alpha\beta}e^{(\kappa)}_{\alpha\beta} \tag{4.9.46}$$

then, with the aid of the constitutive equations, we get

$$W_{12} = W_{21} \tag{4.9.47}$$

On the other hand, by using the strain–displacement relation and the equations of equilibrium, we find that

$$W_{\rho\kappa} = t^{(\rho)}_{\beta\alpha}u^{(\kappa)}_{\alpha,\beta} = \left(t^{(\rho)}_{\beta\alpha}u^{(\kappa)}_{\alpha}\right)_{,\beta} + f^{(\rho)}_{\alpha}u^{(\kappa)}_{\alpha} \tag{4.9.48}$$

If we integrate this relation over Σ and use the divergence theorem, then we obtain

$$\int_{\Sigma} W_{\rho\kappa}da = \int_{\partial\Sigma} t^{(\rho)}_{\beta\alpha}n_{\beta}u^{(\kappa)}_{\alpha}ds + \int_{\Sigma} f^{(\rho)}_{\alpha}u^{(\kappa)}_{\alpha}da \tag{4.9.49}$$

By Equations 4.9.47 and 4.9.49, we arrive at the following reciprocity relation

$$\int_{\Sigma} f^{(1)}_{\alpha}u^{(2)}_{\alpha}da + \int_{\partial\Sigma} t^{(1)}_{\beta\alpha}n_{\beta}u^{(2)}_{\alpha}ds = \int_{\Sigma} f^{(2)}_{\alpha}u^{(1)}_{\alpha}da + \int_{\partial\Sigma} t^{(2)}_{\beta\alpha}n_{\beta}u^{(1)}_{\alpha}ds \tag{4.9.50}$$

In the case of the plane strain of orthotropic bodies, the elastic potential is given by

$$2W_0 = A_{11}e^2_{11} + A_{22}e^2_{22} + 2A_{12}e_{11}e_{22} + 4A_{66}e^2_{12} \tag{4.9.51}$$

It follows from relations 4.9.4 that W_0 is a positive definite quadratic form. As in Section 1.5, we find that

$$2\int_{\Sigma} W_0 da = \int_{\Sigma} f_{\alpha}u_{\alpha}da + \int_{\partial\Sigma} t_{\beta\alpha}n_{\beta}u_{\alpha}ds \tag{4.9.52}$$

Thus we are led to the following theorem, the proof of which is strictly analogous to that given in Section 1.5.2.

Theorem 4.9.2 *Assume that relations* 4.9.4 *holds. Then*

(*i*) *The first boundary-value problem has at most one solution*

(*ii*) *Any two solutions of the second boundary-value problem are equal modulo a rigid displacement.*

Let Σ^+ be a domain in $\mathbb{R}^2$ bounded by a simple closed C^2-curve L, and $\Sigma^- = \mathbb{R}^2 \backslash \overline{\Sigma}^+$. Let $u = (u_1, u_2)$ and $v = (u_1', u_2')$ be two vector fields on Σ^+ such that $u, v \in C^2(\Sigma^+) \cap C^1(\overline{\Sigma}^+)$. The reciprocity relation 4.9.50 leads to

$$\begin{aligned}&\int_{\Sigma^+} \left[uD\left(\frac{\partial}{\partial x}\right) v - vD\left(\frac{\partial}{\partial x}\right) u \right] da \\ &\quad = \int_L \left[uH\left(\frac{\partial}{\partial x}, n_x\right) v - vH\left(\frac{\partial}{\partial x}, n_x\right) u \right] ds\end{aligned} \tag{4.9.53}$$

From Equation 4.9.52, we get

$$2\int_{\Sigma^+} W_0 da = -\int_{\Sigma_+} uD\left(\frac{\partial}{\partial x}\right) uda + \int_L uH\left(\frac{\partial}{\partial x}, n_x\right) uds \tag{4.9.54}$$

Let $\Sigma(y;\varepsilon)$ be the sphere with the center in y and radius ε. Let $y \in \Sigma^+$ and let ε be so small that $\Sigma(y;\varepsilon)$ be entirely contained in Σ^+. Then the relation 4.9.53 can be applied for the region $\Sigma^+ \backslash \Sigma(y;\varepsilon)$ to a regular vector field $u = (u_1, u_2)$ and to vector field $v(x) = \Gamma^{(s)}(x, y)$, $(s = 1, 2)$. We obtain the following representation of Somigliana type

$$\begin{aligned}u(y) &= \int_L \left\{ \Gamma^*(x,y) H\left(\frac{\partial}{\partial x}, n_x\right) u(x) - \left[H\left(\frac{\partial}{\partial x}, n_x\right) \Gamma(x,y)\right]^* u(x) \right\} ds_x \\ &\quad - \int_{\Sigma^+} \Gamma^*(x,y) D\left(\frac{\partial}{\partial x}\right) u(x) da_x\end{aligned} \tag{4.9.55}$$

In view of Equations 4.9.28 and 4.9.42, the relation 4.9.55 implies that

$$\begin{aligned}u(x) &= \int_L \left[\Gamma(x,y) H\left(\frac{\partial}{\partial y}, n_y\right) u(y) - \Lambda(x,y) u(y) \right] ds_y \\ &\quad - \int_{\Sigma^+} \Gamma(x,y) D\left(\frac{\partial}{\partial y}\right) u(y) da_y\end{aligned} \tag{4.9.56}$$

4.9.4 Existence Results

In what follows, we restrict our attention to the equation

$$D\left(\frac{\partial}{\partial x}\right) u = 0 \tag{4.9.57}$$

In this case, Equation 4.9.54 becomes

$$\int_L uH\left(\frac{\partial}{\partial x}, n_x\right) uds = 2\int_{\Sigma^+} W_0 ds \tag{4.9.58}$$

We say that the vector field $u = (u_1, u_2)$ is a regular solution of Equation 4.9.57 in Σ^+ if the formula 4.9.58 can be applied to u, and if u satisfies Equation 4.9.57 in Σ^+.

Let $x \in \Sigma^-$. We describe around x a circle C_R of sufficiently large radius R, containing the region Σ^+. We denote by Σ_R the region bounded by L and C_R. From Equations 4.9.54 and 4.9.57, we get

$$\int_{L+C_R} uH\left(\frac{\partial}{\partial x}, n_x\right)uds = 2\int_{\Sigma_R} W_0 da \tag{4.9.59}$$

If u satisfies the condition

$$\lim_{R\to\infty} R\int_0^{2\pi} uH\left(\frac{\partial}{\partial x}, n_x\right)ud\theta = 0 \tag{4.9.60}$$

then from Equation 4.9.59, we obtain

$$\int_L uH\left(\frac{\partial}{\partial x}, n_x\right)uds = -2\int_{\Sigma_R} W_0 da \tag{4.9.61}$$

We say that the vector field u is a regular solution of Equation 4.9.57 in Σ^- if formula 4.9.61 can be applied to u in Σ^-, and if u satisfies Equation 4.9.57 in Σ^- and the condition 4.9.60.

We consider the following boundary-value problems:

Interior problems. To find a regular solution in Σ^+ of Equation 4.9.57 satisfying one of the conditions

$$\lim_{x\to y} u(x) = f_1(y) \tag{I_1}$$

$$\lim_{x\to y} H\left(\frac{\partial}{\partial x}, n_x\right)u(x) = f_2(y) \tag{I_2}$$

where $x \in \Sigma^+, y \in L$, and f_1 and f_2 are prescribed vector fields.

Exterior problems. To find a regular solution in Σ^- of Equation 4.9.57 satisfying one of the conditions

$$\lim_{x\to y} u(x) = f_3(y) \tag{E_1}$$

$$\lim_{x\to y} H\left(\frac{\partial}{\partial x}, n_x\right)u(x) = f_4(y) \tag{E_2}$$

where $x \in \Sigma^-, y \in L$, and f_3 and f_4 are given.

We assume that f_1 and f_3 are Hölder continuously differentiable on L, and f_2 and f_4 are Hölder continuous on L.

We denote by (I_α^0) and (E_α^0) the homogeneous problems corresponding to (I_α) and (E_α), respectively. We introduce the potential of a single layer

$$V(x;\rho) = \int_L \Gamma(x,y)\rho(y)ds_y \tag{4.9.62}$$

and the potential of a double layer

$$W(x;\nu) = \int_L \Lambda(x,y)\nu(y)ds_y \tag{4.9.63}$$

where $\rho = (\rho_1, \rho_2)$ is Hölder continuous on L and $\nu = (\nu_1, \nu_2)$ is Hölder continuously differentiable on L. As in the classical theory of potentials [55,175], we have the following results.

Theorem 4.9.3 *The potential of a single layer is continuous on* $\mathbb{R}^2$.

Theorem 4.9.4 *The potential of a double layer has finite limits when the point x tends to $y \in L$ from both within and without, and these limits are respectively equal to*

$$\begin{aligned} W^+(y;\nu) &= -\frac{1}{2}\nu(y) + \int_L \Lambda(y,z)\nu(z)ds_z \\ W^-(y;\nu) &= \frac{1}{2}\nu(y) + \int_L \Lambda(y,z)\nu(z)ds_z \end{aligned} \tag{4.9.64}$$

Theorem 4.9.5 *$H(\partial/\partial x, n_x)V(x;\rho)$ tends to finite limits as the point x tends to the boundary point $y \in L$ from within or without, and these limits are respectively equal to*

$$\begin{aligned} \left[H\left(\frac{\partial}{\partial y}, n_y\right)V(y;\rho)\right]^+ &= \frac{1}{2}\rho(y) + \int_L \left[H\left(\frac{\partial}{\partial y}, n_y\right)\Gamma(y,z)\right]\rho(z)ds_z \\ \left[H\left(\frac{\partial}{\partial y}, n_y\right)V(y;\rho)\right]^- &= -\frac{1}{2}\rho(y) + \int_L \left[H\left(\frac{\partial}{\partial y}, n_y\right)\Gamma(y,z)\right]\rho(z)ds_z \end{aligned} \tag{4.9.65}$$

Theorem 4.9.6 *The potentials $V(x;\rho)$ and $W(x;\nu)$ satisfy Equation* 4.9.57 *on* $\Sigma^+ \cup \Sigma^-$.

We seek the solutions of the problems (I_1) and (E_1) in the form of a double-layer potential and the solutions of the problems (I_2) and (E_2) in the form of a single-layer potential. In view of Theorems 4.9.4 and 4.9.5, we obtain for the unknown densities the following singular integral equations

$$-\frac{1}{2}\nu(y) + \int_L \Lambda(y,z)\nu(z)ds_z = f_1(y) \tag{I_1}$$

$$\frac{1}{2}\rho(y) + \int_L \Lambda^*(y,z)\rho(z)ds_z = f_2(y) \tag{I_2}$$

$$\frac{1}{2}\nu(y) + \int_L \Lambda(y,z)\nu(z)ds_z = f_3(y) \tag{E_1}$$

$$-\frac{1}{2}\rho(y) + \int_L \Lambda^*(y,z)\nu(z)ds_z = f_4(y) \tag{E_2}$$

where $y \in L$. The homogeneous equations corresponding to equations (I_1), (I_2), (E_1), and (E_2) for $f_s = 0$, $(s = 1,2,3,4)$, will be denoted by (I_1^0), (I_2^0),

(E_1^0), and (E_2^0), respectively. The equations (I_1) and (E_2), (I_2) and (E_1) are pairwise mutually associate equations.

If we introduce the notations

$$\sigma = x_1 - y_1 + i(x_2 - y_2), \qquad r = [(x_1-y_1)^2 + (x_2-y_2)^2]^{1/2}, \qquad \mathbf{r} = (x_\alpha - y_\alpha)\mathbf{e}_\alpha$$

then we can write

$$\frac{\partial}{\partial s_y}\ln\sigma_k = \frac{\partial}{\partial s_y}\ln\frac{\sigma_k}{r} + \frac{\partial}{\partial s_y}\ln r = \frac{\partial}{\partial s_y}\ln r + \frac{i-\alpha_k}{\sigma\sigma_k} r\cos(\mathbf{r},\mathbf{n}_y) - \frac{i}{r}\cos(\mathbf{r},\mathbf{n}_y) \tag{4.9.66}$$

We note that

$$\frac{\partial}{\partial s_y}\ln r ds_y = \frac{dr}{r} = \frac{dt}{t-t_0} - id\theta \tag{4.9.67}$$

where t and t_0 are the affixes of the points y and x.

Taking into account Equations 4.9.66, 4.9.67, and 4.9.44 and pointing out the characteristic part of the singular operator, the system (I_1) can be written in the form

$$\nu(t_0) + \frac{1}{\pi}\begin{Vmatrix} -\kappa A & \kappa C \\ -\kappa B & \kappa A \end{Vmatrix}\int_L \frac{\nu(t)}{t-t_0}dt + \mathcal{K}(t_0) = -2f_1(t_0) \tag{4.9.68}$$

It is not difficult to prove that the index of the system 4.9.68 is zero [194]. Thus, the system (I_1) is a system of singular integral equations for which Fredholm's basic theorems are valid (cf. [196,242]). We note that the index of the system (I_2) is also zero [194].

Let us consider the problems (I_1) and (E_2). The homogeneous equations (I_1^0) and (E_2^0),

$$-\frac{1}{2}\nu(y) + \int_L \Lambda(y,z)\nu(z)ds_z = 0 \tag{I_1^0}$$

$$-\frac{1}{2}\rho(y) + \int_L \Lambda^*(z,y)\rho(z)ds_z = 0 \tag{E_2^0}$$

have only trivial solutions. We assume the opposite and suppose that ρ^0 is a solution of equation (E_2^0), not equal to zero. Then the single-layer potential

$$V(x;\rho^0) = \int_L \Gamma(x,y)\rho^0(y)ds_y$$

satisfies the condition

$$\left[H\left(\frac{\partial}{\partial y}, n_y\right)V(y;\rho^0)\right]^- = 0, \quad y \in L \tag{4.9.69}$$

When x tends to a point at infinity and y remains fixed on L, then $u_\alpha^{(\beta)}$ tends at infinity as $\delta_{\alpha\beta} \ln r$. If the density $\rho = (\rho_1, \rho_2)$ of the potential of a single layer satisfies the conditions

$$\int_L \rho_\alpha ds = 0, \quad (\alpha = 1, 2) \tag{4.9.70}$$

then the potential $V(x; \rho)$ satisfies the asymptotic relations

$$V = O(r^{-1}), \qquad \frac{\partial V}{\partial R} = O(r^{-2}) \text{ as } r \to \infty \tag{4.9.71}$$

where R is an arbitrary direction.

As in classical theory of potentials [55,172], we find that

$$\int_L H_\alpha \left(\frac{\partial}{\partial x}, n_x \right) \Gamma^{(\beta)}(x, y) ds_x = -\zeta(y)\delta_{\alpha\beta} \tag{4.9.72}$$

where

$$\zeta(y) = \begin{cases} 1, & y \in \Sigma^+ \\ \frac{1}{2}, & y \in L \\ 0, & y \in \Sigma^- \end{cases}$$

If we multiply the equation (E_2^0) by ds_y and integrate on L, on the basis of Equation 4.9.72, we obtain

$$\int_L \rho_\alpha^0(y) ds_y = 0, \quad (\alpha = 1, 2)$$

so that the potential $V(x; \rho^0)$ satisfies the relations 4.9.71. This fact implies that $V(x; \rho^0)$ satisfies the relation 4.9.60. Thus, we conclude that (i) $V(x; \rho^0)$ satisfies Equation 4.9.57 on Σ^- and the condition 4.9.69 on L; (ii) the formula 4.9.61 can be applied to $V(x; \rho^0)$; and (iii) $V(x; \rho^0)$ satisfies the asymptotic relations 4.9.71. It follows that

$$V(x; \rho^0) = 0 \text{ on } \Sigma^- \tag{4.9.73}$$

According to the continuity of the single-layer potential, we have

$$[V(x; \rho^0)]^+ = 0 \text{ on } L$$

Taking into account that $V(x; \rho^0)$ satisfies Equation 4.9.57 on Σ^+, from the uniqueness theorem, we get

$$V(x; \rho^0) = 0 \text{ on } \Sigma^+ \tag{4.9.74}$$

It follows from Equations 4.9.65, 4.9.73, and 4.9.74 that

$$\rho^0(y) = \left[H\left(\frac{\partial}{\partial y}, n_y\right)V(y;\rho^0)\right]^+ - \left[H\left(\frac{\partial}{\partial y}, n_y\right)V(y;\rho^0)\right]^- = 0$$

Thus, our statement concerning the equation (E_2^0) is valid.

Since the equations (I_1^0) and (E_2^0) form an associate set of integral equations, (I_1^0) has also no nontrivial solution. We note that from Equation 4.9.72 and the equation (E_2), with $f_4 = (f_{41}, f_{42})$, we obtain

$$-\int_L \rho_\alpha(y)ds_y = \int_L f_{4\alpha}ds, \quad (\alpha = 1, 2)$$

We have derived the following results.

Theorem 4.9.7 *The problem (I_1) has solution for any Hölder continuously differentiable vector field f_1. This solution is unique and can be expressed by a double-layer potential.*

Theorem 4.9.8 *The problem (E_2) can be solved if and only if*

$$\int_L f_{4\alpha}ds = 0, \quad (\alpha = 1, 2)$$

We now consider the equations (I_2^0) and (E_1^0). We note that the vector field

$$\omega(x) = (c_1 - c_3x_2, c_2 + c_3x_1)$$

where c_i are arbitrary constants, satisfies the boundary-value problem

$$D\left(\frac{\partial}{\partial x}\right)\omega(x) = 0, \quad x \in \Sigma^+, \qquad H\left(\frac{\partial}{\partial x}, n_x\right)\omega(x) = 0 \text{ on } L \qquad (4.9.75)$$

From Equation 4.9.56, we obtain

$$\omega(x) = -\int_L \Lambda(x, y)\omega(y)ds_y, \quad x \in \Sigma^+ \qquad (4.9.76)$$

Passing to the limit in Equation 4.9.76 as the point x approaches the boundary point $x_0 \in L$ from within, according to Equation 4.9.64, we get

$$\frac{1}{2}\omega(x_0) + \int_L \Lambda(x_0, y)\omega(y)ds_y = 0$$

Hence, the matrix $\omega(x)$ satisfies the equation (E_1^0). Clearly, the vector fields

$$\omega^{(1)} = (1, 0), \qquad \omega^{(2)} = (0, 1), \qquad \omega^{(3)} = (-x_2, x_1)$$

are linearly independent solutions of the equation (E_1^0). According to the second Fredholm theorem, the equation (I_2^0) has at least three linearly independent solutions $v^{(i)}$, $(i = 1, 2, 3)$. It is not difficult to prove that $v^{(i)}$ form a complete system of linearly independent solutions of the equation (I_2^0) [194]. This fact implies the completeness of the associate system $(\omega^{(1)}, \omega^{(2)}, \omega^{(3)})$. Hence, the necessary and sufficient conditions to solve the equation (I_2) have the form

$$\int_L \omega^{(j)}(x) f_2(x) ds_x = 0, \quad (j = 1, 2, 3) \tag{4.9.77}$$

If we take $f_2 = (\tilde{t}_1, \tilde{t}_2)$, then the condition 4.9.77 can be written in the form

$$\int_L \tilde{t}_\alpha ds = 0, \qquad \int_L (x_1 \tilde{t}_2 - x_2 \tilde{t}_1) ds = 0 \tag{4.9.78}$$

Thus, we have proved the following theorem.

Theorem 4.9.9 *The problem (I_2) can be solved if and only if the conditions 4.9.78 hold. The solution can be represented as a single-layer potential and is determined within an additive rigid-displacement vector field.*

In the same manner, we can study the problem (E_1). The method of potentials has been applied to study the plane strain problems for cylinders composed of different homogeneous and anisotropic materials [14,194,285,320]. The generalized plane strain problem for homogeneous elastic solids was investigated by this method in various works (see, for example, Ref. 35).

4.10 Deformation of Elastic Cylinders Composed of Nonhomogeneous and Anisotropic Materials

The results presented in Section 3.6 for elastic cylinders composed of different nonhomogeneous and isotropic materials can be extended to the case of anisotropic bodies. In this section, we study Saint-Venant's problem when cylinder B is composed of different nonhomogeneous and anisotropic materials. The results presented in this section have been established in Ref. 152.

We assume that B_ρ is occupied by an anisotropic material with the elastic coefficients $C_{ijkl}^{(\rho)}$. We consider nonhomogeneous bodies characterized by

$$C_{ijkl}^{(\rho)} = C_{ijkl}^{(\rho)}(x_1, x_2), \quad (x_1, x_2) \in A_\rho \tag{4.10.1}$$

The elastic coefficients are supposed to belong to C^∞, and the elastic potential corresponding to the material which occupies B_ρ is assumed to be a positive definite quadratic form. Saint-Venant's problem consists in the determination of a displacement vector field $\mathbf{u} \in C^2(B_1) \cap C^2(B_2) \cap C^1(\overline{B}_1) \cap C^1(\overline{B}_2) \cap C^0(B)$

that satisfies Equations 1.1.1, 1.1.2, and 1.1.8 on B_ρ, the conditions 3.1.1 on the surface of separation Π_0, the conditions for $x_3 = 0$, and the boundary conditions 1.3.1.

4.10.1 Generalized Plane Strain Problem for Composed Cylinders

We assume that cylinder B is composed of different nonhomogeneous and anisotropic materials which occupy the domains B_1 and B_2 introduced in Section 3.1. For the generalized plane strain of this cylinder, the displacement vector field has the form

$$u_j = u_j(x_1, x_2), \quad (x_1, x_2) \in \Sigma_1$$

We assume that the cylinder is subject to body forces $f_k^{(\rho)}$ and to surface tractions $p_k^{(\rho)}$, associated to domain B_ρ, and suppose that $f_k^{(\rho)}$ and $\widetilde{t}_k^{(\rho)}$ are independent of x_3. We restrict our attention to Neumann problem since this problem is involved in the solution of Saint-Venant's problem. In what follows, we assume that $f_i^{(\rho)}$ and $\widetilde{t}_i^{(\rho)}$ belong to C^∞.

The basic equations of the generalized plane strain problem consist of the constitutive equations

$$t_{i\alpha} = C_{i\alpha k\beta}^{(\rho)} u_{k,\beta} \tag{4.10.2}$$

and the equations of equilibrium

$$t_{\alpha i,\alpha} + f_i^{(\rho)} = 0 \tag{4.10.3}$$

on A_ρ. The conditions on the surface of separation reduce to

$$[u_i]_1 = [u_i]_2, \qquad [t_{\alpha i}]_1 n_\alpha^0 = [t_{\alpha i}]_2 n_\alpha^0 \text{ on } \Gamma_0 \tag{4.10.4}$$

The conditions on the lateral boundary take the form

$$[t_{\alpha i} n_\alpha]_\rho = \widetilde{t}_i^{(\rho)} \text{ on } \Gamma_\rho \tag{4.10.5}$$

Following Fichera [88], a solution $u_k \in C^\infty(\overline{A}_1) \cap C^\infty(\overline{A}_2) \cap C^0(\Sigma_1)$ of the generalized plane strain problem exists if and only if

$$\begin{gathered}\sum_{\rho=1}^{2} \left[\int_{A_\rho} f_i^{(\rho)} da + \int_{\Gamma_\rho} \widetilde{t}_i^{(\rho)} ds \right] = 0 \\ \sum_{\rho=1}^{2} \left[\int_{A_\rho} \varepsilon_{\alpha\beta} x_\alpha f_\beta^{(\rho)} da + \int_{\Gamma_\rho} \varepsilon_{\alpha\beta} x_\alpha \widetilde{t}_\beta^{(\rho)} ds \right] = 0\end{gathered} \tag{4.10.6}$$

It is easy to show that if the conditions 4.10.4 are replaced by

$$[u_i]_1 = [u_i]_2, \qquad [t_{\alpha i}]_1 n_\alpha^0 = [t_{\alpha i}]_2 n_\alpha^0 + g_i \text{ on } \Gamma_0 \tag{4.10.7}$$

where g_k are C^∞ functions, then the conditions 4.10.6 are replaced by

$$\sum_{\rho=1}^{2}\left[\int_{A_\rho} f_i^{(\rho)}da+\int_{\Gamma_\rho}\widetilde{t}_i^{(\rho)}ds\right]+\int_{\Gamma_0} g_i ds=0$$
$$\sum_{\rho=1}^{2}\left[\int_{A_\rho}\varepsilon_{\alpha\beta}x_\alpha f_\beta^{(\rho)}da+\int_{\Gamma_\rho}\varepsilon_{\alpha\beta}x_\alpha\widetilde{t}_\beta^{(\rho)}ds\right]+\int_{\Gamma_0}\varepsilon_{\alpha\beta}x_\alpha g_\beta ds=0 \tag{4.10.8}$$

We introduce the generalized plane strain problems $P^{(s)}$, $(s=1,2,3,4)$, characterized by the constitutive equations

$$\pi_{i\alpha}^{(s)}=C_{i\alpha k\beta}^{(\rho)}v_{k,\beta}^{(s)} \text{ on } A_\rho, \quad (s=1,2,3,4), \tag{4.10.9}$$

the equilibrium equations

$$\pi_{\alpha i,\alpha}^{(\beta)}+(C_{i\alpha 33}^{(\rho)}x_\beta)_{,\alpha}=0, \quad (\beta=1,2)$$
$$\pi_{\alpha i,\alpha}^{(3)}+C_{i\alpha 33,\alpha}^{(\rho)}=0 \tag{4.10.10}$$
$$\pi_{\alpha i,\alpha}^{(4)}-\varepsilon_{\eta\beta}(C_{i\kappa\eta 3}^{(\rho)}x_\beta)_{,\kappa}=0 \text{ on } A_\rho$$

and the following conditions

$$[v_i^{(s)}]_1=[v_i^{(s)}]_2, \qquad [\pi_{\alpha i}^{(s)}]_1 n_\alpha^0=[\pi_{\alpha i}^{(s)}]_2 n_\alpha^0+g_i^{(s)} \text{ on } \Gamma_0 \tag{4.10.11}$$

$$[\pi_{\alpha i}^{(\beta)}n_\alpha]_\rho=-C_{i\alpha 33}^{(\rho)}x_\beta n_\alpha, \quad (\beta=1,2), \qquad [\pi_{\alpha i}^{(3)}n_\alpha]_\rho=-C_{i\alpha 33}^{(\rho)}n_\alpha$$
$$[\pi_{\alpha i}^{(4)}n_\alpha]_\rho=\varepsilon_{\eta\beta}C_{i\alpha\eta 3}^{(\rho)}x_\beta n_\alpha \text{ on } \Gamma_\rho \tag{4.10.12}$$

where

$$g_i^{(\beta)}=(C_{i\alpha 33}^{(2)}-C_{i\alpha 33}^{(1)})x_\beta n_\alpha, \qquad g_i^{(3)}=(C_{i\alpha 33}^{(2)}-C_{i\alpha 33}^{(1)})n_\alpha$$
$$g_i^{(4)}=\varepsilon_{\beta\eta}(C_{i\alpha\eta 3}^{(2)}-C_{i\alpha\eta 3}^{(1)})x_\beta n_\alpha \tag{4.10.13}$$

It is easy to verify that the necessary and sufficient conditions 4.10.8 for the existence of the solution are satisfied for each boundary-value problem $P^{(s)}$.

4.10.2 Extension, Bending by Terminal Couples, and Torsion

We assume that the loading applied on the end Σ_1 is statically equivalent to a force $\mathbf{F}=F_3\mathbf{e}_3$ and a moment $\mathbf{M}=M_k\mathbf{e}_k$. Thus, the conditions for $x_3=0$ have the form

$$\int_{\Sigma_1} t_{\alpha 3}da=0 \tag{4.10.14}$$

$$\int_{\Sigma_1} t_{33}da=-F_3, \qquad \int_{\Sigma_1} x_\alpha t_{33}da=\varepsilon_{\alpha\beta}M_\beta$$
$$\int_{\Sigma_1}\varepsilon_{\alpha\beta}x_\alpha t_{\beta 3}da=-M_3 \tag{4.10.15}$$

The problem consists in the finding of a displacement vector field that satisfies Equations 1.1.1, 1.1.2, and 1.1.8 on B_ρ, the conditions 3.1.1 on Π_0, and the boundary conditions 1.3.1, 4.10.14, and 4.10.15.

We seek the solution in the form

$$\begin{aligned} u_\alpha &= -\frac{1}{2}a_\alpha x_3^2 - a_4 \varepsilon_{\alpha\beta} x_\beta x_3 + \sum_{s=1}^{4} a_s v_\alpha^{(s)} \\ u_3 &= (a_1 x_1 + a_2 x_2 + a_3) x_3 + \sum_{s=1}^{4} a_s v_3^{(s)} \end{aligned} \tag{4.10.16}$$

where $v_k^{(s)}$ are the components of the displacement vector in the problem $P^{(s)}$ and a_j are unknown constants. By Equation 4.10.16 and the constitutive equations, we get

$$t_{ij} = C_{ij33}^{(\rho)}(a_1 x_1 + a_2 x_2 + a_3) - C_{ij\alpha 3}^{(\rho)} \varepsilon_{\alpha\beta} a_4 x_\beta + \sum_{s=1}^{4} a_s \pi_{ij}^{(s)} \text{ on } B_\rho \tag{4.10.17}$$

where $\pi_{i\alpha}^{(s)}$ are given by Equation 4.10.9 and

$$\pi_{33}^{(s)} = C_{33i\beta}^{(\rho)} v_{i,\beta}^{(s)}$$

The equilibrium equations 1.1.8 and the boundary conditions 1.3.1 are satisfied on the basis of the relations 4.10.10 and 4.10.12. The conditions 3.1.1 are satisfied in view of relations 4.10.11.

We can show that, on the basis of the equilibrium equations and the conditions 3.1.1 and 1.3.1, the relation 1.3.57 also holds in the case of composed cylinders. Since t_{33} is independent of x_3, we conclude that the conditions 4.10.14 are identically satisfied. From Equations 4.10.15 and 4.10.17, we obtain the following system for the unknown constants

$$\sum_{s=1}^{4} L_{\alpha s}^0 a_s = \varepsilon_{\alpha\beta} M_\beta, \qquad \sum_{s=1}^{4} L_{3s}^0 a_s = -F_3, \qquad \sum_{s=1}^{4} L_{4s}^0 a_s = -M_3 \tag{4.10.18}$$

where we have used the notations

$$\begin{aligned} L_{\alpha\beta}^0 &= \sum_{\rho=1}^{2} \int_{A_\rho} x_\alpha (C_{3333}^{(\rho)} x_\beta + \pi_{33}^{(\beta)}) da \\ L_{\alpha 3}^0 &= \sum_{\rho=1}^{2} \int_{A_\rho} x_\alpha (C_{3333}^{(\rho)} + \pi_{33}^{(3)}) da, \ L_{3\alpha}^0 = \sum_{\rho=1}^{2} \int_{A_\rho} (C_{3333}^{(\rho)} x_\alpha + \pi_{33}^{(\alpha)}) da \\ L_{\alpha 4}^0 &= \sum_{\rho=1}^{2} \int_{A_\rho} x_\alpha (C_{33\eta 3}^{(\rho)} \varepsilon_{\beta\eta} x_\beta + \pi_{33}^{(4)}) da, \ L_{33}^0 = \sum_{\rho=1}^{2} \int_{A_\rho} (C_{3333}^{(\rho)} + \pi_{33}^{(3)}) da \\ L_{34}^0 &= \sum_{\rho=1}^{2} \int_{A_\rho} (C_{33\alpha 3}^{(\rho)} \varepsilon_{\beta\alpha} x_\beta + \pi_{33}^{(4)}) da \end{aligned} \tag{4.10.19}$$

$$L_{4\alpha}^{0} = \sum_{\rho=1}^{2} \int_{A_\rho} \varepsilon_{\eta\beta} x_\eta (C_{\beta 333}^{(\rho)} x_\alpha + \pi_{\beta 3}^{(\alpha)}) da$$

$$L_{43}^{0} = \sum_{\rho=1}^{2} \int_{A_\rho} \varepsilon_{\eta\beta} x_\eta (C_{\beta 333}^{(\rho)} + \pi_{\beta 3}^{(3)}) da$$

$$L_{44}^{0} = \sum_{\rho=1}^{2} \int_{A_\rho} \varepsilon_{\eta\beta} x_\eta (C_{\beta 3\nu 3}^{(\rho)} \varepsilon_{\lambda\nu} x_\lambda + \pi_{\beta 3}^{(4)}) da$$

Let us prove that the system 4.10.18 uniquely determines the constants a_k, $(k = 1, 2, 3, 4)$. We have assumed that the elastic potentials

$$W^{(\rho)}(\mathbf{u}) = \frac{1}{2} C_{ijrs}^{(\rho)} e_{ij}(\mathbf{u}) e_{rs}(\mathbf{u}) = \frac{1}{2} [t_{ij}(\mathbf{u}) e_{ij}(\mathbf{u})]_\rho$$

are positive definite quadratic forms in the variables $e_{rs}(\mathbf{u})$. Let $\mathbf{u}'$ and $\mathbf{u}''$ be displacement vector fields that satisfy Equations 1.1.1, 1.1.2, and 1.1.8 on B_ρ, and the conditions 3.1.1 on Π_0. If we denote

$$W^{(\rho)}(\mathbf{u}', \mathbf{u}'') = \frac{1}{2} C_{ijrs}^{(\rho)} e_{ij}(\mathbf{u}') e_{rs}(\mathbf{u}'')$$

then we can see that the relations 3.6.22, 3.6.23, and 3.6.25 hold.

The relations 4.10.16 and 4.10.17 can be written in the form

$$u_i = \sum_{s=1}^{4} a_s u_i^{(s)}, \qquad t_{ij} = \sum_{s=1}^{4} a_s t_{ij}^{(s)} \tag{4.10.20}$$

where

$$\begin{aligned} u_\alpha^{(\beta)} &= -\frac{1}{2} x_3^2 \delta_{\alpha\beta} + v_\alpha^{(\beta)}, \qquad u_3^{(\beta)} = x_3 x_\beta + v_3^{(\beta)}, \qquad u_i^{(3)} = \delta_{i3} x_3 + v_i^{(3)} \\ u_\alpha^{(4)} &= \varepsilon_{\beta\alpha} x_\beta x_3 + v_\alpha^{(4)}, \qquad u_3^{(4)} = v_3^{(4)} \\ t_{ij}^{(\alpha)} &= C_{ij33}^{(\rho)} x_\alpha + \pi_{ij}^{(\alpha)}, \qquad t_{ij}^{(3)} = C_{ij33}^{(\rho)} + \pi_{ij}^{(3)}, \qquad t_{ij}^{(4)} = \pi_{ij}^{(4)} - C_{ij\alpha 3}^{(\rho)} \varepsilon_{\alpha\beta} x_\beta \end{aligned} \tag{4.10.21}$$

on A_ρ. It follows from Equations 3.6.24 and 4.10.20 that

$$U(\mathbf{u}) = \sum_{r,s=1}^{4} \Lambda_{rs} a_r a_s \tag{4.10.22}$$

where

$$\Lambda_{rs} = \sum_{\rho=1}^{2} \int_{B_\rho} W^{(\rho)}(\mathbf{u}^{(r)}, \mathbf{u}^{(s)}) dv, \quad (r, s = 1, 2, 3, 4) \tag{4.10.23}$$

Clearly, we have

$$\begin{aligned} & t_{\alpha i,\alpha}^{(s)} = 0 \text{ on } A_\rho, \qquad [t_{\alpha i}^{(s)}]_1 n_\alpha^0 = [t_{\alpha i}^{(s)}]_2 n_\alpha^0 \text{ on } \Gamma_0 \\ & [t_{\alpha i}^{(s)} n_\alpha]_\rho = 0 \text{ on } \Gamma_\rho \end{aligned} \tag{4.10.24}$$

On the basis of Equations 4.10.24, we obtain

$$\int_{\Sigma_1} t_{\alpha 3}^{(s)} da = 0, \quad (s = 1, 2, 3, 4) \tag{4.10.25}$$

Let us apply the relations 3.6.23 and 3.6.24 to the displacement fields $u_i^{(s)}$ $(s = 1, 2, 3, 4)$. In view of Equations 4.10.23 and 4.10.17, we find that

$$2\Lambda_{rs} = hL_{rs}^0, \quad (r, s = 1, 2, 3, 4) \tag{4.10.26}$$

In view of Equations 4.10.22 and 4.10.26,

$$\det(L_{rs}^0) > 0 \tag{4.10.27}$$

In view of the relation 4.10.27, we conclude that the system 4.10.18 determines the constants a_k, $(k = 1, 2, 3, 4)$.

4.10.3 Flexure

We assume now that $\mathbf{F} = F_\alpha \mathbf{e}_\alpha$ and $\mathbf{M} = \mathbf{0}$. The conditions on the end located at $x_3 = 0$ are

$$\int_{\Sigma_1} t_{\alpha 3} da = -F_\alpha \tag{4.10.28}$$

$$\int_{\Sigma_1} t_{33} da = 0, \qquad \int_{\Sigma_1} x_\alpha t_{33} da = 0, \qquad \int_{\Sigma_1} \varepsilon_{\alpha\beta} x_\alpha t_{\beta 3} da = 0 \tag{4.10.29}$$

The flexure problem consists in the determination of a displacement vector field that satisfies Equations 1.1.1, 1.1.2, and 1.1.8 on B_ρ, the conditions 3.1.1 on Π_0, and the boundary conditions 1.3.1, 4.10.28, and 4.10.29. We seek the solution in the form

$$\begin{aligned} u_\alpha &= -\frac{1}{2}a_\alpha x_3^2 - a_4 \varepsilon_{\alpha\beta} x_\beta x_3 - \frac{1}{6} b_\alpha x_3^3 - \frac{1}{2} b_4 \varepsilon_{\alpha\beta} x_\beta x_3^2 \\ &\quad + \sum_{s=1}^{4} (a_s + b_s x_3) v_\alpha^{(s)} + v_\alpha(x_1, x_2) \\ u_3 &= (a_1 x_1 + a_2 x_2 + a_3) x_3 + \frac{1}{2}(b_1 x_1 + b_2 x_2 + b_3) x_3^2 \\ &\quad + \sum_{s=1}^{4} (a_s + b_s x_3) v_3^{(s)} + v_3(x_1, x_2) \end{aligned} \tag{4.10.30}$$

where $v_i^{(s)}$ are the components of displacement vector in the problem $P^{(s)}$, v_k are unknown functions, and a_s, b_s, $(s = 1, 2, 3, 4)$, are unknown constants.

It follows from Equations 1.1.9 and 4.10.30 that

$$t_{ij} = C^{(\rho)}_{ij33}[a_1x_1 + a_2x_2 + a_3 + (b_1x_1 + b_2x_2 + b_3)x_3] \\ - C^{(\rho)}_{ij\alpha 3}(a_4 + b_4x_3)\varepsilon_{\alpha\beta}x_\beta \\ + \sum_{s=1}^{4}(a_s + b_sx_3)\pi^{(s)}_{ij} + \sigma_{ij} + k^{(\rho)}_{ij} \text{ on } B_\rho \tag{4.10.31}$$

where

$$\sigma_{ij} = C^{(\rho)}_{ijk\alpha}v_{k,\alpha} \tag{4.10.32}$$

and

$$k^{(\rho)}_{ij} = \sum_{s=1}^{4} C^{(\rho)}_{ijk3}b_sv^{(s)}_k \tag{4.10.33}$$

With the aid of notations 4.10.21, the stress tensor can be written in the form

$$t_{ij} = \sum_{s=1}^{4}(a_s + x_3b_s)t^{(s)}_{ij} + \sigma_{ij} + k^{(\rho)}_{ij} \text{ on } B_\rho \tag{4.10.34}$$

In view of Equations 4.10.10, the equilibrium equations reduce to

$$\sigma_{\alpha i,\alpha} + F^{(\rho)}_i = 0 \text{ on } A_\rho \tag{4.10.35}$$

where

$$F^{(\rho)}_i = k^{(\rho)}_{i\alpha,\alpha} + \sum_{s=1}^{4} b_st^{(s)}_{i3} \tag{4.10.36}$$

On the basis of the relations 4.10.12, the conditions 1.3.1 become

$$[\sigma_{\alpha i}n_\alpha]_\rho = q^{(\rho)}_i \text{ on } L_\rho \tag{4.10.37}$$

where we have used the notations

$$q^{(\rho)}_i = -k^{(\rho)}_{\alpha i}n_\alpha \tag{4.10.38}$$

By Equations 4.10.31 and 4.10.10, we find that the conditions 3.1.1 reduce to

$$[v_i]_1 = [v_i]_2, \qquad [\sigma_{\alpha i}]_1n^0_\alpha = [\sigma_{\alpha i}]_2n^0_\alpha + p_i \text{ on } \Gamma_0 \tag{4.10.39}$$

where

$$p_i = (k^{(2)}_{i\alpha} - k^{(1)}_{i\alpha})n^0_\alpha \tag{4.10.40}$$

Thus, the functions v_i are the components of the displacement vector field in the generalized plane strain problem 4.10.32, 4.10.35, 4.10.37, and 4.10.39. The necessary and sufficient conditions to solve this problem are

$$\sum_{\rho=1}^{2}\left(\int_{A_\rho} F^{(\rho)}_i da + \int_{\Gamma_\rho} q^{(\rho)}_i ds\right) + \int_{\Gamma_0} p_i ds = 0 \\ \sum_{\rho=1}^{2}\left(\int_{A_\rho} \varepsilon_{\alpha\beta}x_\alpha F^{(\rho)}_\beta da + \int_{\Gamma_\rho} \varepsilon_{\alpha\beta}x_\alpha q^{(\rho)}_\beta ds\right) + \int_{\Gamma_0} \varepsilon_{\alpha\beta}x_\alpha p_\beta ds = 0 \tag{4.10.41}$$

The first two conditions 4.10.41 are satisfied on the basis of the relations 4.10.36, 4.10.38, 4.10.40, and 4.10.25. From the remaining conditions, we obtain

$$\sum_{s=1}^{4} L_{rs}^{0} b_s = 0, \quad (r = 3, 4) \tag{4.10.42}$$

where L_{rs}^{0} are defined in Equations 4.10.19.

By using Equations 4.10.31, 1.3.57, and 4.10.19, the conditions 4.10.28 reduce to

$$\sum_{s=1}^{4} L_{\alpha s}^{0} b_s = -F_\alpha \tag{4.10.43}$$

In view of relation 4.10.27, the system 4.10.42 and 4.10.43 uniquely determines the constants b_k, $(k = 1, 2, 3, 4)$. As the conditions 4.10.41 are satisfied, we can assume that the functions v_i are known.

It follows from Equations 4.10.31 and 4.10.29 that the constants a_s, $(s = 1, 2, 3, 4)$, satisfy the equations

$$\sum_{s=1}^{4} L_{rs}^{0} a_s = d_r \tag{4.10.44}$$

where

$$d_\alpha = -\sum_{\rho=1}^{2} \int_{A_\rho} x_\alpha \big(\sigma_{33} + k_{33}^{(\rho)}\big) da, \qquad d_3 = -\sum_{\rho=1}^{2} \int_{A_\rho} \big(\sigma_{33} + k_{33}^{(\rho)}\big) da$$

$$d_4 = -\sum_{\rho=1}^{2} \int_{A_\rho} \varepsilon_{\alpha\beta} x_\alpha \big(\sigma_{\beta 3} + k_{\beta 3}^{(\rho)}\big) da$$

Clearly, the system 4.10.44 can always be solved for a_r, $(r = 1, 2, 3, 4)$. Thus, the flexure problem is solved.

4.11 Cylinders Composed of Different Orthotropic Materials

The deformation of cylinders composed of different orthotropic and homogeneous elastic materials has been studied in various works [28,175,205,339]. In this section, we present the solution of Saint-Venant's problem when cylinder B is composed of different nonhomogeneous and orthotropic materials. We denote by $A_{ij}^{(\rho)}$ the elastic coefficients 4.8.1 of the material which occupies the domain B_ρ, and assume that

$$A_{ij}^{(\rho)} = A_{ij}^{(\rho)}(x_1, x_2), \quad (i, j = 1, 2, \ldots, 6), \quad (x_1, x_2) \in A_\rho \tag{4.11.1}$$

Saint-Venant's problem consists in the determination of a vector field $\mathbf{u} \in C^2(B_1) \cap C^2(B_2) \cap C^1(\overline{B}_1) \cap C^1(\overline{B}_2) \cap C^0(B)$ that satisfies Equations 1.1.1, 4.8.2, and 1.1.8 on B_ρ, the conditions 3.1.1 on the surface of separation Π_0, the conditions for $x_3 = 0$, and the boundary conditions 1.3.1.

4.11.1 Plane Strain Problem

The plane strain problem for homogeneous and orthotropic cylinders has been studied in Section 4.9. When cylinder B is composed of different materials, the equations of the plane strain problem consist of the equations of equilibrium

$$t_{\beta\alpha,\beta} + f_\alpha^{(\rho)} = 0 \tag{4.11.2}$$

the constitutive equations

$$t_{11} = A_{11}^{(\rho)} e_{11} + A_{12}^{(\rho)} e_{22}, \qquad t_{22} = A_{12}^{(\rho)} e_{11} + A_{22}^{(\rho)} e_{22}, \quad t_{12} = 2A_{66}^{(\rho)} e_{12} \tag{4.11.3}$$

and the geometrical equations

$$2e_{\alpha\beta} = u_{\beta,\alpha} + u_{\alpha,\beta} \tag{4.11.4}$$

on A_ρ. The conditions on the surface of separation become

$$[u_\alpha]_1 = [u_\alpha]_2, \qquad [t_{\alpha\beta}]_1 n_\alpha^0 = [t_{\alpha\beta}]_2 n_\alpha^0 \text{ on } \Gamma_0 \tag{4.11.5}$$

In the case of the second boundary-value problem, the boundary conditions are

$$[t_{\alpha\beta} n_\alpha]_\rho = \widetilde{t}_\beta^{(\rho)} \text{ on } \Gamma_\rho \tag{4.11.6}$$

We suppose that the body forces $f_\alpha^{(\rho)}$ and the surface forces $\widetilde{t}_\alpha^{(\rho)}$ are independent of x_3 and are prescribed functions of class C^∞. We continue to assume that $A_{rs}^{(\rho)}$ belongs to C^∞ and that the elastic potential corresponding to the material which occupies B_ρ is positive definite. If the domains A_1 and A_2 satisfy some conditions of regularity, then the second boundary-value problem has a solution $u_\alpha \in C^\infty(A_1 \cup \Gamma_1) \cap C^\infty(A_2 \cup \Gamma_2) \cap C^0(\Sigma_1)$ if and only if the functions $f_\alpha^{(\rho)}$ and $\widetilde{t}_\alpha^{(\rho)}$ satisfy the conditions 3.6.7 (cf. [88]). In what follows, we will have occasion to consider the boundary-value problem characterized by Equations 4.11.2, 4.11.3, and 4.11.4 on A_ρ, the boundary conditions 4.11.6, and the conditions

$$[u_\alpha]_1 = [u_\alpha]_2, \qquad [t_{\alpha\beta}]_1 n_\beta^0 = [t_{\alpha\beta}]_2 n_\beta^0 + g_\alpha \text{ on } \Gamma_0 \tag{4.11.7}$$

where g_α are prescribed functions of class C^∞. In this case, the necessary and sufficient conditions for the existence of the solution are given by the relations 3.6.9.

We introduce three plane strain problems $Q^{(k)}$, $(k = 1, 2, 3)$, characterized by the equations of equilibrium

$$t_{\beta\alpha,\beta}^{(k)} + F_{(k)\alpha}^{(\rho)} = 0 \tag{4.11.8}$$

the constitutive equations

$$t_{11}^{(k)} = A_{11}^{(\rho)} e_{11}^{(k)} + A_{12}^{(\rho)} e_{22}^{(k)}, \qquad t_{22}^{(k)} = A_{12}^{(\rho)} e_{11}^{(k)} + A_{22}^{(\rho)} e_{22}^{(k)}, \qquad t_{12}^{(k)} = 2A_{66}^{(\rho)} e_{12}^{(k)} \tag{4.11.9}$$

and the geometrical equations

$$2e_{\alpha\beta}^{(k)} = u_{\alpha,\beta}^{(k)} + u_{\beta,\alpha}^{(k)} \tag{4.11.10}$$

on A_ρ, where

$$\begin{aligned} &F_{(1)1}^{(\rho)} = (A_{13}^{(\rho)} x_1)_{,1}, \qquad F_{(1)2}^{(\rho)} = (A_{23}^{(\rho)} x_1)_{,2}, \qquad F_{(2)1}^{(\rho)} = (A_{13}^{(\rho)} x_2)_{,1} \\ &F_{(2)2}^{(\rho)} = (A_{23}^{(\rho)} x_2)_{,2}, \qquad F_{(3)1}^{(\rho)} = A_{13,1}^{(\rho)}, \qquad F_{(3)2}^{(\rho)} = A_{23,2}^{(\rho)} \text{ on } A_\rho \end{aligned} \tag{4.11.11}$$

To Equations 4.11.8, 4.11.9, and 4.11.10, we add the conditions

$$[u_\alpha^{(k)}]_1 = [u_\alpha^{(k)}]_2, \qquad [t_{\beta\alpha}^{(k)}]_1 n_\beta^0 = [t_{\beta\alpha}^{(k)}]_2 n_\beta^0 + G_\alpha^{(k)} \text{ on } \Gamma_0 \tag{4.11.12}$$

and

$$[t_{\beta\alpha}^{(k)} n_\beta]_\rho = \widetilde{T}_{(k)\alpha}^{(\rho)} \text{ on } \Gamma_\rho \tag{4.11.13}$$

where

$$\begin{aligned} &G_1^{(1)} = (A_{13}^{(2)} - A_{13}^{(1)}) x_1 n_1^0, \qquad G_2^{(1)} = (A_{23}^{(2)} - A_{23}^{(1)}) x_1 n_2^0 \\ &G_1^{(2)} = (A_{13}^{(2)} - A_{13}^{(1)}) x_2 n_1^0, \qquad G_2^{(2)} = (A_{23}^{(2)} - A_{23}^{(1)}) x_2 n_2^0 \\ &G_1^{(3)} = (A_{13}^{(2)} - A_{13}^{(1)}) n_1^0, \qquad G_2^{(3)} = (A_{23}^{(2)} - A_{23}^{(1)}) n_2^0 \\ &\widetilde{T}_{(1)1}^{(\rho)} = -A_{13}^{(\rho)} x_1 n_1, \qquad \widetilde{T}_{(1)2}^{(\rho)} = -A_{23}^{(\rho)} x_1 n_2, \qquad \widetilde{T}_{(2)1}^{(\rho)} = -A_{13}^{(\rho)} x_2 n_1 \\ &\widetilde{T}_{(2)2}^{(\rho)} = -A_{23}^{(\rho)} x_2 n_2, \qquad \widetilde{T}_{(3)1}^{(\rho)} = -A_{13}^{(\rho)} n_1, \qquad \widetilde{T}_{(3)2}^{(\rho)} = -A_{23}^{(\rho)} n_2 \end{aligned} \tag{4.11.14}$$

It is easy to prove that the necessary and sufficient conditions 3.6.9 for the existence of the solution are satisfied for each boundary-value problem $Q^{(k)}$, $(k = 1, 2, 3)$. We shall assume that the functions $u_\alpha^{(k)}$ are known.

4.11.2 Extension and Bending of Composed Cylinders

Let the loading applied at the end Σ_1 be statically equivalent to the force $\mathbf{F} = F_3 \mathbf{e}_3$ and the moment $\mathbf{M} = M_\alpha \mathbf{e}_\alpha$. The problem of extension and bending consists in the determination of a displacement vector field that satisfies Equations 1.1.1, 4.8.2, and 1.1.8 on B_ρ, the conditions 3.1.1 on Π_0, the boundary conditions 1.3.1 on Π, and the conditions 3.3.1, and 3.3.2 on Σ_1. We try to solve the problem assuming that

$$u_\alpha = -\frac{1}{2} c_\alpha x_3^2 + \sum_{j=1}^{3} c_j u_\alpha^{(j)}, \qquad u_3 = (c_\rho x_\rho + c_3) x_3 \tag{4.11.15}$$

where c_j are unknown constants and $u_\alpha^{(j)}$ are the components of the displacement vector in the problem $Q^{(j)}$, $(j = 1, 2, 3)$. By Equations 4.11.15 and 4.11.10, we obtain

$$e_{\alpha\beta} = \sum_{j=1}^{2} c_j e_{\alpha\beta}^{(j)}, \qquad e_{\alpha 3} = 0, \qquad e_{33} = c_\rho x_\rho + c_3$$

It follows from Equations 4.8.2 and 4.11.3 that

$$\begin{aligned}
t_{11} &= A_{13}^{(\rho)}(c_1 x_1 + c_2 x_2 + c_3) + \sum_{j=1}^{3} c_j t_{11}^{(j)} \\
t_{22} &= A_{23}^{(\rho)}(c_1 x_1 + c_2 x_2 + c_3) + \sum_{j=1}^{3} c_j t_{22}^{(j)} \\
t_{12} &= \sum_{j=1}^{3} c_j t_{12}^{(j)}, \qquad t_{\alpha 3} = 0 \\
t_{33} &= A_{33}^{(\rho)}(c_1 x_1 + c_2 x_2 + c_3) + \sum_{j=1}^{3} c_j \left(A_{13}^{(\rho)} e_{11}^{(j)} + A_{23}^{(\rho)} e_{22}^{(j)}\right) \text{ on } A_\rho
\end{aligned} \tag{4.11.16}$$

The equilibrium equations and the boundary conditions 1.3.1 are satisfied on the basis of the relations 4.11.8, 4.11.11, 4.11.13, and 4.11.14. The conditions 3.1.1 on the surface of separation are satisfied in view of the relations 4.11.12 and 4.11.14. The conditions 3.3.1 are identically satisfied. From Equations 3.3.2 and 4.11.16, we obtain the following equations for the unknown constants,

$$\Gamma_{\alpha j} c_j = \varepsilon_{\alpha\beta} M_\beta, \qquad \Gamma_{3j} c_j = -F_3 \tag{4.11.17}$$

where

$$\begin{aligned}
\Gamma_{\alpha\beta} &= \sum_{\rho=1}^{2} \int_{A_\rho} x_\alpha \left(A_{33}^{(\rho)} x_\beta + A_{13}^{(\rho)} e_{11}^{(\beta)} + A_{23}^{(\rho)} e_{22}^{(\beta)}\right) da \\
\Gamma_{\alpha 3} &= \sum_{\rho=1}^{2} \int_{A_\rho} x_\alpha \left(A_{33}^{(\rho)} + A_{13}^{(\rho)} e_{11}^{(3)} + A_{23}^{(\rho)} e_{22}^{(3)}\right) da \\
\Gamma_{3\alpha} &= \sum_{\rho=1}^{2} \int_{A_\rho} \left(A_{33}^{(\rho)} x_\alpha + A_{13}^{(\rho)} e_{11}^{(\alpha)} + A_{23}^{(\rho)} e_{22}^{(\alpha)}\right) da \\
\Gamma_{33} &= \sum_{\rho=1}^{2} \int_{A_\rho} \left(A_{33}^{(\rho)} + A_{13}^{(\rho)} e_{11}^{(3)} + A_{23}^{(\rho)} e_{22}^{(3)}\right) da
\end{aligned} \tag{4.11.18}$$

As in Section 4.3, we can prove that $\Gamma_{ij} = \Gamma_{ji}$ and that $\det(\Gamma_{ij}) \neq 0$. The system 4.11.17 determines the constants c_1, c_2, and c_3.

4.11.3 Flexure and Torsion

We assume now that $\mathbf{F} = F_\alpha \mathbf{e}_\alpha$ and $\mathbf{M} = M_3 \mathbf{e}_3$. The conditions on the end Σ_1 are given by Equations 1.4.1, 1.3.21, and 1.3.22. The problem consists in the finding of the functions u_k that satisfy Equations 1.1.1, 4.8.2, and 1.1.8 on B_ρ, the conditions 3.1.1 on Π_0, the boundary conditions 1.3.1, and the conditions 1.4.1, 1.3.21, and 1.3.22 on Σ_1. We seek the solution in the form

$$\begin{aligned} u_\alpha &= -\frac{1}{6} d_\alpha x_3^3 - \tau \varepsilon_{\alpha\beta} x_\beta x_3 + x_3 \sum_{j=1}^{3} d_j u_\alpha^{(j)} \\ u_3 &= \frac{1}{2}(d_1 x_1 + d_2 x_2 + d_3) x_3^2 + \Psi(x_1, x_2) \end{aligned} \tag{4.11.19}$$

where $u_\alpha^{(j)}$ are the solutions of the problems $Q^{(j)}$, Ψ is an unknown function, and d_k and τ are unknown constants. By Equations 4.11.19, 1.1.1, and 4.8.2, we obtain

$$\begin{aligned} t_{11} &= A_{13}^{(\rho)}(d_1 x_1 + d_2 x_2 + d_3) x_3 + x_3 \sum_{j=1}^{3} d_j t_{11}^{(j)} \\ t_{22} &= A_{23}^{(\rho)}(d_1 x_1 + d_2 x_2 + d_3) x_3 + x_3 \sum_{j=1}^{3} d_j t_{22}^{(j)}, \qquad t_{12} = x_3 \sum_{j=1}^{3} d_j t_{12}^{(j)} \\ t_{33} &= A_{33}^{(\rho)}(d_1 x_1 + d_2 x_2 + d_3) x_3 + x_3 \sum_{j=1}^{3} d_j (A_{13}^{(\rho)} e_{11}^{(j)} + A_{23}^{(\rho)} e_{22}^{(j)}) \\ t_{23} &= A_{44}^{(\rho)} \left(\Psi_{,2} + \tau x_1 + \sum_{j=1}^{3} d_j u_2^{(j)} \right) \\ t_{31} &= A_{55}^{(\rho)} \left(\Psi_{,1} - \tau x_2 + \sum_{j=1}^{3} d_j u_1^{(j)} \right) \text{ on } B_\rho \end{aligned} \tag{4.11.20}$$

The conditions 1.3.21 are identically satisfied. If we use Equations 4.11.20, 4.11.8, 4.11.11, and 4.11.14, then we find that the equations of equilibrium 1.1.8 and the conditions 3.1.1 and 1.3.1 are satisfied if the function Ψ satisfies the equation

$$\left(A_{55}^{(\rho)} \Psi_{,1}\right)_{,1} + \left(A_{44}^{(\rho)} \Psi_{,2}\right)_{,2} = -q^{(\rho)} \text{ on } A_\rho \tag{4.11.21}$$

and the conditions

$$\begin{aligned} [\Psi]_1 = [\Psi]_2,\ & \left[A_{55}^{(1)} \Psi_{,1} n_1^0 + A_{44}^{(1)} \Psi_{,2} n_2^0\right]_1 \\ &= \left[A_{55}^{(2)} \Psi_{,1} n_1^0 + A_{44}^{(2)} \Psi_{,2} n_2^0\right]_2 + h \text{ on } \Gamma_0 \\ A_{55}^{(\rho)} \Psi_{,1} n_1 + A_{44}^{(\rho)} & \Psi_{,2} n_2 = \kappa^{(\rho)} \text{ on } \Gamma_\rho \end{aligned} \tag{4.11.22}$$

where

$$
\begin{aligned}
q^{(\rho)} &= A_{33}^{(\rho)}(d_1x_1 + d_2x_2 + d_3) - \tau\big[\big(A_{55}^{(\rho)}x_2\big)_{,1} - \big(A_{44}^{(\rho)}x_1\big)_{,2}\big] \\
&\quad + \sum_{j=1}^{3} d_j\big[\big(A_{55}^{(\rho)}u_1^{(j)}\big)_{,1} + \big(A_{44}^{(\rho)}u_2^{(j)}\big)_{,2} + A_{13}^{(\rho)}e_{11}^{(j)} + A_{23}^{(\rho)}e_{22}^{(j)}\big] \\
h &= \tau\big[\big(A_{55}^{(1)} - A_{55}^{(2)}\big)x_2n_1^0 - \big(A_{44}^{(1)} - A_{44}^{(2)}\big)x_1n_2^0\big] \\
&\quad + \sum_{j=1}^{3} d_j\big[\big(A_{55}^{(2)} - A_{55}^{(1)}\big)u_1^{(j)}n_1^0 + \big(A_{44}^{(2)} - A_{44}^{(1)}\big)u_2^{(j)}n_2^0\big] \\
\kappa^{(\rho)} &= \tau\big(A_{55}^{(\rho)}x_2n_1 - A_{44}^{(\rho)}x_1n_2\big) \\
&\quad - \sum_{j=1}^{3} d_j\big[A_{55}^{(\rho)}u_1^{(j)}n_1 + A_{44}^{(\rho)}u_2^{(j)}n_2\big]
\end{aligned}
\tag{4.11.23}
$$

The necessary and sufficient condition for the existence of the solution of the boundary-value problem 4.11.21 and 4.11.22 is (cf. [55,88])

$$
\sum_{\rho=1}^{2}\left(\int_{A_\rho} q^{(\rho)}da + \int_{\Gamma_\rho} \kappa^{(\rho)}ds\right) + \int_{\Gamma_0} hds = 0 \tag{4.11.24}
$$

Substituting the relations 4.11.23 into Equation 4.11.24, we get

$$
\Gamma_{3j}d_j = 0 \tag{4.11.25}
$$

where Γ_{3j} are defined by Equations 4.11.18. In view of Equations 1.3.57, 4.11.20, and 4.11.18, we find that the conditions 1.4.1 reduce to

$$
\Gamma_{\alpha j}d_j = -F_\alpha \tag{4.11.26}
$$

The system 4.11.25 and 4.11.26 uniquely determines the constants d_k.

Let us introduce the function $\varphi \in C^2(A_1) \cap C^2(A_2) \cap C^1(\overline{A}_1) \cap C^1(\overline{A}_2) \cap C^0(\Sigma_1)$ which satisfies the equation

$$
\big(A_{55}^{(\rho)}\varphi_{,1}\big)_{,1} + \big(A_{44}^{(\rho)}\varphi_{,2}\big)_{,2} = \big(A_{55}^{(\rho)}x_2\big)_{,1} - \big(A_{44}^{(\rho)}x_1\big)_{,2} \text{ on } A_\rho \tag{4.11.27}
$$

and the conditions

$$
\begin{aligned}
[\varphi]_1 = [\varphi]_2, \qquad \big[A_{55}^{(1)}\varphi_{,1}n_1^0 + A_{44}^{(1)}\varphi_{,2}n_2^0\big]_1 &= \big[A_{55}^{(2)}\varphi_{,1}n_1^0 + A_{44}^{(2)}\varphi_{,2}n_2^0\big]_2 \\
&\quad + \big(A_{55}^{(1)} - A_{55}^{(2)}\big)x_2n_1^0 - \big(A_{44}^{(1)} - A_{44}^{(2)}\big)x_1n_2^0 \text{ on } \Gamma_0 \\
A_{55}^{(\rho)}\varphi_{,1}n_1 + A_{44}^{(\rho)}\varphi_{,2}n_2 &= A_{55}^{(\rho)}x_2n_1 - A_{44}^{(\rho)}x_1n_2 \text{ on } \Gamma_\rho
\end{aligned}
\tag{4.11.28}
$$

It is not difficult to verify that the necessary and sufficient condition for the existence of the function φ is satisfied. We introduce the function χ by

$$
\Psi = \tau\varphi + \chi \tag{4.11.29}
$$

In view of Equations 4.11.21, 4.11.22, 4.11.23, and 4.11.27, we find that the function χ satisfies the equation

$$\left(A_{55}^{(\rho)}\chi_{,1}\right)_{,1} + \left(A_{44}^{(\rho)}\chi_{,2}\right)_{,2} = g^{(s)} \text{ on } A_\rho \tag{4.11.30}$$

and the conditions

$$[\chi]_1 = [\chi]_2, \quad \left[A_{55}^{(1)}\chi_{,1}n_1^0 + A_{44}^{(1)}\chi_{,2}n_2^0\right]_1 = \left[A_{55}^{(2)}\chi_{,1}n_1^0 + A_{44}^{(2)}\chi_{,2}n_2^0\right]_2 + f \text{ on } \Gamma_0$$
$$A_{55}^{(\rho)}\chi_{,1}n_1 + A_{44}^{(\rho)}\chi_{,2}n_2 = k^{(\rho)} \text{ on } \Gamma_\rho \tag{4.11.31}$$

where

$$g^{(\rho)} = -A_{33}^{(\rho)}(d_1x_1 + d_2x_2 + d_3) - \sum_{j=1}^{3} d_j\Big[\left(A_{55}^{(\rho)}u_1^{(j)}\right)_{,1} + \left(A_{44}^{(\rho)}u_2^{(j)}\right)_{,2}$$
$$+ A_{13}^{(\rho)}e_{11}^{(j)} + A_{23}^{(\rho)}e_{22}^{(j)}\Big]$$
$$f = \sum_{j=1}^{3} d_j\left[\left(A_{55}^{(2)} - A_{55}^{(1)}\right)u_1^{(j)}n_1^0 + \left(A_{44}^{(2)} - A_{44}^{(1)}\right)u_2^{(j)}n_2^0\right]$$
$$k^{(\rho)} = -\sum_{j=1}^{3} d_j\left(A_{55}^{(\rho)}u_1^{(j)}n_1 + A_{44}^{(\rho)}u_2^{(j)}n_2\right) \tag{4.11.32}$$

It follows from Equations 4.11.20, 4.11.29, and 1.3.22 that

$$D^*\tau = -M_3 - \mathfrak{M} \tag{4.11.33}$$

where D^* is the torsional rigidity defined by

$$D^* = \sum_{\rho=1}^{2}\int_{A_\rho}\left[A_{44}^{(\rho)}x_1(\varphi_{,2} + x_1) - A_{55}^{(\rho)}x_2(\varphi_{,1} - x_2)\right]da \tag{4.11.34}$$

and $\mathfrak{M}$ is given by

$$\mathfrak{M} = \sum_{\rho=1}^{2}\int_{A_\rho}\left[x_1A_{44}^{(\rho)}\chi_{,2} - x_2A_{55}^{(\rho)}\chi_{,1} + \sum_{j=1}^{3} d_j\left(A_{44}^{(\rho)}x_1u_2^{(j)} - A_{55}^{(\rho)}x_2u_1^{(j)}\right)\right]da \tag{4.11.35}$$

The constant τ is determined by Equation 4.11.33.

If $F_\alpha = 0$, then from Equations 4.11.25 and 4.11.26, we obtain $d_k = 0$, so that $\chi = 0$ and $\mathfrak{M} = 0$. In this case, from Equation 4.11.19, we find the solution of the torsion problem.

4.12 Exercises

4.12.1 A homogeneous and orthotropic elastic cylinder occupies the domain

$$B = \left\{ x : \frac{x_1^2}{a^2} + \frac{x_2^2}{b^2} < 1, \quad 0 < x_3 < h \right\}, \qquad (a > 0, b > 0)$$

Investigate the torsion of the cylinder.

4.12.2 Study the flexure of an elliptical right cylinder made of a homogeneous and orthotropic elastic material.

4.12.3 Investigate the torsion of a right cylinder of rectangular cross section, composed of two homogeneous orthotropic elastic materials.

4.12.4 Study extension, bending, and torsion of a circular cylinder $B = \{x : x_1^2 + x_2^2 < a^2, 0 < x_3 < h\}$ made from a nonhomogeneous orthotropic material with the following constitutive coefficients

$$A_{ij} = A_{ij}^* e^{-\alpha r}, \quad \alpha > 0, \quad (i, j = 1, 2, \ldots, 6)$$

where A_{ij}^* and α are prescribed constants, and $r = (x_1^2 + x_2^2)^{1/2}$.

4.12.5 Study the deformation of a circular cylinder made of a homogeneous and orthotropic elastic material when the lateral boundary is subjected to a constant pressure.

4.12.6 Investigate the Almansi–Michell problem for homogeneous and orthotropic elastic cylinders.

4.12.7 Study the extension, bending, and torsion of an anisotropic elastic cylinder for the case when the medium is homogeneous and has a plane of elastic symmetry, normal to the axis of cylinder.

4.12.8 A homogeneous and orthotropic elliptical cylinder is in equilibrium in the absence of body forces. The cylinder is subjected on the lateral surface to the tractions $\widetilde{t}_1 = 0, \widetilde{t}_2 = 0, \widetilde{t}_3 = Px_3$, where P is a given constant. Investigate the deformation of the body.

4.12.9 Investigate the Almansi problem for inhomogeneous and orthotropic elastic cylinders.

Chapter 5

Cosserat Elastic Continua

5.1 Basic Equations

In a remarkable study, E. Cosserat and F. Cosserat [54] gave a systematic development of the mechanics of continuous media in which each point has the six degree of freedom of a rigid body. The orientation of a given particle of such a medium can be represented mathematically by the values of three mutually perpendicular unit vectors which Ericksen and Truesdell [76] called directors. In the 1960s, the subject matter was reopened in the works [81,110,228]. These early theories were discussed in Refs. 85,193,332. The Cosserat elastic continuum has been used as model for bones and for engineering materials like concrete and other composites (see [85] and references therein).

In this section, we present the basic equations of the linear theory of an elastic Cosserat medium. This theory is usually called *theory of micropolar elasticity* (cf. [83,254]). An account of the historical developments of the theory as well as references to various contributions may be found in the works by Eringen and Kafadar [84], Nowacki [255], Dyszlewicz [74], and Eringen [85]. In Chapters 5 and 6, we present a study of the deformation of right cylinders made of a Cosserat elastic material. The particular rod theory based on the concept of a Cosserat curve is not considered here. The reader interested in this subject will find an account in Ref. 284.

As remarked above, a Cosserat medium is a continuum in which each point has the degrees of freedom of a rigid body. Thus, the deformation of such a body is described by

$$\mathbf{u} = \mathbf{u}(\mathbf{x}, t), \qquad \boldsymbol{\varphi} = \boldsymbol{\varphi}(\mathbf{x}, t), \quad (\mathbf{x}, t) \in B \times I \tag{5.1.1}$$

where $\mathbf{u}$ is the displacement vector field, $\boldsymbol{\varphi}$ is the microrotation vector field, and I is a given interval of time. We consider an arbitrary region $\mathcal{P}$ of the continuum, bounded by a surface $\partial\mathcal{P}$ at time t, and we suppose that P is the corresponding region at time t_0, bounded by the surface ∂P. We postulate the energy balance in the form [85,332]

$$\begin{aligned}\int_P \rho_0(\dot{u}_i\ddot{u}_i + Y_{ij}\dot{\varphi}_i\ddot{\varphi}_j + \dot{e})dv \\ = \int_P \rho_0(\Phi_i\dot{u}_i + G_i\dot{\varphi}_i)dv + \int_{\partial P}(s_i\dot{u}_i + m_i\dot{\varphi}_i)da\end{aligned} \tag{5.1.2}$$

where ρ_0 is the density in the reference configuration, Y_{ij} are coefficients of inertia, e is the internal energy per unit mass, $\mathbf{\Phi}$ is the body force per unit mass, $\mathbf{G}$ is the body couple per unit mass, $\mathbf{s}$ is the stress vector associated with the surface $\partial\mathcal{P}$ but measured per unit area of the surface ∂P, $\mathbf{m}$ is the couple stress vector associated with the surface $\partial\mathcal{P}$ but measured per unit area of ∂P, and a superposed dot denotes the material derivative with respect to the time. We suppose that the body has arrived at a given state at a time t through some prescribed motion. Following Green and Rivlin [109], we consider a second motion which differs from the given motion only by a constant superposed rigid body translational velocity, the body occupying the same position at time t, and we assume that e, $\mathbf{\Phi}$, $\mathbf{G}, \mathbf{s}$, and $\mathbf{m}$ are unaltered by such superposed rigid velocity. If we use Equation 5.1.2 with $\dot{u}_i$ replaced by $\dot{u}_i + a_i$, where a_i is an arbitrary constant, we obtain

$$\int_P \rho_0 \ddot{u}_i dv = \int_P f_i dv + \int_{\partial P} s_i da \tag{5.1.3}$$

where $f_i = \rho_0 \Phi_i$. From Equation 5.1.3, by the usual methods, we obtain

$$s_i = t_{ji} n_j \tag{5.1.4}$$

and

$$t_{ji,j} + f_i = \rho_0 \ddot{u}_i \tag{5.1.5}$$

In view of Equations 5.1.4 and 5.1.5, the relation 5.1.2 reduces to

$$\int_P \rho_0 (\dot{e} + Y_{ij}\dot{\varphi}_i \ddot{\varphi}_j) dv = \int_P (g_i \dot{\varphi}_i + t_{ji}\dot{u}_{i,j}) dv + \int_{\partial P} m_i \dot{\varphi}_i da \tag{5.1.6}$$

where $g_i = \rho_0 G_i$. With an argument similar to that used in obtaining Equation 5.1.4, from Equation 5.1.6, we obtain

$$(m_i - m_{ji} n_j)\dot{\varphi}_i = 0 \tag{5.1.7}$$

where m_{ji} is the couple stress tensor. If we use Equation 5.1.7 in Equation 5.1.6 and apply the resulting equation to an arbitrary region, then we find the local form of the conservation of energy

$$\dot{W} = (m_{ji,j} + g_i - \rho_0 Y_{ij} \ddot{\varphi}_j)\dot{\varphi}_i + t_{ji}\dot{u}_{i,j} + m_{ji}\dot{\varphi}_{i,j} \tag{5.1.8}$$

where $W = \rho_0 e$. Let us now consider a motion of the body which differs from the given motion only by a superposed uniform rigid body angular velocity, the body occupying the same position at time t, and let us assume that $W, t_{ij}, m_{ij}, g_i - \rho_0 Y_{ij}\ddot{\varphi}_j$ are unaltered by such motion. In this case, $\dot{\varphi}_i$ are replaced by $\dot{\varphi}_i + b_i$, where b_i are arbitrary constants, and $\dot{u}_i$ are replaced by $\dot{u}_i + \varepsilon_{ijk} b_j x_k$, where ε_{ijk} is the alternating symbol. Equation 5.1.8 holds when $\dot{u}_{i,j}$ is replaced by $\dot{u}_{i,j} + \varepsilon_{jik} b_k$ and $\dot{\varphi}_i$ by $\dot{\varphi}_i + b_i$. It follows that

$$\dot{W} = (m_{ji,j} + g_i - \rho_0 Y_{ij}\ddot{\varphi}_j)(\dot{\varphi}_i + b_i) + t_{ji}(\dot{u}_{i,j} + \varepsilon_{jir} b_r) + m_{ji}\dot{\varphi}_{i,j}$$

With the help of Equation 5.1.8 and the arbitrariness of the constants b_k, we obtain

$$m_{ji,j} + \varepsilon_{irs}t_{rs} + g_i = \rho_0 Y_{ij}\ddot{\varphi}_j \tag{5.1.9}$$

By Equations 5.1.8 and 5.1.9, we get

$$\dot{W} = t_{ij}\dot{e}_{ij} + m_{ij}\dot{\kappa}_{ij} \tag{5.1.10}$$

where

$$e_{ij} = u_{j,i} + \varepsilon_{jik}\varphi_k, \qquad \kappa_{ij} = \varphi_{j,i} \tag{5.1.11}$$

We restrict our attention to the linear theory of elastic materials where the independent constitutive variables are e_{ij} and κ_{ij}. It is simple to see that e_{ij} and κ_{ij} are invariant under superposed rigid-body motions. We assume that W, t_{ij}, m_{ij}, and m_i are functions of e_{ij}, κ_{ij}, and x_m, consistent with the assumption of the linear theory, and that there is no internal constraint. From Equation 5.1.7, we obtain

$$m_i = m_{ji}n_j \tag{5.1.12}$$

On the basis of constitutive equations, from Equation 5.1.10, we find that

$$t_{ij} = \frac{\partial W}{\partial e_{ij}}, \qquad m_{ij} = \frac{\partial W}{\partial \kappa_{ij}} \tag{5.1.13}$$

In the linear theory, and assuming that the initial body is free from stresses and couple stresses, we have

$$W = \frac{1}{2}A_{ijrs}e_{ij}e_{rs} + B_{ijrs}e_{ij}\kappa_{rs} + \frac{1}{2}C_{ijrs}\kappa_{ij}\kappa_{rs} \tag{5.1.14}$$

where A_{ijrs}, B_{ijrs}, and C_{ijrs} are smooth functions on B which satisfy the symmetry relations

$$A_{ijrs} = A_{rsij}, \qquad C_{ijrs} = C_{rsij} \tag{5.1.15}$$

In the case of homogeneous bodies, the constitutive coefficients A_{ijrs}, B_{ijrs}, and C_{ijrs} are constants. By Equations 5.1.13 and 5.1.14, we find the following constitutive equations

$$\begin{aligned} t_{ij} &= A_{ijrs}e_{rs} + B_{ijrs}\kappa_{rs} \\ m_{ij} &= B_{rsij}e_{rs} + C_{ijrs}\kappa_{rs} \end{aligned} \tag{5.1.16}$$

In the case of an isotropic and centrosymmetric material, the constitutive equations 5.1.16 become

$$\begin{aligned} t_{ij} &= \lambda e_{rr}\delta_{ij} + (\mu + \kappa)e_{ij} + \mu e_{ji} \\ m_{ij} &= \alpha\kappa_{rr}\delta_{ij} + \beta\kappa_{ji} + \gamma\kappa_{ij} \end{aligned} \tag{5.1.17}$$

where $\lambda, \mu, \kappa, \alpha, \beta$, and γ are constitutive coefficients.

If we assume that W is a positive definite quadratic form in the variables e_{ij} and κ_{ij}, then we find that the constitutive coefficients of an isotropic body satisfy the inequalities [85]

$$\begin{aligned} &3\lambda + 2\mu + \kappa > 0, \qquad 2\mu + \kappa > 0, \qquad \kappa > 0 \\ &3\alpha + \beta + \gamma > 0, \qquad \gamma + \beta > 0, \qquad \gamma - \beta > 0 \end{aligned} \tag{5.1.18}$$

In what follows, we restrict our attention to the linear theory of equilibrium. The basic equations of the theory of elastostatics consist of the equations of equilibrium

$$t_{ji,j} + f_i = 0, \qquad m_{ji,j} + \varepsilon_{irs} t_{rs} + g_i = 0 \text{ on } B \tag{5.1.19}$$

the constitutive equations 5.1.16, and the geometrical equations 5.1.11. To these equations, we adjoin boundary conditions. In the *first boundary-value problem*, the boundary conditions are

$$u_i = \widetilde{u}_i, \qquad \varphi_i = \widetilde{\varphi}_i \text{ on } \partial B \tag{5.1.20}$$

where $\widetilde{u}_i$ and $\widetilde{\varphi}_i$ are given. The *second boundary-value problem* is characterized by the boundary conditions

$$t_{ji} n_j = \widetilde{t}_i, \qquad m_{ji} n_j = \widetilde{m}_i \text{ on } \partial B \tag{5.1.21}$$

where $\widetilde{t}_i$ and $\widetilde{m}_i$ are prescribed functions.

We assume that (i) B is a bounded regular region; (ii) f_i and g_i are continuous on $\overline{B}$; (iii) A_{ijrs}, B_{ijrs}, and C_{ijrs} are smooth on $\overline{B}$ and satisfy the conditions 5.1.15; (iv) $\widetilde{t}_i$ and $\widetilde{m}_i$ are piecewise regular on ∂B; and (v) $\widetilde{u}_i$ and $\widetilde{\varphi}_i$ are continuous on ∂B.

The first boundary-value problem consists in finding the functions $u_i, \varphi_i \in C^2(B) \cap C^0(\overline{B})$ that satisfy the Equations 5.1.11, 5.1.16, and 5.1.19 on B, and the boundary conditions 5.1.20 on ∂B. The second boundary-value problem consists in the determination of the functions $u_i, \varphi_i \in C^2(B) \cap C^1(\overline{B})$ that satisfy the Equations 5.1.11, 5.1.16, and 5.1.19 on B, and the boundary conditions 5.1.21.

The existence and uniqueness of solutions of the boundary-value problems of linear elastostatics have been studied in various works [126,164,196]. We recall the following existence result.

Theorem 5.1.1 *Assume that W is positive definite and that the hypotheses (i)–(iv) hold. Then the second boundary-value problem has solution if and only if*

$$\int_B f_i dv + \int_{\partial B} \widetilde{t}_i da = 0, \qquad \int_B (\varepsilon_{irs} x_r f_s + g_i) dv + \int_{\partial B} (\varepsilon_{irs} x_r \widetilde{t}_s + \widetilde{m}_i) da = 0 \tag{5.1.22}$$

Any two solutions of the second boundary-value problem are equal, modulo a rigid deformation.

We note that a rigid deformation has the form

$$u_i = a_i + \varepsilon_{ijr} b_j x_r, \qquad \varphi_i = b_i$$

where a_j and b_j are constants.

We assume for the remainder of this chapter that the material is homogeneous and isotropic, and that the elastic potential W is positive definite.

5.2 Plane Strain

With a view toward deriving a solution of Saint-Venant's problem, we present some results concerning the plane deformation of homogeneous and isotropic elastic cylinders. Throughout this section, we assume that the region B refers to the interior of the right cylinder considered in Section 1.2. We suppose that the vector fields $\mathbf{f}, \mathbf{g}, \widetilde{\mathbf{u}}, \widetilde{\varphi}, \widetilde{\mathbf{t}}$, and $\widetilde{\mathbf{m}}$ are independent of the axial coordinate, and parallel to the x_1, x_2-plane. The state of plane strain of the cylinder B is characterized by

$$u_\alpha = u_\alpha(x_1, x_2), \qquad u_3 = 0, \qquad \varphi_\alpha = 0, \qquad \varphi_3 = \varphi_3(x_1, x_2), \quad (x_1, x_2) \in \Sigma_1 \tag{5.2.1}$$

These restrictions, in conjunction with the constitutive equations, imply that the stress tensor and couple stress tensor are independent of the axial coordinate. It follows from Equations 5.1.11 and 5.2.1 that

$$e_{\alpha\beta} = u_{\beta,\alpha} + \varepsilon_{\beta\alpha}\varphi_3, \qquad \kappa_{\alpha 3} = \varphi_{3,\alpha} \tag{5.2.2}$$

The constitutive equations 5.1.17 show that nonzero components of the stress tensor and couple stress tensor are $t_{\alpha\beta}, m_{\alpha 3}, t_{33}$, and $m_{3\alpha}$. Further,

$$t_{\alpha\beta} = \lambda e_{\rho\rho}\delta_{\alpha\beta} + (\mu + \kappa)e_{\alpha\beta} + \mu e_{\beta\alpha}, \qquad m_{\alpha 3} = \gamma\kappa_{\alpha 3} \tag{5.2.3}$$

The equations of equilibrium 5.1.19 reduce to

$$t_{\beta\alpha,\beta} + f_\alpha = 0, \qquad m_{\alpha 3,\alpha} + \varepsilon_{\alpha\beta} t_{\alpha\beta} + g_3 = 0 \tag{5.2.4}$$

on Σ_1. The nonzero surface loads acting at a regular point x on the curve Γ are given by

$$s_\alpha = t_{\beta\alpha} n_\beta, \qquad m_3 = m_{\alpha 3} n_\alpha \tag{5.2.5}$$

where $n_\alpha = \cos(n_x, x_\alpha)$ and $\mathbf{n}_x$ is the unit vector of the outward normal to Γ at x.

By Equations 5.2.2 and 5.2.3, we can express the equations of equilibrium of homogeneous and isotropic solids in terms of the functions u_α and φ_3,

$$\begin{aligned}(\mu+\kappa)\Delta u_\alpha + (\lambda+\mu)u_{\beta,\beta\alpha} + \kappa\varepsilon_{\alpha\beta}\varphi_{3,\beta} &= -f_\alpha \\ \gamma\Delta\varphi_3 + \kappa\varepsilon_{\alpha\beta}u_{\beta,\alpha} - 2\kappa\varphi_3 &= -g_3\end{aligned} \tag{5.2.6}$$

on Σ_1. In the case of the first boundary-value problem, the boundary conditions are

$$u_\alpha = \widetilde{u}_\alpha, \qquad \varphi_3 = \widetilde{\varphi}_3 \text{ on } \Gamma \tag{5.2.7}$$

where $\widetilde{u}_\alpha$ and $\widetilde{\varphi}_3$ are prescribed functions. The second boundary-value problem is characterized by the boundary conditions

$$t_{\beta\alpha}n_\beta = \widetilde{t}_\alpha, \qquad m_{\alpha 3}n_\alpha = \widetilde{m}_3 \text{ on } \Gamma \tag{5.2.8}$$

where $\widetilde{t}_\alpha$ and $\widetilde{m}_3$ are given.

The functions t_{33} and $m_{3\alpha}$ can be determined after the functions u_α and φ_3 are found.

5.2.1 Polar Coordinates

In the solution of various boundary-value problems, it is convenient to employ the polar coordinates (r,θ), such that $x_1 = r\cos\theta, x_2 = r\sin\theta$. The equations of equilibrium can be written in the form

$$\begin{aligned}&\frac{\partial t_{rr}}{\partial r} + \frac{1}{r}\frac{\partial t_{\theta r}}{\partial \theta} + \frac{1}{r}(t_{rr} - t_{\theta\theta}) + f_r = 0 \\ &\frac{\partial t_{r\theta}}{\partial r} + \frac{1}{r}\frac{\partial t_{\theta\theta}}{\partial \theta} + \frac{1}{r}(t_{r\theta} + t_{\theta r}) + f_\theta = 0 \\ &\frac{\partial m_{rz}}{\partial r} + \frac{1}{r}\frac{\partial m_{\theta z}}{\partial \theta} + \frac{1}{r}m_{rz} + t_{r\theta} - t_{\theta r} + g_3 = 0\end{aligned} \tag{5.2.9}$$

on Σ_1, where $t_{rr}, t_{\theta\theta}, t_{r\theta}$, and $t_{\theta r}$ are the physical components of the stress tensor, m_{rz} and $m_{\theta z}$ are physical components of couple stress tensor, and f_r and f_θ are the physical components of the body force. The constitutive equations become

$$\begin{aligned}&t_{rr} = (\lambda+2\mu+\kappa)e_{rr} + \lambda e_{\theta\theta}, \qquad t_{r\theta} = (\mu+\kappa)e_{r\theta} + \mu e_{\theta r} \\ &t_{\theta\theta} = \lambda e_{rr} + (\lambda+2\mu+\kappa)e_{\theta\theta}, \qquad t_{\theta r} = (\mu+\kappa)e_{\theta r} + \mu e_{r\theta} \\ &m_{rz} = \gamma\kappa_{rz}, \qquad m_{\theta z} = \gamma\kappa_{\theta z}\end{aligned} \tag{5.2.10}$$

where

$$\begin{aligned}&e_{rr} = \frac{\partial u_r}{\partial r}, \qquad e_{\theta\theta} = \frac{1}{r}\left(\frac{\partial u_\theta}{\partial \theta} + u_r\right), \qquad e_{r\theta} = \frac{\partial u_\theta}{\partial r} - \varphi_3 \\ &e_{\theta r} = \frac{1}{r}\left(\frac{\partial u_r}{\partial \theta} - u_\theta\right) + \varphi_3, \qquad \kappa_{\theta z} = \frac{1}{r}\frac{\partial \varphi_3}{\partial \theta}, \qquad \kappa_{rz} = \frac{\partial \varphi_3}{\partial r}\end{aligned} \tag{5.2.11}$$

Here, u_r and u_θ are the physical components of the displacement vector field, so that $u_r + iu_\theta = (u_1 + iu_2)e^{-i\theta}$. The equations of equilibrium 5.2.9 can be expressed in terms of the functions u_r, u_θ, and φ_3. Thus we obtain

$$
\begin{aligned}
&(\lambda+2\mu+\kappa)\left(\frac{\partial^2 u_r}{\partial r^2}+\frac{1}{r}\frac{\partial u_r}{\partial r}-\frac{1}{r^2}u_r+\frac{1}{r}\frac{\partial^2 u_\theta}{\partial r\partial\theta}-\frac{1}{r^2}\frac{\partial u_\theta}{\partial\theta}\right)\\
&\quad-(\mu+\kappa)\frac{1}{r}\left(\frac{\partial^2 u_\theta}{\partial r\partial\theta}+\frac{1}{r}\frac{\partial u_\theta}{\partial\theta}-\frac{1}{r}\frac{\partial^2 u_r}{\partial\theta^2}\right)+\kappa\frac{1}{r}\frac{\partial\varphi_3}{\partial\theta}+f_r=0\\
&(\lambda+2\mu+\kappa)\left(\frac{1}{r^2}\frac{\partial^2 u_\theta}{\partial\theta^2}+\frac{1}{r}\frac{\partial^2 u_r}{\partial r\partial\theta}+\frac{1}{r^2}\frac{\partial u_r}{\partial\theta}\right)-(\mu+\kappa)\\
&\quad\times\left(\frac{1}{r}\frac{\partial^2 u_r}{\partial r\partial\theta}-\frac{1}{r^2}\frac{\partial u_r}{\partial\theta}-\frac{\partial^2 u_\theta}{\partial r^2}-\frac{1}{r}\frac{\partial u_\theta}{\partial r}+\frac{1}{r^2}u_\theta\right)-\kappa\frac{\partial\varphi_3}{\partial r}+f_\theta=0\\
&\gamma\left(\frac{\partial^2\varphi_3}{\partial r^2}+\frac{1}{r}\frac{\partial\varphi_3}{\partial r}+\frac{1}{r^2}\frac{\partial^2\varphi_3}{\partial\theta^2}\right)+\kappa\left(\frac{\partial u_\theta}{\partial r}+\frac{1}{r}u_\theta-\frac{1}{r}\frac{\partial u_r}{\partial\theta}-2\varphi_3\right)+g_3=0
\end{aligned}
\tag{5.2.12}
$$

Let $\widetilde{t}_r$ and $\widetilde{t}_\theta$ be the physical components of the given traction vector. Then the conditions 5.2.8 can be expressed in the form

$$
\begin{aligned}
t_{rr}n_r + t_{\theta r}n_\theta = \widetilde{t}_r, \qquad t_{r\theta}n_r + t_{\theta\theta}n_\theta = \widetilde{t}_\theta\\
m_{rz}n_r + m_{\theta z}n_\theta = \widetilde{m}_3 \text{ on } \Gamma
\end{aligned}
\tag{5.2.13}
$$

where n_r and n_θ are physical components of the vector $\mathbf{n}$.

The plane strain problems for homogeneous and isotropic bodies can be studied with the aid of the method of functions of complex variables [8]. In this section, we use the method of potentials [140,195] to derive existence and uniqueness results.

5.2.2 Solution of Field Equations

We now give a representation of solutions of Equations 5.2.6. We introduce the operator

$$\Omega = \frac{1}{b}\Delta\Delta(\Delta - \kappa^2) \tag{5.2.14}$$

where

$$b^{-1} = \gamma(\mu+\kappa)(\lambda+2\mu+\kappa), \qquad k = \left[\frac{\kappa(2\mu+\kappa)}{\gamma(\mu+\kappa)}\right]^{1/2} \tag{5.2.15}$$

We note that the relations 5.1.18 imply that b and k^2 are strictly positive.

Theorem 5.2.1 *Let*

$$
\begin{aligned}
u_\alpha &= (\lambda+2\mu+\kappa)\Delta(\gamma\Delta - 2\kappa)G_\alpha - [\gamma(\lambda+\mu)\Delta\\
&\quad -\kappa(2\lambda+2\mu+\kappa)]G_{\beta,\beta\alpha} + \kappa(\lambda+2\mu+\kappa)\varepsilon_{\beta\alpha}\Delta G_{3,\beta}\\
\varphi_3 &= (\lambda+2\mu+\kappa)[\kappa\varepsilon_{\alpha\beta}\Delta G_{\alpha,\beta} + (\mu+\kappa)\Delta\Delta G_3]
\end{aligned}
\tag{5.2.16}
$$

where G_j are fields of class C^6 on Σ_1 that satisfy the equations

$$\Omega G_\alpha = -f_\alpha, \qquad \Omega G_3 = -g_3 \tag{5.2.17}$$

Then u_α and φ_3 satisfy Equations 5.2.6 on Σ_1.

Proof. In view of Equations 5.2.16, we find that

$$\begin{aligned}
&(\mu+\kappa)\Delta u_\alpha + (\lambda+\mu)u_{\beta,\beta\alpha} + \kappa\varepsilon_{\alpha\beta}\varphi_{3,\beta}\\
&\quad = (\mu+\kappa)(\lambda+2\mu+\kappa)\Delta\Delta(\gamma\Delta-2\kappa)G_\alpha - (\mu+\kappa)[\gamma(\lambda+\mu)\Delta\Delta\\
&\qquad - \kappa(2\lambda+2\mu+\kappa)\Delta]G_{\beta,\beta\alpha} + \kappa(\lambda+2\mu+\kappa)(\mu+\kappa)\varepsilon_{\beta\alpha}\Delta\Delta G_{3,\beta}\\
&\qquad + (\lambda+\mu)(\lambda+2\mu+\kappa)\Delta(\gamma\Delta-2\kappa)G_{\beta,\beta\alpha} - (\lambda+\mu)[\gamma(\lambda+\mu)\Delta\\
&\qquad - \kappa(2\lambda+2\mu+\kappa)]\Delta G_{\beta,\beta\alpha} + \kappa(\lambda+2\mu+\kappa)[\kappa\Delta G_{\alpha,\beta\beta}\\
&\qquad - \kappa\Delta G_{\beta,\alpha\beta} + \varepsilon_{\alpha\beta}(\mu+\kappa)\Delta\Delta G_{3,\beta}]\\
&\quad = \gamma(\mu+\kappa)(\lambda+2\mu+\kappa)\Delta\Delta(\Delta-k^2)G_\alpha
\end{aligned} \tag{5.2.18}$$

Similarly, we obtain

$$\begin{aligned}
&\gamma\Delta\varphi_3 + \kappa\varepsilon_{\alpha\beta}u_{\beta,\alpha} - 2\kappa\varphi_3\\
&\quad = \gamma(\lambda+2\mu+\kappa)[\kappa\varepsilon_{\alpha\beta}\Delta\Delta G_{\alpha,\beta} + (\mu+\kappa)\Delta\Delta\Delta G_3]\\
&\qquad + \kappa(\lambda+2\mu+\kappa)\Delta(\gamma\Delta-2\kappa)\varepsilon_{\alpha\beta}G_{\beta,\alpha} + \kappa^2(\lambda+2\mu+\kappa)\Delta G_{3,\alpha\alpha}\\
&\qquad - 2\kappa(\lambda+2\mu+\kappa)[\kappa\varepsilon_{\alpha\beta}\Delta G_{\alpha,\beta} + (\mu+\kappa)\Delta\Delta G_3]\\
&\quad = \gamma(\lambda+2\mu+\kappa)(\mu+\kappa)(\Delta-k^2)\Delta\Delta G_3
\end{aligned} \tag{5.2.19}$$

In view of Equation 5.2.17, from Equations 5.2.18 and 5.2.19, we obtain the desired result. □

5.2.3 Fundamental Solutions

We use Theorem 5.2.1 to establish the fundamental solutions of Equations 5.2.6. First, we assume that

$$f_1 = \delta(x-y), \qquad f_2 = 0, \qquad g_3 = 0$$

where $\delta(\cdot)$ is the Dirac measure and $y(y_\alpha)$ is a fixed point. In this case, we take in Equations 5.2.16, $G_1 = f, G_2 = 0$, and $G_3 = 0$. From Equations 5.2.17, it follows that the function f satisfies the equation

$$\Delta\Delta(\Delta-k^2)f = -b\delta(x-y) \tag{5.2.20}$$

In general, if $f_\alpha = \delta_{\alpha\beta}\delta(x-y), g_3 = 0$, then we take $G_\alpha = f\delta_{\alpha\beta}$, $G_3 = 0$, where f is a solution of Equation 5.2.20. In this case, from Equations 5.2.16, we obtain the functions $u_\alpha^{(\beta)}$ and $\varphi_3^{(\beta)}$. If we assume that

$$f_\alpha = 0, \qquad g_3 = \delta(x-y)$$

then we take $G_\alpha = 0, G_3 = f$, where f satisfies Equation 5.2.20. We denote by $u_\alpha^{(3)}$ and $\varphi_3^{(3)}$ the functions resulting from Equations 5.2.16 when $G_\alpha = 0$ and $G_3 = f$. Thus, we obtain

$$\begin{aligned} u_\alpha^{(\beta)} &= \delta_{\alpha\beta}(\lambda + 2\mu + \kappa)\Delta(\gamma\Delta - 2\kappa)f - [\gamma(\lambda + \mu)\Delta - \kappa(2\lambda + 2\mu + \kappa)]f_{,\alpha\beta} \\ \varphi_3^{(\beta)} &= (\lambda + 2\mu + \kappa)\kappa\varepsilon_{\beta\alpha}\Delta f_{,\alpha} \\ u_\alpha^{(3)} &= \kappa(\lambda + 2\mu + \kappa)\varepsilon_{\beta\alpha}\Delta f_{,\beta} \\ \varphi_3^{(3)} &= (\lambda + 2\mu + \kappa)(\mu + \kappa)\Delta\Delta f \end{aligned} \tag{5.2.21}$$

The functions $u_\alpha^{(j)}$ and $\varphi_3^{(j)}$ represent the fundamental solutions of the system 5.2.6.

Let us study Equation 5.2.20. We note that if the functions H_k satisfy the equations

$$\Delta\Delta H_1 = S, \qquad \Delta H_2 = S, \qquad (\Delta - k^2)H_3 = S$$

then the solution of the equation

$$\Delta\Delta(\Delta - k^2)H = S$$

can be written in the form

$$H = -\frac{1}{k^4}(k^2 H_1 + H_2 - H_3) \tag{5.2.22}$$

If $S = -b\delta(x - y)$, then

$$\begin{gathered} H_1 = -\frac{b}{8\pi}r^2 \ln r, \qquad H_2 = -\frac{b}{2\pi}\ln r, \qquad H_3 = \frac{b}{2\pi}K_0(kr) \\ r = [(x_1 - y_1)^2 + (x_2 - y_2)^2]^{1/2} \end{gathered} \tag{5.2.23}$$

where K_n is the modified Bessel function of order n. It follows from Equations 5.2.22 and 5.2.23 that the solution of Equation 5.2.20 is given by

$$f = \frac{b}{8\pi k^4}\left[k^2 r^2 \ln r + 4 \ln r + 4K_0(kr)\right] \tag{5.2.24}$$

Let us note that

$$\begin{aligned} \frac{\partial^2}{\partial x_\alpha \partial x_\beta} &= (x_\alpha - y_\alpha)(x_\beta - y_\beta)r^{-2}\frac{d^2}{dr^2} \\ &\quad + [\delta_{\alpha\beta}r^2 - (x_\alpha - y_\alpha)(x_\beta - y_\beta)]r^{-3}\frac{d}{dr} \end{aligned} \tag{5.2.25}$$

$$\frac{d}{dr}K_0(kr) = -kK_1(kr), \qquad \frac{d^2}{dr^2}K_0(kr) = k^2 K_0(kr) + kr^{-1}K_1(kr)$$

Moreover, for $x \neq y$, we have

$$\begin{aligned} f_{,\alpha} &= \frac{b}{8\pi k^4}(x_\alpha - y_\alpha)\{k^2(1+2\ln r) + 4r^{-2}[1 - krK_1(kr)]\} \\ f_{,\alpha\beta} &= \frac{b}{8\pi k^4}\{k^2[2\delta_{\alpha\beta}\ln r + \delta_{\alpha\beta} + 2r^{-2}(x_\alpha - y_\alpha)(x_\beta - y_\beta)] \\ &\quad + 4r^{-4}[r^2\delta_{\alpha\beta} - 2(x_\alpha - y_\alpha)(x_\beta - y_\beta)][1 - rkK_1(kr)] \\ &\quad + 4k^2r^{-2}(x_\alpha - y_\alpha)(x_\beta - y_\beta)K_0(kr)\} \\ \Delta f &= \frac{b}{2\pi k^2}[1 + K_0(kr) + \ln r], \qquad \Delta\Delta f = \frac{b}{2\pi}K_0(kr) \end{aligned} \tag{5.2.26}$$

We have the following expansions in series

$$\begin{aligned} K_0(x) &= -\ln x - \frac{1}{4}x^2\ln x - \frac{1}{64}x^4\ln x - \cdots \\ K_1(x) &= \frac{1}{x} + \frac{1}{2}x\ln x + \frac{1}{16}x^3\ln x + \cdots \end{aligned} \tag{5.2.27}$$

Let $\Gamma(x,y)$ be the matrix of fundamental solutions

$$\Gamma(x,y) = \|\Gamma_{mn}(x,y)\|_{3\times 3} \tag{5.2.28}$$

where

$$\Gamma_{\alpha\beta} = u_\alpha^{(\beta)}, \qquad \Gamma_{\alpha 3} = u_\alpha^{(3)}, \qquad \Gamma_{3k} = \varphi_3^{(k)} \tag{5.2.29}$$

We note that

$$\Gamma(x,y) = \Gamma^*(y,x) \tag{5.2.30}$$

We write A^* for the transpose of A. Let us denote by $\Gamma^{(k)}$, $(k = 1,2,3)$, the columns of the matrix $\Gamma(x,y)$.

It follows from Equations 5.2.21 and 5.2.24 that

$$\Gamma = -\frac{1}{2\pi}\begin{Vmatrix} d & 0 & 0 \\ 0 & d & 0 \\ 0 & 0 & \gamma^{-1} \end{Vmatrix}\ln r + \Omega, \qquad d = \frac{\lambda + 3\mu + \kappa}{2(\lambda + 2\mu + \kappa)(\mu + \kappa)} \tag{5.2.31}$$

where we have pointed out the terms with singularities.

We introduce the matricial differential operator

$$D\left(\frac{\partial}{\partial x}\right) = \left\|D_{ij}\left(\frac{\partial}{\partial x}\right)\right\|_{3\times 3} \tag{5.2.32}$$

where

$$\begin{aligned} D_{\alpha\beta}\left(\frac{\partial}{\partial x}\right) &= (\mu + \kappa)\delta_{\alpha\beta}\Delta + (\lambda + \mu)\frac{\partial^2}{\partial x_\alpha \partial x_\beta} \\ D_{\alpha 3}\left(\frac{\partial}{\partial x}\right) &= \kappa\varepsilon_{\alpha\beta}\frac{\partial}{\partial x_\beta} \\ D_{3\beta}\left(\frac{\partial}{\partial x}\right) &= \kappa\varepsilon_{\rho\beta}\frac{\partial}{\partial x_\rho}, \qquad D_{33}\left(\frac{\partial}{\partial x}\right) = \gamma\Delta - 2\kappa \end{aligned} \tag{5.2.33}$$

The system 5.2.6 can be written in matricial form. As in Ref. 195, the vector $\mathbf{v} = (v_1, v_2, \ldots, v_m)$ shall be considered as a column matrix so that the product of the matrix $A = \|a_{ij}\|_{m\times m}$ and the vector $\mathbf{v}$ is an m-dimensional vector. The vector $\mathbf{v}$ multiplied by the matrix A will denote the matrix product between the row matrix $\|v_1, v_2, \ldots, v_m\|$ and the matrix A. We denote

$$u = (u_1, u_2, \varphi_3), \qquad F = -(f_1, f_2, g_3) \tag{5.2.34}$$

Equations 5.2.6 can be written in the form

$$D\left(\frac{\partial}{\partial x}\right) u = F \tag{5.2.35}$$

We introduce the matricial operator

$$T\left(\frac{\partial}{\partial x}, n_x\right) = \left\| T_{ij}\left(\frac{\partial}{\partial x}, n_x\right)\right\|_{3\times 3} \tag{5.2.36}$$

where

$$\begin{aligned} T_{\alpha\beta}\left(\frac{\partial}{\partial x}, n_x\right) &= (\mu+\kappa)\delta_{\alpha\beta}\frac{\partial}{\partial n_x} + \left(\lambda n_\alpha \frac{\partial}{\partial x_\beta} + \mu n_\beta \frac{\partial}{\partial x_\alpha}\right) \\ T_{\alpha 3}\left(\frac{\partial}{\partial x}, n_x\right) &= \kappa\varepsilon_{\alpha\beta}n_\beta, \qquad T_{3\alpha}\left(\frac{\partial}{\partial x}, n_x\right) = 0, \qquad T_{33}\left(\frac{\partial}{\partial x}, n_x\right) = \gamma\frac{\partial}{\partial n_x} \end{aligned} \tag{5.2.37}$$

If we denote

$$\mathcal{T} = (t_1, t_2, m_3) \tag{5.2.38}$$

then the relations 5.2.5 can be written as

$$\mathcal{T} = T\left(\frac{\partial}{\partial x}, n_x\right) u \tag{5.2.39}$$

Let $T_i(\partial/\partial x, n_x)$ be the row matrix with the elements $T_{ij}(\partial/\partial x, n_x)$. The relations 5.2.5 become

$$t_\alpha = T_\alpha\left(\frac{\partial}{\partial x}, n_x\right) u, \qquad m_3 = T_3\left(\frac{\partial}{\partial x}, n_x\right) u \tag{5.2.40}$$

We denote by $\Lambda(x,y)$ the matrix obtained from $T(\partial/\partial x, n_x)\Gamma(x,y)$ by interchanging the rows and columns and replacing x by y, that is,

$$\Lambda(x,y) = \left[T\left(\frac{\partial}{\partial y}, n_y\right)\Gamma(y,x)\right]^* \tag{5.2.41}$$

We can verify that

$$D\left(\frac{\partial}{\partial x}\right)\Lambda(x,y) = 0 \quad \text{for } x \neq y \tag{5.2.42}$$

It follows from Equations 5.2.21, 5.2.24, 5.2.28, and 5.2.36 that

$$\Lambda = M + Z, \qquad M = \|M_{ij}\|_{3\times 3}, \qquad Z = \|Z_{ij}\|_{3\times 3} \tag{5.2.43}$$

where

$$\begin{aligned}
&M_{11} = M_{22} = M_{33} = -\frac{1}{2\pi}\frac{\partial}{\partial n_y}(\ln r)\\
&M_{12} = -M_{21} = -\frac{c}{2\pi}\frac{d}{ds_y}(\ln r), \qquad M_{\alpha 3} = M_{3\alpha} = 0\\
&Z_{ij} = O(\ln r) \quad \text{as } r \to 0\\
&c = \frac{2\mu^2 + \mu\kappa - \lambda\kappa}{2(\lambda + 2\mu + \kappa)(\mu + \kappa)}, \qquad \frac{d}{ds_x} = \frac{\partial}{\partial x_2}n_1 - \frac{\partial}{\partial x_1}n_2
\end{aligned} \tag{5.2.44}$$

If $x \neq y$, then each column $\Gamma^{(j)}(x,y)$, $(j=1,2,3)$, of the matrix $\Gamma(x,y)$ satisfies at x the homogeneous system 5.2.6, that is,

$$D\left(\frac{\partial}{\partial x}\right)\Gamma(x,y) = 0 \tag{5.2.45}$$

5.2.4 Somigliana Relations

We consider two states of plane strain defined on the domain Σ and characterized by the displacements $u_\alpha^{(\rho)}$, the microrotations $\varphi_3^{(\rho)}$, the strain measures $e_{\alpha\beta}^{(\rho)}$ and $\kappa_{\alpha 3}^{(\rho)}$, the components of the stress tensor $t_{\alpha\beta}^{(\rho)}$, and the components of the couple stress tensor $m_{\alpha 3}^{(\rho)}$, $(\rho = 1,2)$. We assume that the state $A^{(\rho)} = \{u_\alpha^{(\rho)}, \varphi_3^{(\rho)}, e_{\alpha\beta}^{(\rho)}, \kappa_{\alpha 3}^{(\rho)}, t_{\alpha\beta}^{(\rho)}, m_{\alpha 3}^{(\rho)}\}$ corresponds to the body loads $I^{(\rho)} = \{f_\alpha^{(\rho)}, g_3^{(\rho)}\}$. We denote

$$t_\alpha^{(\rho)} = t_{\beta\alpha}^{(\rho)} n_\beta, \qquad m_3^{(\rho)} = m_{\alpha 3}^{(\rho)} n_\alpha \tag{5.2.46}$$

In what follows, we shall use the following reciprocal theorem.

Theorem 5.2.2 *If $A^{(\rho)}$ are elastic states corresponding to the body loads $I^{(\rho)}$, then*

$$\begin{aligned}
&\int_\Sigma \left(f_\alpha^{(1)} u_\alpha^{(2)} + g_3^{(1)} \varphi_3^{(2)}\right) da + \int_{\partial\Sigma} \left(t_\alpha^{(1)} u_\alpha^{(2)} + m_3^{(1)} \varphi_3^{(2)}\right) ds\\
&= \int_\Sigma \left(f_\alpha^{(2)} u_\alpha^{(1)} + g_3^{(2)} \varphi_3^{(1)}\right) da + \int_{\partial\Sigma} \left(t_\alpha^{(2)} u_\alpha^{(1)} + m_3^{(2)} \varphi_3^{(1)}\right) ds
\end{aligned} \tag{5.2.47}$$

Proof. We introduce the notation

$$2W_{\nu\eta} = t_{\alpha\beta}^{(\nu)} e_{\alpha\beta}^{(\eta)} + m_{\alpha 3}^{(\nu)} \kappa_{\alpha 3}^{(\eta)} \tag{5.2.48}$$

It follows from Equations 5.2.3 that

$$W_{12} = W_{21} \tag{5.2.49}$$

On the other hand, from Equations 5.2.2 and 5.2.4, we get

$$2W_{\nu\eta} = f_\alpha^{(\nu)} u_\alpha^{(\eta)} + g_3^{(\nu)} \varphi_3^{(\eta)} + \left(t_{\beta\alpha}^{(\nu)} u_\alpha^{(\eta)} + m_{\beta 3}^{(\nu)} \varphi_3^{(\eta)}\right)_{,\beta} \tag{5.2.50}$$

so that

$$2\int_\Sigma W_{\nu\eta} da = \int_\Sigma \left(f_\alpha^{(\nu)} u_\alpha^{(\eta)} + g_3^{(\nu)} \varphi_3^{(\eta)}\right) da + \int_{\partial\Sigma} \left(t_\alpha^{(\nu)} u_\alpha^{(\eta)} + m_3^{(\nu)} \varphi_3^{(\eta)}\right) ds \tag{5.2.51}$$

By Equations 5.2.49 and 5.2.51, we obtain the desired result. □

The elastic potential $\widetilde{W}$ in the case of the plane strain is defined by

$$2\widetilde{W} = \lambda e_{\nu\nu} e_{\rho\rho} + (\mu + \kappa) e_{\alpha\beta} e_{\alpha\beta} + \mu e_{\beta\alpha} e_{\alpha\beta} + \gamma \kappa_{\alpha 3} \kappa_{\alpha 3} \tag{5.2.52}$$

Theorem 5.2.3 *Assume that $\widetilde{W}$ is a positive definite quadratic form. Then*

(i) *The first boundary-value problem has at most one solution*

(ii) *Any two solutions of the second boundary-value problem are equal modulo a plane rigid deformation*

Proof. It follows from Equations 5.2.2, 5.2.4, and 5.2.52 that

$$2\widetilde{W} = t_{\alpha\beta} e_{\alpha\beta} + m_{\alpha 3} \kappa_{\alpha 3} = f_\alpha u_\alpha + g_3 \varphi_3 + (t_{\beta\alpha} u_\alpha + m_{\beta 3} \varphi_3)_{,\beta}$$

If we integrate this relation over Σ and use the divergence theorem, then we obtain

$$2\int_\Sigma \widetilde{W} da = \int_\Sigma (f_\alpha u_\alpha + g_3 \varphi_3) da + \int_{\partial\Sigma} (t_{\beta\alpha} n_\beta u_\alpha + m_{\beta 3} n_\beta \varphi_3) ds \tag{5.2.53}$$

Let (u'_α, φ'_3) and $(u''_\alpha, \varphi''_3)$ be two solutions of a boundary-value problem, and $u^0_\alpha = u'_\alpha - u''_\alpha$, $\varphi^0_3 = \varphi'_3 - \varphi''_3$. Clearly, $(u^0_\alpha, \varphi^0_3)$ is a solution corresponding to $f_\alpha = 0, g_3 = 0$, and to null boundary data. Since $\widetilde{W}$ is positive definite, from Equation 5.2.53, we obtain

$$u^0_{\beta,\alpha} + \varepsilon_{\beta\alpha} \varphi^0_3 = 0, \qquad \varphi^0_{3,\alpha} = 0$$

so that

$$u^0_1 = c_1 - c_3 x_2, \qquad u^0_2 = c_2 + c_3 x_1, \qquad \varphi^0_3 = c_3 \tag{5.2.54}$$

where c_k are arbitrary constants. In the first boundary-value problem, we find that $c_k = 0$. □

Let Σ^+ be a domain in the x_1, x_2-plane, bounded by a simple closed C^2-curve L, and let Σ^- be the complementary of $\Sigma^+ \cup L$ to the entire plane. Let $u = (u_1, u_2, \varphi_3)$ and $v = (u_1', u_2', \varphi_3')$ be two vector fields on Σ^+ such that $u, v \in C^2(\Sigma^+) \cap C^1(\overline{\Sigma}^+)$. The reciprocity relation 5.2.47 leads to

$$\int_{\Sigma^+} \left[uD\left(\frac{\partial}{\partial x}\right) v - vD\left(\frac{\partial}{\partial x}\right) u \right]$$
$$= \int_L \left[uH\left(\frac{\partial}{\partial x}, n_x\right) v - vH\left(\frac{\partial}{\partial x}, n_x\right) u \right] ds \qquad (5.2.55)$$

From Equation 5.2.53, we get

$$2 \int_{\Sigma^+} \widetilde{W} da = - \int_{\Sigma^+} uD\left(\frac{\partial}{\partial x}\right) u da + \int_L uH\left(\frac{\partial}{\partial x}, n_x\right) u ds \qquad (5.2.56)$$

Let $\Sigma(y; \varepsilon)$ be the sphere with the center in y and radius ε. Let $y \in \Sigma^+$ and let ε be so small that $\Sigma(y; \varepsilon)$ be entirely contained in Σ^+. Then the relation 5.2.55 can be applied for the region $\Sigma^+ \backslash \Sigma(y; \varepsilon)$ to a regular vector field $u = (u_1, u_2, \varphi_3)$ and to vector field $v(x) = \Gamma^{(s)}(x, y)$, $(s = 1, 2, 3)$. We obtain the following representation of Somigliana type

$$u(y) = \int_L \left\{ \Gamma^*(x, y) H\left(\frac{\partial}{\partial x}, n_x\right) u(x) - \left[H\left(\frac{\partial}{\partial x}, n_x\right) \Gamma(x, y) \right]^* u(x) \right\} ds_x$$
$$- \int_{\Sigma^+} \Gamma^*(x, y) D\left(\frac{\partial}{\partial x}\right) u(x) da_x \qquad (5.2.57)$$

In view of Equations 5.2.30 and 5.2.41, the relation 5.2.57 implies that

$$u(x) = \int_L \left[\Gamma(x, y) H\left(\frac{\partial}{\partial y}, n_y\right) u(y) - \Lambda(x, y) u(y) \right] ds_y$$
$$- \int_{\Sigma^+} \Gamma(x, y) D\left(\frac{\partial}{\partial y}\right) u(y) da_y \qquad (5.2.58)$$

5.2.5 Existence Theorems

In what follows, we restrict our attention to the equation

$$D\left(\frac{\partial}{\partial x}\right) u = 0 \qquad (5.2.59)$$

In this case, Equation 5.2.56 becomes

$$\int_L uH\left(\frac{\partial}{\partial x}, n_x\right) u ds = 2 \int_{\Sigma^+} \widetilde{W} da \qquad (5.2.60)$$

We say that the vector field $u = (u_1, u_2, \varphi_3)$ is a regular solution of Equation 5.2.59 in Σ^+ if the formula 5.2.60 can be applied to u and if it satisfies Equation 5.2.59 in Σ^+.

Let $x \in \Sigma^-$. We describe around x a circle C_R of sufficiently large radius R, containing the region Σ^+. We denote by Σ_R the region bounded by L and C_R. From Equations 5.2.53 and 5.2.59, we get

$$\int_{L+C_R} uH\left(\frac{\partial}{\partial x}, n_x\right)uds = 2\int_{\Sigma_R} \widetilde{W}da \tag{5.2.61}$$

If u satisfies the condition

$$\lim_{R\to\infty} R\int_0^{2\pi} uH\left(\frac{\partial}{\partial x}, n_x\right)ud\theta = 0 \tag{5.2.62}$$

then from Equation 5.2.61, we obtain

$$\int_L uH\left(\frac{\partial}{\partial x}, n_x\right)uds = -2\int_{\Sigma_R} \widetilde{W}da \tag{5.2.63}$$

We say that the vector field u is a regular solution of Equation 5.2.59 in Σ^- if the formula 5.2.63 can be applied to u in Σ^- and if u satisfies Equation 5.2.59 in Σ^- and the condition 5.2.62.

We consider the following boundary-value problems.

Interior problems. To find a regular solution in Σ^+ of Equation 5.2.59 satisfying one of the conditions

$$\lim_{x\to y} u(x) = f_1(y) \tag{I_1}$$

$$\lim_{x\to y} H\left(\frac{\partial}{\partial x}, n_x\right)u(x) = f_2(y) \tag{I_2}$$

where $x \in \Sigma^+, y \in L$, and f_1 and f_2 are prescribed vector fields.

Exterior problems. To find a regular solution in Σ^- of Equation 5.2.59 satisfying one of the conditions

$$\lim_{x\to y} u(x) = f_3(y) \tag{E_1}$$

$$\lim_{x\to y} H\left(\frac{\partial}{\partial x}, n_x\right)u(x) = f_4(y) \tag{E_2}$$

where $x \in \Sigma^-$, $y \in L$, and f_3 and f_4 are given.

We assume that f_1 and f_3 are Hölder continuously differentiable on L, and f_2 and f_4 are Hölder continuous on L.

We denote by (I_α^0) and (E_α^0) the homogeneous problems corresponding to (I_α) and (E_α), respectively. We introduce the potential of a single layer

$$V(x;\rho) = \int_L \Gamma(x, y)\rho(y)ds_y \tag{5.2.64}$$

and the potential of a double layer

$$W(x;\nu) = \int_L \Lambda(x,y)\nu(y)ds_y \tag{5.2.65}$$

where $\rho = (\rho_1, \rho_2, \rho_3)$ is Hölder continuous on L and $\nu = (\nu_1, \nu_2, \nu_3)$ is Hölder continuously differentiable on L. As in the classical theory [55], we have the following results.

Theorem 5.2.4 *The potential of a single layer is continuous on* $\mathbb{R}^2$.

Theorem 5.2.5 *The potential of a double layer has finite limits when point x tends to $y \in L$ from both within and without, and these limits are respectively equal to*

$$\begin{aligned} W^+(y;\nu) &= -\frac{1}{2}\nu(y) + \int_L \Lambda(y,z)\nu(z)ds_z \\ W^-(y;\nu) &= \frac{1}{2}\nu(y) + \int_L \Lambda(y,z)\nu(z)ds_z \end{aligned} \tag{5.2.66}$$

the integrals being conceived in the sense of Cauchy's principal value.

Theorem 5.2.6 *$H(\partial/\partial x, n_x)V(x;\rho)$ tends to finite limits as point x tends to the boundary point $y \in L$, from within or without, and these limits are respectively equal to*

$$\begin{aligned} \left[H\left(\frac{\partial}{\partial y}, n_y\right)V(y;\rho)\right]^+ &= \frac{1}{2}\rho(y) + \int_L \left[H\left(\frac{\partial}{\partial y}, n_y\right)\Gamma(y,z)\right]\rho(z)ds_z \\ \left[H\left(\frac{\partial}{\partial y}, n_y\right)V(y;\rho)\right]^- &= -\frac{1}{2}\rho(y) + \int_L \left[H\left(\frac{\partial}{\partial y}, n_y\right)\Gamma(y,z)\right]\rho(z)ds_z \end{aligned} \tag{5.2.67}$$

Theorem 5.2.7 *The potentials $V(x;\rho)$ and $W(x;\nu)$ satisfy Equation* 5.2.59 *on* $\Sigma^+ \cup \Sigma^-$.

We seek the solutions of the problems (I_1) and (E_1) in the form of a double-layer potential and the solutions of the problems (I_2) and (E_2) in the form of a single-layer potential. In view of Theorems 5.2.5 and 5.2.6, we obtain for the unknown densities the following singular integral equations

$$-\frac{1}{2}\nu(y) + \int_L \Lambda(y,z)\nu(z)ds_z = f_1(y) \tag{I_1}$$

$$\frac{1}{2}\rho(y) + \int_L \Lambda^*(z,y)\rho(z)ds_z = f_2(y) \tag{I_2}$$

$$\frac{1}{2}\nu(y) + \int_L \Lambda(y,z)\nu(z)ds_z = f_3(y) \tag{E_1}$$

$$-\frac{1}{2}\rho(y) + \int_L \Lambda^*(z,y)\rho(z)ds_z = f_4(y) \tag{E_2}$$

where $y \in L$. The homogeneous equations corresponding to equations (I_1), (I_2), (E_1), and (E_2) for $f_s = 0$, ($s = 1, 2, 3, 4$), will be denoted by (I_1^0), (I_2^0), (E_1^0), and (E_2^0), respectively. The equations (I_1) and (E_2), and (I_2) and (E_1) are piecewise mutually associate equations.

We note that

$$\frac{d\ln r}{ds_z}ds_z = \frac{dr}{r} = \frac{dt}{t - t_0} - id\theta \tag{5.2.68}$$

where t and t_0 are the affixes of the points z and y. Taking into account Equations 5.2.43 and 5.2.68 and pointing out the characteristic part of the singular operator [242], the system (I_1) can be written in the form

$$\nu(t_0) + \frac{1}{\pi}\begin{Vmatrix} 0 & c & 0 \\ -c & 0 & 0 \\ 0 & 0 & 0 \end{Vmatrix} \int_L \frac{\nu(t)}{t - t_0}dt + \mathcal{K}\nu(t_0) = -2f_1(t_0) \tag{5.2.69}$$

Let us denote by $[h(t)]_L$ the increment of the function h as the point t describes once the curve L in the direction leaving the domain Σ^+ on the left. The index of the system 5.2.69 is

$$n = \frac{1}{2\pi}\left[\arg\left(\frac{\det \mathcal{D}}{\det \mathcal{S}}\right)\right]_L$$

where

$$\mathcal{D} = \|d_{ij}\|_{3\times 3}, \qquad \mathcal{S} = \|s_{ij}\|_{3\times 3}$$
$$d_{mn} = s_{mn} = 1 \text{ for } m = n, \qquad d_{21} = -d_{12} = s_{12} = -s_{21} = ic$$
$$d_{\alpha 3} = d_{3\alpha} = 0, \qquad s_{\alpha 3} = s_{3\alpha} = 0$$

Since in our case we have $n = 0$, the system (I_1) is a system of singular integral equations for which Fredholm's basic theorems are valid [196]. In a similar way, we can prove that the index of the system (I_2) is zero.

Let us consider the problems (I_1) and (E_2). The homogeneous equations (I_1^0) and (E_2^0)

$$-\frac{1}{2}\nu(y) + \int_L \Lambda(y,z)\nu(z)ds_z = 0 \tag{I_1^0}$$

$$-\frac{1}{2}\rho(y) + \int_L \Lambda^*(z,y)\rho(z)ds_z = 0 \tag{E_2^0}$$

have only trivial solutions. We assume the opposite and suppose that ρ^0 is a solution of equation (E_2^0), not equal to zero. Then, the single-layer potential

$$V(x;\rho^0) = \int_L \Gamma(x,y)\rho^0(y)ds_y$$

satisfies the condition

$$\left[H\left(\frac{\partial}{\partial y}, n_y\right) V(y;\rho^0)\right]^- = 0, \quad y \in L \tag{5.2.70}$$

When x tends to a point at infinity and y remains fixed on L then $u_\alpha^{(\beta)}$ tends at infinity as $\delta_{\alpha\beta} \ln r$. If the density $\rho = (\rho_1, \rho_2, \rho_3)$ of the potential of a single layer satisfies the conditions

$$\int_L \rho_\alpha ds = 0, \quad (\alpha = 1, 2) \tag{5.2.71}$$

then the potential $V(x;\rho)$ satisfies the asymptotic relations

$$V = O(r^{-1}), \qquad \frac{\partial V}{\partial R} = O(r^{-2}) \text{ as } r \to \infty \tag{5.2.72}$$

where $\mathbf{R}$ is an arbitrary direction. As in classical theory of elasticity, we have

$$\int_L H_\alpha\left(\frac{\partial}{\partial x}, n_x\right) \Gamma^{(\beta)}(x, y) ds_x = -\zeta(y)\delta_{\alpha\beta} \tag{5.2.73}$$

where

$$\zeta(u) = \begin{cases} 1, & y \in \Sigma^+, \\ \frac{1}{2}, & y \in L, \\ 0, & y \in \Sigma^- \end{cases}$$

If we multiply the equation (E_2^0) by ds_y and integrate on L, on the basis of Equation 5.2.73, we obtain

$$\int_L \rho_\alpha^0(y) ds_y = 0, \quad (\alpha = 1, 2)$$

so that the potential $V(x;\rho^0)$ satisfies Equation 5.2.72. This fact implies that $V(x;\rho^0)$ satisfies the relation 5.2.62. Thus, we conclude that (i) $V(x;\rho^0)$ satisfies Equation 5.2.59 on Σ^- and the condition 5.2.70 on L; (ii) the formula 5.2.63 can be applied to $V(x;\rho^0)$; and (iii) $V(x;\rho^0)$ satisfies the asymptotic relations 5.2.72. It follows that

$$V(x;\rho^0) = 0 \text{ on } \Sigma^- \tag{5.2.74}$$

According to the continuity of the single-layer potential, we have

$$[V(x;\rho^0)]^+ = 0 \text{ on } L$$

Taking into account that $V(x;\rho^0)$ satisfies Equation 5.2.59 on Σ^+, from the uniqueness theorem, we get

$$V(x;\rho^0) = 0 \text{ on } \Sigma^+ \tag{5.2.75}$$

It follows from Equations 5.2.67, 5.2.74, and 5.2.75 that

$$\rho^0(y) = \left[H\left(\frac{\partial}{\partial y}, n_y\right) V(y; \rho^0)\right]^+ - \left[H\left(\frac{\partial}{\partial y}, n_y\right) V(y; \rho^0)\right]^- = 0$$

Thus, our statement concerning the equation (E_2^0) is valid.

Since the equations (I_1^0) and (E_2^0) form an associate set of integral equations, (I_1^0) has also no nontrivial solution. We note that from Equation 5.2.73 and the equation (E_2), with $f_4 = (f_{41}, f_{42}, f_{43})$, we obtain

$$-\int_L \rho_\alpha(y) ds_y = \int_L f_{4\alpha} ds, \quad (\alpha = 1, 2)$$

Thus we obtain the following results.

Theorem 5.2.8 *Problem (I_1) has solution for any Hölder continuously differentiable vector field f_1. This solution is unique and can be expressed by a double-layer potential.*

Theorem 5.2.9 *Problem (E_2) can be solved if and only if*

$$\int_L f_{4\alpha} ds = 0, \quad (\alpha = 1, 2)$$

We now consider the equations (I_2^0) and (E_1^0). We note that the vector field

$$\omega(x) = (c_1 - c_3 x_2, c_2 + c_3 x_1, c_3)$$

where c_i are arbitrary constants, satisfies the equations

$$D\left(\frac{\partial}{\partial x}\right)\omega(x) = 0, \quad x \in \Sigma^+, \qquad H\left(\frac{\partial}{\partial x}, n_x\right)\omega(x) = 0 \text{ on } L \tag{5.2.76}$$

From Equation 5.2.58, we obtain

$$\omega(x) = -\int_L \Lambda(x, y)\omega(y) ds_y, \quad x \in \Sigma^+ \tag{5.2.77}$$

Passing to the limit in Equation 5.2.77 as the point x approaches the boundary point $x_0 \in L$ from within, according to Equation 5.2.66, we get

$$\frac{1}{2}\omega(x_0) + \int_L \Lambda(x_0, y)\omega(y) ds_y = 0$$

Hence, the matrix $\omega(x)$ satisfies the equation (E_1^0). Clearly, the vector fields

$$\omega^{(1)} = (1, 0, 0), \qquad \omega^{(2)} = (0, 1, 0), \qquad \omega^{(3)} = (-x_2, x_1, 1)$$

are linearly independent solutions of the equation (E_1^0). According to the second Fredholm theorem, the equation (I_2^0) has at least three linearly independent solutions $v^{(i)}$, $(i = 1, 2, 3)$. As in classical theory [194], we can prove that $v^{(i)}$ forms a complete system of linearly independent solutions of the equation (I_2^0). This fact implies the completeness of the associate system $(\omega^{(1)}, \omega^{(2)}, \omega^{(3)})$. Hence, the necessary and sufficient conditions to solve the equation (I_2) have the form

$$\int_L \omega^{(j)}(x) f_2(x) ds_x = 0, \quad (j = 1, 2, 3) \tag{5.2.78}$$

If we take $f_2 = (\tilde{t}_1, \tilde{t}_2, \tilde{m})$, then the conditions 5.2.78 can be written in the form

$$\int_L \tilde{t}_\alpha ds = 0, \qquad \int_L (x_1 \tilde{t}_2 - x_2 \tilde{t}_1 + \tilde{m}) ds = 0 \tag{5.2.79}$$

Thus, we have the following result.

Theorem 5.2.10 *Problem (I_2) can be solved if and only if the conditions 5.2.79 hold. The solution can be represented as a single-layer potential and is determined within an additive plane rigid deformation.*

As in classical theory, we can study the problem (E_1). These results have been established in Ref. 140.

On the basis of Theorem 5.2.10, we find that the second boundary-value problem has solution if and only if

$$\begin{gathered} \int_{\Sigma_1} f_\alpha da + \int_\Gamma \tilde{t}_\alpha ds = 0 \\ \int_{\Sigma_1} (\varepsilon_{\alpha\beta} x_\alpha f_\beta + g_3) da + \int_\Gamma (\varepsilon_{\alpha\beta} x_\alpha \tilde{t}_\beta + \tilde{m}_3) ds = 0 \end{gathered} \tag{5.2.80}$$

5.3 Saint-Venant's Problem for Cosserat Cylinders

In this section, we study the Saint-Venant's problem within the linear theory of Cosserat elastic bodies. We show that the method of Section 1.7 can be extended to derive a solution of Saint-Venant's problem. Minimum principles characterizing the solutions of extension, bending, torsion, and flexure problems are presented in Section 5.4. These principles lead to a solution of Truesdell's problem for Cosserat cylinders.

Saint-Venant's problem for Cosserat elastic bodies has been studied in various works [85,141,143,154,188,338].

5.3.1 Preliminaries

We assume for the remainder of this chapter that the domain B is occupied by a homogeneous and isotropic material. We denote by u the six-dimensional vector field on B, defined by $u = (u_1, u_2, u_3, \varphi_1, \varphi_2, \varphi_3) = (u_i, \varphi_i)$, where u_j are the components of the displacement vector field, and φ_k are the components of the microrotation vector field. Let us denote the strain measures associated with u by $e_{ij}(u)$ and $\kappa_{ij}(u)$, that is

$$e_{ij}(u) = u_{j,i} + \varepsilon_{jik}\varphi_k, \qquad \kappa_{ij}(u) = \varphi_{j,i} \tag{5.3.1}$$

We note that $e_{ij}(u) = 0$, $\kappa_{ij}(u) = 0$ if and only if $u_i = a_i + \varepsilon_{ijk}b_jx_k$, $\varphi_i = b_i$, where a_k and b_k are arbitrary constants. Let

$$\mathcal{R}^* = \left\{u^0 : u^0 = \left(u_i^0, \varphi_i^0\right),\ u_i^0 = a_i + \varepsilon_{ijk}b_jx_k,\ \varphi_i^0 = b_i\right\} \tag{5.3.2}$$

where a_i and b_i are constants. If $u \in \mathcal{R}^*$, then u is called a rigid deformation. We denote by $t_{ij}(u)$ and $m_{ij}(u)$ the components of the stress tensor and couple stress tensor, associated with u. In the case of isotropic and homogeneous bodies, we have

$$\begin{aligned} t_{ij}(u) &= \lambda e_{rr}(u)\delta_{ij} + (\mu+\kappa)e_{ij}(u) + \mu e_{ji}(u) \\ m_{ij}(u) &= \alpha\kappa_{rr}(u)\delta_{ij} + \beta\kappa_{ji}(u) + \gamma\kappa_{ij}(u) \end{aligned} \tag{5.3.3}$$

where $\lambda, \mu, \kappa, \alpha, \beta$, and γ are given constants. Over the past decade, a determination of the constitutive constants has become possible (see [30,259,260] and references therein).

We call a six-dimensional vector field u an equilibrium vector field for B if $u \in C^1(\overline{B}) \cap C^2(B)$ and

$$(t_{ji}(u))_{,j} = 0, \qquad (m_{ji}(u))_{,j} + \varepsilon_{ijk}t_{jk}(u) = 0 \tag{5.3.4}$$

hold on B.

Let $s_i(u)$ and $m_i(u)$ be the components of the stress vector and couple stress vector at regular points of ∂B, corresponding to the stress tensor $t_{ij}(u)$ and couple stress tensor $m_{ij}(u)$ defined on $\overline{B}$, that is,

$$s_i(u) = t_{ji}(u)n_j, \qquad m_i(u) = m_{ji}(u)n_j \tag{5.3.5}$$

The elastic potential corresponding to u is given by

$$\begin{aligned} 2W(u) &= \lambda e_{rr}(u)e_{ss}(u) + (\mu+\kappa)e_{ij}(u)e_{ij}(u) + \mu e_{ij}(u)e_{ji}(u) \\ &\quad + \alpha\kappa_{rr}(u)\kappa_{ss}(u) + \beta\kappa_{ij}(u)\kappa_{ji}(u) + \gamma\kappa_{ij}(u)\kappa_{ij}(u) \end{aligned} \tag{5.3.6}$$

We assume that the elastic potential is a positive definite quadratic form in the variables $e_{ij}(u)$ and $\kappa_{ij}(u)$.

The strain energy $U(u)$ corresponding to u is defined by

$$U(u) = \int_B W(u)dv \tag{5.3.7}$$

In the following, two six-dimensional vector fields differing by a rigid deformation will be regarded identical.

The functional $U(\cdot)$ generates the bilinear functional

$$U(u,v) = \frac{1}{2}\int_B [\lambda e_{rr}(u)e_{ss}(v) + (\mu+\kappa)e_{ij}(u)e_{ij}(v) + \mu e_{ij}(u)e_{ji}(v) + \alpha\kappa_{rr}(u)\kappa_{ss}(v) + \beta\kappa_{ij}(u)\kappa_{ji}(v) + \gamma\kappa_{ij}(u)\kappa_{ij}(v)]dv \qquad (5.3.8)$$

The set of smooth vector fields u over $\overline{B}$ can be made into a real vector space with the inner product

$$\langle u, v\rangle = 2U(u,v) \qquad (5.3.9)$$

This inner product generates the energy norm $\|u\|_e^2 = \langle u, u\rangle$. As in Theorem 5.2.2, we can prove that for any equilibrium vector fields $u = (u_i, \varphi_i)$ and $v = (v_i, \psi_i)$, one has

$$\langle u, v\rangle = \int_{\partial B} [v_i s_i(u) + \psi_i m_i(u)]da \qquad (5.3.10)$$

which implies the reciprocity relation

$$\int_{\partial B} [u_i s_i(v) + \varphi_i m_i(v)]da = \int_{\partial B} [v_i s_i(u) + \psi_i m_i(u)]da \qquad (5.3.11)$$

We assume that the region B from here on refers to the interior of the right cylinder defined in Section 1.2. We consider the equilibrium problem of cylinder B which, in the absence of body forces and body couples, is subjected to surface forces and surface couples prescribed over its ends and is free from lateral loading. Thus, the problem consists in the determination of an equilibrium six-dimensional vector field u on B, subject to the requirements

$$\begin{aligned} &s_i(u) = 0, \qquad m_i(u) = 0 \text{ on } \Pi \\ &s_i(u) = \widetilde{t}_i^{(\alpha)}, \qquad m_i(u) = \widetilde{m}_i^{(\alpha)} \text{ on } \Sigma_\alpha, \quad (\alpha = 1,2) \end{aligned} \qquad (5.3.12)$$

where $\widetilde{t}_i^{(\alpha)}$ and $\widetilde{m}_i^{(\alpha)}$, $(\alpha = 1,2)$, are prescribed functions. We assume that the hypotheses of Theorem 5.1.1 hold. It follows from Equations 5.3.10 and 5.1.22 that the necessary and sufficient conditions for the existence of a solution to this problem are given by

$$\begin{aligned} &\int_{\Sigma_1} \widetilde{t}_i^{(1)} da + \int_{\Sigma_2} \widetilde{t}_i^{(2)} da = 0 \\ &\int_{\Sigma_1} (\varepsilon_{ijk}x_j\widetilde{t}_k^{(1)} + \widetilde{m}_i^{(1)})da + \int_{\Sigma_2} (\varepsilon_{ijk}x_j\widetilde{t}_k^{(2)} + \widetilde{m}_i^{(2)})da = 0 \end{aligned} \qquad (5.3.13)$$

In the formulation of Saint-Venant, the conditions 5.3.12 are replaced by

$$\begin{aligned} &s_i(u) = 0, \qquad m_i(u) = 0 \text{ on } \Pi \\ &R_i(u) = F_i, \qquad H_i(u) = M_i \end{aligned} \qquad (5.3.14)$$

where $\mathbf{F}$ and $\mathbf{M}$ are given vectors representing the resultant of surface forces and the resultant moment about O of the surface forces and surface couples acting on Σ_1. Here, $R_i(\cdot)$ and $H_i(\cdot)$ are the linear functionals defined by

$$\begin{aligned} R_i(u) &= -\int_{\Sigma_1} t_{3i}(u)da \\ H_\alpha(u) &= -\int_{\Sigma_1} [\varepsilon_{\alpha\beta}x_\beta t_{33}(u) + m_{3\alpha}(u)]da \\ H_3(u) &= -\int_{\Sigma_1} [\varepsilon_{\alpha\beta}x_\alpha t_{3\beta}(u) + m_{33}(u)]da \end{aligned} \tag{5.3.15}$$

Saint-Venant's problem for Cosserat elastic cylinders consists in the determination of an equilibrium vector field $u = (u_i, \varphi_i)$ on B that satisfies the conditions 5.3.14. Let $K(\mathbf{F}, \mathbf{M})$ be the class of solutions to this problem. We continue to denote by $K_I(F_3, M_1, M_2, M_3)$ the set of all solutions of the extension, bending, and torsion problems (the problem (P_1)), and by $K_{II}(F_1, F_2)$ the set of all solutions of the flexure problem (the problem (P_2)).

From the conditions of equilibrium of cylinder B, we obtain

$$\begin{aligned} &\int_{\Sigma_2} t_{3i}(u)da = -R_i(u) \\ &\int_{\Sigma_2} [x_\alpha t_{33}(u) + \varepsilon_{\beta\alpha} m_{3\beta}(u)]da = \varepsilon_{\alpha\beta} H_\beta(u) - hR_\alpha(u) \\ &\int_{\Sigma_2} [\varepsilon_{\alpha\beta}x_\alpha t_{3\beta}(u) + m_{33}(u)]da = -H_3(u) \end{aligned} \tag{5.3.16}$$

We denote by Λ the set of all equilibrium vector fields u that satisfy the conditions

$$s_i(u) = 0, \qquad m_i(u) = 0 \text{ on } \Pi$$

Theorem 5.3.1 *If $u \in \Lambda$ and $u_{,3} \in C^1(\overline{B}) \cap C^2(B)$ then $u_{,3} \in \Lambda$ and*

$$\mathbf{R}(u_{,3}) = \mathbf{0}, \qquad H_\alpha(u_{,3}) = \varepsilon_{\alpha\beta} R_\beta(u), \qquad H_3(u_{,3}) = 0 \tag{5.3.17}$$

Proof. The first assertion follows at once from the fact that $t_{ij}(u_{,3}) = (t_{ij}(u))_{,3}$, $m_{ij}(u_{,3}) = (m_{ij}(u))_{,3}$ and hypotheses. From Equations 5.3.4, we arrive at

$$\begin{aligned} t_{3i}(u_{,3}) &= (t_{3i}(u))_{,3} = -(t_{\alpha i}(u))_{,\alpha} \\ \varepsilon_{\alpha\beta}x_\beta t_{33}(u_{,3}) + m_{3\alpha}(u_{,3}) &= (m_{3\alpha}(u))_{,3} + \varepsilon_{\alpha\beta}x_\beta(t_{33}(u))_{,3} \\ &= -(m_{\rho\alpha}(u))_{,\rho} - \varepsilon_{\alpha ij}t_{ij}(u) - \varepsilon_{\alpha\beta}x_\beta(t_{\rho 3}(u))_{,\rho} \\ &= -(m_{\rho\alpha}(u))_{,\rho} - \varepsilon_{\alpha ij}t_{ij}(u) - \varepsilon_{\alpha\beta}[(x_\beta t_{\rho 3}(u))_{,\rho} - t_{\beta 3}] \\ &= -[m_{\rho\alpha}(u) + \varepsilon_{\alpha\beta}x_\beta t_{\rho 3}(u)]_{,\rho} - \varepsilon_{\rho\alpha}t_{3\rho}(u) \\ \varepsilon_{\alpha\beta}x_\alpha t_{3\beta}(u_{,3}) + m_{33}(u_{,3}) &= -\varepsilon_{\alpha\beta}x_\alpha(t_{\rho\beta}(u))_{,\rho} - (m_{\rho 3}(u))_{,\rho} - \varepsilon_{\alpha\beta}t_{\alpha\beta}(u) \\ &= -[\varepsilon_{\alpha\beta}x_\alpha t_{\rho\beta}(u) + m_{\rho 3}(u)]_{,\rho} \end{aligned} \tag{5.3.18}$$

Using the divergence theorem, Equations 5.3.15 and 5.3.18, we obtain

$$\begin{aligned}
\mathbf{R}(u_{,3}) &= \int_\Gamma \mathbf{s}(u)ds \\
H_\alpha(u_{,3}) &= \int_\Gamma [\varepsilon_{\alpha\beta}x_\beta s_3(u) + m_\alpha(u)]ds + \varepsilon_{\alpha\rho}R_\rho(u) \\
H_3(u_{,3}) &= \int_\Gamma [\varepsilon_{\alpha\beta}x_\alpha s_\beta(u) + m_3(u)]ds
\end{aligned}$$

The desired result is now immediate. □

Theorem 5.3.1 has the following immediate consequences.

Corollary 5.3.1 *If* $u \in K_I(F_3, M_1, M_2, M_3)$ *and* $u_{,3} \in C^1(\overline{B}) \cap C^2(B)$, *then* $u_{,3} \in \Lambda$ *and* $\mathbf{R}(u_{,3}) = \mathbf{0}, \mathbf{H}(u_{,3}) = \mathbf{0}$.

Corollary 5.3.2 *If* $u \in K_{II}(F_1, F_2)$ *and* $u_{,3} \in C^1(\overline{B}) \cap C^2(B)$, *then* $u_{,3} \in K_I(0, F_2, -F_1, 0)$.

The above results will be used to establish a solution of Saint-Venant's problem.

We note that in Ref. 21, Berglund extended Toupin's version of Saint-Venant's principle to the case of Cosserat elastic cylinders.

5.3.2 Extension, Bending, and Torsion

Corollary 5.3.1 allows us to establish a method to derive a solution to the problem (P_1). Let $\mathcal{A}^*$ be the class of solutions to the Saint-Venant's problem corresponding to $\mathbf{F} = \mathbf{0}$ and $\mathbf{M} = \mathbf{0}$. In view of definition 5.3.2, it follows that $\mathcal{R}^* \subset \mathcal{A}^*$. We note that if $u \in K_I(F_3, M_1, M_2, M_3)$ and $u_{,3} \in C^1(\overline{B}) \cap C^2(B)$, then by Corollary 5.3.1, $u_{,3} \in \mathcal{A}^*$. It is natural to seek a solution v of the problem (P_1) such that $v_{,3}$ is a rigid deformation.

Theorem 5.3.2 *Let S be the set of all vector fields $u \in C^1(\overline{B}) \cap C^2(B)$ such that $u_{,3} \in \mathcal{R}^*$. Then there exists a vector field $v \in S$ which is solution of the problem (P_1).*

Proof. Let $v \in C^1(\overline{B}) \cap C^2(B)$, $v = (v_i, \omega_i)$, such that

$$v_{,3} = (\alpha_i + \varepsilon_{ijk}\beta_j x_k, \beta_i)$$

where α_i and β_i are constants. We find

$$\begin{aligned}
v_\alpha &= -\frac{1}{2}a_\alpha x_3^2 - a_4\varepsilon_{\alpha\beta}x_\beta x_3 + w_\alpha(x_1, x_2) \\
v_3 &= (a_\rho x_\rho + a_3)x_3 + w_3(x_1, x_2) \\
\omega_\alpha &= \varepsilon_{\alpha\beta}a_\beta x_3 + \chi_\alpha(x_1, x_2), \qquad \omega_3 = a_4 x_3 + \chi_3(x_1, x_2)
\end{aligned} \tag{5.3.19}$$

modulo a rigid deformation. Here $w = (w_i, \chi_i)$ is an arbitrary vector field independent of x_3, and we have used the notations $a_\alpha = \varepsilon_{\rho\alpha}\beta_\rho, a_3 = \alpha_3, a_4 = \beta_3$. Now we prove that the functions w_i and χ_i, and the constants a_s, $(s = 1, 2, 3, 4)$, can be determined so that $v \in K_I(F_3, M_1, M_2, M_3)$. By Equations 5.3.1 and 5.3.19,

$$\begin{aligned}
e_{\alpha\beta}(v) &= e_{\alpha\beta}(w^0), & e_{3\alpha}(v) &= -\varepsilon_{\alpha\beta}(a_4 x_\beta + \chi_\beta) \\
e_{\alpha 3}(v) &= e_{\alpha 3}(v'), & e_{33}(v) &= a_\rho x_\rho + a_3 \\
\kappa_{\alpha\beta}(v) &= \kappa_{\alpha\beta}(w'), & \kappa_{3\alpha}(v) &= \varepsilon_{\alpha\beta} a_\beta \\
\kappa_{\alpha 3}(v) &= \kappa_{\alpha 3}(w^0), & \kappa_{33}(v) &= a_4
\end{aligned}$$

where

$$w^0 = (w_1, w_2, 0, 0, 0, \chi_3), \qquad w' = (0, 0, w_3, \chi_1, \chi_2, 0) \tag{5.3.20}$$

Clearly,

$$\begin{aligned}
&t_{\alpha\beta}(v) = \lambda(a_\rho x_\rho + a_3)\delta_{\alpha\beta} + T_{\alpha\beta}(w^0), \qquad t_{\alpha 3}(v) = P_\alpha(w') - \mu a_4 \varepsilon_{\alpha\rho} x_\rho \\
&t_{3\alpha}(v) = Q_\alpha(w') + (\mu + \kappa) a_4 \varepsilon_{\beta\alpha} x_\beta \\
&t_{33}(v) = (\lambda + 2\mu + \kappa)(a_\rho x_\rho + a_3) + \lambda e_{\rho\rho}(w^0) \\
&m_{\nu\eta}(v) = \alpha a_4 \delta_{\nu\eta} + H_{\nu\eta}(w'), \qquad m_{\alpha 3}(v) = \beta \varepsilon_{\alpha\rho} a_\rho + M_{\alpha 3}(w^0) \\
&m_{3\alpha}(v) = \gamma \varepsilon_{\alpha\rho} a_\rho + \beta \chi_{3,\alpha}, \qquad m_{33}(v) = (\alpha + \beta + \gamma) a_4 + \alpha \chi_{\rho,\rho}
\end{aligned} \tag{5.3.21}$$

where

$$\begin{aligned}
&T_{\alpha\beta}(w^0) = \lambda e_{\rho\rho}(w^0)\delta_{\alpha\beta} + (\mu + \kappa) e_{\alpha\beta}(w^0) + \mu e_{\beta\alpha}(w^0) \\
&M_{\alpha 3}(w^0) = \gamma \kappa_{\alpha 3}(w^0), \qquad P_\alpha(w') = (\mu + \kappa) w_{3,\alpha} + \kappa \varepsilon_{\alpha\beta} \chi_\beta \\
&Q_\alpha(w') = \mu w_{3,\alpha} + \kappa \varepsilon_{\beta\alpha} \chi_\beta, \qquad H_{\nu\eta}(w') = \alpha \chi_{\rho,\rho} \delta_{\eta\nu} + \beta \chi_{\nu,\eta} + \gamma \chi_{\eta,\nu}
\end{aligned} \tag{5.3.22}$$

We introduce the following notations

$$w_3 = a_4 \varphi, \qquad \chi_\alpha = a_4 \psi_\alpha, \qquad \widehat{w} = (0, 0, \varphi, \psi_1, \psi_2, 0) \tag{5.3.23}$$

Clearly, $w = w^0 + a_4 \widehat{w}$. Let $\mathcal{T}$ be the set of all vector fields $\widehat{w} \in C^1(\overline{B}) \cap C^2(B)$ such that $\widehat{w} = (0, 0, \varphi, \psi_1, \psi_2, 0)$. We introduce the operators L_i on $\mathcal{T}$ defined by

$$\begin{aligned}
L_\nu \widehat{w} &= \gamma \Delta \psi_\nu + (\alpha + \beta)\psi_{\rho,\rho\nu} + \kappa \varepsilon_{\nu\beta} \varphi_{,\beta} - 2\kappa \psi_\nu \\
L_3 \widehat{w} &= (\mu + \kappa)\Delta \varphi + \kappa \varepsilon_{\alpha\beta} \psi_{\beta,\alpha}
\end{aligned} \tag{5.3.24}$$

With the help of Equations 5.3.21, 5.3.23, and 5.3.24, the equations of equilibrium and the conditions on the lateral boundary reduce to

$$\begin{aligned}
&\left(T_{\beta\alpha}(w^0)\right)_{,\beta} + f_\alpha^0 = 0, \qquad \left(M_{\rho 3}(w^0)\right)_{,\rho} + \varepsilon_{\alpha\beta} T_{\alpha\beta}(w^0) = 0 \text{ on } \Sigma_1 \\
&T_{\beta\alpha}(w^0) n_\beta = t_\alpha^0, \qquad M_{\alpha 3}(w^0) n_\alpha = m_3^0 \text{ on } \Gamma
\end{aligned} \tag{5.3.25}$$

and

$$L_i\widehat{w} = h_i \text{ on } \Sigma_1, \qquad N_i\widehat{w} = \zeta_i \text{ on } \Gamma \tag{5.3.26}$$

where

$$\begin{aligned} f_\alpha^0 &= \lambda a_\alpha, \quad t_\alpha^0 = -\lambda(a_\rho x_\rho + a_3)n_\alpha, \quad m_3^0 = \beta\varepsilon_{\rho\alpha}a_\rho n_\alpha \\ h_\alpha &= \kappa x_\alpha, \quad h_3 = 0, \quad \zeta_\nu = -\alpha n_\nu, \quad \zeta = \mu\varepsilon_{\alpha\beta}x_\beta n_\alpha \end{aligned} \tag{5.3.27}$$

and

$$\begin{aligned} N_\nu\widehat{w} &= (\alpha\psi_{\rho,\rho}\delta_{\eta\nu} + \beta\psi_{\eta,\nu} + \gamma\psi_{\nu,\eta})n_\eta \\ N_3\widehat{w} &= (\mu+\kappa)\frac{\partial\varphi}{\partial n} + \kappa\varepsilon_{\alpha\beta}\psi_\beta n_\alpha \end{aligned} \tag{5.3.28}$$

From Equations 5.3.20, 5.3.22, and 5.3.25, we conclude that w^0 is characterized by a plane strain problem (cf. Section 5.2). It is easy to verify that the necessary and sufficient conditions to solve the boundary-value problem 5.3.25 are satisfied. Thus, the boundary-value problem 5.3.25 has solutions for any constants a_1, a_2, and a_3. We denote by $w^{(i)} = (u_1^{(i)}, u_2^{(i)}, 0, 0, 0, \varphi_3^{(i)})$, $(i = 1, 2, 3)$, a solution of the boundary-value problem 5.3.25 when $a_j = \delta_{ij}$. We can write

$$w^0 = \sum_{i=1}^{3} a_i w^{(i)} \tag{5.3.29}$$

where $w^{(i)}$ are characterized by the equations

$$\begin{aligned} &\left(T_{\beta\alpha}(w^{(\rho)})\right)_{,\beta} + \lambda\delta_{\alpha\rho} = 0, \qquad \left(T_{\beta\alpha}(w^{(3)})\right)_{,\beta} = 0 \\ &\left(M_{\rho 3}(w^{(i)})\right)_{,\rho} + \varepsilon_{\alpha\beta}T_{\alpha\beta}(w^{(i)}) = 0 \text{ on } \Sigma_1 \end{aligned} \tag{5.3.30}$$

and the boundary conditions

$$\begin{aligned} T_{\beta\alpha}(w^{(\rho)})n_\beta &= -\lambda x_\rho n_\alpha, \qquad T_{\beta\alpha}(w^{(3)})n_\beta = -\lambda n_\alpha \\ M_{\alpha 3}(w^{(\rho)})n_\alpha &= \beta\varepsilon_{\rho\alpha}n_\alpha, \qquad M_{\alpha 3}(w^{(3)})n_\alpha = 0 \text{ on } \Gamma \end{aligned} \tag{5.3.31}$$

In what follows, we shall assume that the vector fields $w^{(i)}$, $(i = 1, 2, 3)$, are known.

We consider now the boundary-value problem defined by

$$L_i\widehat{w} = \eta_i \text{ on } \Sigma_1, \qquad N_i\widehat{w} = \rho_i \text{ on } \Gamma \tag{5.3.32}$$

where η_i and ρ_i are C^∞ functions. It is known (cf. [141,154]) that the boundary-value problem 5.3.32 has a solution $\widehat{w} \in C^1(\overline{\Sigma}_1) \cap C^2(\Sigma_1)$ if and only if

$$\int_{\Sigma_1} \eta_3 - \int_\Gamma \rho_3 ds = 0 \tag{5.3.33}$$

The necessary and sufficient condition for the existence of a solution of the boundary-value problem 5.3.26 is satisfied.

By Equations 5.3.19, 5.3.20, 5.3.23, and 5.3.29, we see that the vector field v can be written in the form

$$v = \sum_{j=1}^{4} a_j v^{(j)} \tag{5.3.34}$$

where the vector fields $v^{(j)} = (v_i^{(j)}, \omega_i^{(j)})$, $(j = 1, 2, 3, 4)$, are defined by

$$\begin{aligned} &v_\alpha^{(\beta)} = -\frac{1}{2}x_3^2\delta_{\alpha\beta} + u_\alpha^{(\beta)}, \qquad v_\alpha^{(3)} = u_\alpha^{(3)}, \qquad v_\alpha^{(4)} = \varepsilon_{\beta\alpha}x_\beta x_3 \\ &v_3^{(\beta)} = x_\beta x_3, \qquad v_3^{(3)} = x_3, \qquad v_3^{(4)} = \varphi, \qquad \omega_\alpha^{(\beta)} = \varepsilon_{\alpha\beta}x_3 \\ &\omega_\alpha^{(3)} = 0, \qquad \omega_\alpha^{(4)} = \psi_\alpha, \qquad \omega_3^{(i)} = \varphi_3^{(i)}, \qquad \omega_3^{(4)} = x_3 \end{aligned} \tag{5.3.35}$$

Clearly, $v^{(j)} \in \Lambda$, $(j = 1, 2, 3, 4)$. The relations 5.3.34 and 5.3.35 lead to

$$\begin{aligned} &v_\alpha = -\frac{1}{2}x_3^2 a_\alpha + a_4\varepsilon_{\beta\alpha}x_\beta x_3 + \sum_{i=1}^{3} a_i u_\alpha^{(i)}, \qquad v_3 = (a_\rho x_\rho + a_3)x_3 + a_4\varphi \\ &\omega_\alpha = \varepsilon_{\alpha\beta}a_\beta x_3 + a_4\psi_a, \qquad \omega_3 = a_4 x_3 + \sum_{i=1}^{3} a_i\varphi_3^{(i)} \end{aligned} \tag{5.3.36}$$

By Equations 5.3.21, 5.3.23, and 5.3.29, we arrive at

$$t_{ij}(v) = \sum_{s=1}^{4} a_s t_{ij}(v^{(s)}), \qquad m_{ij}(v) = \sum_{s=1}^{4} a_s m_{ij}(v^{(s)}) \tag{5.3.37}$$

where

$$\begin{aligned} &t_{\alpha\beta}(v^{(\rho)}) = \lambda x_\rho\delta_{\alpha\beta} + T_{\alpha\beta}(w^{(\rho)}), \qquad t_{\alpha\beta}(v^{(3)}) = \lambda\delta_{\alpha\beta} + T_{\alpha\beta}(w^{(3)}) \\ &t_{\alpha\beta}(v^{(4)}) = 0, \qquad t_{\alpha 3}(v^{(i)}) = 0, \qquad t_{\alpha 3}(v^{(4)}) = P_\alpha(\widehat{w}) - \mu\varepsilon_{\alpha\beta}x_\beta \\ &t_{3\alpha}(v^{(i)}) = 0, \qquad t_{3\alpha}(v^{(4)}) = Q_\alpha(\widehat{w}) + (\mu + \kappa)\varepsilon_{\beta\alpha}x_\beta \\ &t_{33}(v^{(\rho)}) = (\lambda + 2\mu + \kappa)x_\rho + \lambda u_{\alpha,\alpha}^{(\rho)}, \qquad t_{33}(v^{(3)}) = \lambda + 2\mu + \kappa + \lambda u_{\alpha,\alpha}^{(3)} \\ &t_{33}(v^{(4)}) = 0, \qquad m_{\nu\eta}(v^{(i)}) = 0, \qquad m_{\nu\eta}(v^{(4)}) = \alpha\delta_{\nu\eta} + H_{\nu\eta}(\widehat{w}) \\ &m_{\alpha 3}(v^{(\rho)}) = \beta\varepsilon_{\alpha\rho} + M_{\alpha 3}(w^{(\rho)}), \qquad m_{\alpha 3}(v^{(3)}) = M_{\alpha 3}(w^{(3)}), \qquad m_{\alpha 3}(v^{(4)}) = 0 \\ &m_{3\alpha}(v^{(4)}) = 0, \qquad m_{33}(v^{(i)}) = 0, \qquad m_{33}(v^{(4)}) = \alpha + \beta + \gamma + \alpha\psi_{\rho,\rho} \\ &m_{3\alpha}(v^{(\rho)}) = \gamma\varepsilon_{\alpha\rho} + \beta\varphi_{3,\alpha}^{(\rho)}, \qquad m_{3\alpha}(v^{(3)}) = \beta\varphi_{3,\alpha}^{(3)} \end{aligned} \tag{5.3.38}$$

The conditions on the end Σ_1 are

$$R_\alpha(v) = 0, \qquad R_3(v) = F_3, \qquad H_i(v) = M_i \tag{5.3.39}$$

Since $v_{,3} \in \mathcal{R}^*$, by Theorem 5.3.1, we find that $R_\alpha(v) = 0$. The other conditions from Equations 5.3.39 furnish the following system for the constants $a_1, a_2, a_3,$

and a_4

$$D_{\alpha j}a_j = \varepsilon_{\alpha\rho}M_\rho, \qquad D_{3j}a_j = -F_3, \qquad Da_4 = -M_3 \tag{5.3.40}$$

where

$$\begin{aligned}
D_{\alpha\eta} &= \int_{\Sigma_1} \{x_\alpha[(\lambda+2\mu+\kappa)x_\eta + \lambda u^{(\eta)}_{\nu,\nu}] - \beta\varepsilon_{\alpha\rho}\varphi^{(\eta)}_{3,\rho} + \gamma\delta_{\alpha\eta}\}da \\
D_{\alpha 3} &= \int_{\Sigma_1} \{x_\alpha(\lambda+2\mu+\kappa+\lambda u^{(3)}_{\nu,\nu}) - \beta\varepsilon_{\alpha\rho}\varphi^{(3)}_{3,\rho}\}da \\
D_{3\alpha} &= \int_{\Sigma_1} [(\lambda+2\mu+\kappa)x_\alpha + \lambda u^{(\alpha)}_{\nu,\nu}]da \\
D_{33} &= \int_{\Sigma} [\lambda+2\mu+\kappa+\lambda u^{(3)}_{\rho,\rho}]da \\
D &= \int_{\Sigma_1} [\mu\varepsilon_{\alpha\beta}x_\alpha\varphi_{,\beta} + \kappa x_\alpha\psi_\alpha + (\mu+\kappa)x_\rho x_\rho + \alpha\psi_{\rho,\rho} + \alpha+\beta+\gamma]da
\end{aligned} \tag{5.3.41}$$

We note that the constants D_{ij} and D can be calculated after the functions $\{u^{(i)}_\alpha, \varphi^{(i)}_3\}$, $(i = 1, 2, 3)$, and $(\varphi, \psi_1, \psi_2)$ are found.

Let us prove that the system 5.3.40 can always be solved for a_1, a_2, a_3, and a_4. In the view of Equations 5.3.7 and 5.3.34,

$$U(v) = \frac{1}{2}\sum_{i,j=1}^{4} \langle v^{(i)}, v^{(j)}\rangle a_i a_j$$

Since $W(v)$ is positive definite and $v^{(i)}$ is not a rigid deformation, it follows that

$$\det\langle v^{(i)}, v^{(j)}\rangle \neq 0, \quad (i, j = 1, 2, 3, 4) \tag{5.3.42}$$

By Equations 5.3.10, 5.3.11, 5.3.35, 5.3.38, and $v^{(i)} \in \Lambda$, $(i = 1, 2, 3, 4)$,

$$\begin{aligned}
\langle v^{(\alpha)}, v^{(\beta)}\rangle &= \int_{\partial B} \left[v^{(\alpha)}_j s_j(v^{(\beta)}) + \omega^{(\alpha)}_j m_j(v^{(\beta)})\right]da \\
&= -\frac{1}{2}h^2\int_{\Sigma_2} t_{3\alpha}(v^{(\beta)})da + hD_{\alpha\beta} \\
\langle v^{(\alpha)}, v^{(3)}\rangle &= hD_{\alpha 3}, \qquad \langle v^{(3)}, v^{(3)}\rangle = hD_{33} \\
\langle v^{(i)}, v^{(4)}\rangle &= 0, \qquad \langle v^{(4)}, v^{(4)}\rangle = hD
\end{aligned}$$

If we use the relations $v^{(i)} \in \Lambda$ and $v^{(i)}_{,3} \in \mathcal{R}^*$, then by Theorem 5.3.1 and Equations 5.3.16, we find that $R_\alpha(v^{(i)}) = 0$. Thus,

$$\langle v^{(i)}, v^{(j)}\rangle = hD_{ij}, \qquad \langle v^{(i)}, v^{(4)}\rangle = 0, \qquad \langle v^{(4)}, v^{(4)}\rangle = hD \tag{5.3.43}$$

It follows from Equations 5.3.11, 5.3.42, and 5.3.43 that $D_{ij} = D_{ji}$ and

$$\det(D_{ij}) \neq 0, \qquad D \neq 0 \tag{5.3.44}$$

Thus, the system 5.3.40 uniquely determines the constants a_1, a_2, a_3, and a_4. □

Remark 1. The proof of this theorem offers a constructive procedure to obtain a solution of the extension–bending–torsion problem. This solution has the form 5.3.36 where the functions $u_\alpha^{(i)}, \varphi_3^{(i)}$, $(i = 1, 2, 3)$, are solutions of the plane strain problems 5.3.30 and 5.3.31, the set of functions $(\varphi, \psi_1, \psi_2)$ is characterized by the boundary-value problem 5.3.26, and the constants a_1, a_2, a_3, and a_4 are determined by Equations 5.3.40.

Remark 2. The functions $u_\alpha^{(3)}$ and $\varphi_3^{(3)}$ can be determined in the following way. The corresponding equilibrium equations and boundary conditions are satisfied if one choose

$$T_{\beta\alpha}(w^{(3)}) = -\lambda\delta_{\alpha\beta}, \qquad M_{\alpha 3}(w^{(3)}) = 0$$

Since λ is constant, the above functions satisfy the compatibility conditions [83]. By the constitutive equations,

$$u_{1,1}^{(3)} = u_{2,2}^{(3)} = -\nu, \qquad u_{1,2}^{(3)} + \varphi_3^{(3)} = u_{2,1}^{(3)} - \varphi_3^{(3)} = 0, \qquad \varphi_{3,\alpha}^{(3)} = 0$$

where $\nu = \lambda(2\lambda + 2\mu + \kappa)^{-1}$. The integration of these equations yields

$$u_\alpha^{(3)} = -\nu x_\alpha, \qquad \varphi_3^{(3)} = 0$$

modulo a plane rigid displacement.

From Equations 5.3.41, we get

$$D_{\alpha 3} = D_{3\alpha} = AEx_\alpha^0, \qquad D_{33} = EA \tag{5.3.45}$$

where A is the area of the cross section, x_α^0 are the coordinates of the centroid of Σ_1 and

$$E = (2\mu + \kappa)(3\lambda + 2\mu + \kappa)/(2\lambda + 2\mu + \kappa)$$

We note that we established the relations $D_{3\alpha} = AEx_\alpha^0$ without recourse to the determination of $u_\nu^{(\rho)}$.

Remark 3. If the rectangular cartesian coordinate frame is chosen in such a way that the origin O coincides with the centroid of the cross section Σ_1, then the problems of extension and bending can be treated independently one of the other.

In view of Equations 5.3.36, 5.3.40, and 5.3.45, if $x_\alpha^0 = 0$, then we find the following solutions:

1. *Extension solution* $(F_\alpha = 0, M_i = 0)$

$$u_\alpha = -a_3\nu x_\alpha, \qquad u_3 = a_3 x_3, \qquad \varphi_i = 0$$

where

$$EAa_3 = -F_3$$

2. *Bending solution* $(F_i = 0, M_3 = 0)$

$$\begin{aligned} u_\alpha &= -\frac{1}{2}a_\alpha x_3^2 + \sum_{\rho=1}^{2} a_\rho u_\alpha^{(\rho)}, \qquad u_3 = a_\beta x_\beta x_3 \\ \varphi_\alpha &= \varepsilon_{\alpha\beta} a_\beta x_3, \qquad \varphi_3 = \sum_{\rho=1}^{2} a_\rho \varphi_3^{(\rho)} \end{aligned} \tag{5.3.46}$$

where the functions $u_\alpha^{(\rho)}, \varphi_3^{(\rho)}$, $(\rho = 1, 2)$, are solutions of the corresponding plane strain problems from Equations 5.3.30 and 5.3.31, and the constants a_1 and a_2 are determined by

$$D_{\alpha\beta} a_\beta = \varepsilon_{\alpha\eta} M_\eta$$

3. *Torsion solution* $(F_i = 0, M_\alpha = 0)$

$$u_\alpha = \varepsilon_{\beta\alpha} a_4 x_\beta x_3, \qquad u_3 = a_4 \varphi, \qquad \varphi_\alpha = a_4 \psi_\alpha, \qquad \varphi_3 = a_4 x_3 \tag{5.3.47}$$

where the torsion functions φ, ψ_1, and ψ_2 are characterized by the boundary-value problem 5.3.26 and a_4 is given by

$$Da_4 = -M_3$$

D is the torsional rigidity for micropolar cylinders.

The solution of the torsion problem for a circular cylinder has been presented in Refs. 188 and 338 (see the solution of Exercise 5.7.3). The extension and bending of a circular cylinder has been studied in Refs. 188–190 (see the solution of Exercise 5.7.2).

5.3.3 Flexure

By a solution of flexure problem, we mean a vector field $u \in \Lambda$ that satisfies the conditions

$$R_\alpha(u) = F_\alpha, \qquad R_3(u) = 0, \qquad H_i(u) = 0 \tag{5.3.48}$$

Let $\widehat{a} = (a_1, a_2, a_3, a_4)$. We denote, for the remainder of this chapter, by $v\{\widehat{a}\}$ the vector field v defined by Equation 5.3.36.

With the help of Corollaries 5.3.1 and 5.3.2 and Theorem 5.3.2, we are led to seek a solution of the flexure problem in the form

$$u = \int_0^{x_3} v\{\widehat{b}\}dx_3 + v\{\widehat{c}\} + w' \tag{5.3.49}$$

where $\widehat{b} = (b_1, b_2, b_3, b_4)$ and $\widehat{c} = (c_1, c_2, c_3, c_4)$ are two constant four-dimensional vectors, and $w' = (w_i', \chi_i')$ is a vector independent of x_3 such that $w' \in C^1(\overline{\Sigma}_1) \cap C^2(\Sigma_1)$.

Theorem 5.3.3 *Let Y be the set of all vector fields of the form 5.3.49. Then there exists a vector field $u^0 \in Y$ which is solution of the problem (P_2).*

Proof. Let $u^0 \in Y$. Next, we prove that the vector field $w' = (w_i', \chi_i')$ and the constants b_i, c_i, $(i = 1, 2, 3, 4)$, can be determined so that $u^0 \in K_{II}(F_1, F_2)$. First, we determine the vector $\widehat{b}$. Thus, if $u^0 \in K_{II}(F_1, F_2)$, then by Corollary 5.3.2 and Equation 5.3.49,

$$v\{\widehat{b}\} \in K_I(0, F_2, -F_1, 0) \tag{5.3.50}$$

In view of Equations 5.3.40 and 5.3.50, we obtain

$$D_{\alpha j} b_j = -F_\alpha, \qquad D_{3j} b_j = 0, \qquad b_4 = 0 \tag{5.3.51}$$

This system determines b_1, b_2, and b_3. From Equations 5.3.36, 5.3.49, and 5.3.51, we find that

$$\begin{aligned}
u_\alpha^0 &= -\frac{1}{6} b_\alpha x_3^3 - \frac{1}{2} c_\alpha x_3^2 - c_4 \varepsilon_{\alpha\beta} x_\beta x_3 + \sum_{i=1}^{3} (b_i x_3 + c_i) u_\alpha^{(i)} + w_\alpha' \\
u_3^0 &= \frac{1}{2}(b_\rho x_\rho + b_3) x_3^2 + (c_\rho x_\rho + c_3) x_3 + c_4 \varphi + w_3' \\
\varphi_\alpha^0 &= \frac{1}{2} \varepsilon_{\alpha\beta} b_\beta x_3^2 + \varepsilon_{\alpha\beta} c_\beta x_3 + c_4 \psi_\alpha + \chi_\alpha' \\
\varphi_3^0 &= c_4 x_3 + \sum_{i=1}^{3} (b_i x_3 + c_i) \varphi_3^{(i)} + \chi_3'
\end{aligned} \tag{5.3.52}$$

where $(u_\alpha^{(i)}, \varphi_3^{(i)})$, $(i = 1, 2, 3)$, are characterized by Equations 5.3.30 and 5.3.31. It follows from Equations 5.3.1, 5.3.3, and 5.3.52 that

$$\begin{aligned}
t_{\alpha\beta}(u^0) &= \lambda[(b_\rho x_\rho + b_3) x_3 + c_\rho x_\rho + c_3] \delta_{\alpha\beta} \\
&\quad + \sum_{i=1}^{3} (b_i x_3 + c_i) T_{\alpha\beta}(w^{(i)}) + T_{\alpha\beta}(\omega^0) \\
t_{33}(u^0) &= (\lambda + 2\mu + \kappa)[(b_\rho x_\rho + b_3) x_3 + c_\rho x_\rho + c_3] \\
&\quad + \lambda \sum_{i=1}^{3} (b_i x_3 + c_i) u_{\rho,\rho}^{(i)} + \lambda w_{\rho,\rho}'
\end{aligned}$$

$$t_{\alpha 3}(u^0) = P_\alpha(\overline{w}) + c_4[P_\alpha(\widehat{w}) + \mu\varepsilon_{\beta\alpha}x_\beta] + \mu\sum_{i=1}^{3} b_i u_\alpha^{(i)}$$
$$t_{3\alpha}(u^0) = [Q_\alpha(\overline{w}) + (\mu+\kappa)\varepsilon_{\beta\alpha}x_\beta] + (\mu+\kappa)\sum_{i=1}^{3} b_i u_\alpha^{(i)}$$
$$m_{\lambda\nu}(u^0) = H_{\lambda\nu}(\overline{w}) + c_4[\alpha\delta_{\lambda\nu} + H_{\lambda\nu}(\widehat{w})] + \alpha\delta_{\lambda\nu}\sum_{i=1}^{3} b_i\varphi_3^{(i)} \quad (5.3.53)$$
$$m_{33}(u^0) = \alpha(c_4\psi_{\rho,\rho} + \chi'_{\rho,\rho}) + (\alpha+\beta+\gamma)\left(c_4 + \sum_{i=1}^{3} b_i\varphi_3^{(i)}\right)$$
$$m_{\alpha 3}(u^0) = \beta\varepsilon_{\alpha\nu}(b_\nu x_3 + c_\nu) + \sum_{i=1}^{3}(b_i x_3 + c_i)M_{\alpha 3}(w^{(i)}) + M_{\alpha 3}(\omega^0)$$
$$m_{3\alpha}(u^0) = \gamma\varepsilon_{\alpha\nu}(b_\nu x_3 + c_\nu) + \beta\sum_{i=1}^{3}(b_i x_3 + c_i)\varphi_{3,\alpha}^{(i)} + \beta\chi'_{3,\alpha}$$

where we have used the notations

$$\omega^0 = (w'_1, w'_2, 0, 0, 0, \chi'_3), \qquad \overline{w} = (0, 0, w'_3, \chi'_1, \chi'_2, 0)$$

If we substitute Equation 5.3.53 into equations of equilibrium, we find, with the aid of Equations 5.3.26 and 5.3.30, that

$$(T_{\beta\alpha}(\omega^0))_{,\beta} = 0, \qquad (M_{\rho 3}(\omega^0))_{,\rho} + \varepsilon_{\alpha\beta}T_{\alpha\beta}(\omega^0) = 0 \text{ on } \Sigma_1 \quad (5.3.54)$$

and

$$L_i\overline{w} = \xi_i \text{ on } \Sigma_1 \quad (5.3.55)$$

where

$$\xi_\nu = -\gamma\varepsilon_{\nu\rho}b_\rho - \sum_{i=1}^{3} b_i\left[(\alpha+\beta)\varphi_{3,\nu}^{(i)} - \varepsilon_{\nu\beta}\kappa u_\beta^{(i)}\right]$$
$$\xi_3 = -(\lambda+2\mu+\kappa)(b_\rho x_\rho + b_3) - (\lambda+\mu)\sum_{i=1}^{3} b_i u_{\rho,\rho}^{(i)}$$

In view of Equations 5.3.26 and 5.3.31, the conditions on the lateral boundary reduce to

$$T_{\beta\alpha}(\omega^0)n_\beta = 0, \qquad M_{\alpha 3}(\omega^0)n_\alpha = 0 \text{ on } \Gamma \quad (5.3.56)$$

and

$$N_\rho\overline{w} = -\alpha n_\rho\sum_{i=1}^{3} b_i\varphi_3^{(i)}, \qquad N_3\overline{w} = -\mu n_\alpha\sum_{i=1}^{3} b_i u_\alpha^{(i)} \text{ on } \Gamma \quad (5.3.57)$$

The relations 5.3.54 and 5.3.56 constitute a plane strain problem corresponding to null data. We conclude that $w'_\alpha = 0$ and $\chi'_3 = 0$. The necessary and sufficient condition for the existence of a solution to the boundary-value problem 5.3.55 and 5.3.57 reduces to

$$D_{3i}b_i = 0$$

This condition is satisfied on the basis of Equations 5.3.51. Thus, the functions w_3' and χ_α' are characterized by the boundary-value problem 5.3.55 and 5.3.57. The conditions $R_\alpha(u) = F_\alpha$ are satisfied by Equations 5.3.50 and 5.3.51 and Theorem 5.3.1. The conditions $R_3(u) = 0$, $\mathbf{H}(u) = \mathbf{0}$ reduce to

$$D_{ij}c_j = 0 \tag{5.3.58}$$

and

$$\begin{aligned} Dc_4 = -\int_\Sigma \Bigg\{ & \varepsilon_{\alpha\beta}x_\alpha \left[\mu w_{3,\beta}' + \varepsilon_{\nu\beta}\kappa\chi_\nu' + (\mu+\kappa)\sum_{i=1}^{3} b_i u_\beta^{(i)}\right] \\ & + (\alpha+\beta+\gamma)\sum_{i=1}^{3} b_i\varphi_3^{(i)} + \alpha\chi_{\rho,\rho}'\Bigg\} da \end{aligned} \tag{5.3.59}$$

By Equations 5.3.44 and 5.3.58, we conclude that $c_i = 0$. The constant c_4 is given by Equation 5.3.59. □

The flexure problem for a circular cylinder was investigated in Refs. 189 and 190.

Remark 4. The above theorem offers a constructive procedure to obtain a solution of the flexure problem. This solution has the form 5.3.52 where $w_\alpha' = \chi_3' = 0, c_i = 0$, the functions $u_\alpha^{(i)}, \varphi_3^{(i)}$, $(i = 1, 2, 3)$, are solutions of the plane strain problems 5.3.30 and 5.3.31, the functions φ and ψ_α are characterized by the boundary-value problem 5.3.36, the functions w_3' and χ_α' are characterized by the boundary-value problem 5.3.55 and 5.3.57, and the constants b_i and c_4 are determined by Equations 5.3.51 and 5.3.59.

Remark 5. If the rectangular cartesian coordinate frame is chosen in such a way that the origin O coincides with the centroid of the cross section Σ_1, then Equation 5.3.45 implies $D_{3\alpha} = 0$. It follows from Equation 5.3.51 that $b_3 = 0$. In this case, Equation 5.3.52 yields the following solution of the flexure problem

$$\begin{aligned} u_\alpha^0 &= -\frac{1}{6}b_\alpha x_3^3 + c_4\varepsilon_{\beta\alpha}x_\beta x_3 + x_3\sum_{\rho=1}^{2} b_\rho u_\alpha^{(\rho)} \\ u_3^0 &= \frac{1}{2}(b_1x_1 + b_2x_2)x_3^2 + c_4\varphi + w_3' \\ \varphi_\alpha^0 &= \frac{1}{2}\varepsilon_{\alpha\beta}b_\beta x_3^2 + c_4\psi_\alpha + \chi_\alpha' \\ \varphi_3^0 &= c_4x_3 + x_3\sum_{\rho=1}^{2} b_\rho\varphi_3^{(\rho)} \end{aligned} \tag{5.3.60}$$

where the constants b_α are determined by

$$D_{\alpha\beta} b_\beta = -F_\alpha$$

and c_4 is given by Equation 5.3.59.

The stress tensor and couple stress tensor are

$$t_{\alpha\beta}(u^0) = \lambda b_\rho x_\rho x_3 \delta_{\alpha\beta} + x_3 \sum_{\rho=1}^{2} b_\rho T_{\alpha\beta}(w^{(\rho)})$$

$$t_{33}(u^0) = (\lambda + 2\mu + \kappa) b_\rho x_\rho x_3 + x_3 \lambda \sum_{\rho=1}^{2} b_\rho u_{\nu,\nu}^{(\rho)}$$

$$t_{\alpha 3}(u^0) = P_\alpha(\overline{w}) + c_4[P_\alpha(\widehat{w}) + \mu \varepsilon_{\beta\alpha} x_\beta] + \mu \sum_{\rho=1}^{2} b_\rho u_\alpha^{(\rho)}$$

$$t_{3\alpha}(u^0) = Q_\alpha(\overline{w}) + c_4[Q_\alpha(\widehat{w}) + \varepsilon_{\beta\alpha}(\mu + \kappa) x_\beta] + (\mu + \kappa) \sum_{\rho=1}^{2} b_\rho u_\alpha^{(\rho)}$$

$$m_{\lambda\nu}(u^0) = H_{\lambda\nu}(\overline{w}) + c_4[H_{\lambda\nu}(\widehat{w}) + \alpha \delta_{\lambda\nu}] + \alpha \delta_{\lambda\nu} \sum_{\rho=1}^{2} b_\rho \varphi_3^{(\rho)}$$

$$m_{33}(u^0) = \alpha(c_4 \psi_{\rho,\rho} + \chi'_{\rho,\rho}) + (\alpha + \beta + \gamma)\left(c_4 + \sum_{\rho=1}^{2} b_\rho \varphi_3^{(\rho)}\right)$$

$$m_{\alpha 3}(u^0) = \beta \varepsilon_{\alpha\nu} x_3 b_\nu + x_3 \sum_{\rho=1}^{2} b_\rho M_{\alpha 3}(w^{(\rho)})$$

$$m_{3\alpha}(u^0) = \gamma \varepsilon_{\alpha\nu} b_\nu x_3 + \beta x_3 \sum_{\rho=1}^{2} b_\rho \varphi_{3,\alpha}^{(\rho)}$$

Remark 6. If we replace Equations 5.3.58 and 5.3.59 by

$$D_{\alpha j} c_j = \varepsilon_{\alpha\rho} M_\rho, \qquad D_{3j} c_j = -F_3$$

$$Dc_4 = -M_3 - \int_{\Sigma_1} \left\{ \varepsilon_{\alpha\beta} x_\alpha \left[\mu w'_{3,\beta} + \varepsilon_{\nu\beta} \kappa \chi'_\nu + (\mu + \kappa) \sum_{i=1}^{3} b_i u_\beta^{(i)} \right] \right.$$
$$\left. + (\alpha + \beta + \gamma) \sum_{i=1}^{3} b_i \varphi_3^{(i)} + \alpha \chi'_{\rho,\rho} \right\} da$$

then the vector field u^0 defined by Equation 5.3.52 belongs to $K(\mathbf{F}, \mathbf{M})$.

Remark 7. The plane problems 5.3.30 and 5.3.31 can be reduced to plane strain problems without body loads. Let us introduce the functions $u_\alpha^{*(\eta)}$,

$\varphi_3^{*(\eta)}$, $(\eta = 1, 2)$, by

$$u_1^{*(1)} = u_1^{(1)} + \frac{1}{2}\nu(x_1^2 - x_2^2), \qquad u_2^{*(1)} = u_2^{(1)} + \nu x_1 x_2, \qquad \varphi_3^{*(1)} = \varphi_3^{(1)} + \nu x_2$$
$$u_1^{*(2)} = u_1^{(2)} + \nu x_1 x_2, \qquad u_2^{*(2)} = u_2^{(2)} - \frac{1}{2}\nu(x_1^2 - x_2^2), \qquad \varphi_3^{*(2)} = \varphi_3^{(2)} - \nu x_1 \tag{5.3.61}$$

where

$$\nu = \lambda/(2\lambda + 2\mu + \kappa) \tag{5.3.62}$$

We define $e_{\alpha\beta}^{*(\eta)}, \kappa_{\alpha 3}^{*(\eta)}, t_{\alpha\beta}^{*(\eta)}$, and $m_{\alpha 3}^{*(\eta)}$ by

$$e_{\alpha\beta}^{*(\eta)} = u_{\beta,\alpha}^{*(\eta)} + \varepsilon_{\beta\alpha}\varphi_3^{*(\eta)}, \qquad \kappa_{\alpha 3}^{*(\eta)} = \varphi_{3,\alpha}^{*(\eta)}$$
$$t_{\alpha\beta}^{*(\eta)} = \lambda e_{\rho\rho}^{*(\eta)}\delta_{\alpha\beta} + (\mu + \kappa)e_{\alpha\beta}^{*(\eta)} + \mu e_{\beta\alpha}^{*(\eta)}, \qquad m_{\alpha 3}^{*(\eta)} = \gamma\kappa_{\alpha 3}^{*(\eta)}, \qquad (\eta = 1, 2) \tag{5.3.63}$$

It follows from Equations 5.3.1, 5.3.29, 5.3.61, and 5.3.62 that

$$e_{\alpha\beta}(w^{(\eta)}) = e_{\alpha\beta}^{*(\eta)} - \nu\delta_{\alpha\beta}x_\eta, \qquad \kappa_{\alpha 3}(w^{(\eta)}) = \kappa_{\alpha 3}^{*(\eta)} + \varepsilon_{\alpha\eta}\nu$$
$$T_{\alpha\beta}(w^{(\eta)}) = t_{\alpha\beta}^{*(\eta)} - \lambda\delta_{\alpha\beta}x_\eta, \qquad M_{\alpha 3}(w^{(\eta)}) = m_{\alpha 3}^{*(\eta)} + \gamma\varepsilon_{\alpha\eta}\nu \tag{5.3.64}$$

By Equations 5.3.30, 5.3.31, and 5.3.64, we obtain the equations

$$t_{\beta\alpha,\beta}^{*(\eta)} = 0, \qquad m_{\alpha 3,\alpha}^{*(\eta)} + \varepsilon_{\alpha\beta}t_{\alpha\beta}^{*(\eta)} = 0 \text{ on } \Sigma_1 \tag{5.3.65}$$

and the boundary conditions

$$t_{\beta\alpha}^{*(\eta)}n_\beta = 0, \qquad m_{\alpha 3}^{*(\eta)}n_\alpha = (\beta + \gamma\nu)\varepsilon_{\eta\alpha}n_\alpha \text{ on } \Gamma \tag{5.3.66}$$

We denote by $\mathfrak{M}^{(\eta)}$, $(\eta = 1, 2)$, the plane strain problem characterized by Equations 5.3.63 and 5.3.65 on Σ_1, and the boundary conditions 5.3.66 on Γ. If we substitute Equation 5.3.61 into Equation 5.3.36 and use the relations $u_\alpha^{(3)} = -\nu x_\alpha$, $\varphi_3^{(3)} = 0$, then the solution of the problem of extension and bending can be written in the form

$$u_1 = -\frac{1}{2}a_1\left[x_3^2 + \nu(x_1^2 - x_2^2)\right] - a_2\nu x_1 x_2 - a_3\nu x_1 + a_1 u_1^{*(1)} + a_2 u_1^{*(2)}$$
$$u_2 = -a_1\nu x_1 x_2 - \frac{1}{2}a_2\left[x_3^2 - \nu(x_1^2 - x_2^2)\right] - a_3\nu x_2 + a_1 u_2^{*(1)} + a_2 u_2^{*(2)}$$
$$u_3 = (a_1 x_1 + a_2 x_2 + a_3)x_3, \qquad \varphi_\alpha = \varepsilon_{\alpha\beta}a_\beta x_3$$
$$\varphi_3 = -a_1\nu x_2 + a_2\nu x_1 + a_1\varphi_3^{*(1)} + a_2\varphi_3^{*(2)} \tag{5.3.67}$$

where $\{u_\alpha^{*(\eta)}, \varphi_3^{*(\eta)}\}$ is the solution of the problem $\mathfrak{M}^{(\eta)}$. Similarly, on the basis of Equations 5.3.61 and 5.3.52, we can express the solution of the flexure problem in terms of the solutions of the problems $\mathfrak{M}^{(\eta)}$.

5.4 Minimum Principles

In this section, we present minimum strain-energy characterizations of the solutions obtained in Section 5.3. We assume that the origin O coincides with the centroid of Σ_1. Since in the extension solution the microrotation vector vanishes, we renounce to study this solution. First, we study the bending problem. We denote by A_I the set of all equilibrium vector fields u that satisfy the conditions

$$\begin{gathered} s_i(u) = 0, \qquad m_i(u) = 0 \text{ on } \Pi, \qquad t_{3\rho}(u) = 0, \\ m_{33}(u) = 0 \text{ on } \Sigma_\beta, \qquad H_\alpha(u) = M \end{gathered} \tag{5.4.1}$$

Theorem 5.4.1 *Let v be the solution 5.3.46 of the bending problem, corresponding to a couple of moment $\mathbf{M}(M_1, M_2, 0)$. Then*

$$U(v) \leq U(u)$$

for every $u \in A_I$, and equality holds only if $u = v$ (modulo a rigid deformation).

Proof. We note that $v = (v_i, \omega_i) \in A_I$. Let $u \in A_I$ and define $u' = u - v$. Then u' is an equilibrium vector field that satisfies

$$\begin{gathered} s_i(u') = 0, \qquad m_i(u') = 0 \text{ on } \Pi, \qquad t_{3\rho}(u') = 0, \\ m_{33}(u') = 0 \text{ on } \Sigma_\beta, \qquad H_\alpha(u') = 0 \end{gathered} \tag{5.4.2}$$

We can write,

$$U(u) = U(u') + U(v) + \langle u', v \rangle. \tag{5.4.3}$$

It follows from Equations 5.3.10, 5.3.16, 5.3.46, and 5.4.2 that

$$\begin{aligned} \langle u', v \rangle &= \int_{\partial B} [v_i s_i(u') + \omega_i m_i(u')] da \\ &= \int_{\Sigma_2} [v_i t_{3i}(u') + \omega_i m_{3i}(u')] da - \int_{\Sigma_1} [v_i t_{3i}(u') + \omega_i m_{3i}(u')] da \\ &= h \int_{\Sigma_2} [(a_1 x_1 + a_2 x_2) t_{33}(u') + \varepsilon_{\alpha\beta} a_\beta m_{3\alpha}(u')] da \\ &= h a_\alpha [\varepsilon_{\alpha\beta} H_\beta(u') - h R_\alpha(u')] = 0 \end{aligned} \tag{5.4.4}$$

From Equations 5.4.3 and 5.4.4, we see that $U(u) \geq U(v)$, and $U(u) = U(v)$ only if u' is a rigid deformation. □

We denote by A_{II} the set of all equilibrium vector fields u that satisfy conditions

$$\begin{gathered} s_i(u) = 0, \qquad m_i(u) = 0 \text{ on } \Pi, \qquad t_{33}(u) = 0, \\ m_{3\alpha}(u) = 0 \text{ on } \Sigma_\beta, \qquad H_3(u) = M_3 \end{gathered} \tag{5.4.5}$$

Theorem 5.4.2 *Let v be the solution* 5.3.47 *of the torsion problem corresponding to the scalar torque M_3. Then*

$$U(v) \leq U(u)$$

for every $u \in A_{II}$, and equality holds only if $u = v$.

Proof. We consider $u \in A_{II}$ and $v = (v_i, \omega_i)$. Since $v \in A_{II}$, it follows that the field $u' = u - v$ is an equilibrium vector field that satisfies

$$\begin{aligned} &s_i(u') = 0, \qquad m_i(u') = 0 \text{ on } \Pi, \qquad t_{33}(u') = 0 \\ &m_{3\alpha}(u') = 0 \text{ on } \Sigma_\rho, \qquad H_3(u') = 0 \end{aligned} \tag{5.4.6}$$

In view of Equations 5.3.10, 5.3.16, 5.3.47, and 5.4.6 we obtain

$$\langle u', v \rangle = a_4 h \int_{\Sigma_2} [\varepsilon_{\beta\alpha} x_\beta t_{3\alpha}(u') + m_{33}(u')] da = -a_4 h H_3(u') = 0$$

Thus,

$$U(u) - U(v) = U(u - v)$$

The conclusion is now immediate. □

Let $\mathcal{E}$ denote the set of all equilibrium vector fields u that satisfy the conditions

$$\begin{aligned} &u_{,3} \in C^1(\overline{B}) \cap C^2(B), \qquad s_i(u) = 0, \qquad m_i(u) = 0 \text{ on } \Pi \\ &[t_{3\alpha}(u_{,3})](x_1, x_2, 0) = [t_{3\alpha}(u_{,3})](x_1, x_2, h) \\ &[m_{33}(u_{,3})](x_1, x_2, 0) = [m_{33}(u_{,3})](x_1, x_2, h) \ \ (x_1, x_2) \in \Sigma_1 \\ &R_\alpha(u) = F_\alpha \end{aligned}$$

Theorem 5.4.3 *Let u^0 be the solution* 5.3.60 *of the flexure problem corresponding to the loads F_1 and F_2. Then*

$$U(u^0_{,3}) \leq U(u_{,3})$$

for every $u \in \mathcal{E}$, and equality holds only if $u_{,3} = u^0_{,3}$.

Proof. We assume that $u \in \mathcal{E}$. Since $u^0 \in \mathcal{E}$, it follows that the vector field u' defined by $u' = u - u^0$ is an equilibrium displacement field that satisfies

$$\begin{aligned} &u'_{,3} \in C^1(\overline{B}) \cap C^2(B), \qquad s_i(u') = 0, \qquad m_i(u') = 0 \text{ on } \Pi \\ &[t_{3\alpha}(u'_{,3})](x_1, x_2, 0) = [t_{3\alpha}(u'_{,3})](x_1, x_2, h) \\ &[m_{33}(u'_{,3})](x_1, x_2, 0) = [m_{33}(u'_{,3})](x_1, x_2, h) \quad (x_1, x_2) \in \Sigma_1, \quad R_\alpha(u') = 0 \end{aligned} \tag{5.4.7}$$

We can write,

$$U(u_{,3}) = U(u'_{,3}) + U(u^0_{,3}) + \langle u'_{,3}, u^0_{,3} \rangle$$

From Equations 5.3.10, 5.3.16, 5.3.60, and 5.4.7,

$$\langle u'_{,3}, u^0_{,3}\rangle = \int_{\partial B} [u^0_{i,3} s_i(u'_{,3}) + \varphi^0_{i,3} m_i(u'_{,3})]\, da = -\frac{1}{2} b_\alpha h^2 \int_{\Sigma_2} t_{3\alpha}(u'_{,3})\, da$$
$$+ h b_\alpha \int_{\Sigma_2} [x_\alpha t_{33}(u'_{,3}) + \varepsilon_{\beta\alpha} m_{3\beta}(u'_{,3})]\, da$$
$$= \frac{1}{2} b_\alpha h^2 R_\alpha(u'_{,3}) + h b_\alpha [\varepsilon_{\alpha\beta} H_\beta(u'_{,3}) - h R_\alpha(u'_{,3})]$$
$$= -\frac{1}{2} b_\alpha h^2 R_\alpha(u'_{,3}) + h b_\alpha \varepsilon_{\alpha\beta} H_\beta(u'_{,3})$$

In view of Theorem 5.3.1 and Equation 5.4.7, we find

$$\langle u'_{,3}, u^0_{,3}\rangle = 0$$

so that

$$U(u_{,3}) = U(u'_{,3}) + U(u^0_{,3})$$

The conclusion is now immediate. □

5.5 Global Strain Measures

In this section, we study Truesdell's problem for Cosserat elastic cylinders.

We first consider Truesdell's problem for the torsion of Cosserat elastic cylinders. We denote by T the set of all solutions of the torsion problem corresponding to the scalar torque M_3. We have to solve the following problem: to define the functional $\tau(\cdot)$ on T such that

$$M_3 = D\tau(u) \ \text{ for every } u \in T \tag{5.5.1}$$

Let T_0 be the set of all equilibrium vector fields u that satisfy the conditions

$$s_i(u) = 0, \qquad m_i(u) = 0 \text{ on } \Pi, \qquad t_{33}(u) = 0, \qquad m_{3\alpha}(u) = 0 \text{ on } \Sigma_\beta$$
$$R_\alpha(u) = 0, \qquad H_3(u) = M_3 \tag{5.5.2}$$

If $u \in T_0$, then $R_3(u) = 0$, $H_\alpha(u) = 0$, so that $u \in T$. We define the real function

$$\xi \to \|u - \xi v^{(4)}\|^2_e$$

where $u \in T_0$ and $v^{(4)}$ is given by Equation 5.3.35. This function attains its minimum at

$$\gamma(u) = \langle u, v^{(4)}\rangle / \|v^{(4)}\|^2_e \tag{5.5.3}$$

Let us prove that $\gamma(u) = \tau(u)$ for every $u \in T_0$. By Equations 5.3.10, 5.3.35, and 5.5.2, we find that

$$\langle u, v^{(4)}\rangle = h H_3(u) \tag{5.5.4}$$

In view of the relations 5.3.38, we obtain

$$\|v^{(4)}\|_e^2 = hD \tag{5.5.5}$$

where D is defined in Equation 5.3.41. Thus, from Equations 5.5.3, 5.5.4, and 5.5.5, we arrive at

$$H_3(u) = D\gamma(u) \tag{5.5.6}$$

From Equations 5.5.1 and 5.5.6, we see that $\tau(u) = \gamma(u)$ for each $u \in T_0$. On the other hand, by Equations 5.3.10, 5.3.11, and 5.3.35, we find that

$$\langle u, v^{(4)} \rangle = N(u) \tag{5.5.7}$$

where

$$\begin{aligned} N(u) = &\int_{\Sigma_2} \left\{ u_\alpha \left[\mu\varphi_{,\alpha} + \kappa\varepsilon_{\beta\alpha}\psi_\beta + \frac{1}{2}\varepsilon_{\beta\alpha}(2\mu+\kappa)x_\beta \right] \right. \\ &\left. + \varphi_3(\alpha\psi_{\rho,\rho} + \beta + \gamma) \right\} da \\ &- \int_{\Sigma_1} \left\{ u_\alpha \left[\mu\varphi_{,\alpha} + \kappa\varepsilon_{\beta\alpha}\psi_\beta + \frac{1}{2}\varepsilon_{\beta\alpha}(2\mu+\kappa)x_\beta \right] \right. \\ &\left. + \varphi_3(\alpha\psi_{\rho,\rho} + \beta + \gamma) \right\} da \end{aligned}$$

In view of Equations 5.5.3, 5.5.5, and 5.5.7, we get

$$\tau(u) = \frac{1}{hD} N(u) \quad \text{for each } u \in T_0$$

This relation defines the generalized twist on the subclass T_0 of solutions to the torsion problem. By Equation 5.5.1, we interpret the right-hand side of the above relation as the global measure of strain appropriate to torsion.

In what follows we assume that the rectangular cartesian coordinate is chosen in such a way that the origin O coincides with the centroid of the cross section Σ_1.

Truesdell's problem can be set also for the flexure. Thus we are led to the following problem: to define the functionals $\eta_\alpha(\cdot)$ on $K_{II}(F_1, F_2)$ such that

$$D_{\alpha\rho}\eta_\rho(u) = -F_\alpha \tag{5.5.8}$$

for each $u \in K_{II}(F_1, F_2)$.

We denote by G the set of all equilibrium vector fields u that satisfy the conditions

$$\begin{aligned} &u_{,3} \in C^1(\overline{B}) \cap C^2(B), \qquad s_i(u) = 0, \qquad m_i(u) = 0 \text{ on } \Pi \\ &[t_{3\alpha}(u_{,3})](x_1, x_2, 0) = [t_{3\alpha}(u_{,3})](x_1, x_2, h) \\ &[m_{33}(u_{,3})](x_1, x_2, 0) = [m_{33}(u_{,3})](x_1, x_2, h), \ (x_1, x_2) \in \Sigma_1 \\ &R_\alpha(u) = F_\alpha, \qquad R_3(u) = 0, \qquad \mathbf{H}(u) = \mathbf{0} \end{aligned} \tag{5.5.9}$$

If $u \in G$, then $u \in K_{II}(F_1, F_2)$. Let us consider the real function f defined by

$$f(\xi_1, \xi_2) = 2U\big(u_{,3} - \xi_1 v^{(1)} - \xi_2 v^{(2)}\big) \tag{5.5.10}$$

where $u \in G$ and $v^{(\rho)}$, $(\rho = 1, 2)$, are given by Equations 5.3.35. By Equations 5.3.43 and 5.5.10,

$$f = hD_{\alpha\beta}\xi_\alpha\xi_\beta - 2\xi_\alpha\langle u_{,3}, v^{(\alpha)}\rangle + \langle u_{,3}, u_{,3}\rangle$$

Since $D_{\alpha\beta}$ is positive definite, f will be a minimum at $(\rho_1(u), \rho_2(u))$ if and only if $(\rho_1(u), \rho_2(u))$ is the solution of the following system of equations

$$hD_{\alpha\beta}\rho_\beta(u) = \langle u_{,3}, v^{(\alpha)}\rangle \tag{5.5.11}$$

Let us prove that $\rho_\alpha(u) = \eta_\alpha(u)$, $(\alpha = 1, 2)$, for every $u \in G$. By Equations 5.3.10, 5.3.16, 5.3.35, and 5.5.9, we obtain

$$\begin{aligned}\langle u_{,3}, v^{(\alpha)}\rangle &= \int_{\partial B}\big[v_i^{(\alpha)} s_i(u_{,3}) + \omega_i^{(\alpha)} m_i(u_{,3})\big]da \\ &= -\frac{1}{2}h^2 R_\alpha(u_{,3}) + h\varepsilon_{\alpha\beta}H_\beta(u_{,3})\end{aligned} \tag{5.5.12}$$

By Equation 5.5.12 and Theorem 5.3.1, we find

$$\langle u_{,3}, v^{(\alpha)}\rangle = -hR_\alpha(u) \tag{5.5.13}$$

It follows from Equations 5.5.11 and 5.5.13 that

$$D_{\alpha\beta}\rho_\beta(u) = -R_\alpha(u) \tag{5.5.14}$$

Thus, from Equations 5.5.8, 5.5.9, and 5.5.14, we conclude that $\eta_\alpha(u) = \rho_\alpha(u)$, $(\alpha = 1, 2)$, for each $u \in G$.

On the other hand, by Equations 5.3.10, 5.3.11, and 5.3.38, we find

$$\langle u_{,3}, v^{(\alpha)}\rangle = \int_{\partial B}\big[u_{i,3}t_{3i}(v^{(\alpha)}) + \varphi_{i,3}m_{3i}(v^{(\alpha)})\big]da = S_\alpha(u) \tag{5.5.15}$$

where

$$\begin{aligned}S_\rho(u) = &\int_{\Sigma_2}\big\{u_{3,3}\big[(\lambda + 2\mu + \kappa)x_\rho + \lambda u_{\nu,\nu}^{(\rho)}\big] + \varphi_{\alpha,3}\big[\gamma\varepsilon_{\alpha\rho} + \beta\varphi_{3,\alpha}^{(\rho)}\big]\big\}da \\ &- \int_{\Sigma_1}\big\{u_{3,3}\big[(\lambda + 2\mu + \kappa)x_\rho + \lambda u_{\nu,\nu}^{(\rho)}\big] + \varphi_{\alpha,3}\big[\gamma\varepsilon_{\alpha\rho} + \beta\varphi_{3,\alpha}^{(\rho)}\big]\big\}da\end{aligned}$$

for each $u = (u_i, \varphi_i) \in G$.

From Equations 5.5.11 and 5.5.15, we get

$$D_{\alpha\beta}\eta_\beta(u) = \frac{1}{h}S_\alpha(u), \quad (\alpha = 1, 2)$$

for every $u \in G$. This system defines $\eta_\alpha(\cdot)$ on the subclass G of solutions to the flexure problem. We can interpret $\eta_\alpha(u)$ as the global measures of strain appropriate to flexure, associated with $u \in G$.

Truesdell's problem can be set and solved also for extension and bending.

5.6 Theory of Loaded Cosserat Cylinders

Now we consider that the body force **f** and the body couple **g** are prescribed on B. By an equilibrium vector field on B corresponding to the body loads $\{\mathbf{f},\mathbf{g}\}$ we mean a six-dimensional vector field $u \in C^1(\overline{B}) \cap C^2(B)$ that satisfies the equations

$$(t_{ji}(u))_{,j} + f_i = 0, \qquad (m_{ji}(u))_{,j} + \varepsilon_{ijk}t_{jk}(u) + g_i = 0 \tag{5.6.1}$$

on B. We assume that the conditions 5.3.14 are replaced by

$$s_i(u) = p_i, \qquad m_i(u) = k_i \text{ on } \Pi, \qquad \mathbf{R}(u) = \mathbf{F}, \qquad \mathbf{H}(u) = \mathbf{M} \tag{5.6.2}$$

where **p** and **k** are prescribed vector fields, and **F** and **M** are prescribed vectors. The problem of loaded cylinder consists in finding an equilibrium vector field on B that corresponds to the body loads $\{\mathbf{f},\mathbf{g}\}$ and satisfies the conditions 5.6.2.

When $\mathbf{f},\mathbf{g},\mathbf{p}$, and **k** are independent of the axial coordinate, we refer to this problem as Almansi–Michell problem. We denote by (P_3) the Almansi–Michell problem corresponding to the system of loads $\{\mathbf{F},\mathbf{M},\mathbf{f},\mathbf{g},\mathbf{p},\mathbf{k}\}$. Let $K_{III}(\mathbf{F},\mathbf{M},\mathbf{f},\mathbf{g},\mathbf{p},\mathbf{k})$ denote the class of solutions to the problem (P_3).

Theorem 5.6.1 *If $u \in C^1(\overline{B}) \cap C^2(B)$, then*

$$\begin{aligned}
R_i(u_{,3}) &= \int_{\partial\Sigma_1} s_i(u)ds - \int_{\Sigma_1} (t_{ji}(u))_{,j}da \\
H_\alpha(u_{,3}) &= \int_{\partial\Sigma_1} [\varepsilon_{\alpha\beta}x_\beta s_3(u) + m_\alpha(u)]ds - \int_{\Sigma_1} [\varepsilon_{\alpha\beta}x_\beta(t_{j3}(u))_{,j} \\
&\quad + (m_{j\alpha}(u))_{,j} + \varepsilon_{\alpha rs}t_{rs}(u)]da + \varepsilon_{\alpha\beta}R_\beta(u) \\
H_3(u_{,3}) &= \int_{\partial\Sigma_1} [\varepsilon_{\alpha\beta}x_\alpha s_\beta(u) + m_3(u)]ds - \int_{\Sigma_1} [\varepsilon_{\alpha\beta}x_\alpha(t_{j\beta}(u))_{,j} \\
&\quad + (m_{j3}(u))_{,j} + \varepsilon_{\alpha\beta}t_{\alpha\beta}(u)]da
\end{aligned}$$

The proof of this theorem is analogous to that given for Theorem 5.3.1.

Let us consider the problem (P_3). Theorem 5.6.1 has the following consequence.

Corollary 5.6.1 *If $u \in K_{III}(\mathbf{F},\mathbf{M},\mathbf{f},\mathbf{g},\mathbf{p},\mathbf{k})$ and $\mathbf{u}_{,3} \in C^1(\overline{B}) \cap C^2(B)$, then $u_{,3} \in K(\mathbf{G},\mathbf{Z})$ where*

$$\begin{aligned}
\mathbf{G} &= \int_\Gamma \mathbf{p}ds + \int_{\Sigma_1} \mathbf{f}da \\
Z_\alpha &= \int_\Gamma (\varepsilon_{\alpha\beta}x_\beta p_3 + k_\alpha)ds + \int_{\Sigma_1} (\varepsilon_{\alpha\beta}x_\beta f_3 + g_\alpha)da + \varepsilon_{\alpha\beta}F_\beta \qquad (5.6.3) \\
Z_3 &= \int_\Gamma (\varepsilon_{\alpha\beta}x_\alpha p_\beta + k_3)ds + \int_{\Sigma_1} (\varepsilon_{\alpha\beta}x_\alpha f_\beta + g_3)da
\end{aligned}$$

With the help of Corollary 5.6.1 and Equation 5.3.49, we are led to seek a solution of the problem (P_3) in the form

$$u = \int_0^{x_3}\int_0^{x_3} v\{\widehat{b}\}dx_3dx_3 + \int_0^{x_3} v\{\widehat{c}\}dx_3 + v\{\widehat{d}\} + x_3u' + u^0 \tag{5.6.4}$$

where $\widehat{b}, \widehat{c}$, and $\widehat{d}$ are unknown constant vectors, u' and u^0 are unknown vector fields independent of x_3, and $v\{\widehat{a}\}$ is defined by Equations 5.3.36.

Theorem 5.6.2 *Let V be the set of all vector fields of the form* 5.6.4. *Then there exists a vector field $\widehat{u} \in V$ which is solution of the problem (P_3).*

Proof. Let us determine $\widehat{b}, \widehat{c}, \widehat{d}, u'$, and u^0 such that $\widehat{u} \in K_{III}(\mathbf{F}, \mathbf{M}, \mathbf{f}, \mathbf{g}, \mathbf{p}, \mathbf{k})$. If $\widehat{u} \in K_{III}(\mathbf{F}, \mathbf{M}, \mathbf{f}, \mathbf{g}, \mathbf{p}, \mathbf{k})$, then by Corollaries 5.3.1, 5.6.1, and Equation 5.6.4,

$$\int_0^{x_3} v\{\widehat{b}\}dx_3 + v\{\widehat{c}\} + u' \in K(\mathbf{G}, \mathbf{Z})$$

By Theorem 5.3.3 and Equation 5.3.51, we obtain

$$D_{\alpha j}b_j = -G_\alpha, \qquad D_{3j}b_j = 0, \qquad b_4 = 0 \tag{5.6.5}$$

and $u' = (0, 0, \chi, \chi_1, \chi_2, 0)$ is characterized by

$$\begin{aligned}
L_\nu u' &= -\gamma\varepsilon_{\nu\rho}b_\rho - \sum_{i=1}^{3} b_i\left[(\alpha+\beta)\varphi_{3,\nu}^{(i)} - \kappa\varepsilon_{\nu\beta}u_\beta^{(i)}\right] \\
L_3u' &= -(\lambda+2\mu+\kappa)(b_\rho x_\rho + b_3) - (\lambda+\mu)\sum_{i=1}^{3} b_i u_{\rho,\rho}^{(i)} \text{ on } \Sigma_1 \\
N_\rho u' &= -\alpha n_\rho \sum_{i=1}^{3} b_i\varphi_3^{(i)}, \qquad N_3u' = -\mu n_\alpha \sum_{i=1}^{3} b_i u_\alpha^{(i)} \text{ on } \Gamma
\end{aligned} \tag{5.6.6}$$

Moreover, the constant vector $\widehat{c}$ is determined by

$$\begin{aligned}
D_{\alpha j}c_j &= \varepsilon_{\alpha\rho}G_\rho, \qquad D_{3j}c_j = -G_3 \\
Dc_4 &= -Z_3 - \int_{\Sigma_1}\left\{\varepsilon_{\alpha\beta}x_\alpha\left[\mu\chi_{,\beta} + \kappa\varepsilon_{\nu\beta}\chi_\nu + (\mu+\kappa)\sum_{i=1}^{3} b_i u_\beta^{(i)}\right]\right. \\
&\qquad \left. + (\alpha+\beta+\gamma)\sum_{i=1}^{3} b_i\varphi_3^{(i)} + \alpha\chi_{\rho,\rho}\right\}da
\end{aligned} \tag{5.6.7}$$

From Equations 5.6.4 and 5.6.5, we get

$$\begin{aligned}
\widehat{u}_\alpha &= -\frac{1}{24}b_\alpha x_3^4 - \frac{1}{6}c_\alpha x_3^3 - \frac{1}{2}d_\alpha x_3^2 - \frac{1}{2}c_4\varepsilon_{\alpha\beta}x_\beta x_3^2 \\
&\quad - d_4\varepsilon_{\alpha\beta}x_\beta x_3 + \sum_{i=1}^{3}\left(d_j + c_j x_3 + \frac{1}{2}b_j x_3^2\right)u_\alpha^{(i)} + w_\alpha \\
\widehat{u}_3 &= \frac{1}{6}(b_\rho x_\rho + b_3)x_3^3 + \frac{1}{2}(c_\rho x_\rho + c_3)x_3^2 + (d_\rho x_\rho + d_3)x_3 \\
&\quad + (c_4 x_3 + d_4)\varphi + x_3\chi + \Psi \\
\widehat{\varphi}_\alpha &= \varepsilon_{\alpha\beta}\left(\frac{1}{6}b_\beta x_3^3 + \frac{1}{2}c_\beta x_3^2 + d_\beta x_3\right) + (c_4 x_3 + d_4)\psi_\alpha + x_3\chi_\alpha + \Psi_\alpha \\
\widehat{\varphi}_3 &= \sum_{i=1}^{3}\left(\frac{1}{2}b_i x_3^2 + c_i x_3 + d_i\right)\varphi_3^{(i)} + \frac{1}{2}c_4 x_3^2 + d_4 x_3 + w_3
\end{aligned} \tag{5.6.8}$$

where $u^0 = (w_1, w_2, \Psi, \Psi_1, \Psi_2, w_3)$. The constitutive equations imply that

$$\begin{aligned}
t_{\alpha\beta}(\widehat{u}) &= \lambda\left[\frac{1}{2}(b_\rho x_\rho + b_3)x_3^2 + (c_\rho x_\rho + c_3)x_3 + d_\rho x_\rho + d_3\right]\delta_{\alpha\beta} \\
&\quad + \lambda(\chi + c_4\varphi)\delta_{\alpha\beta} + \sum_{i=1}^{3}\left(\frac{1}{2}b_i x_3^2 + c_i x_3 + d_i\right)T_{\alpha\beta}(w^{(i)}) + T_{\alpha\beta}(\omega^0) \\
t_{33}(\widehat{u}) &= (\lambda + 2\mu + \kappa)\left[d_\rho x_\rho + d_3 + (c_\rho x_\rho + c_3)x_3 + \frac{1}{2}(b_\rho x_\rho + b_3)x_3^2\right] \\
&\quad + (\lambda + 2\mu + \kappa)(\chi + c_4\varphi) + \lambda\sum_{i=1}^{3}\left(\frac{1}{2}b_i x_3^2 + c_i x_3 + d_i\right)u_{\rho,\rho}^{(i)} + \lambda w_{\alpha,\alpha} \\
t_{\alpha 3}(\widehat{u}) &= P_\alpha(\omega) + x_3 P_\alpha(u') + (d_4 + c_4 x_3)[P_\alpha(\widehat{w}) + \mu\varepsilon_{\beta\alpha}x_\beta] \\
&\quad + \mu\sum_{i=1}^{3}(c_i + b_i x_3)u_\alpha^{(i)} \\
t_{3\alpha}(\widehat{u}) &= Q_\alpha(\omega) + x_3 Q_\alpha(u') + (d_4 + c_4 x_3)[Q_\alpha(\widehat{w}) + (\mu + \kappa)\varepsilon_{\beta\alpha}x_\beta] \\
&\quad + (\mu + \kappa)\sum_{i=1}^{3}(c_i + b_i x_3)u_\alpha^{(i)} \\
m_{\lambda\nu}(\widehat{u}) &= H_{\lambda\nu}(\omega) + x_3 H_{\lambda\nu}(u') + (d_4 + c_4 x_3)[H_{\lambda\nu}(\widehat{w}) + \delta_{\lambda\nu}] \\
&\quad + \alpha\delta_{\lambda\nu}\sum_{i=1}^{3}(c_i + b_i x_3)\varphi_3^{(i)} \\
m_{33}(\widehat{u}) &= (\alpha + \beta + \gamma)\left[d_4 + c_4 x_3 + \sum_{i=1}^{3}(c_i + b_i x_3)\varphi_3^{(i)}\right] \\
&\quad + \alpha(d_4 + c_4 x_3)\psi_{\rho,\rho} + \alpha(\Psi_{\rho,\rho} + x_3\chi_{\rho,\rho})
\end{aligned}$$

$$
\begin{aligned}
m_{\alpha 3}(\widehat{u}) &= \beta\varepsilon_{\alpha\nu}\left(d_\nu + c_\nu x_3 + \frac{1}{2}b_\nu x_3^2\right) + \beta(\chi_\alpha + c_4\psi_\alpha) \\
&\quad + \sum_{i=1}^{3}\left(d_i + c_i x_3 + \frac{1}{2}b_i x_3^2\right)M_{\alpha 3}(w^{(i)}) + M_{\alpha 3}(\omega^0) \\
m_{3\alpha}(\widehat{u}) &= \gamma\varepsilon_{\alpha\nu}\left(d_\nu + c_\nu x_3 + \frac{1}{2}b_\nu x_3^2\right) + \gamma(\chi_\alpha + c_4\psi_\alpha) \\
&\quad + \beta\sum_{i=1}^{3}\left(d_i + c_i x_3 + \frac{1}{2}b_i x_3^2\right)\varphi_{3,\alpha}^{(i)} + \beta w_{3,\alpha}
\end{aligned} \tag{5.6.9}
$$

where $\omega^0 = (w_1, w_2, 0, 0, 0, w_3)$, $\omega = (0, 0, \Psi, \Psi_1, \Psi_2, 0)$.

The equations of equilibrium and the conditions on the lateral boundary reduce to

$$
\begin{aligned}
&\left(T_{\beta\alpha}(\omega^0)\right)_{,\beta} + h_\alpha = 0 \\
&\left(M_{\alpha 3}(\omega^0)\right)_{,\alpha} + \varepsilon_{\alpha\beta}T_{\alpha\beta}(\omega^0) + g = 0 \text{ on } \Sigma_1 \\
&T_{\beta\alpha}(\omega^0)n_\beta = p_\alpha^0, \qquad M_{\alpha 3}(\omega^0)n_\alpha = q^0 \text{ on } \Gamma
\end{aligned} \tag{5.6.10}
$$

and

$$
L_i\omega = \gamma_i \text{ on } \Sigma_1, \qquad N_i\omega = \rho_i \text{ on } \Gamma, \tag{5.6.11}
$$

where

$$
\begin{aligned}
h_\alpha &= \lambda(\chi + c_4\varphi)_{,\alpha} + Q_\alpha(u') + c_4[Q_\alpha(u') + (\mu+\kappa)\varepsilon_{\beta\alpha}x_\beta] \\
&\quad + (\mu+\kappa)\sum_{i=1}^{3} b_i w_\alpha^{(i)} + f_\alpha \\
g &= \beta(\chi_\alpha + c_4\psi_\alpha)_{,\alpha} + (\alpha+\beta+\gamma)\left(c_4 + \sum_{i=1}^{3} b_i\varphi_3^{(i)}\right) \\
&\quad + \alpha(\chi_\rho + c_4\psi_\rho)_{,\rho} + g_3 \\
p_\alpha^0 &= p_\alpha - \lambda(\chi + c_4\varphi)n_\alpha, \qquad q^0 = k_3 - \beta(\chi_\alpha + c_4\psi_\alpha)n_\alpha \\
\gamma_\nu &= -\sum_{i=1}^{3} c_i\left[\alpha\varphi_{3,\nu}^{(i)} + \beta\varphi_{3,\nu}^{(i)} - \kappa\varepsilon_{\nu\beta}u_\beta^{(i)}\right] - \gamma\varepsilon_{\nu\beta}c_\beta - g_\nu \\
\gamma_3 &= -(\lambda+\mu)\sum_{i=1}^{3} c_i w_{\alpha,\alpha}^{(i)} - (\lambda+2\mu+\kappa)(c_\rho x_\rho + c_3) - f_3 \\
\rho_\nu &= k_\nu - n_\nu\alpha\sum_{i=1}^{3} c_i\varphi_3^{(i)}, \qquad \rho_3 = p_3 - \mu\sum_{i=1}^{3} c_i w_\alpha^{(i)} n_\alpha
\end{aligned} \tag{5.6.12}
$$

With the help of Equations 5.6.5, 5.6.7, and 5.6.12, the divergence theorem, and Theorem 5.6.2, we get

$$\int_{\Sigma_1} h_\alpha da + \int_\Gamma p_\alpha^0 ds = G_\alpha - R_\alpha(\widehat{u}_{,3}) = G_\alpha - \varepsilon_{\beta\alpha} H_\beta(\widehat{u}_{,33})$$
$$= G_\alpha + D_{\alpha i} b_i = 0$$
$$\int_{\Sigma_1} (\varepsilon_{\alpha\beta} x_\alpha h_\beta + g) da + \int_\Gamma (\varepsilon_{\alpha\beta} x_\alpha p_\beta^0 + q^0) ds = Z_3 - H_3(\widehat{u}_{,3}) = 0$$
$$\int_{\Sigma_1} \gamma_3 da - \int_\Gamma \rho_3 ds = -\int_{\Sigma_1} f_3 da - \int_\Gamma p_3 ds - D_{3j} c_j = 0$$

We conclude that the necessary and sufficient conditions to solve the boundary-value problems 5.6.10 and 5.6.11 are satisfied.

It follows from Equations 5.3.15, 5.6.7, and 5.6.9 that

$$H_\alpha(\widehat{u}_{,3}) = \varepsilon_{\beta\alpha} D_{\beta i} c_i = \int_\Gamma (\varepsilon_{\alpha\beta} x_\beta p_3 + k_\alpha) ds + \int_{\Sigma_1} (\varepsilon_{\alpha\beta} x_\beta f_3 + g_\alpha) da + \varepsilon_{\alpha\beta} F_\beta$$

By Theorem 5.6.1,

$$H_\alpha(\widehat{u}_{,3}) = \int_\Gamma (\varepsilon_{\alpha\beta} x_\beta p_3 + k_\alpha) ds + \int_{\Sigma_1} (\varepsilon_{\alpha\beta} x_\beta f_3 + g_\alpha) da + \varepsilon_{\alpha\beta} R_\beta(\widehat{u})$$

The last two relations imply that $R_\alpha(\widehat{u}) = F_\alpha$.

The conditions $R_3(\widehat{u}) = F_3$ and $\mathbf{H}(\widehat{u}) = \mathbf{M}$ reduce to

$$D_{ij} d_j = r_i$$
$$Dd_4 = -M_3 - \int_{\Sigma_1} \left\{ \varepsilon_{\alpha\beta} x_\alpha \left[\mu \Psi_{,\beta} + \kappa \varepsilon_{\nu\beta} \Psi_\nu + (\mu + \kappa) \sum_{i=1}^3 c_i w_\beta^{(i)} \right] + (\alpha + \beta + \gamma) \sum_{i=1}^3 c_i \varphi_3^{(i)} + \alpha \Psi_{\nu,\nu} \right\} da \qquad (5.6.13)$$

where

$$r_\alpha = \varepsilon_{\alpha\beta} M_\beta - \int_{\Sigma_1} \{ x_\alpha [\lambda w_{\rho,\rho} + (\lambda + 2\mu + \kappa)(\chi + c_4 \varphi)] - \varepsilon_{\alpha\beta} [\gamma(\chi_\beta + c_4 \psi_\beta) + \beta w_{3,\beta}] \} da$$
$$r_3 = -F_3 - \int_{\Sigma_1} [(\lambda + 2\mu + \kappa)(\chi + c_4 \varphi) + \lambda w_{\alpha,\alpha}] da$$

The vector $\widehat{d}$ is defined by Equation 5.6.13. □

Next, we study the Almansi problem. Let u^* be an equilibrium vector field on B which corresponds to the body loads $\{\mathbf{f} = \mathbf{f}^* x_3^n, \mathbf{g} = \mathbf{g}^* x_3^n\}$, and satisfies the conditions

$$s_i(u^*) = p_i^* x_3^n, \qquad m_i(u^*) = k_i^* x_3^n \text{ on } \Pi, \qquad \mathbf{R}(u^*) = \mathbf{0}, \qquad \mathbf{H}(u^*) = \mathbf{0} \qquad (5.6.14)$$

where $\mathbf{f}^*, \mathbf{g}^*, \mathbf{p}^*$, and $\mathbf{k}^*$ are prescribed vector fields independent of x_3, and n is a positive integer or zero. Let u be an equilibrium vector field on B which corresponds to the body loads $\{\mathbf{f} = \mathbf{f}^* x_3^{n+1}, \mathbf{g} = \mathbf{g}^* x_3^{n+1}\}$ and satisfies the conditions

$$s_i(u) = p_i^* x_3^{n+1}, \qquad m_i(u) = k_i^* x_3^{n+1} \text{ on } \Pi, \qquad \mathbf{R}(u) = \mathbf{0}, \qquad \mathbf{H}(u) = \mathbf{0} \tag{5.6.15}$$

As in Section 2.3, we can prove that Almansi problem reduces to the finding a vector field u once the vector field u^* is known. Moreover, we are led to seek the vector field u in the form

$$u = (n+1)\left[\int_0^{x_3} u^* dx_3 + v\{\widehat{a}\} + w\right] \tag{5.6.16}$$

where $\widehat{a} = (a_1, a_2, a_3, a_4)$ is an unknown four-dimensional vector and w is an unknown vector field independent of x_3. From Equation 5.6.16 and the constitutive equations, we have

$$\begin{aligned} t_{ij}(u) &= (n+1)\left[\int_0^{x_3} t_{ij}(u^*)dx_3 + \sum_{r=1}^{4} a_r t_{ij}(v^{(r)}) + t_{ij}(w) + k_{ij}\right] \\ m_{ij}(u) &= (n+1)\left[\int_0^{x_3} m_{ij}(u^*)dx_3 + \sum_{r=1}^{4} a_r m_{ij}(v^{(r)}) + m_{ij}(w) + h_{ij}\right] \end{aligned} \tag{5.6.17}$$

where

$$\begin{aligned} k_{\alpha\beta} &= \lambda\delta_{\alpha\beta} u_3^*(x_1, x_2, 0), & k_{33} &= (\lambda + 2\mu + \kappa)u_3^*(x_1, x_2, 0) \\ k_{\alpha 3} &= \mu u_\alpha^*(x_1, x_2, 0), & k_{3\alpha} &= (\mu + \kappa)u_\alpha^*(x_1, x_2, 0) \\ h_{\eta\nu} &= \alpha\delta_{\eta\nu}\varphi_3^*(x_1, x_2, 0), & h_{33} &= (\alpha + \beta + \gamma)\varphi_3^*(x_1, x_2, 0) \\ h_{\alpha 3} &= \beta\varphi_\alpha^*(x_1, x_2, 0), & h_{3\alpha} &= \gamma\varphi_\alpha^*(x_1, x_2, 0) \end{aligned}$$

The equations of equilibrium and the conditions on the lateral boundary reduce to

$$\begin{aligned} &\left(T_{\beta\alpha}(\omega)\right)_{,\beta} + E_\alpha = 0 \\ &\left(M_{\rho 3}(\omega)\right)_{,\rho} + \varepsilon_{\alpha\beta} T_{\alpha\beta}(\omega) + J = 0 \text{ on } \Sigma_1 \\ &T_{\beta\alpha}(\omega) n_\beta = p'_\alpha, \qquad M_{\alpha 3}(\omega) n_\alpha = q' \text{ on } \Gamma \end{aligned} \tag{5.6.18}$$

and

$$L_i \omega^* = \zeta_i \text{ on } \Sigma_1, \qquad N_i \omega^* = \xi_i \text{ on } \Gamma \tag{5.6.19}$$

where

$$\begin{aligned} &w = (v_1, v_2, v_3, \chi_1, \chi_2, \chi_3), \qquad \omega = (v_1, v_2, 0, 0, 0, \chi_3) \\ &\omega^* = (0, 0, v_3, \chi_1, \chi_2, 0), \qquad E_\alpha = k_{\rho\alpha,\rho} + [t_{3\alpha}(u^*)](x_1, x_2, 0) \\ &J = h_{\alpha 3,\alpha} + [m_{33}(u^*)](x_1, x_2, 0), \qquad p'_\alpha = -k_{\rho\alpha} n_\rho, \; q' = -h_{\rho 3} n_\rho \\ &\zeta_\alpha = -h_{\rho\alpha,\rho} - [m_{3\alpha}(u^*)](x_1, x_2, 0) \\ &\zeta_3 = -k_{\rho 3,\rho} + [t_{33}(u^*)](x_1, x_2, 0), \qquad \xi_\alpha = -h_{\rho\alpha} n_\rho, \qquad \xi_3 = -k_{\rho 3} n_\rho \end{aligned} \tag{5.6.20}$$

From Equation 5.6.20, we get

$$\int_{\Sigma_1} E_\alpha da + \int_\Gamma p'_\alpha ds = -R_\alpha(u^*) = 0$$

$$\int_{\Sigma_1} (\varepsilon_{\alpha\beta} x_\alpha E_\beta + J) da + \int_\Gamma (\varepsilon_{\alpha\beta} x_\alpha p'_\beta + q') ds = -H_3(u^*) = 0$$

$$\int_{\Sigma_1} \zeta_3 da - \int_\Gamma \xi_3 ds = -R_3(u^*) = 0$$

Thus, the necessary and sufficient conditions to solve the boundary-value problems 5.6.18 and 5.6.19 are satisfied. We shall assume that the functions v_i and χ_i are known.

In view of Theorem 5.6.1, we have $R_\alpha(u) = \varepsilon_{\beta\alpha} H_\beta[(n+1)u^*] = 0$. The conditions $R_3(u) = 0$, $\mathbf{H}(u) = \mathbf{0}$ reduce to

$$D_{\alpha j} a_j = -\int_{\Sigma_1} [x_\alpha(k_{33} + t_{33}(w)) - \varepsilon_{\alpha\rho}(h_{3\rho} + m_{3\rho}(w))] da$$

$$D_{3j} a_j = -\int_{\Sigma_1} [k_{33} + t_{33}(w)] da$$

$$D a_4 = -\int_{\Sigma_1} \{\varepsilon_{\alpha\beta} x_\alpha [k_{3\beta} + t_{3\beta}(w)] + m_{33}(w) + h_{33}\} da$$

This system can always be solved for a_1, a_2, a_3, and a_4.

The problems of Almansi and Michell for Cosserat elastic bodies have been studied in Refs. 155 and 287 using the semi-inverse method.

5.7 Exercises

5.7.1 A homogeneous and isotropic Cosserat elastic material occupies a right cylinder B with the cross section $\Sigma_1 = \{x : x_1^2 + x_2^2 < a^2, x_3 = 0\}$, $(a > 0)$. The body is in equilibrium in the absence of body forces and body couples. Investigate the plane strain of the cylinder when the lateral boundary is subjected to the loading

$$\widetilde{t}_\alpha = 0, \qquad \widetilde{m}_3 = q_1 n_1 + q_2 n_2$$

where q_α are prescribed constants.

5.7.2 Study the extension and bending of a homogeneous and isotropic Cosserat elastic cylinder that occupies the domain $B = \{x : x_1^2 + x_2^2 < a^2, 0 < x_3 < h\}$, $(a > 0)$.

5.7.3 Study the torsion problem of a right circular cylinder of radius a, made of a homogeneous and isotropic Cosserat elastic material.

5.7.4 Investigate the torsion of a homogeneous and isotropic Cosserat elastic cylinder with square cross section.

5.7.5 Investigate the deformation of a homogeneous and isotropic Cosserat elastic circular cylinder which is subjected to a temperature field independent of the axial coordinate.

5.7.6 Study the Saint-Venant's problem for a homogeneous and hemitropic Cosserat elastic right cylinder.

5.7.7 A homogeneous and isotropic Cosserat elastic continuum occupies the domain $B = \{x : x_1^2 + x_2^2 < a^2,\, a < x_3 < h\}$, $(a > 0)$. Study the extension of cylinder B which is subjected to a uniform pressure on the lateral surface.

5.7.8 Investigate the torsion of a homogeneous and orthotropic Cosserat elastic cylinder.

5.7.9 Study the problem of loaded cylinders in the theory of homogeneous and hemitropic Cosserat elastic solids.

Chapter 6

Nonhomogeneous Cosserat Cylinders

6.1 Plane Strain Problems

In this chapter, we study the deformation of nonhomogeneous Cosserat elastic cylinders, when the constitutive coefficients are independent of the axial coordinate. In the first part of the chapter, we consider the case of isotropic bodies and assume that

$$\lambda = \lambda(x_1, x_2), \qquad \mu = (x_1, x_2), \ldots, \qquad \gamma = \gamma(x_1, x_2), \quad (x_1, x_2) \in \Sigma_1 \tag{6.1.1}$$

We suppose that the domain Σ_1 is C^∞-smooth [88], and that the elastic coefficients belong to C^∞. The basic equations of the plane strain, parallel to the x_1, x_2-plane, consist of the equations of equilibrium

$$t_{\beta\alpha,\beta} + f_\alpha = 0, \qquad m_{\alpha 3,\alpha} + \varepsilon_{\alpha\beta} t_{\alpha\beta} + g_3 = 0 \tag{6.1.2}$$

the constitutive equations

$$t_{\alpha\beta} = \lambda e_{\rho\rho}\delta_{\alpha\beta} + (\mu + \kappa)e_{\alpha\beta} + \mu e_{\beta\alpha}, \qquad m_{\alpha 3} = \gamma\kappa_{\alpha 3} \tag{6.1.3}$$

and the geometrical equations

$$e_{\alpha\beta} = u_{\beta,\alpha} + \varepsilon_{\beta\alpha}\varphi_3, \qquad \kappa_{\alpha 3} = \varphi_{3,\alpha} \tag{6.1.4}$$

on Σ_1. We restrict our attention to the second boundary-value problem, so that we consider the boundary conditions

$$t_{\beta\alpha} n_\beta = \widetilde{t}_\alpha, \qquad m_{\alpha 3} n_\alpha = \widetilde{m}_3 \text{ on } \Gamma \tag{6.1.5}$$

We assume that $f_\alpha, g_3, \widetilde{t}_\alpha$, and $\widetilde{m}_3$ are functions of class C^∞, and that the elastic potential $\widetilde{W}$ is a positive definite quadratic form in the variables $e_{\alpha\beta}$ and $\kappa_{\alpha 3}$.

The second boundary-value problem consists in the determination of the functions u_α and φ_3 of class $C^2(\Sigma_1) \cap C^1(\overline{\Sigma}_1)$ that satisfy Equations 6.1.2, 6.1.3, and 6.1.4 on Σ_1 and the boundary conditions 6.1.5 on Γ, when the constitutive coefficients are given by Equation 6.1.1.

We note the following existence result (cf. [88,142,147]) which holds under the above assumptions of regularity.

Theorem 6.1.1 *The second boundary-value problem has solutions belonging to $C^\infty(\overline{\Sigma}_1)$ if and only if the functions $f_\alpha, g_3, \widetilde{t}_\alpha$, and $\widetilde{m}_3$ satisfy the conditions* 5.2.80.

We denote by $\mathscr{A}^{(1)}$ the plane strain problem characterized by the loading

$$f_\alpha = (\lambda x_1)_{,\alpha}, \qquad g = -\beta_{,2}, \qquad \widetilde{t}_\alpha = -\lambda x_1 n_\alpha, \qquad \widetilde{m}_3 = \beta n_2$$

and by $\mathscr{A}^{(2)}$ the plane strain problem with the loads

$$f_\alpha = (\lambda x_2)_{,\alpha}, \qquad g = \beta_{,1}, \qquad \widetilde{t}_\alpha = -\lambda x_2 n_\alpha, \qquad \widetilde{m}_3 = -\beta n_1$$

Let us denote by $\mathscr{A}^{(3)}$ the plane strain problem where

$$f_\alpha = \lambda_{,\alpha}, \qquad g = 0, \qquad \widetilde{t}_\alpha = -\lambda n_\alpha, \qquad \widetilde{m} = 0$$

In what follows, we denote the components of displacement vector, microrotation vector, strain tensor, stress tensor, and couple-stress tensor from the problem $\mathscr{A}^{(s)}$, $(s = 1, 2, 3)$, by $v_\alpha^{(s)}, \psi_3^{(s)}, \gamma_{\alpha\beta}^{(s)}, \sigma_{\alpha\beta}^{(s)}$, and $\mu_{\alpha 3}^{(s)}$, respectively. Thus, we have

$$\begin{aligned}
&\sigma_{\beta\alpha,\beta}^{(\eta)} + (\lambda x_\eta)_{,\alpha} = 0, \qquad \sigma_{\beta\alpha,\beta}^{(3)} + \lambda_{,\alpha} = 0 \\
&\mu_{\beta 3,\beta}^{(\eta)} + \varepsilon_{\alpha\beta}\sigma_{\alpha\beta}^{(\eta)} + \varepsilon_{\alpha\eta}\beta_{,\alpha} = 0, \qquad \mu_{\beta 3,\beta}^{(3)} + \varepsilon_{\alpha\beta}\sigma_{\alpha\beta}^{(3)} = 0 \\
&\sigma_{\alpha\beta}^{(s)} = \lambda\gamma_{\eta\eta}^{(s)}\delta_{\alpha\beta} + (\mu+\kappa)\gamma_{\alpha\beta}^{(s)} + \mu\gamma_{\beta\alpha}^{(s)}, \qquad \mu_{\alpha 3}^{(s)} = \gamma\psi_{3,\alpha}^{(s)} \\
&\gamma_{\alpha\beta}^{(s)} = v_{\beta,\alpha}^{(s)} + \varepsilon_{\beta\alpha}\psi_3^{(s)} \text{ on } \Sigma_1
\end{aligned} \tag{6.1.6}$$

and the boundary conditions

$$\begin{aligned}
&\sigma_{\beta\alpha}^{(\eta)} n_\beta = -\lambda x_\eta n_\alpha, \qquad \sigma_{\beta\alpha}^{(3)} n_\beta = -\lambda n_\alpha \\
&\mu_{\alpha 3}^{(\eta)} n_\alpha = \varepsilon_{\eta\nu}\beta n_\nu, \qquad \mu_{\alpha 3}^{(3)} n_\alpha = 0 \text{ on } \Gamma
\end{aligned} \tag{6.1.7}$$

The necessary and sufficient conditions 5.2.80 for the existence of the solution are satisfied for each boundary-value problem $\mathscr{A}^{(s)}$. In what follows, we assume that the functions $v_\alpha^{(s)}$ and $\psi_3^{(s)}$ have been determined.

6.2 Saint-Venant's Problem

We assume that the right cylinder B is occupied by an isotropic and non-homogeneous Cosserat elastic material with the constitutive coefficients 6.1.1. We suppose that the elastic potential 5.3.6 is positive definite. In the absence of body forces and body couples, the equilibrium equations are

$$t_{ji,j} = 0, \qquad m_{ji,j} + \varepsilon_{ijk}t_{jk} = 0 \text{ on } B \tag{6.2.1}$$

We suppose that the cylinder is free of lateral loading, so that we have the conditions

$$t_{\alpha i}n_\alpha = 0, \qquad m_{\alpha i}n_\alpha = 0 \text{ on } \Pi \tag{6.2.2}$$

Let the loading applied on the end Σ_1 be statically equivalent to a force $\mathbf{F} = F_k\mathbf{e}_k$ and a moment $\mathbf{M} = M_k\mathbf{e}_k$. Thus, for $x_3 = 0$ we have the following conditions

$$\int_{\Sigma_1} t_{3\alpha}da = -F_\alpha \tag{6.2.3}$$

$$\int_{\Sigma_1} t_{33}da = -F_3 \tag{6.2.4}$$

$$\int_{\Sigma_1} (x_\alpha t_{33} - \varepsilon_{\alpha\beta}m_{3\beta})da = \varepsilon_{\alpha\beta}M_\beta \tag{6.2.5}$$

$$\int_{\Sigma_1} (\varepsilon_{\alpha\beta}x_\alpha t_{3\beta} + m_{33})da = -M_3 \tag{6.2.6}$$

Saint-Venant's problem consists in the finding of the functions u_i and φ_i that satisfy Equations 5.1.11, 5.1.17, and 6.2.1 on B, the conditions 6.2.2 on Π, and the conditions for $x_3 = 0$. As in the classical theory of elasticity, the problem will be reduced to the study of plane problems.

6.2.1 Extension and Bending of Cosserat Cylinders

We suppose that the resultant force and the resultant moment about O of the loads acting on Σ_1 are given by $\mathbf{F} = F_3\mathbf{e}_3$ and $\mathbf{M} = M_\alpha\mathbf{e}_\alpha$, respectively. In this case, the conditions on the end Σ_1 reduce to

$$\int_{\Sigma_1} t_{3\alpha}da = 0, \qquad \int_{\Sigma_1} (\varepsilon_{\alpha\beta}x_\alpha t_{3\beta} + m_{33})da = 0 \tag{6.2.7}$$

$$\int_{\Sigma_1} t_{33}da = -F_3, \qquad \int_{\Sigma_1} (x_\alpha t_{33} - \varepsilon_{\alpha\beta}m_{3\beta})da = \varepsilon_{\alpha\beta}M_\beta \tag{6.2.8}$$

The problem of extension and bending consists in the determination of the displacements u_k and microrotations φ_k that satisfy Equations 5.1.11, 5.1.17, and 6.2.1 on B and the boundary conditions 6.2.2, 6.2.7, and 6.2.8, when the constitutive coefficients are prescribed functions of the form 6.1.1.

We seek the solution of the problem in the form

$$\begin{aligned} u_\alpha &= -\frac{1}{2}a_\alpha x_3^2 + \sum_{s=1}^{3} a_s v_\alpha^{(s)}, \qquad u_3 = (a_1x_1 + a_2x_2 + a_3)x_3 \\ \varphi_\alpha &= \varepsilon_{\alpha\beta}a_\beta x_3, \qquad \varphi_3 = \sum_{s=1}^{3} a_s\psi_3^{(s)} \end{aligned} \tag{6.2.9}$$

where $v_\alpha^{(s)}$ and $\psi_3^{(s)}$ are the solutions of the problems $\mathscr{A}^{(s)}$, and a_s are unknown constants.

From Equations 5.1.11, 5.1.17, and 6.2.9, we obtain

$$\begin{aligned}
t_{\alpha\beta} &= \lambda(a_1x_1 + a_2x_2 + a_3)\delta_{\alpha\beta} + \sum_{s=1}^{3} a_s\sigma_{\alpha\beta}^{(s)} \\
t_{33} &= (\lambda + 2\mu + \kappa)(a_1x_1 + a_2x_2 + a_3) + \lambda\sum_{s=1}^{3} a_s\gamma_{\alpha\alpha}^{(s)} \\
t_{\alpha3} &= t_{3\beta} = 0, \qquad m_{\alpha\beta} = m_{33} = 0 \\
m_{\alpha3} &= \varepsilon_{\alpha\nu}\beta a_\nu + \sum_{s=1}^{3} a_s\mu_{\alpha3}^{(s)}, \qquad m_{3\alpha} = \varepsilon_{\alpha\nu}\gamma a_\nu + \sum_{s=1}^{3} a_s\mu_{3\alpha}^{(s)}
\end{aligned} \tag{6.2.10}$$

Using Equations 6.1.6, 6.1.7, and 6.2.1, we see that the equilibrium equations 6.2.1 and the boundary conditions 6.2.2 are satisfied. It follows from Equation 6.2.1 that the conditions 6.2.7 are identically satisfied. The conditions 6.2.8 lead to the following system for the unknown constants a_1, a_2, and a_3,

$$A_{rs}a_s = B_r \tag{6.2.11}$$

where

$$\begin{aligned}
A_{\alpha\beta} &= \int_{\Sigma_1} \{x_\alpha[(\lambda + 2\mu + \kappa)x_\beta + \lambda\gamma_{\eta\eta}^{(\beta)}] - \varepsilon_{\alpha\lambda}(\varepsilon_{\lambda\beta}\gamma + \mu_{3\lambda}^{(\beta)})\}da \\
A_{\alpha3} &= \int_{\Sigma_1} \{x_\alpha[\lambda + 2\mu + \kappa + \lambda\gamma_{\eta\eta}^{(3)}] - \varepsilon_{\alpha\lambda}\mu_{3\lambda}^{(3)}\}da \\
A_{3\alpha} &= \int_{\Sigma_1} [(\lambda + 2\mu + \kappa)x_\alpha + \lambda\gamma_{\eta\eta}^{(\alpha)}]da \\
A_{33} &= \int_{\Sigma_1} [\lambda + 2\mu + \kappa + \lambda\gamma_{\eta\eta}^{(3)}]da, \ \ B_\alpha = \varepsilon_{\alpha\beta}M_\beta, \ \ B_3 = -F_3
\end{aligned} \tag{6.2.12}$$

As in Section 5.3 we can prove that

$$\det(A_{rs}) \neq 0 \tag{6.2.13}$$

It follows that the system 6.2.11 uniquely determines the constants a_s. Thus, the solution is given by Equations 6.2.9 where $\{v_\alpha^{(s)}, \psi_3^{(s)}\}$ is the solution of the problem $\mathscr{A}^{(s)}$, and a_j are given by Equations 6.2.11.

6.2.2 Torsion

Let us suppose that $\mathbf{F} = \mathbf{0}$ and $\mathbf{M} = M_3\mathbf{e}_3$. Then the conditions on the end Σ_1 become

$$\int_{\Sigma_1} t_{3\alpha}da = 0 \tag{6.2.14}$$

$$\int_{\Sigma_1} t_{33}da = 0, \qquad \int_{\Sigma_1} (x_\alpha t_{33} - \varepsilon_{\alpha\beta}m_{3\beta})da = 0 \tag{6.2.15}$$

$$\int_{\Sigma_1} (\varepsilon_{\alpha\beta}x_\alpha t_{3\beta} + m_{33})da = -M_3 \tag{6.2.16}$$

The problem of torsion consists in the determination of the functions $u_i, \varphi_i \in C^2(B) \cap C^1(\overline{B})$ that satisfy Equations 5.1.11, 5.1.17, and 6.2.1 on B, the conditions for $x_3 = 0$ and the boundary conditions 6.2.2. We seek the solution of this problem in the form

$$\begin{aligned} u_\alpha &= \varepsilon_{\beta\alpha}\tau x_\beta x_3, \qquad & u_3 &= \tau\Phi(x_1, x_2) \\ \varphi_\alpha &= \tau\Phi_\alpha(x_1, x_2), \qquad & \varphi_3 &= \tau x_3 \end{aligned} \tag{6.2.17}$$

where Φ and Φ_α are unknown functions and τ is an unknown constant.

Let $V = (G, G_1, G_2)$ be an ordered triplet of functions G, G_1, and G_2 defined on Σ_1. We introduce the notations

$$\begin{aligned} T_\alpha V &= (\mu + \kappa)G_{,\alpha} + \kappa\varepsilon_{\alpha\beta}G_\beta, \qquad S_\alpha V = \mu G_{,\alpha} + \kappa\varepsilon_{\beta\alpha}G_\beta \\ M_{\nu\rho}V &= \alpha G_{\eta,\eta}\delta_{\nu\rho} + \beta G_{\nu,\rho} + \gamma G_{\rho,\nu} \\ \mathcal{L}_\alpha V &= (M_{\beta\alpha}V)_{,\beta} + \varepsilon_{\alpha\beta}(T_\beta V - S_\beta V) \\ \mathcal{L}_3 V &= (T_\alpha V)_{,\alpha} \\ \mathcal{N}_\alpha V &= (M_{\beta\alpha}V)n_\beta, \qquad \mathcal{N}_3 V = n_\alpha T_\alpha V \end{aligned} \tag{6.2.18}$$

By Equations 5.1.11, 5.1.17, and 6.2.17, we obtain

$$\begin{aligned} & t_{\alpha\beta} = 0, \qquad t_{33} = 0, \qquad t_{\alpha 3} = \tau(T_\alpha\Lambda + \mu\varepsilon_{\beta\alpha}x_\beta) \\ & t_{3\alpha} = \tau[S_\alpha\Lambda + (\mu + \kappa)\varepsilon_{\beta\alpha}x_\beta], \qquad m_{\eta\nu} = (M_{\eta\nu}\Lambda + \alpha\delta_{\eta\nu}) \\ & m_{\alpha 3} = m_{3\alpha} = 0, \qquad m_{33} = \tau(\alpha\Phi_{\nu,\nu} + \alpha + \beta + \gamma) \end{aligned} \tag{6.2.19}$$

where $\Lambda = (\Phi, \Phi_1, \Phi_2)$. The equilibrium equations 6.2.1 reduce to

$$\mathcal{L}_\nu\Lambda = x_\nu\kappa - \alpha_{,\nu}, \qquad \mathcal{L}_3\Lambda = \varepsilon_{\alpha\beta}(\mu x_\beta)_{,\alpha} \text{ on } \Sigma_1 \tag{6.2.20}$$

The boundary conditions 6.2.2 become

$$\mathcal{N}_\nu\Lambda = -\alpha n_\nu, \qquad \mathcal{N}_3\Lambda = \mu\varepsilon_{\alpha\beta}x_\beta n_\alpha \text{ on } \Gamma \tag{6.2.21}$$

Let us consider the boundary-value problem

$$\mathcal{L}_i V = \xi_i \text{ on } \Sigma_1, \qquad \mathcal{N}_i V = \zeta_i \text{ on } \Gamma \tag{6.2.22}$$

where ξ_i and ζ_i are C^∞ functions. We have the following result (cf. [137]).

Theorem 6.2.1 *The boundary problem 6.2.22 has solutions belonging to $C^\infty(\Sigma_1)$ if and only if*

$$\int_{\Sigma_1} \xi_3 da = \int_\Gamma \zeta_3 ds \tag{6.2.23}$$

The necessary and sufficient condition 6.2.23 for the existence of the solution of the boundary-value problem 6.2.20 and 6.2.21 is satisfied. In what follows, we assume that the functions Φ and Φ_α are known.

From Equations 6.2.16 and 6.2.19, we obtain

$$\tau D^* = -M_3 \tag{6.2.24}$$

where

$$D^* = \int_{\Sigma_1} (\varepsilon_{\alpha\beta} x_\alpha \{\mu \Phi_{,\beta} + \varepsilon_{\nu\beta}[\kappa \Phi_\nu + (\mu+\kappa) x_\nu]\} + \alpha \Phi_{\nu,\nu} + \alpha + \beta + \gamma) da \tag{6.2.25}$$

The positive definiteness of the elastic potential implies that $D^* > 0$, so that the relation 6.2.24 determines the constant τ.

The conditions 6.2.14 are satisfied on the basis of the equations of equilibrium and the boundary conditions. Thus, for the first of 6.2.14 we have

$$\begin{aligned} \int_{\Sigma_1} t_{31} da &= \int_{\Sigma_1} (t_{13} - m_{\alpha 3,\alpha}) da = \int_{\Sigma_1} (t_{13} + x_1 t_{\alpha 3,\alpha} - m_{\alpha 2,\alpha}) da \\ &= \int_{\Sigma_1} [(x_1 t_{\alpha 3})_{,\alpha} - m_{\alpha 2,\alpha}] da = \int_{\Gamma} (x_1 t_{\alpha 3} n_\alpha - m_{\alpha 2} n_\alpha) ds = 0 \end{aligned}$$

In a similar way we can prove that the second condition of 6.2.14 is satisfied. We conclude that Equation 6.2.17, where (Φ, Φ_1, Φ_2) satisfies the boundary-value problem 6.2.20 and 6.2.21 and τ is given by Equation 6.2.24, is a solution of the torsion problem. This solution was established in Ref. 137.

6.2.3 Flexure

We assume that the loading applied on Σ_1 is statically equivalent to the force $\mathbf{F} = F_\alpha \mathbf{e}_\alpha$ and the moment $\mathbf{M} = \mathbf{0}$. The conditions on Σ_1 are given by

$$\int_{\Sigma_1} t_{3\alpha} da = -F_\alpha \tag{6.2.26}$$

$$\int_{\Sigma_1} t_{33} da = 0, \qquad \int_{\Sigma_1} (x_\alpha t_{33} - \varepsilon_{\alpha\beta} m_{3\beta}) da = 0 \tag{6.2.27}$$

$$\int_{\Sigma_1} (\varepsilon_{\alpha\beta} x_\alpha t_{3\beta} + m_{33}) da = 0 \tag{6.2.28}$$

The flexure problem consists in the solving of Equations 5.1.11, 5.1.17, and 6.2.1 on B with the boundary conditions for $x_3 = 0$ and 6.2.2. On the basis of Theorem 5.3.3, we try to solve the problem assuming that

$$\begin{aligned} u_\alpha &= -\frac{1}{6} b_\alpha x_3^3 + x_3 \sum_{s=1}^{3} b_s v_\alpha^{(s)} + \varepsilon_{\beta\alpha} \tau x_\beta x_3 \\ u_3 &= \frac{1}{2}(b_1 x_1 + b_2 x_2 + b_3) x_3^2 + \tau \Phi + \Psi(x_1, x_2) \\ \varphi_\alpha &= \frac{1}{2} \varepsilon_{\alpha\beta} b_\beta x_3^2 + \tau \Phi_\alpha + \Psi_\alpha(x_1, x_2) \\ \varphi_3 &= x_3 \sum_{s=1}^{3} b_s \psi_3^{(s)} + \tau x_3 \end{aligned} \tag{6.2.29}$$

where Φ and Φ_α satisfy the boundary-value problem 6.2.20 and 6.2.21, $v_\alpha^{(s)}$ and $\psi_3^{(s)}$ are the solutions of the problems $\mathscr{A}^{(s)}$, Ψ and Ψ_α are unknown functions, and b_k and τ are unknown constants. From Equations 5.1.11, 5.1.17, and 6.2.29, we obtain

$$
\begin{aligned}
t_{\alpha\beta} &= \lambda x_3(b_1x_1 + b_2x_2 + b_3)\delta_{\alpha\beta} + x_3\sum_{s=1}^{3} b_s\sigma_{\alpha\beta}^{(s)} \\
t_{33} &= (\lambda + 2\mu + \kappa)(b_1x_1 + b_2x_2 + b_3)x_3 + \lambda x_3\sum_{s=1}^{3} b_s\gamma_{\alpha\alpha}^{(s)} \\
t_{\alpha 3} &= \tau(T_\alpha\Lambda + \mu\varepsilon_{\beta\alpha}x_\beta) + T_\alpha\omega + \mu\sum_{s=1}^{3} b_s v_\alpha^{(s)} \\
t_{3\alpha} &= \tau[S_\alpha\Lambda + \varepsilon_{\beta\alpha}(\mu+\kappa)x_\beta] + S_\alpha\omega + (\mu+\kappa)\sum_{s=1}^{3} b_s v_\alpha^{(s)} \\
m_{\eta\nu} &= \tau(M_{\eta\nu}\Lambda + \alpha\delta_{\eta\nu}) + M_{\eta\nu}\omega + \alpha\delta_{\eta\nu}\sum_{s=1}^{3} b_s\psi_3^{(s)} \\
m_{\alpha 3} &= \beta\varepsilon_{\alpha\rho}b_\rho x_3 + x_3\sum_{s=1}^{3} b_s\mu_{\alpha 3}^{(s)} \\
m_{3\alpha} &= \gamma\varepsilon_{\alpha\beta}b_\beta x_3 + x_3\sum_{s=1}^{3} b_s\mu_{3\alpha}^{(s)} \\
m_{33} &= (\alpha+\beta+\gamma)\left(\tau + \sum_{s=1}^{3} b_s\psi_3^{(s)}\right) + \alpha(\tau\Phi_{\nu,\nu} + \Psi_{\nu,\nu})
\end{aligned}
\tag{6.2.30}
$$

where $\Lambda = (\Phi, \Phi_1, \Phi_2)$ and $\omega = (\Psi, \Psi_1, \Psi_2)$.

With the help of the relations 6.1.6, 6.1.7, 6.2.20, and 6.2.21 we see that the equilibrium equations 6.2.1 and the boundary conditions 6.2.2 reduce to

$$
\begin{aligned}
\mathcal{L}_\nu\omega &= -\gamma\varepsilon_{\nu\beta}b_\beta - \sum_{s=1}^{3} b_s\left[\left(\alpha\psi^{(s)}\right)_{,\nu} + \mu_{3\nu}^{(s)} - \varepsilon_{\nu\beta}\kappa v_\beta^{(s)}\right] \\
\mathcal{L}_3\omega &= -(\lambda + 2\mu + \kappa)(b_1x_1 + b_2x_2 + b_3) - \sum_{s=1}^{3} b_s\left[\lambda\gamma_{\alpha\alpha}^{(s)} + \left(\mu v_\alpha^{(s)}\right)_{,\alpha}\right] \text{ on } \Sigma_1
\end{aligned}
\tag{6.2.31}
$$

and

$$
\mathscr{N}_\nu\omega = -\alpha n_\nu\sum_{s=1}^{3} b_s\psi_3^{(s)}, \qquad \mathscr{N}_3\omega = -\mu n_\alpha\sum_{s=1}^{3} b_s v_\alpha^{(s)} \text{ on } \Gamma \tag{6.2.32}
$$

The condition 6.2.23 for the existence of the solution of the boundary-value problem 6.2.31 and 6.2.32 takes the form

$$
A_{3s}b_s = 0 \tag{6.2.33}
$$

where A_{3s} are given by Equations 6.2.12.

Let us impose the conditions 6.2.26. With the aid of Equations 6.2.1 and 6.2.2, we can write

$$\int_{\Sigma_1} t_{31} da = \int_{\Sigma_1} (t_{13} - m_{j2,j} + x_1 t_{i3,i}) da = \int_{\Gamma} (x_1 t_{\alpha 3} n_\alpha - m_{\alpha 2} n_\alpha) ds$$
$$+ \int_{\Sigma_1} (x_1 t_{33} - m_{32})_{,3} da = \int_{\Sigma_1} (x_1 t_{33} - m_{32})_{,3} da \qquad (6.2.34)$$

In a similar way, we obtain

$$\int_{\Sigma_1} t_{32} da = \int_{\Sigma_1} (x_2 t_{33} + m_{31})_{,3} da \qquad (6.2.35)$$

By Equations 6.2.29, 6.2.34, and 6.2.35, the conditions 6.2.26 reduce to

$$A_{\alpha s} b_s = -F_\alpha \qquad (6.2.36)$$

The system 6.2.33 and 6.2.36 uniquely determines the constants b_k. From Equations 6.2.16 and 6.2.30, we obtain

$$\tau D^* = -\int_{\Sigma_1} \left\{ \varepsilon_{\alpha\beta} x_\alpha \left[\mu \Psi_{,\beta} + \varepsilon_{\nu\beta} \kappa \Psi_\nu + (\mu + \kappa) \sum_{s=1}^{3} b_s v_\beta^{(s)} \right] \right.$$
$$\left. + (\alpha + \beta + \gamma) \sum_{s=1}^{3} b_s \psi_3^{(s)} + \alpha \Psi_{\nu,\nu} \right\} da$$

where D^* is given by Equation 6.2.25. The above relation determines the constant τ.

The conditions 6.2.27 are satisfied on the basis of Equation 6.2.30. Thus, the solution of the flexure problem has the form 6.2.29.

The results presented in this section generalize the results established in Ref. 149 for the classical theory of elasticity.

6.3 Problems of Almansi and Michell

In this section, we study the problem of loaded cylinders made of non-homogeneous and isotropic Cosserat elastic materials. We assume that the constitutive coefficients are independent of the axial coordinate.

6.3.1 Uniformly Loaded Cylinders

We study first the Almansi–Michell problem stated in Section 5.6. In this case the equilibrium equations are given by Equation 5.1.19, where

$$f_i = f_i^{(0)}(x_1, x_2), \qquad g_i = g_i^{(0)}(x_1, x_2), \quad (x_1, x_2) \in \Sigma_1 \qquad (6.3.1)$$

Here, $f_i^{(0)}$ and $g_i^{(0)}$ are prescribed functions. The conditions on the lateral surface have the form

$$t_{\alpha i}n_\alpha = \widetilde{t}_i^{(0)}, \qquad m_{\alpha i}n_\alpha = \widetilde{m}_i^{(0)} \text{ on } \Pi \tag{6.3.2}$$

where $\widetilde{t}_i^{(0)}$ and $\widetilde{m}_i^{(0)}$ are independent of the axial coordinate.

The Almansi–Michell problem consists in the finding of the functions $u_i, \varphi_i \in C^2(B) \cap C^1(\overline{B}_1)$ that satisfy Equations 5.1.11, 5.1.17, and 5.1.19 on B, the conditions 6.3.2 on Π, and the conditions on the end Σ_1, when the body loads, the constitutive coefficients and $\widetilde{t}_i^{(0)}, \widetilde{m}_i^{(0)}$ are independent of x_3. We seek the solution in the form

$$\begin{aligned}
u_\alpha &= -\frac{1}{2}a_\alpha x_3^2 - \frac{1}{6}b_\alpha x_3^3 - \frac{1}{24}c_\alpha x_3^4 + \varepsilon_{\beta\alpha}\left(\tau_1 x_3 + \frac{1}{2}\tau_2 x_3^2\right)x_\beta \\
&\quad + \sum_{s=1}^{3}\left(a_s + b_s x_3 + \frac{1}{2}c_s x_3^2\right)v_\alpha^{(s)} + v_\alpha(x_1,x_2) \\
u_3 &= (a_1x_1 + a_2x_2 + a_3)x_3 + \frac{1}{2}(b_1x_1 + b_2x_2 + b_3)x_3^2 \\
&\quad + \frac{1}{6}(c_1x_1 + c_2x_2 + c_3)x_3^3 + (\tau_1 + x_3\tau_2)\Phi + \Psi(x_1,x_2) + x_3\chi(x_1,x_2) \\
\varphi_\alpha &= \varepsilon_{\alpha\beta}\left(a_\beta x_3 + \frac{1}{2}b_\beta x_3^2 + \frac{1}{6}c_\beta x_3^3\right) + (\tau_1 + x_3\tau_2)\Phi_\alpha \\
&\quad + \Psi_\alpha(x_1,x_2) + x_3\chi_\alpha(x_1,x_2) \\
\varphi_3 &= \sum_{s=1}^{3}\left(a_s + x_3 b_s + \frac{1}{2}c_s x_3^2\right)\psi_3^{(s)} + \tau_1 x_3 + \frac{1}{2}\tau_2 x_3^2 + w(x_1,x_2)
\end{aligned} \tag{6.3.3}$$

where $v_\alpha^{(s)}, \psi_3^{(s)}$ are the solutions of the problems $\mathscr{A}^{(s)}$, Φ and Φ_α satisfy the boundary-value problem 6.2.20 and 6.2.21; $\Psi, \Psi_\alpha, \chi, \chi_\alpha, v_\alpha$, and w are unknown functions, and a_i, b_i, c_i, τ_1, and τ_2 are unknown constants. From Equations 5.1.11, 5.1.17, and 6.3.3, we obtain

$$\begin{aligned}
t_{\alpha\beta} &= \lambda\Big[a_1x_1 + a_2x_2 + a_3 + (b_1x_1 + b_2x_2 + b_3)x_3 \\
&\quad + \frac{1}{2}(c_1x_1 + c_2x_2 + c_3)x_3^2\Big]\delta_{\alpha\beta} \\
&\quad + \lambda(\chi + \tau_2\Phi)\delta_{\alpha\beta} + \sum_{s=1}^{3}\left(a_s + b_s x_3 + \frac{1}{2}c_s x_3^2\right)\sigma_{\alpha\beta}^{(s)} + \sigma_{\alpha\beta} \\
t_{33} &= (\lambda + 2\mu + \kappa)\Big[a_1x_1 + a_2x_2 + a_3 + (b_1x_1 + b_2x_2 + b_3)x_3 \\
&\quad + \frac{1}{2}(c_1x_1 + c_2x_2 + c_3)x_3^2\Big] + (\lambda + 2\mu + \kappa)(\chi + \tau_2\Phi) \\
&\quad + \lambda\sum_{s=1}^{3}\left(a_s + b_s x_3 + \frac{1}{2}c_s x_3^2\right)\gamma_{\alpha\alpha}^{(s)} + \lambda\gamma_{\alpha\alpha}
\end{aligned}$$

$$
\begin{aligned}
t_{\alpha 3} &= T_\alpha \Omega + x_3 T_\alpha V + (\tau_1 + \tau_2 x_3)(T_\alpha \Lambda + \mu \varepsilon_{\beta\alpha} x_\beta) \\
&\quad + \mu \sum_{s=1}^{3} (b_s + c_s x_3) v_\alpha^{(s)} \\
t_{3\alpha} &= S_\alpha \Omega + x_3 S_\alpha V + (\tau_1 + \tau_2 x_3)[S_\alpha \Lambda + (\mu + \kappa)\varepsilon_{\beta\alpha} x_\beta] \\
&\quad + (\mu + \kappa) \sum_{s=1}^{3} (b_s + c_s x_3) v_\alpha^{(s)} \\
m_{\lambda\nu} &= M_{\lambda\nu} \Omega + x_3 M_{\lambda\nu} V + (\tau_1 + \tau_2 x_3)(M_{\lambda\nu} \Lambda + \delta_{\lambda\nu}) \\
&\quad + \alpha \delta_{\lambda\nu} \sum_{s=1}^{3} (b_s + c_s x_3) \psi_3^{(s)} \\
m_{33} &= (\alpha + \beta + \gamma) \left[\tau_1 + \tau_2 x_3 + \sum_{s=1}^{3} (b_s + c_s x_3) \psi^{(s)} \right] \\
&\quad + \alpha(\tau_1 + \tau_2 x_3)\Phi_{\lambda,\lambda} + \alpha(\Psi_{\lambda,\lambda} + x_3 \chi_{\lambda,\lambda}) \\
m_{\alpha 3} &= \beta \varepsilon_{\alpha\nu} \left(a_\nu + b_\nu x_3 + \frac{1}{2} c_\nu x_3^2 \right) + \beta(\chi_\alpha + \tau_2 \Phi_\alpha) \\
&\quad + \sum_{s=1}^{3} \left(a_s + b_s x_3 + \frac{1}{2} c_s x_3^2 \right) \mu_{\alpha 3}^{(s)} + \mu_{\alpha 3} \\
m_{3\alpha} &= \gamma \varepsilon_{\alpha\nu} \left(a_\nu + b_\nu x_3 + \frac{1}{2} c_\nu x_3^2 \right) + \gamma(\chi_\alpha + \tau_2 \Phi_\alpha) \\
&\quad + \sum_{s=1} \left(a_s + b_s x_3 + \frac{1}{2} c_s x_3^2 \right) \mu_{3\alpha}^{(s)} + \mu_{3\alpha}
\end{aligned} \tag{6.3.4}
$$

where we have used the notations $\Lambda = (\Phi, \Phi_1, \Phi_2)$, $\Omega = (\Psi, \Psi_1, \Psi_2)$, $V = (\chi, \chi_1, \chi_2)$ and

$$
\begin{aligned}
&\sigma_{\alpha\beta} = \lambda \gamma_{\nu\nu} \delta_{\alpha\beta} + (\mu + \kappa)\gamma_{\alpha\beta} + \mu \gamma_{\beta\alpha} \\
&\mu_{\alpha 3} = \gamma w_{,\alpha}, \qquad \mu_{3\alpha} = \beta w_{,\alpha}, \qquad \gamma_{\alpha\beta} = v_{\beta,\alpha} + \varepsilon_{\beta\alpha} w
\end{aligned} \tag{6.3.5}
$$

With the help of Equations 6.1.6, 6.2.20, and 6.3.4, the equilibrium equations 5.1.19 reduce to

$$
\sigma_{\beta\alpha,\beta} + H_\alpha = 0, \qquad \mu_{\alpha 3,\alpha} + \varepsilon_{\alpha\beta} \sigma_{\alpha\beta} + H = 0 \tag{6.3.6}
$$

$$
\mathcal{L}_i \Omega = G_i \tag{6.3.7}
$$

$$
\mathcal{L}_i V = K_i \tag{6.3.8}
$$

on Σ_1, where

$$H_\alpha = [\lambda(\chi + \tau_2\Phi)]_{,\alpha} + S_\alpha V + \tau_2[S_\alpha \Lambda + (\mu+\kappa)\varepsilon_{\beta\alpha}x_\beta] + (\mu+\kappa)\sum_{s=1}^{3} c_\alpha v_\alpha^{(s)} + f_\alpha^{(0)}$$

$$H = [\beta(\chi_\alpha + \tau_2\Phi_\alpha)]_{,\alpha} + (\alpha+\beta+\gamma)\left(\tau_2 + \sum_{s=1}^{3} c_s\varphi^{(s)}\right) + \alpha(\chi_\nu + \tau_2\Phi_\nu)_{,\nu} + g_3^{(0)}$$

$$G_\nu = -\sum_{s=1}^{3} b_s\left[\left(\alpha\psi_3^{(s)}\right)_{,\nu} + \mu_{3\nu}^{(s)} - \varepsilon_{\nu\beta}\kappa v_\beta^{(s)}\right] - \gamma\varepsilon_{\nu\beta}b_\beta - g_\nu^{(0)} \tag{6.3.9}$$

$$G_3 = -\sum_{s=1}^{3} b_s\left[\lambda\gamma_{\alpha\alpha}^{(s)} + \left(\mu v_\alpha^{(s)}\right)_{,\alpha}\right] - (\lambda+2\mu+\kappa)(b_1x_1 + b_2x_2 + b_3) - f_3^{(0)}$$

$$K_\nu = -\sum_{s=1}^{3} c_s\left[\left(\alpha\psi_3^{(s)}\right)_{,\nu} + \mu_{3\nu}^{(s)} - \varepsilon_{\nu\beta}\kappa v_\beta^{(s)}\right] - \gamma\varepsilon_{\nu\beta}c_\beta$$

$$K_3 = -\sum_{s=1}^{3} c_s\left[\lambda\gamma_{\alpha\alpha}^{(s)} + \left(\mu v_\alpha^{(s)}\right)_{,\alpha}\right] - (\lambda+2\mu+\kappa)(c_1x_1 + c_2x_2 + c_3)$$

Using Equations 6.1.7, 6.2.21, and 6.3.4, the conditions 6.3.2 become

$$\sigma_{\alpha\beta}n_\alpha = S_\beta, \qquad \mu_{\alpha 3}n_\alpha = S \tag{6.3.10}$$

$$\mathscr{N}_i\Omega = N_i \tag{6.3.11}$$

$$\mathscr{N}_i V = P_i \tag{6.3.12}$$

on Γ, where

$$S_\beta = \tilde{t}_\beta^{(0)} - \lambda(\chi + \tau_2\Phi)n_\beta, \qquad S = \tilde{m}_3^{(0)} - \beta(\chi_\alpha + \tau_2\Phi_\alpha)n_\alpha$$

$$N_\nu = \tilde{m}_\nu^{(0)} - n_\nu\alpha\sum_{s=1}^{3} b_s\psi_3^{(s)}, \qquad N_3 = \tilde{t}_3^{(0)} - \mu\sum_{s=1}^{3} b_s v_\alpha^{(s)}n_\alpha \tag{6.3.13}$$

$$P_\nu = -\alpha n_\nu\sum_{s=1}^{3} c_s\psi_3^{(s)}, \qquad P_3 = -\mu\sum_{s=1}^{3} c_s v_\alpha^{(s)}n_\alpha$$

From Equations 6.3.5, 6.3.6, and 6.3.10, it follows that the functions v_α and w satisfy the equations and the boundary conditions in a plane strain problem. The necessary and sufficient conditions 5.2.80 for the existence of the solution of this problem are

$$\int_{\Sigma_1} H_\alpha + \int_\Gamma S_\alpha ds = 0$$

$$\int_{\Sigma_1} (\varepsilon_{\alpha\beta}x_\alpha H_\beta + H)da + \int_\Gamma (\varepsilon_{\alpha\beta}x_\alpha S_\beta + S)ds = 0 \tag{6.3.14}$$

Using Equations 6.3.9 and 6.3.13 and the divergence theorem, we obtain

$$\int_{\Sigma_1} H_\alpha da + \int_\Gamma S_\alpha ds = \int_{\Sigma_1} f_\alpha^{(0)} da + \int_\Gamma \widetilde{t}_\alpha^{(0)} ds + \int_{\Sigma_1} t_{3\alpha,3} da$$
$$\int_{\Sigma_1} (\varepsilon_{\alpha\beta} x_\alpha H_\beta + H) da + \int_\Gamma (\varepsilon_{\alpha\beta} x_\alpha S_\beta + S) ds = \int_{\Sigma_1} (\varepsilon_{\alpha\beta} x_\alpha f_\beta^{(0)} + g_3^{(0)}) da$$
$$+ \int_\Gamma (\varepsilon_{\alpha\beta} x_\alpha \widetilde{t}_\beta^{(0)} + \widetilde{m}_3^{(0)}) ds + \int_{\Sigma_1} (\varepsilon_{\alpha\beta} x_\alpha t_{3\beta,3} + m_{33,3}) da \qquad (6.3.15)$$

With the help of equilibrium equations 5.1.19 and 6.3.1, we have

$$t_{31,3} = (t_{13} - m_{j2,j})_{,3} = (t_{13} - m_{j2,j} + x_1 t_{s3,s})_{,3}$$
$$= x_1 t_{33,33} - m_{32,33} + (x_1 t_{\alpha 3})_{,\alpha 3}$$
$$t_{32,3} = x_2 t_{33,33} + m_{31,33} + (x_2 t_{\alpha 3})_{,\alpha 3}$$

so that, taking into account the conditions 6.3.2, we obtain

$$\int_{\Sigma_1} t_{3\alpha,3} da = \int_{\Sigma_1} (x_\alpha t_{33,33} - \varepsilon_{\alpha\beta} m_{3\beta,33}) da \qquad (6.3.16)$$

Substituting Equations 6.3.4 into Equation 6.3.16, we find that

$$\int_{\Sigma_1} t_{3\alpha,3} da = A_{\alpha i} c_i$$

so that the condition $6.3.14_1$ can be written in the form

$$A_{\alpha i} c_i = -\int_{\Sigma_1} f_\alpha^{(0)} da - \int_\Gamma \widetilde{t}_\alpha^{(0)} ds \qquad (6.3.17)$$

Let us consider the boundary-value problem 6.3.8 and 6.3.12. The necessary and sufficient condition for the existence of the solution of this problem is

$$\int_{\Sigma_1} K_3 da - \int_\Gamma P_3 ds = 0 \qquad (6.3.18)$$

Using Equations 6.2.12, 6.3.9, and 6.3.10, from Equation 6.3.18 we obtain

$$A_{3i} c_i = 0 \qquad (6.3.19)$$

In view of Equation 6.2.13, the system 6.3.17 and 6.3.19 uniquely determines the constants c_i. Let us consider now the boundary-value problem 6.3.7 and 6.3.11. The necessary and sufficient condition for the existence of the solution of this problem is

$$\int_{\Sigma_1} G_3 da - \int_\Gamma N_3 ds = 0 \qquad (6.3.20)$$

By using Equations 6.2.12, 6.3.9, and 6.3.13, the condition 6.3.20 reduces to

$$A_{3s}b_s = -\int_{\Sigma_1} f_3^{(0)} da - \int_{\Gamma} \widetilde{t}_3^{(0)} ds \tag{6.3.21}$$

Let us impose the conditions 6.2.3. We can write

$$\begin{aligned}
\int_{\Sigma_1} t_{31} da &= \int_{\Sigma_1} (t_{13} - m_{j2,j} - g_2^{(0)}) da \\
&= \int_{\Sigma_1} [t_{13} - m_{j2,j} + x_1(t_{s3,s} + f_3^{(0)}) - g_2^{(0)}] da \\
&= \int_{\Sigma_1} [(x_1 t_{\alpha 3})_{,\alpha} - m_{\alpha 2,\alpha} + x_1 t_{33,3} - m_{32,3} - g_2^{(0)}] da \\
&= \int_{\Gamma} (x_1 \widetilde{t}_3^{(0)} - \widetilde{m}_2^{(0)}) ds + \int_{\Sigma_1} (x_1 f_3^{(0)} - g_2^{(0)}) da \qquad (6.3.22) \\
&\quad + \int_{\Sigma_1} (x_1 t_{33,3} - m_{32,3}) da \\
\int_{\Sigma_1} t_{32} da &= \int_{\Sigma_1} (x_2 f_3^{(0)} + g_1^{(0)}) da + \int_{\Gamma} (x_2 \widetilde{t}_3^{(0)} + \widetilde{m}_1^{(0)}) ds \\
&\quad + \int_{\Sigma_1} (x_2 t_{33,3} + m_{31,3}) da
\end{aligned}$$

By using Equations 6.3.4 and 6.3.22, the conditions 6.2.3 become

$$A_{\alpha s} b_s = -F_\alpha - \int_{\Sigma_1} (x_\alpha f_3^{(0)} + \varepsilon_{\beta\alpha} g_\beta^{(0)}) da - \int_{\Gamma} (x_\alpha \widetilde{t}_3^{(0)} + \varepsilon_{\beta\alpha} \widetilde{m}_\beta^{(0)}) ds \tag{6.3.23}$$

Equations 6.3.21 and 6.3.23 determine the constants b_s. In what follows we assume that the functions χ, χ_α, Ψ, and Ψ_α, and the constants b_i and c_i are known.

With the help of Equations $6.3.15_2$ and 6.3.4, the condition $6.3.15_2$ reduces to

$$\begin{aligned}
\tau_2 D^* &= -\int_{\Sigma_1} (\varepsilon_{\alpha\beta} x_\alpha f_\beta^{(0)} + g_3^{(0)}) da + \int_{\Gamma} (\varepsilon_{\alpha\beta} x_\alpha \widetilde{t}_\beta^{(0)} + \widetilde{m}_3^{(0)}) ds \\
&\quad - \int_{\Sigma_1} \left\{ \varepsilon_{\alpha\beta} x_\alpha \left[\mu \chi_{,\beta} + \varepsilon_{\nu\beta} \kappa \chi_\nu + (\mu + \kappa) \sum_{s=1}^{3} c_s v_\beta^{(s)} \right] \right. \\
&\quad \left. + (\alpha + \beta + \gamma) \sum_{s=1}^{3} c_s \psi_3^{(s)} + \alpha \chi_{\nu,\nu} \right\} da
\end{aligned}$$

where D^* is given by Equation 6.2.25. The above relation permits the determination of the constant τ_2. From Equations 6.2.4, 6.2.5, and 6.3.4, we obtain

$$A_{ij} a_j = C_i \tag{6.3.24}$$

where

$$C_\alpha = \varepsilon_{\alpha\beta} M_\beta - \int_\Sigma \{x_\alpha[\lambda\gamma_{\nu\nu} + (\lambda + 2\mu + \kappa)(\chi + \tau_2\Phi)] - \varepsilon_{\alpha\beta}[\gamma(\chi_\beta + \tau_2\Phi_\beta) + \mu_{3\beta}]\}da$$

$$C_3 = -F_3 - \int_\Sigma [(\lambda + 2\mu + \kappa)(\chi + \tau_2\Phi) + \lambda\gamma_{\alpha\alpha}]da$$

Equations 6.3.24 determine the constants a_j. By Equations 6.2.6 and 6.3.4, we find that

$$\tau_1 D^* = -M_3 - \int_\Sigma \left\{ \varepsilon_{\alpha\beta} x_\alpha \left[\mu\Psi_{,\beta} + \varepsilon_{\nu\beta}\kappa\Psi_\nu + (\mu+\kappa)\sum_{s=1}^{3} b_s v_\beta^{(s)} \right] + (\alpha+\beta+\gamma)\sum_{s=1}^{3} b_s \psi^{(s)} + \alpha\Psi_{\nu,\nu} \right\} da$$

This relation determines the constant τ_1. The Almansi–Michell problem is therefore solved.

6.3.2 Almansi's Problem

We assume that $f_i, g_i, \widetilde{t}_i$, and $\widetilde{m}_i$ are polynomials of degree r in the axial coordinate, namely

$$\begin{aligned} f_i &= \sum_{k=0}^{r} F_{ik}(x_1, x_2)x_3^k, & g_i &= \sum_{k=0}^{r} G_{ik}(x_1, x_2)x_3^k \\ \widetilde{t}_i &= \sum_{k=0}^{r} p_{ik}(x_1, x_2)x_3^k, & \widetilde{m}_i &= \sum_{k=0}^{r} q_{ik}(x_1, x_2)x_3^k \end{aligned} \tag{6.3.25}$$

where F_{ik}, G_{ik}, p_{ik}, and q_{ik} are prescribed functions of class C^∞. In the case of nonhomogeneous Cosserat cylinders, the problem of Almansi consists in finding of the functions $u_i, \varphi_i \in C^2(B) \cap C^1(\overline{B})$ that satisfy the Equations 5.1.11, 5.1.17, and 5.1.19 on B, the conditions

$$t_{\alpha i} n_\alpha = \widetilde{t}_i, \qquad m_{\alpha i} n_\alpha = \widetilde{m}_i \text{ on } \Pi \tag{6.3.26}$$

and the conditions on the end Σ_1, when $f_i, g_i, \widetilde{t}_i$, and $\widetilde{m}_i$ are given by Equation 6.3.25 and the constitutive coefficients have the form 6.1.1. As in Section 2.3, the Almansi problem can be reduced to the following problem: to find the functions $u_i, \varphi_i \in C^2(B) \cap C^1(\overline{B})$ which satisfy the equations

$$\begin{aligned} &t_{ji,j} + \mathscr{F}_i(x_1, x_2)x_3^{n+1} = 0, \qquad m_{ji,j} + \varepsilon_{irs}t_{rs} + \mathcal{H}_i(x_1, x_2)x_3^{n+1} = 0 \\ &t_{ij} = \lambda e_{rr}\delta_{ij} + (\mu+\kappa)e_{ij} + \mu e_{ji} \\ &m_{ij} = \alpha\varphi_{r,r}\delta_{ij} + \beta\varphi_{i,j} + \gamma\varphi_{j,i}, \qquad e_{ij} = u_{j,i} + \varepsilon_{jir}\varphi_r \text{ on } B \end{aligned} \tag{6.3.27}$$

and the boundary conditions

$$\int_{\Sigma_1} t_{3i} da = 0, \qquad \int_{\Sigma_1} (\varepsilon_{ijk} x_j t_{3k} + m_{3i}) da = 0 \tag{6.3.28}$$

$$t_{\alpha i} n_\alpha = p_i(x_1, x_2) x_3^{n+1}, \qquad m_{\alpha i} n_\alpha = q_i(x_1, x_2) x_3^{n+1} \text{ on } \Pi \tag{6.3.29}$$

when the solution of the equations

$$\begin{aligned} & t^*_{ji,j} + \mathscr{F}_i(x_1, x_2) x_3^n = 0, \qquad m^*_{ji,j} + \varepsilon_{irs} t^*_{rs} + \mathcal{H}_i(x_1, x_2) x_3^n = 0 \\ & t^*_{ij} = \lambda e^*_{rr} \delta_{ij} + (\mu + \kappa) e^*_{ij} + \mu e^*_{ji} \\ & m^*_{ij} = \alpha \varphi^*_{r,r} \delta_{ij} + \beta \varphi^*_{i,j} + \gamma \varphi^*_{j,i}, \qquad e^*_{ij} = u^*_{j,i} + \varepsilon_{jir} \varphi^*_r \end{aligned} \tag{6.3.30}$$

with the conditions

$$\int_{\Sigma_1} t^*_{3i} da = 0, \qquad \int_{\Sigma_1} (\varepsilon_{ijk} x_j t^*_{3k} + m^*_{3i}) da = 0 \tag{6.3.31}$$

$$t^*_{\alpha i} n_\alpha = p_i(x_1, x_2) x_3^n, \qquad m^*_{\alpha i} n_\alpha = q_i(x_1, x_2) x_3^n \text{ on } \Pi \tag{6.3.32}$$

is known. In the above relations, $\mathscr{F}_i, \mathcal{H}_i, p_i$, and q_i are prescribed functions which belong to C^∞. We seek the solution of Almansi problem in the form

$$u_i = (n+1)\left[\int_0^{x_3} u^*_i dx_3 + v_i\right], \qquad \varphi_i = (n+1)\left[\int_0^{x_3} \varphi^*_i dx_3 + \psi_i\right] \tag{6.3.33}$$

where v_i and ψ_i are unknown functions. From Equations 6.3.27 and 6.3.33, we obtain

$$\begin{aligned} t_{ij} &= (n+1)\left[\int_0^{x_3} t^*_{ij} dx_3 + \tau_{ij} + k_{ij}\right] \\ m_{ij} &= (n+1)\left[\int_0^{x_3} m^*_{ij} dx_3 + \mu_{ij} + h_{ij}\right] \end{aligned} \tag{6.3.34}$$

where

$$\begin{aligned} & \tau_{ij} = \lambda \gamma_{rr} \delta_{ij} + (\mu + \kappa) \gamma_{ij} + \mu \gamma_{ji} \\ & \mu_{ij} = \alpha \psi_{r,r} \delta_{ij} + \beta \psi_{i,j} + \gamma \psi_{j,i}, \qquad \gamma_{ij} = v_{j,i} + \varepsilon_{jik} \psi_k \end{aligned} \tag{6.3.35}$$

and

$$\begin{aligned} k_{\alpha\beta} &= \lambda \delta_{\alpha\beta} u^*_3(x_1, x_2, 0), & k_{33} &= (\lambda + 2\mu + \kappa) u^*_3(x_1, x_2, 0) \\ k_{\alpha 3} &= \mu u^*_\alpha(x_1, x_2, 0), & k_{3\alpha} &= (\mu + \kappa) u^*_\alpha(x_1, x_2, 0) \\ h_{\eta\nu} &= \alpha \delta_{\eta\nu} \varphi^*_3(x_1, x_2, 0), & h_{33} &= (\alpha + \beta + \gamma) \varphi^*_3(x_1, x_2, 0) \\ h_{\alpha 3} &= \beta \varphi^*_\alpha(x_1, x_2, 0), & h_{3\alpha} &= \gamma \varphi^*_\alpha(x_1, x_2, 0) \end{aligned} \tag{6.3.36}$$

By using Equations 6.3.30, the equilibrium equations reduce to

$$\tau_{ji,j} + Y_i = 0, \qquad \mu_{ji,j} + \varepsilon_{irs} \tau_{rs} + Z_i = 0 \tag{6.3.37}$$

where

$$Y_i = k_{\alpha i,\alpha} + t^*_{3i}(x_1, x_2, 0), \qquad Z_i = h_{\alpha i,\alpha} + m^*_{3i}(x_1, x_2, 0) \tag{6.3.38}$$

Let us note that Y_i and Z_i are independent of the axial coordinate.

With the help of Equations 6.3.32 and 6.3.34, the conditions 6.3.29 become

$$\tau_{\beta i} n_\beta = \rho_i, \qquad \mu_{\beta i} n_\beta = \eta_i \text{ on } \Pi \tag{6.3.39}$$

where

$$\rho_i = -k_{\alpha i} n_\alpha, \qquad \eta_i = -h_{\alpha i} n_\alpha$$

From Equations 6.3.28 and 6.3.31, we obtain

$$\int_{\Sigma_1} \tau_{3i} da = -T_i, \qquad \int_{\Sigma_1} (\varepsilon_{ijk} x_j \tau_{3k} + \mu_{3i}) da = -\Omega_i \tag{6.3.40}$$

where

$$T_i = \int_{\Sigma_1} k_{3i} da, \qquad \Omega_i = \int_{\Sigma_1} (\varepsilon_{ijs} x_j k_{3s} + h_{3i}) da$$

Thus, the functions v_i and ψ_i satisfy Equations 6.3.7 and 6.3.35 on B and the boundary conditions 6.3.39 and 6.3.40. This problem was studied in Section 6.3.1. The solution has the form 6.3.3 in which $c_i = b_i = \tau_2 = 0$, $\chi = \chi_\alpha = 0$. Thus, the considered problem is solved.

The results presented in this chapter were established in Ref. 155.

6.4 Anisotropic Cosserat Cylinders

This section is concerned with the deformation of nonhomogeneous and anisotropic Cosserat elastic cylinders. Throughout this section we consider nonhomogeneous materials where the elastic coefficients are independent of the axial coordinate, namely

$$\begin{gathered} A_{jikl} = A_{ijkl}(x_1, x_2), \qquad B_{ijkl} = B_{ijkl}(x_1, x_2) \\ C_{ijkl} = C_{ijkl}(x_1, x_2), \quad (x_1, x_2) \in \Sigma_1 \end{gathered} \tag{6.4.1}$$

We suppose that the functions A_{ijrs}, B_{ijrs}, and C_{ijrs} belong to C^∞, and that the domain Σ_1 is C^∞-smooth. We consider only a C^∞-*theory* but it is possible to get a classical solution under more general assumptions of regularity [88].

6.4.1 Generalized Plane Strain

We assume that the cylinder B is occupied by an anisotropic elastic material for which the constitutive coefficients are independent of x_3. We define the state of generalized plane strain of the cylinder B to be that state in which the displacement vector and microrotation vector are independent of the axial coordinate,

$$u_i = u_i(x_1, x_2), \qquad \varphi_i = \varphi_i(x_1, x_2), \quad (x_1, x_2) \in \Sigma_1 \tag{6.4.2}$$

This restriction implies that $e_{ij}, \kappa_{ij}, t_{ij}$, and m_{ij} are independent of x_3. We assume that on the lateral surface of the cylinder, there are prescribed stress vector and the couple stress vector, and that the loads are independent of the axial coordinate. In the case of the generalized plane strain, the equations of equilibrium are

$$t_{\alpha i,\alpha} + f_i = 0, \qquad m_{\alpha i,\alpha} + \varepsilon_{ijk} t_{jk} + g_i = 0 \text{ on } \Sigma_1 \tag{6.4.3}$$

The geometrical equations imply that

$$e_{\alpha i} = u_{i,\alpha} + \varepsilon_{i\alpha k}\varphi_k, \qquad e_{3i} = \varepsilon_{i3k}\varphi_k, \qquad \kappa_{\alpha i} = \varphi_{i,\alpha}, \qquad \kappa_{3i} = 0 \tag{6.4.4}$$

The constitutive equations reduce to

$$\begin{gathered} t_{\alpha i} = A_{\alpha ijk} e_{jk} + B_{\alpha i\beta j}\kappa_{\beta j}, \qquad t_{3\alpha} = A_{3\alpha jk} e_{jk} + B_{3\alpha\beta j}\kappa_{\beta j} \\ m_{\alpha i} = B_{jk\alpha i} e_{jk} + C_{\alpha i\beta j}\kappa_{\beta j} \end{gathered} \tag{6.4.5}$$

and

$$t_{33} = A_{33jk} e_{jk} + B_{33\beta j}\kappa_{\beta j}, \qquad m_{3i} = B_{jk3i} e_{jk} + C_{3i\beta j}\kappa_{\beta j} \tag{6.4.6}$$

On the lateral surface of the cylinder, we have the boundary conditions

$$t_{\alpha i} n_\alpha = \widetilde{t}_i, \qquad m_{\alpha i} n_\alpha = \widetilde{m}_i \text{ on } \Gamma \tag{6.4.7}$$

The generalized plane strain problem consists in the determination of the functions u_i and φ_i which satisfy Equations 6.4.3, 6.4.4, and 6.4.5 on Σ_1 and the boundary conditions 6.4.7 on Γ. The functions t_{33} and m_{3i} can be calculated from Equations 6.4.6 after the components u_i and φ_i have been determined.

The conditions of equilibrium of the cylinder B can be written in the form

$$\begin{gathered} \int_{\Sigma_1} f_i da + \int_\Gamma \widetilde{t}_i ds = 0 \\ \int_{\Sigma_1} (\varepsilon_{\alpha\beta} x_\alpha f_\beta + g_3) da + \int_\Gamma (\varepsilon_{\alpha\beta} x_\alpha \widetilde{t}_\beta + \widetilde{m}_3) ds = 0 \end{gathered} \tag{6.4.8}$$

and

$$\begin{gathered} \int_{\Sigma_1} (x_2 f_3 + g_1) da + \int_\Gamma (x_2 \widetilde{t}_3 + \widetilde{m}_1) ds - \int_{\Sigma_1} t_{32} da = 0 \\ \int_{\Sigma_1} (x_1 f_3 - g_2) da + \int_\Gamma (x_1 \widetilde{t}_3 - \widetilde{m}_2) ds - \int_{\Sigma_1} t_{31} da = 0 \end{gathered} \tag{6.4.9}$$

The conditions 6.4.9 are identically satisfied on the basis of the relations 6.4.3 and 6.4.7. Thus, for the first of Equation 6.4.9, we have

$$\begin{aligned}\int_{\Sigma_1} t_{32}da &= \int_{\Sigma_1}(t_{23}+g_1+m_{\alpha 1,\alpha})da\\ &= \int_{\Sigma_1}[t_{23}+x_2(t_{\alpha 3,\alpha}+f_3)+m_{\alpha 1,\alpha}+g_1]da\\ &= \int_{\Sigma_1}[(x_2t_{\alpha 3})_{,\alpha}+x_2f_3+g_1+m_{\alpha 1,\alpha}]da\\ &= \int_{\Gamma}(x_2t_{\alpha 3}+m_{\alpha 1})n_\alpha ds+\int_{\Sigma_1}(x_2f_3+g_1)da\\ &= \int_{\Gamma}(x_2\widetilde{t}_3+\widetilde{m}_1)ds+\int_{\Sigma_1}(x_2f_3+g_1)da\end{aligned}$$

In a similar way, we can prove that the second condition 6.4.9 is satisfied.

The elastic potential in the case of generalized plane strain is given by

$$\begin{aligned}2W^0 &= A_{\alpha ijk}e_{\alpha i}e_{jk}+A_{3\alpha jk}e_{3\alpha}e_{jk}+B_{\alpha i\beta j}e_{\alpha i}\kappa_{\beta j}\\ &\quad + B_{3\alpha\beta j}e_{3\alpha}\kappa_{\beta j}+B_{jk\alpha i}e_{jk}\kappa_{\alpha i}+C_{\alpha i\beta j}\kappa_{\alpha i}\kappa_{\beta j}\end{aligned}$$

We suppose that W^0 is a positive definite quadratic form in the variables $e_{\alpha i}, e_{3\alpha}$, and $\kappa_{\alpha i}$. We recall the following result (cf. [88,154]).

Theorem 6.4.1 *The generalized plane strain problem has a solution belonging to $C^\infty(\overline{\Sigma}_1)$ if and only if the functions $f_i, g_3, \widetilde{t}_i$, and $\widetilde{m}_3$ satisfy the conditions* 6.4.8.

In what follows we will use four special problems $C^{(s)}$, $(s = 1, 2, 3, 4)$, of generalized plane strain. The problems $C^{(s)}$ correspond to the systems of loading $f_i^{(s)}, g_i^{(s)}, \widetilde{t}_i^{(s)}, \widetilde{m}_i^{(s)}$, where

$$\begin{aligned}f_i^{(\beta)} &= (A_{\alpha i33}x_\beta+\varepsilon_{\nu\beta}B_{\alpha i3\nu})_{,\alpha}, \qquad f_i^{(3)} = A_{\alpha i33,\alpha}\\ f_i^{(4)} &= (A_{\alpha i3\nu}\varepsilon_{\beta\nu}x_\beta+B_{\alpha i33})_{,\alpha}\\ g_i^{(\beta)} &= (B_{33\alpha i}x_\beta+\varepsilon_{\nu\beta}C_{\alpha i3\nu})_{,\alpha}+\varepsilon_{ijk}(A_{jk33}x_\beta+\varepsilon_{\nu\beta}B_{jk3\nu})\\ g_i^{(3)} &= B_{33\alpha i,\alpha}+\varepsilon_{ijk}A_{jk33}\\ g_i^{(4)} &= (B_{3\nu\alpha i}\varepsilon_{\beta\nu}x_\beta+C_{\alpha i33})_{,\alpha}+\varepsilon_{ijk}(A_{jk3\nu}\varepsilon_{\beta\nu}x_\beta+B_{jk33})\\ \widetilde{t}_i^{(\beta)} &= -(A_{\alpha i33}x_\beta+\varepsilon_{\nu\beta}B_{\alpha i3\nu})n_\alpha, \qquad \widetilde{t}_i^{(3)} = -A_{\alpha i33}n_\alpha\\ \widetilde{t}_i^{(4)} &= -(A_{\alpha i3\nu}\varepsilon_{\beta\nu}x_\beta+B_{\alpha i33})n_\alpha\\ \widetilde{m}_i^{(\beta)} &= -(B_{33\alpha i}x_\beta+\varepsilon_{\nu\beta}C_{\alpha i3\nu})n_\alpha, \qquad \widetilde{m}_i^{(3)} = -B_{33\alpha i}n_\alpha\\ m_i^{(4)} &= -(B_{3\nu\alpha i}\varepsilon_{\beta\nu}x_\beta+C_{\alpha i33})n_\alpha\end{aligned} \tag{6.4.10}$$

We denote by $u_i^{(s)}$ and $\varphi_i^{(s)}$, respectively, the components of the displacement vector and the components of the microrotation vector from the problem $C^{(s)}$.

The problem $C^{(s)}$ is characterized by the equations

$$\begin{gathered}
t_{\alpha i,\alpha}^{(s)} + f_i^{(s)} = 0, \qquad m_{\alpha i,\alpha}^{(s)} + \varepsilon_{ijk} t_{jk}^{(s)} + g_i^{(s)} = 0 \\
t_{\alpha i}^{(s)} = A_{\alpha ijk} e_{jk}^{(s)} + B_{\alpha i\beta j} \kappa_{\beta j}^{(s)}, \qquad t_{3\alpha}^{(s)} = A_{3\alpha jk} e_{jk}^{(s)} + B_{3\alpha\beta j} \kappa_{\beta j}^{(s)} \\
m_{\alpha i}^{(s)} = B_{jk\alpha i} e_{jk}^{(s)} + C_{\alpha i\beta j} \kappa_{\beta j}^{(s)} \\
e_{\alpha i}^{(s)} = u_{i,\alpha}^{(s)} + \varepsilon_{i\alpha k} \varphi_k^{(s)}, \qquad e_{3i}^{(s)} = \varepsilon_{i3k} \varphi_k^{(s)}, \qquad \kappa_{\alpha i}^{(s)} = \varphi_{i,\alpha}^{(s)}
\end{gathered} \tag{6.4.11}$$

on Σ_1, and the boundary conditions

$$t_{\alpha i}^{(s)} n_\alpha = \tilde{t}_i^{(s)}, \qquad m_{\alpha i}^{(s)} n_\alpha = \tilde{m}_i^{(s)} \text{ on } \Gamma \tag{6.4.12}$$

We denote

$$t_{33}^{(s)} = A_{33jk} e_{jk}^{(s)} + B_{33\beta j} \kappa_{\beta j}^{(s)}, \qquad m_{3i}^{(s)} = B_{jk3i} e_{jk}^{(s)} + C_{3i\beta j} \kappa_{\beta j}^{(s)}$$

It is a simple matter to see that the necessary and sufficient conditions 6.4.8 for the existence of the solution are satisfied for each boundary-value problem $C^{(s)}$. In what follows we assume that the functions $u_i^{(s)}$ and $\varphi_i^{(s)}$ are known.

6.4.2 Extension, Bending, and Torsion

Let us study the deformation of the cylinder B when the loading applied on the end Σ_1 is statically equivalent to a force $\mathbf{F} = F_3 \mathbf{e}_3$ and a moment $\mathbf{M} = M_3 \mathbf{e}_3$. The problem consists in the solving of Equations 5.1.11, 5.1.16, and 6.2.1 on B, with the conditions 4.10.14, 4.10.15, and 6.2.20. We seek the solution in the form

$$\begin{gathered}
u_\alpha = -\frac{1}{2} a_\alpha x_3^2 + \varepsilon_{\beta\alpha} a_4 x_3 x_\beta + \sum_{s=1}^{4} a_s u_\alpha^{(s)} \\
u_3 = (a_1 x_1 + a_2 x_2 + a_3) x_3 + \sum_{s=1}^{4} a_s u_3^{(s)} \\
\varphi_\alpha = \varepsilon_{\alpha\beta} a_\beta x_3 + \sum_{s=1}^{4} a_s \varphi_\alpha^{(s)}, \qquad \varphi_3 = a_4 x_3 + \sum_{s=1}^{4} a_s \varphi_3^{(s)}
\end{gathered} \tag{6.4.13}$$

where $u_i^{(s)}$ and $\varphi_i^{(s)}$ are the solutions of the problems $C^{(s)}$ and a_s are unknown constants. From Equations 5.1.11, 5.1.16, and 6.4.13, we obtain

$$t_{ij} = \sum_{s=1}^{4} a_s \tau_{ij}^{(s)}, \qquad m_{ij} = \sum_{s=1}^{4} a_s \mu_{ij}^{(s)} \tag{6.4.14}$$

where

$$\tau_{ij}^{(\alpha)} = t_{ij}^{(\alpha)} + A_{ij33}x_\alpha + \varepsilon_{\nu\alpha}B_{ij3\nu}, \qquad \tau_{ij}^{(3)} = t_{ij}^{(3)} + A_{ij33}$$
$$\tau_{ij}^{(4)} = t_{ij}^{(4)} + A_{ij3\nu}\varepsilon_{\beta\nu}x_\beta + B_{ij33}, \qquad \mu_{ij}^{(\alpha)} = m_{ij}^{(\alpha)} + B_{33ij}x_\alpha + \varepsilon_{\nu\alpha}C_{ij3\nu}$$
$$\mu_{ij}^{(3)} = m_{ij}^{(3)} + B_{33ij}, \qquad \mu_{ij}^{(4)} = m_{ij}^{(4)} + B_{3\nu ij}\varepsilon_{\beta\nu}x_\beta + C_{ij33} \tag{6.4.15}$$

The equilibrium equations 6.2.1 and the boundary conditions 6.2.2 are satisfied on the basis of the relations 6.4.11 and 6.4.12. As in Section 6.2, we can prove that the conditions 4.10.14 are identically satisfied. From 4.10.15 and 6.4.14, we find that

$$\sum_{s=1}^{4}\mathcal{D}_{\alpha s}a_s = \varepsilon_{\alpha\beta}M_\beta, \qquad \sum_{s=1}^{4}\mathcal{D}_{3s}a_s = -F_3, \qquad \sum_{s=1}^{4}\mathcal{D}_{4s}a_s = -M_3 \tag{6.4.16}$$

where

$$\mathcal{D}_{\alpha s} = \int_{\Sigma_1}(x_\alpha\tau_{33}^{(s)} + \varepsilon_{\beta\alpha}\mu_{3\beta}^{(s)})da, \qquad \mathcal{D}_{3s} = \int_{\Sigma_1}\tau_{33}^{(s)}da$$
$$\mathcal{D}_{4s} = \int_{\Sigma_1}(\varepsilon_{\alpha\beta}x_\alpha\tau_{3\beta}^{(s)} + \mu_{33}^{(s)})da \tag{6.4.17}$$

As in Section 4.3, we can prove that

$$\det(\mathcal{D}_{rs}) \neq 0 \tag{6.4.18}$$

so that the system 6.4.16 uniquely determines the constants a_s, $(s = 1, 2, 3, 4)$.

6.4.3 Flexure

The problem of flexure consists in the determination of a solution of the Equations 5.1.11, 5.1.16, and 6.2.1 on B which satisfies the conditions 6.2.2 and the conditions for $x_3 = 0$. We seek the solution in the form

$$u_\alpha = -\frac{1}{2}a_\alpha x_3^2 - \frac{1}{6}b_\alpha x_3^3 + \varepsilon_{\beta\alpha}\left(a_4x_3 + \frac{1}{2}b_4x_3^2\right)x_\beta$$
$$+\sum_{s=1}^{4}(a_s + b_sx_3)u_\alpha^{(s)} + v_\alpha(x_1, x_2)$$
$$u_3 = (a_1x_1 + a_2x_2 + a_3)x_3 + \frac{1}{2}(b_1x_1 + b_2x_2 + b_3)x_3^2$$
$$+\sum_{s=1}^{4}(a_s + b_sx_3)u_3^{(s)} + v_3(x_1, x_2) \tag{6.4.19}$$
$$\varphi_\alpha = \varepsilon_{\alpha\beta}\left(a_\beta x_3 + \frac{1}{2}b_\beta x_3^2\right) + \sum_{s=1}^{4}(a_s + b_sx_3)\varphi_\alpha^{(s)} + \psi_\alpha(x_1, x_2)$$
$$\varphi_3 = a_4x_3 + \frac{1}{2}b_4x_3^2 + \sum_{s=1}^{4}(a_s + b_sx_3)\varphi_3^{(s)} + \psi_3(x_1, x_2)$$

where $u_i^{(s)}$ and $\varphi_i^{(s)}$ are the solutions of the problems $C^{(s)}$, v_i and ψ_i are unknown functions, and a_k and b_k, $(k = 1, 2, 3, 4)$, are unknown constants.

From Equations 5.1.11, 5.1.17, and 6.4.19, we obtain

$$t_{ij} = \sum_{s=1}^{4} (a_s + b_s x_3)\tau_{ij}^{(s)} + \tau_{ij} + K_{ij}$$
$$m_{ij} = \sum_{s=1}^{4} (a_s + b_s x_3)\mu_{ij}^{(s)} + \mu_{ij} + H_{ij} \tag{6.4.20}$$

where $\tau_{ij}^{(s)}$ and $\mu_{ij}^{(s)}$ are given by Equations 6.4.15, τ_{ij} and μ_{ij} are defined by

$$\tau_{ij} = A_{ijrs}\gamma_{rs} + B_{ijrs}\nu_{rs}, \qquad \mu_{ij} = B_{rsij}\gamma_{rs} + C_{ijrs}\nu_{rs}$$
$$\gamma_{\alpha i} = v_{i,\alpha} + \varepsilon_{i\alpha k}\psi_k, \qquad \gamma_{3i} = \varepsilon_{i3k}\psi_k, \qquad \nu_{\alpha i} = \psi_{i,\alpha}, \qquad \nu_{3i} = 0 \tag{6.4.21}$$

and

$$K_{ij} = \sum_{s=1}^{4} b_s\big(A_{ij3k}u_k^{(s)} + B_{ij3k}\varphi_k^{(s)}\big), \qquad H_{ij} = \sum_{s=1}^{4} b_s\big(B_{3kij}u_k^{(s)} + C_{ij3k}\varphi_k^{(s)}\big) \tag{6.4.22}$$

With the help of Equations 6.4.11 and 6.4.20, the equilibrium equations 6.2.1 reduce to

$$\tau_{\alpha i,\alpha} + Q_i = 0, \qquad \mu_{\alpha i,\alpha} + \varepsilon_{ijk}\tau_{jk} + G_i = 0 \text{ on } \Sigma_1 \tag{6.4.23}$$

where

$$Q_i = K_{\alpha i,\alpha} + \sum_{s=1}^{4} b_s \tau_{3i}^{(s)}, \qquad G_i = H_{\alpha i,\alpha} + \varepsilon_{ijk}K_{jk} + \sum_{s=1}^{4} b_s \mu_{3i}^{(s)} \tag{6.4.24}$$

In view of the relations 6.4.20 and 6.4.12, the conditions on the lateral surface become

$$\tau_{\alpha i}n_\alpha = p_i, \qquad \mu_{\alpha i}n_\alpha = q_i \text{ on } \Gamma \tag{6.4.25}$$

where

$$p_i = -K_{\alpha i}n_\alpha, \qquad q_i = -H_{\alpha i}n_\alpha \tag{6.4.26}$$

Thus, the functions v_i and ψ_i are the components of the displacement vector and the components of the microrotation vector in the generalized plane strain problem 6.4.21, 6.4.23, and 6.4.25. The necessary and sufficient conditions to solve this problem are

$$\int_{\Sigma_1} Q_i da + \int_\Gamma p_i ds = 0$$
$$\int_{\Sigma_1} (\varepsilon_{\alpha\beta}x_\alpha Q_\beta + G_3)da + \int_\Gamma (\varepsilon_{\alpha\beta}x_\alpha p_\beta + q_3)ds = 0 \tag{6.4.27}$$

It is a simple matter to see that we have

$$\tau_{\alpha i,\alpha}^{(s)} = 0, \qquad \mu_{\alpha i,\alpha}^{(s)} + \varepsilon_{ijk}\tau_{jk}^{(s)} = 0 \text{ on } \Sigma_1 \tag{6.4.28}$$

and

$$\tau_{\alpha i}^{(s)} n_\alpha = 0, \qquad \mu_{\alpha i}^{(s)} n_\alpha = 0 \text{ on } \Gamma \tag{6.4.29}$$

Using Equations 6.4.28 and 6.4.29, we find that

$$\begin{aligned}\int_{\Sigma_1} \tau_{3\alpha}^{(s)} da &= \int_{\Sigma_1} \left[\tau_{\alpha 3}^{(s)} + \varepsilon_{\beta\alpha}\mu_{\rho\beta,\rho}^{(s)}\right] da = \int_{\Sigma_1} \left[\tau_{\alpha 3}^{(s)} + x_\alpha \tau_{\nu 3,\nu}^{(s)} + \varepsilon_{\beta\alpha}\mu_{\rho\beta,\rho}^{(s)}\right] da \\ &= \int_{\Sigma_1} \left[\left(x_\alpha \tau_{\nu 3}^{(s)}\right)_{,\nu} + \varepsilon_{\beta\alpha}\mu_{\rho\beta,\rho}^{(s)}\right] da = \int_\Gamma \left[x_\alpha \tau_{\nu 3}^{(s)} n_\nu + \varepsilon_{\beta\alpha}\mu_{\rho\beta}^{(s)} n_\rho\right] ds = 0\end{aligned} \tag{6.4.30}$$

By Equations 6.4.24, 6.4.26, and 6.4.30, we find that the first two conditions 6.4.27 are identically satisfied. From the remaining conditions, we get

$$\sum_{s=1}^{4} \mathcal{D}_{rs} b_s = 0, \quad (r = 3, 4) \tag{6.4.31}$$

where $\mathcal{D}_{rs}$ are given by Equations 6.4.17. Taking into account the equilibrium equations and the boundary conditions 6.2.2, we obtain

$$\begin{aligned}\int_{\Sigma_1} t_{3\alpha} da &= \int_{\Sigma_1} (t_{\alpha 3} + \varepsilon_{\beta\alpha} m_{j\beta,j}) da = \int_{\Sigma_1} (t_{\alpha 3} + x_\alpha t_{i3,i} + \varepsilon_{\beta\alpha} m_{j\beta,j}) da \\ &= \int_\Gamma (x_\alpha t_{\nu 3} + \varepsilon_{\beta\alpha} m_{\nu\beta}) n_\nu ds + \int_{\Sigma_1} (x_\alpha t_{33,3} + \varepsilon_{\beta\alpha} m_{3\beta,3}) da \\ &= \int_{\Sigma_1} (x_\alpha t_{33} + \varepsilon_{\beta\alpha} m_{3\beta})_{,3} da\end{aligned} \tag{6.4.32}$$

Using Equations 6.4.20 and 6.4.32, the conditions 6.2.26 reduce to

$$\sum_{s=1}^{4} \mathcal{D}_{\alpha s} b_s = -F_\alpha \tag{6.4.33}$$

The system 6.4.31 and 6.4.33 can always be solved for the constants c_s. Thus the conditions 6.4.27 are satisfied. In what follows we assume that the functions v_j and ψ_i have been determined.

In view of Equations 6.4.20, from the conditions 6.2.27 and 6.2.28, we obtain the following equations for the unknown constants a_s

$$\sum_{s=1}^{4} \mathcal{D}_{rs} a_s = d_r, \quad (r = 1, 2, 3, 4) \tag{6.4.34}$$

where

$$d_\alpha = -\int_{\Sigma_1} [x_\alpha(\tau_{33} + K_{33}) - \varepsilon_{\alpha\beta}(\mu_{3\beta} + H_{3\beta})]da$$

$$d_3 = -\int_{\Sigma_1} (\tau_{33} + K_{33})da \tag{6.4.35}$$

$$d_4 = -\int_{\Sigma_1} [\varepsilon_{\alpha\beta}x_\alpha(\tau_{3\beta} + K_{3\beta}) + \mu_{33} + H_{33}]da$$

Equations 6.4.34 uniquely determine the constants a_s, $(s = 1, 2, 3, 4)$, so that the flexure problem is solved.

6.4.4 Uniformly Loaded Cylinders

Let us consider the Almansi–Michell problem for anisotropic elastic bodies. The problem consists in the determination of the functions $u_i, \varphi_i \in C^2(B) \cap C^1(\overline{B})$ that satisfy the Equations 5.1.11, 5.1.16, and 5.1.19 on B, the conditions on the end Σ_1, and the conditions 6.3.2 on Π, when the body loads have the form 6.3.1.

Following Ref. 161, we try to solve the problem assuming that

$$\begin{aligned}
u_\alpha &= -\frac{1}{2}a_\alpha x_3^2 - \frac{1}{6}b_\alpha x_3^3 - \frac{1}{24}c_\alpha x_3^4 + \varepsilon_{\beta\alpha}\left(a_4 x_3 + \frac{1}{2}b_4 x_3^2 + \frac{1}{6}c_4 x_3^3\right)x_\beta \\
&\quad + \sum_{s=1}^{4}\left(a_s + b_s x_3 + \frac{1}{2}c_s x_3^2\right)u_\alpha^{(s)} + w_\alpha(x_1, x_2) + x_3 v_\alpha(x_1, x_2) \\
u_3 &= (a_1 x_1 + a_2 x_2 + a_3)x_3 + \frac{1}{2}(b_1 x_1 + b_2 x_2 + b_3)x_3^2 \\
&\quad + \frac{1}{6}(c_1 x_1 + c_2 x_2 + c_3)x_3^3 \\
&\quad + \sum_{s=1}^{4}\left(a_s + b_s x_3 + \frac{1}{2}c_s x_3^2\right)u_3^{(s)} + w_3(x_1, x_2) + x_3 v_3(x_1, x_2) \\
\varphi_\alpha &= \varepsilon_{\alpha\beta}\left(a_\beta x_3 + \frac{1}{2}b_\beta x_3^2 + \frac{1}{6}c_\beta x_3^3\right) + \sum_{s=1}^{4}\left(a_s + b_s x_3 + \frac{1}{2}c_s x_3^2\right)\varphi_\alpha^{(s)} \\
&\quad + \chi_\alpha(x_1, x_\alpha) + x_3\psi_\alpha(x_1, x_2) \\
\varphi_3 &= a_4 x_3 + \frac{1}{2}b_4 x_3^2 + \frac{1}{6}c_4 x_3^3 + \sum_{s=1}^{4}\left(a_s + b_s x_3 + \frac{1}{2}c_s x_3^2\right)\varphi_3^{(s)} \\
&\quad + \chi_3(x_1, x_2) + x_3\psi_3(x_1, x_2)
\end{aligned} \tag{6.4.36}$$

where $u_i^{(s)}$ and $\varphi_i^{(s)}$ are the solutions of the problems $C^{(s)}$, v_i, ψ_i, w_i, and χ_i are unknown functions, and a_s, b_s, and c_s, $(s = 1, 2, 3, 4)$, are unknown constants. By Equations 5.1.11, 5.1.16, and 6.4.36, we get

$$
\begin{aligned}
t_{ij} &= \sum_{s=1}^{4}\left(a_s + b_s x_3 + \frac{1}{2}c_s x_3^2\right)\tau_{ij}^{(s)} + \tau_{ij} + x_3\sigma_{ij} + k_{ij} + x_3 K_{ij} \\
m_{ij} &= \sum_{s=1}^{4}\left(a_s + b_s x_3 + \frac{1}{2}c_s x_3^2\right)\mu_{ij}^{(s)} + \mu_{ij} + x_3\nu_{ij} + h_{ij} + x_3 H_{ij}
\end{aligned}
\tag{6.4.37}
$$

where $\tau_{ij}^{(s)}$ and $\mu_{ij}^{(s)}$ are given by Equation 6.4.15, τ_{ij} and μ_{ij} are defined by

$$
\begin{aligned}
&\tau_{ij} = A_{ijrs}\xi_{rs} + B_{ijrs}\eta_{rs}, \qquad \mu_{ij} = B_{rsij}\xi_{rs} + C_{ijrs}\eta_{rs} \\
&\xi_{\alpha i} = w_{i,\alpha} + \varepsilon_{i\alpha k}\chi_k, \qquad \xi_{3i} = \varepsilon_{i3k}\chi_k, \qquad \eta_{\alpha i} = \chi_{i,\alpha}
\end{aligned}
\tag{6.4.38}
$$

the functions σ_{ij} and ν_{ij} have the expressions

$$
\begin{aligned}
&\sigma_{ij} = A_{ijrs}\gamma_{rs} + B_{ijrs}\zeta_{rs}, \qquad \nu_{ij} = B_{rsij}\gamma_{rs} + C_{ijrs}\zeta_{rs} \\
&\gamma_{\alpha i} = v_{i,\alpha} + \varepsilon_{i\alpha k}\psi_k, \qquad \gamma_{3i} = \varepsilon_{i3k}\psi_k, \qquad \zeta_{\alpha i} = \psi_{i,\alpha}
\end{aligned}
\tag{6.4.39}
$$

and we have used the notations

$$
\begin{aligned}
k_{ij} &= A_{ij3r}v_r + B_{ij3r}\psi_r + \sum_{s=1}^{4} b_s\left(A_{ij3r}u_r^{(s)} + B_{ij3r}\varphi_r^{(s)}\right) \\
K_{ij} &= \sum_{s=1}^{4} c_s\left[A_{ij3r}u_r^{(s)} + B_{ij3r}\varphi_r^{(s)}\right] \\
h_{ij} &= B_{3rij}v_r + C_{ij3r}\psi_r + \sum_{s=1}^{4} b_s\left[B_{3rij}u_r^{(s)} + C_{ij3r}\varphi_r^{(s)}\right] \\
H_{ij} &= \sum_{s=1}^{4} c_s\left(B_{3rij}u_r^{(s)} + C_{ij3r}\varphi_r^{(s)}\right)
\end{aligned}
\tag{6.4.40}
$$

The equations of equilibrium 5.1.19 reduce to

$$
\begin{aligned}
&\tau_{\alpha i,\alpha} + k_{\alpha i,\alpha} + \sum_{s=1}^{4} b_s\tau_{3i}^{(s)} + \sigma_{3i} + K_{3i} + f_i^{(0)} = 0 \\
&\mu_{\alpha i,\alpha} + \varepsilon_{ijk}\tau_{jk} + g_i^{(0)} + h_{\alpha i,\alpha} + \sum_{s=1}^{4} b_s\mu_{3i}^{(s)} + \nu_{3i} + H_{3i} + \varepsilon_{ijr}k_{jr} = 0
\end{aligned}
\tag{6.4.41}
$$

and

$$
\begin{aligned}
&\sigma_{\alpha i,\alpha} + K_{\alpha i,\alpha} + \sum_{s=1}^{4} c_s\tau_{3i}^{(s)} = 0 \\
&\nu_{\alpha i,\alpha} + \varepsilon_{ijk}\sigma_{jk} + H_{\alpha i,\alpha} + \varepsilon_{ijk}K_{jk} + \sum_{s=1}^{4} c_s\mu_{3i}^{(s)} = 0 \text{ on } \Sigma_1
\end{aligned}
\tag{6.4.42}
$$

The boundary conditions 6.3.2 become

$$\tau_{\alpha i}n_\alpha = P_i, \qquad \mu_{\alpha i}n_\alpha = Q_i \text{ on } \Gamma \tag{6.4.43}$$

and

$$\sigma_{\alpha i}n_\alpha = T_i, \qquad \nu_{\alpha i}n_\alpha = S_i \text{ on } \Gamma \tag{6.4.44}$$

where

$$P_i = \widetilde{t}_i^{(0)} - k_{\alpha i}n_\alpha, \qquad Q_i = \widetilde{m}_i^{(0)} - h_{\alpha i}n_\alpha, \qquad T_i = -K_{\alpha i}n_\alpha, \qquad S_i = -H_{\alpha i}n_\alpha$$

The necessary and sufficient conditions to solve the generalized plane strain problem 6.4.39, 6.4.42, and 6.4.44 reduce to

$$\int_{\Sigma_1}\sum_{s=1}^{4} c_s\tau_{3i}^{(s)}da = 0, \qquad \int_{\Sigma_1}\sum_{s=1}^{4} c_s\left[\varepsilon_{\alpha\beta}x_\alpha\tau_{3\beta}^{(s)} + \mu_{33}^{(s)}\right]da = 0 \tag{6.4.45}$$

By using Equation 6.4.30, it follows that the first two conditions 6.4.45 are satisfied. The remaining conditions imply

$$\sum_{s=1}^{4}\mathcal{D}_{rs}c_s = 0, \quad (r = 3, 4) \tag{6.4.46}$$

The necessary and sufficient conditions for the existence of the solution of the generalized plane strain problem 6.4.38, 6.4.41, and 6.4.43 are

$$\begin{aligned}
&\int_{\Sigma_1} f_i^{(0)}da + \int_\Gamma \widetilde{t}_i^{(0)}ds + \int_{\Sigma_1} t_{3i,3}da = 0\\
&\int_{\Sigma_1}(\varepsilon_{\alpha\beta}x_\alpha f_\beta^{(0)} + g_3^{(0)})da + \int_\Gamma(\varepsilon_{\alpha\beta}x_\alpha\widetilde{t}_\beta^{(0)} + \widetilde{m}_3^{(0)})ds\\
&\qquad + \int_{\Sigma_1}(\varepsilon_{\alpha\beta}x_\alpha t_{3\beta,3} + m_{33,3})da = 0
\end{aligned} \tag{6.4.47}$$

By using Equation 6.3.16, the first two conditions 6.4.47 reduce to

$$\sum_{s=1}^{4}\mathcal{D}_{\alpha s}c_s = -\int_{\Sigma_1} f_\alpha^{(0)}da - \int_\Gamma \widetilde{t}_\alpha^{(0)}ds \tag{6.4.48}$$

The system 6.4.46 and 6.4.48 determines the constants c_s, $(s = 1, 2, 3, 4)$. Thus, the conditions 6.4.45 are satisfied, and in what follows we can assume that the functions v_i and ψ_i are known. The remaining conditions from Equation 6.4.47 become

$$\begin{aligned}
\sum_{s=1}^{4}\mathcal{D}_{3s}b_s &= -\int_{\Sigma_1} f_3^{(0)}da - \int_\Gamma \widetilde{t}_3^{(0)}ds - \int_{\Sigma_1}(\sigma_{33} + K_{33})da\\
\sum_{s=1}^{4}\mathcal{D}_{4s}b_s &= -\int_{\Sigma_1}(\varepsilon_{\alpha\beta}x_\alpha f_\beta^{(0)} + g_3^{(0)})da - \int_\Gamma(\varepsilon_{\alpha\beta}x_\alpha\widetilde{t}_\beta^{(0)} + \widetilde{m}_3^{(0)})ds\\
&\quad - \int_{\Sigma_1}[\varepsilon_{\alpha\beta}x_\alpha(\sigma_{3\beta} + K_{3\beta}) + \nu_{33} + H_{33}]da
\end{aligned} \tag{6.4.49}$$

With the help of Equation 6.3.22, the conditions 6.2.3 take the form

$$\sum_{s=1}^{4} \mathcal{D}_{\alpha s} b_s = -F_\alpha - \int_\Gamma (x_\alpha \widetilde{t}_3^{(0)} + \varepsilon_{\beta\alpha} \widetilde{m}_\beta^{(0)}) ds - \int_{\Sigma_1} (x_\alpha f_3^{(0)} + \varepsilon_{\beta\alpha} g_\beta^{(0)}) da$$
$$- \int_{\Sigma_1} [x_\alpha(\sigma_{33} + K_{33}) + \varepsilon_{\beta\alpha}(\nu_{3\beta} + H_{3\beta})] da \qquad (6.4.50)$$

In view of Equation 6.4.18, the system 6.4.49 and 6.4.50 determines the constants b_s, $(s = 1, 2, 3, 4)$. The conditions 6.4.47 are satisfied so that we can consider that the functions w_i and χ_i are given.

From Equations 6.2.4, 6.2.5, and 6.2.6, we obtain the following system for the constants a_s

$$\sum_{s=1}^{4} \mathcal{D}_{rs} a_s = \zeta_r, \quad (r = 1, 2, 3, 4) \qquad (6.4.51)$$

where

$$\zeta_\alpha = \varepsilon_{\alpha\beta} M_\beta - \int_{\Sigma_1} [x_\alpha(\tau_{33} + k_{33}) + \varepsilon_{\beta\alpha}(\mu_{3\beta} + h_{3\beta})] da$$
$$\zeta_3 = -F_3 - \int_{\Sigma_1} (\tau_{33} + k_{33}) da$$
$$\zeta_4 = -M_3 - \int_{\Sigma_1} [\varepsilon_{\alpha\beta} x_\alpha(\tau_{3\beta} + k_{3\beta}) + \mu_{33} + h_{33}] da$$

On the basis of Equation 6.4.18, from Equation 6.4.51, we can find the constants a_1, a_2, a_3, and a_4.

6.4.5 Recurrence Process

In this case the problem of Almansi reduces to the finding of the functions $u_i, \varphi_i \in C^2(B) \cap C^1(\overline{B})$ that satisfy the equations

$$t_{ji,j} + \mathscr{F}_i(x_1, x_2) x_3^{n+1} = 0, \qquad m_{ji,j} + \varepsilon_{irs} t_{rs} + \mathcal{H}_i(x_1, x_2) x_3^{n+1} = 0$$
$$t_{ij} = A_{ijrs} e_{rs} + B_{ijrs} \kappa_{rs}, \qquad m_{ij} = B_{rsij} e_{rs} + C_{ijrs} \kappa_{rs} \qquad (6.4.52)$$
$$e_{ij} = u_{j,i} + \varepsilon_{jik} \varphi_k, \qquad \kappa_{ij} = \varphi_{j,i} \text{ on } B$$

and the boundary conditions 6.3.28 and 6.3.29, when the solution of the equations

$$t^*_{ji,j} + \mathscr{F}_i(x_1, x_2) x_3^n = 0, \qquad m^*_{ji,j} + \varepsilon_{irs} t^*_{rs} + \mathcal{H}_i(x_1, x_2) x_3^n = 0$$
$$t^*_{ij} = A_{ijrs} e^*_{rs} + B_{ijrs} \kappa^*_{rs}, \qquad m^*_{ij} = B_{rsij} e^*_{rs} + C_{ijrs} \kappa^*_{rs} \qquad (6.4.53)$$
$$e^*_{ij} = u^*_{j,i} + \varepsilon_{jik} \varphi^*_k, \qquad \kappa^*_{ij} = \varphi^*_{j,i} \text{ on } B$$

with the conditions 6.3.31 and 6.3.32, is known. We seek the solution of this problem in the form 6.3.33, where v_i and ψ_i are unknown functions.

By Equations 6.3.33, 6.4.52, and 6.4.53, we get

$$t_{ij} = (n+1)\left(\int_0^{x_3} t_{ij}^* dx_3 + \tau_{ij} + T_{ij}\right)$$
$$m_{ij} = (n+1)\left(\int_0^{x_3} m_{ij}^* dx_3 + \mu_{ij} + M_{ij}\right)$$

where τ_{ij} and μ_{ij} are defined by

$$\tau_{ij} = A_{ijrs}\xi_{rs} + B_{ijrs}\eta_{rs}, \qquad \mu_{ij} = B_{rsij}\xi_{rs} + C_{ijrs}\eta_{rs}$$
$$\xi_{ij} = v_{j,i} + \varepsilon_{jik}\psi_k, \qquad \eta_{ij} = \psi_{j,i}$$

and we have used the notations

$$T_{ij} = A_{ij3r}u_r^*(x_1, x_2, 0) + B_{ij3r}\varphi_r^*(x_1, x_2, 0)$$
$$M_{ij} = B_{3rij}u_r^*(x_1, x_2, 0) + C_{ij3r}\varphi_r^*(x_1, x_2, 0)$$

The equilibrium equations reduce to

$$\tau_{ji,j} + P_i = 0, \qquad \mu_{ji,j} + \varepsilon_{irs}\tau_{rs} + Q_i = 0 \text{ on } B$$

Here we have used the notations

$$P_i = T_{ji,j} + t_{3i}^*(x_1, x_2, 0), \qquad Q_i = M_{ji,j} + m_{3i}^*(x_1, x_2, 0)$$

The conditions on the lateral surface become

$$\tau_{\alpha i}n_\alpha = s_i, \qquad \mu_{\alpha i}n_\alpha = r_i \text{ on } \Pi$$

where

$$s_i = -T_{\alpha i}n_\alpha, \qquad r_i = -M_{\alpha i}n_\alpha$$

One can see that P_i, Q_i, s_i, and r_i are independent of the axial coordinate. In view of Equations 6.3.31, from Equations 6.3.28, we obtain

$$\int_{\Sigma_1} \tau_{3i} da = -\widetilde{A}_i, \qquad \int_{\Sigma_1} (\varepsilon_{ijk}x_j\tau_{3k} + \mu_{3i}) da = -\widetilde{B}_i$$

where

$$\widetilde{A}_i = \int_{\Sigma_1} T_{3i} da, \qquad \widetilde{B}_i = \int_{\Sigma_1} (\varepsilon_{ijk}x_jT_{3k} + M_{3i}) da$$

Thus, for the unknown functions v_i and ψ_i, we have obtained a problem of Almansi–Michell type. The solution of this problem has the form 6.4.36.

6.5 Cylinders Composed of Different Elastic Materials

This section is concerned with the deformation of a cylinder composed of different isotropic Cosserat elastic materials. We now assume that B is a composed cylinder, as described in Section 3.1. We suppose that the domain B_ρ is occupied by an isotropic material with the constitutive coefficients $\lambda^{(\rho)}, \mu^{(\rho)}, \ldots, \gamma^{(\rho)}$ and that

$$\lambda^{(\rho)} = \lambda^{(\rho)}(x_1, x_2), \qquad \mu^{(\rho)} = \mu^{(\rho)}(x_1, x_2), \ldots, \gamma^{(\rho)} = \gamma^{(\rho)}(x_1, x_2) \\ (x_1, x_2) \in A_\rho \tag{6.5.1}$$

We assume that the elastic coefficients belong to C^∞ and that the elastic potential corresponding to the body which occupies B_ρ is a positive definite quadratic form. We can consider B as being occupied by an elastic medium which, in general, has elastic coefficients discontinuous along Π_0.

The functions u_i, φ_i, t_i, and m_i must be continuous in passing from one medium to another so that we have the conditions

$$[u_i]_1 = [u_i]_2, \qquad [\varphi_i]_1 = [\varphi_i]_2 \\ [t_{\beta i}]_1 n^0_\beta = [t_{\beta i}]_2 n^0_\beta, \qquad [m_{\beta i}]_1 n^0_\beta = [m_{\beta i}]_2 n^0_\beta \text{ on } \Pi_0 \tag{6.5.2}$$

where it has been indicated that the expressions in brackets are calculated for the domains B_1 and B_2, respectively. Here, n^0_β are the direction cosines of the vector normal to Π_0, outward to B_1.

6.5.1 Plane Strain Problems

The plane strain problem for Cosserat elastic solids has been introduced in Sections 5.2 and 6.1. Let us consider now the problem of the plane strain associated to the cylinder B, which is occupied by two materials. The equilibrium equations for the plane strain can be written in the form

$$t_{\beta\alpha,\beta} + f^{(\rho)}_\alpha = 0, \qquad m_{\beta 3,\beta} + \varepsilon_{\alpha\beta} t_{\alpha\beta} + g^{(\rho)}_3 = 0 \text{ on } A_\rho \tag{6.5.3}$$

We assume that the functions $f^{(\rho)}_\alpha$ and $g^{(\rho)}_3$ belong to C^∞. The constitutive Equations 6.1.3 lead to

$$t_{\alpha\beta} = \lambda^{(\rho)} e_{\eta\eta}\delta_{\alpha\beta} + (\mu^{(\rho)} + \kappa^{(\rho)}) e_{\alpha\beta} + \mu^{(\rho)} e_{\beta\alpha} \\ m_{\alpha 3} = \gamma^{(\rho)} \varphi_{3,\alpha} \text{ on } A_\rho \tag{6.5.4}$$

Since the displacement vector, the microrotation vector, the stress vector, and the couple-stress vector are continuous in passing one medium to another, in the plane strain we have the conditions

$$[u_\alpha]_1 = [u_\alpha]_2, \qquad [\varphi_3]_1 = [\varphi_3]_2 \\ [t_{\alpha\beta}]_1 n^0_\alpha = [t_{\alpha\beta}]_2 n^0_\alpha, \qquad [m_{\alpha 3}]_1 n^0_\alpha = [m_{\alpha 3}]_2 n^0_\alpha \text{ on } \Gamma_0 \tag{6.5.5}$$

We consider the following boundary conditions

$$[t_{\beta\alpha}n_\beta]_\rho = h_\alpha^{(\rho)}, \qquad [m_{\alpha 3}n_\alpha]_\rho = q^{(\rho)} \text{ on } \Gamma_\rho \tag{6.5.6}$$

where $h_\alpha^{(\rho)}$ and $q^{(\rho)}$ are functions of class C^∞. If the domains A_ρ satisfy some conditions of regularity [88], then the above plane strain problem has a solution if and only if

$$\begin{aligned}
&\sum_{\rho=1}^{2}\left[\int_{A_\rho} f_\alpha^{(\rho)}da + \int_{\Gamma_\rho} h_\alpha^{(\rho)}da\right] = 0\\
&\sum_{\rho=1}^{2}\left[\int_{A_\rho} (\varepsilon_{\alpha\beta}x_\alpha f_\beta^{(\rho)} + g_3^{(\rho)})da + \int_{\Gamma_\rho} (\varepsilon_{\alpha\beta}x_\alpha h_\beta^{(\rho)} + q^{(\rho)})ds\right] = 0
\end{aligned} \tag{6.5.7}$$

In what follows, we assume that the requirements which insure this result are fulfilled. If the conditions 6.5.5 are replaced by

$$\begin{aligned}
[u_\alpha]_1 = [u_\alpha]_2, &\qquad [\varphi_3]_1 = [\varphi_3]_2\\
[t_{\alpha\beta}]_1 n_\alpha^0 = [t_{\alpha\beta}]_2 n_\alpha^0 + p_\beta, &\qquad [m_{\alpha 3}]_1 n_\alpha^0 = [m_{\alpha 3}]_2 n_\alpha^0 + q \text{ on } \Gamma_0
\end{aligned} \tag{6.5.8}$$

where p_α and q are functions of class C^∞, then the conditions 6.5.7 are replaced by

$$\begin{aligned}
&\sum_{\rho=1}^{2}\left[\int_{A_\rho} f_\alpha^{(\rho)}da + \int_{\Gamma_\rho} h_\alpha^{(\rho)}ds\right] + \int_{\Gamma_0} p_\alpha ds = 0\\
&\sum_{\rho=1}^{2}\left[\int_{A_\rho} (\varepsilon_{\alpha\beta}x_\alpha f_\beta^{(\rho)} + g_3^{(\rho)})da + \int_{\Gamma_\rho} (\varepsilon_{\alpha\beta}x_\alpha h_\beta^{(\rho)} + q^{(\rho)})ds\right]\\
&\quad + \int_{\Gamma_0} (\varepsilon_{\alpha\beta}x_\alpha p_\beta + q)ds = 0
\end{aligned} \tag{6.5.9}$$

We will have occasion to use three special problems $\mathcal{E}^{(s)}$, $(s = 1, 2, 3)$, of plane strain. In what follows, we denote by $u_\alpha^{(s)}, \varphi^{(s)}, e_{\alpha\beta}^{(s)}, t_{ij}^{(s)}$, and $m_{ij}^{(s)}$ the solution of the problem $\mathcal{E}^{(s)}$. The problems $\mathcal{E}^{(s)}$ are characterized by the equations of equilibrium

$$\begin{aligned}
&t_{\beta\alpha,\beta}^{(\eta)} + (\lambda^{(\rho)}x_\eta)_{,\alpha} = 0, \qquad t_{\beta\alpha,\beta}^{(3)} + \lambda_{,\alpha}^{(\rho)} = 0, \quad (\eta = 1, 2)\\
&m_{\beta 3,\beta}^{(\eta)} + \varepsilon_{\alpha\beta}t_{\alpha\beta}^{(\eta)} + \varepsilon_{\alpha\eta}\beta_{,\alpha}^{(\rho)} = 0, \qquad m_{\beta 3,\beta}^{(3)} + \varepsilon_{\alpha\beta}t_{\alpha\beta}^{(3)} = 0
\end{aligned} \tag{6.5.10}$$

the constitutive equations

$$t_{\alpha\beta}^{(s)} = \lambda^{(\rho)}e_{\eta\eta}^{(s)}\delta_{\alpha\beta} + (\mu^{(\rho)} + \kappa^{(\rho)})e_{\alpha\beta}^{(s)} + \mu^{(\rho)}e_{\beta\alpha}^{(s)}, \qquad m_{\alpha 3}^{(s)} = \gamma^{(\rho)}\varphi_{3,\alpha}^{(s)} \tag{6.5.11}$$

the geometrical equations

$$e_{\alpha\beta}^{(s)} = u_{\beta,\alpha}^{(s)} + \varepsilon_{\beta\alpha}\varphi_3^{(s)}, \qquad \kappa_{\alpha 3}^{(s)} = \varphi_{3,\alpha}^{(s)} \tag{6.5.12}$$

on A_ρ, and the following conditions

$$\begin{gathered}[u_\alpha^{(s)}]_1 = [u_\alpha^{(s)}]_2, \qquad [\varphi_3^{(s)}]_1 = [\varphi_3^{(s)}]_2 \\ [t_{\beta\alpha}^{(s)}]_1 n_\beta^0 = [t_{\beta\alpha}^{(s)}]_2 n_\beta^0 + P_\alpha^{(s)}, \qquad [m_{\alpha 3}^{(s)}]_1 n_\alpha^0 = [m_{\alpha 3}^{(s)}]_2 n_\alpha^0 + Q^{(s)} \text{ on } \Gamma_0\end{gathered} \tag{6.5.13}$$

$$\begin{aligned}[t_{\beta\alpha}^{(\eta)} n_\beta]_\rho &= -\lambda^{(\rho)} x_\eta n_\alpha, \qquad [t_{\beta\alpha}^{(3)} n_\beta]_\rho = -\lambda^{(\rho)} n_\alpha \\ [m_{\alpha 3}^{(\eta)} n_\alpha]_\rho &= \varepsilon_{\eta\nu} \beta^{(\rho)} n_\nu, \qquad [m_{\alpha 3}^{(3)} n_\alpha]_\rho = 0 \text{ on } \Gamma_\rho\end{aligned} \tag{6.5.14}$$

where we have used the notations

$$\begin{gathered}P_\alpha^{(\eta)} = (\lambda^{(2)} - \lambda^{(1)}) x_\eta n_\alpha^0, \qquad P_\alpha^{(3)} = (\lambda^{(2)} - \lambda^{(1)}) n_\alpha^0 \\ Q^{(\eta)} = \varepsilon_{\eta\alpha}(\beta^{(1)} - \beta^{(2)}) n_\alpha^0, \qquad Q^{(3)} = 0\end{gathered} \tag{6.5.15}$$

The necessary and sufficient conditions for the existence of the solution are satisfied for each boundary-value problem $\mathcal{E}^{(s)}$.

6.5.2 Extension and Bending

The problem of extension and bending for a composed cylinder consists in the solving of the Equations 5.1.11, 5.1.17, and 6.2.1 on B_ρ with the conditions 6.2.2, 6.2.7, 6.2.8, and 6.5.2 when the constitutive coefficients have the form 6.5.1. We try to solve this problem assuming that

$$\begin{gathered}u_\alpha = -\frac{1}{2} a_\alpha x_3^2 + \sum_{s=1}^{3} a_s u_\alpha^{(s)}, \qquad u_3 = (a_1 x_1 + a_2 x_2 + a_3) x_3 \\ \varphi_\alpha = \varepsilon_{\alpha\beta} a_\beta x_3, \qquad \varphi_3 = \sum_{s=1}^{3} a_s \varphi_3^{(s)}\end{gathered} \tag{6.5.16}$$

where $u_\alpha^{(s)}, \varphi_3^{(s)}$ are the solutions of the problems $\mathcal{E}^{(s)}$, and a_s are unknown constants. From Equations 5.1.11, 5.1.17, and 6.5.16, we get

$$\begin{aligned}t_{\alpha\beta} &= \lambda^{(\rho)}(a_1 x_1 + a_2 x_2 + a_3)\delta_{\alpha\beta} + \sum_{s=1}^{3} a_s t_{\alpha\beta}^{(s)} \\ t_{33} &= (\lambda^{(\rho)} + 2\mu^{(\rho)} + \kappa^{(\rho)})(a_1 x_1 + a_2 x_2 + a_3) + \lambda^{(\rho)} \sum_{s=1}^{3} a_s e_{\alpha\alpha}^{(s)} \\ t_{\alpha 3} &= t_{3\alpha} = 0, \qquad m_{\alpha\beta} = m_{33} = 0, \qquad m_{\alpha 3} = \varepsilon_{\alpha\nu}\beta^{(\rho)} a_\nu + \sum_{s=1}^{3} a_s m_{\alpha 3}^{(s)} \\ m_{3\alpha} &= \varepsilon_{\alpha\nu}\gamma^{(\rho)} a_\nu + \sum_{s=1}^{3} a_s m_{3\alpha}^{(s)}\end{aligned} \tag{6.5.17}$$

By using Equations 6.5.13, 6.5.14, and 6.5.15, it follows that the conditions 6.2.2 and 6.5.2 are satisfied. The conditions 6.2.7 are satisfied on the basis of the relations 6.5.17. From Equations 6.2.8 and 6.5.17 we obtain the following system for the unknown constants a_k,

$$Y_{rs}a_s = C_r \tag{6.5.18}$$

where

$$\begin{aligned}
Y_{\alpha\beta} &= \sum_{\rho=1}^{2}\int_{A_\rho}\{x_\alpha[(\lambda^{(\rho)} + 2\mu^{(\rho)} + \kappa^{(\rho)})x_\beta + \lambda^{(\rho)}e_{\eta\eta}^{(\beta)}] \\
&\quad - \varepsilon_{\alpha\lambda}(\varepsilon_{\lambda\beta}\gamma^{(\rho)} + m_{3\lambda}^{(\beta)})\}da \\
Y_{\alpha 3} &= \sum_{\rho=1}^{2}\int_{A_\rho}\{x_\alpha[\lambda^{(\rho)} + 2\mu^{(\rho)} + \kappa^{(\rho)} + \lambda^{(\rho)}e_{\eta\eta}^{(3)}] - \varepsilon_{\alpha\lambda}m_{3\lambda}^{(3)}\}da \\
Y_{3\alpha} &= \sum_{\rho=1}^{2}\int_{A_\rho}[(\lambda^{(\rho)} + 2\mu^{(\rho)} + \kappa^{(\rho)})x_\alpha + \lambda^{(\rho)}e_{\eta\eta}^{(\alpha)}]da \\
Y_{33} &= \sum_{\rho=1}^{2}\int_{A_\rho}[\lambda^{(\rho)} + 2\mu^{(\rho)} + \kappa^{(\rho)} + \lambda^{(\rho)}e_{\eta\eta}^{(3)}]da \\
C_\alpha &= \varepsilon_{\alpha\beta}M_\beta, \qquad C_3 = -F_3
\end{aligned} \tag{6.5.19}$$

Following the procedure from Section 3.6 we can prove that $\det(Y_{rs}) \neq 0$, so that the system 6.5.18 determines the constants a_s.

6.5.3 Torsion

The problem of torsion for a cylinder composed of two materials consists in the finding of the functions u_i and φ_i that satisfy Equations 5.1.11, 5.1.17, and 6.2.1 on B_ρ, the conditions 6.5.2 on the surface of separation Π_0, the conditions on the end Σ_1, and the conditions 6.2.2 on the lateral boundary of the cylinder B. Following Ref. 141 we seek the solution in the form

$$u_\alpha = \varepsilon_{\beta\alpha}\tau x_\beta x_3, \qquad u_3 = \tau\Phi(x_1, x_2), \qquad \varphi_\alpha = \tau\Phi_\alpha(x_1, x_2), \qquad \varphi_3 = \tau x_3 \tag{6.5.20}$$

where Φ, Φ_1, and Φ_2 are unknown functions, and τ is an unknown constant. Let $\widetilde{V} = (G, G_1, G_2)$ be an ordered triplet of functions G, G_1, and G_2. We introduce the notations

$$
\begin{aligned}
T_\alpha^{(\rho)}\widetilde{V} &= (\mu^{(\rho)}+\kappa^{(\rho)})G_{,\alpha}+\kappa^{(\rho)}\varepsilon_{\alpha\beta}G_\beta, \qquad S_\alpha^{(\rho)}\widetilde{V}=\mu^{(\rho)}G_{,\alpha}+\kappa^{(\rho)}\varepsilon_{\beta\alpha}G_\beta\\
M_{\alpha\beta}^{(\rho)}\widetilde{V} &= \alpha^{(\rho)}G_{\nu,\nu}\delta_{\alpha\beta}+\beta^{(\rho)}G_{\alpha,\beta}+\gamma^{(\rho)}G_{\beta,\alpha}\\
\mathcal{L}_\nu^{(\rho)}\widetilde{V} &= \big(M_{\beta\nu}^{(\rho)}\widetilde{V}\big)_{,\beta}+\varepsilon_{\nu\beta}\big(T_\beta^{(\rho)}\widetilde{V}-S_\beta^{(\rho)}\widetilde{V}\big)\\
&= \big(\alpha^{(\rho)}G_{\lambda,\lambda}\big)_{,\nu}+\big(\beta^{(\rho)}G_{\lambda,\nu}\big)_{,\lambda}+\big(\gamma^{(\rho)}G_{\nu,\lambda}\big)_{,\lambda}+\varepsilon_{\nu\beta}\kappa^{(\rho)}G_{,\beta}-2\kappa^{(\rho)}G_\nu\\
\mathcal{L}_3^{(\rho)}\widetilde{V} &= \big(T_\alpha^{(\rho)}\widetilde{V}\big)_{,\alpha}=\big[(\mu^{(\rho)}+\kappa^{(\rho)})G_{,\alpha}\big]_{,\alpha}+\varepsilon_{\alpha\beta}\big(\kappa^{(\rho)}G_\beta\big)_{,\alpha}\\
\mathcal{N}_\nu^{(\rho)}\widetilde{V} &= \big(M_{\alpha\nu}^{(\rho)}\widetilde{V}\big)n_\alpha=\alpha^{(\rho)}G_{\lambda,\lambda}n_\nu+\beta^{(\rho)}G_{\lambda,\nu}n_\lambda+\gamma^{(\rho)}G_{\nu,\lambda}n_\lambda\\
\mathcal{N}_3^{(\rho)}\widetilde{V} &= \big(T_\alpha^{(\rho)}\widetilde{V}\big)n_\alpha=(\mu^{(\rho)}+\kappa^{(\rho)})G_{,\alpha}n_\alpha+\kappa^{(\rho)}\varepsilon_{\alpha\beta}G_\beta n_\alpha
\end{aligned}
\tag{6.5.21}
$$

Taking into account Equations 5.1.11, 5.1.17, and 6.5.20, we obtain

$$
\begin{aligned}
&t_{\alpha\beta}=0, \qquad t_{33}=0, \qquad t_{\alpha 3}=\tau\big[T_\alpha^{(\rho)}\Lambda+\mu^{(\rho)}\varepsilon_{\beta\alpha}x_\beta\big]\\
&t_{3\alpha}=\tau\big[S_\alpha^{(\rho)}\Lambda+(\mu^{(\rho)}+\kappa^{(\rho)})\varepsilon_{\beta\alpha}x_\beta\big], \qquad m_{\eta\nu}=\tau\big[M_{\eta\nu}^{(\rho)}\Lambda+\alpha^{(\rho)}\delta_{\eta\nu}\big]\\
&m_{\alpha 3}=m_{3\alpha}=0, \qquad m_{33}=\tau\big[\alpha^{(\rho)}\Phi_{\nu,\nu}+\alpha^{(\rho)}+\beta^{(\rho)}+\gamma^{(\rho)}\big] \text{ on } A_\rho
\end{aligned}
\tag{6.5.22}
$$

where $\Lambda=(\Phi,\Phi_1,\Phi_2)$. The equations of equilibrium 6.2.1 reduce to

$$
\mathcal{L}_i^{(\rho)}\Lambda=\mathscr{F}_i^{(\rho)} \text{ on } A_\rho \tag{6.5.23}
$$

where

$$
\mathscr{F}_\nu^{(\rho)}=x_\nu\kappa^{(\rho)}-\alpha_{,\nu}^{(\rho)}, \qquad \mathscr{F}_3^{(\rho)}=\varepsilon_{\alpha\beta}(\mu^{(\rho)}x_\beta)_{,\alpha} \tag{6.5.24}
$$

The boundary conditions 6.2.2 become

$$
\mathscr{N}_i^{(\rho)}\Lambda=\sigma_i^{(\rho)} \text{ on } \Gamma_\rho \tag{6.5.25}
$$

where

$$
\sigma_\nu^{(\rho)}=-\alpha^{(\rho)}n_\nu, \qquad \sigma_3^{(\rho)}=\mu^{(\rho)}\varepsilon_{\alpha\beta}x_\beta n_\alpha \tag{6.5.26}
$$

The conditions 6.5.2 reduce to

$$
\big[\mathscr{N}_i^{(1)}\Lambda\big](n^0)-\big[\mathscr{N}_i^{(2)}\Lambda\big](n^0)=k_i \text{ on } \Gamma_0 \tag{6.5.27}
$$

where we have used the notations

$$
k_\nu=(\alpha^{(2)}-\alpha^{(1)})n_\nu^0, \qquad k_3=(\mu^{(1)}-\mu^{(2)})\varepsilon_{\alpha\beta}x_\beta n_\alpha^0 \tag{6.5.28}
$$

and $[\mathscr{N}_i^{(\rho)}](n^0)$ denotes the operator $\mathscr{N}_i^{(\rho)}$ for $n_\alpha=n_\alpha^0$. Following Refs. 88, 137, and 154, the necessary and sufficient condition for the existence of the solution of the boundary-value problem 6.5.23, 6.5.25, and 6.5.27 is

$$\sum_{\rho=1}^{2}\left[\int_{A_\rho}\mathscr{F}_3^{(\rho)}da - \int_{\Gamma_\rho}\sigma_3^{(\rho)}ds\right] - \int_{\Gamma_0}k_3 ds = 0$$

By using Equations 6.5.24, 6.5.26, and 6.5.28, it is easy to see that this condition is satisfied. We assume that the functions Φ and Φ_α are known. Taking into account Equations 6.5.22 it follows that the conditions 6.2.15 are satisfied. As in Section 6.2.2, we can prove that the conditions 6.2.14 are identically satisfied. From Equation 6.2.16 we determine the constant τ. Thus, by using Equations 6.5.22, the condition 6.2.16 reduces to

$$\tau D' = -M_3 \tag{6.5.29}$$

where D' is the torsional rigidity

$$\begin{aligned} D' = \sum_{\rho=1}^{2}\int_{A_\rho}&(\varepsilon_{\alpha\beta}x_\alpha\{\mu^{(\rho)}\Phi_{,\beta} + \varepsilon_{\nu\beta}[\kappa^{(\rho)}\Phi_\nu + (\mu^{(\rho)} + \kappa^{(\rho)})x_\nu]\} \\ &+ \alpha^{(\rho)}\Phi_{\nu,\nu} + \alpha^{(\rho)} + \beta^{(\rho)} + \gamma^{(\rho)})da \end{aligned} \tag{6.5.30}$$

By using the method presented in Section 5.3 we can prove that $D' \neq 0$, so that the relation 6.5.29 determines the constant τ.

6.5.4 Flexure

We suppose that the loading applied on Σ_1 is statically equivalent to the force $\mathbf{F} = F_\alpha \mathbf{e}_\alpha$ and the moment $\mathbf{M} = \mathbf{0}$. The problem consists in the determination of the displacement and microrotation vector fields that satisfy Equations 5.1.11, 5.1.17, and 6.2.1 on B_ρ, the conditions 6.5.2 on the surface of separation, the conditions on the end Σ_1, and the conditions 6.2.2 on the surface Π. Following Ref. 155, we seek the solution in the form

$$\begin{aligned} u_\alpha &= -\frac{1}{6}b_\alpha x_3^3 + x_3\sum_{s=1}^{3}b_s u_\alpha^{(s)} + \varepsilon_{\beta\alpha}\tau x_\beta x_3 \\ u_3 &= \frac{1}{2}(b_1x_1 + b_2x_2 + b_3)x_3^2 + \tau\Phi + \Psi(x_1,x_2) \\ \varphi_\alpha &= \frac{1}{2}\varepsilon_{\alpha\beta}b_\beta x_3^2 + \tau\Phi_\alpha + \Psi_\alpha(x_1,x_2) \\ \varphi_3 &= x_3\sum_{s=1}^{3}b_s\varphi_3^{(s)} + \tau x_3 \end{aligned} \tag{6.5.31}$$

where Φ, Φ_1, and Φ_2 satisfy Equations 6.5.23 and the conditions 6.5.25 and 6.5.27; $u_\alpha^{(s)}$ and $\varphi_3^{(s)}$ are the solutions of the problems $\mathcal{E}^{(s)}$; Ψ, Ψ_1, and Ψ_2

are unknown functions, and b_i and τ are unknown constants. By Equations 5.1.11, 5.1.17, and 6.5.31, we find that

$$
\begin{aligned}
t_{\alpha\beta} &= x_3 \sum_{s=1}^{3} b_s t_{\alpha\beta}^{(s)} + \lambda^{(\rho)} x_3 (b_1 x_1 + b_2 x_2 + b_3) \delta_{\alpha\beta} \\
t_{33} &= (\lambda^{(\rho)} + 2\mu^{(\rho)} + \kappa^{(\rho)})(b_1 x_1 + b_2 x_2 + b_3) x_3 + \lambda^{(\rho)} x_3 \sum_{s=1}^{3} b_s e_{\alpha\alpha}^{(s)} \\
t_{\alpha 3} &= \tau \big[T_\alpha^{(\rho)} \Lambda + \mu^{(\rho)} \varepsilon_{\beta\alpha} x_\beta \big] + T_\alpha^{(\rho)} \Omega + \mu^{(\rho)} \sum_{s=1}^{3} b_s u_\alpha^{(s)} \\
t_{3\alpha} &= \tau \big[S_\alpha^{(\rho)} \Lambda + \varepsilon_{\beta\alpha} (\mu^{(\rho)} + \kappa^{(\rho)}) x_\beta \big] + S_\alpha^{(\rho)} \Omega + (\mu^{(\rho)} + \kappa^{(\rho)}) \sum_{s=1}^{3} b_s u_\alpha^{(s)} \\
m_{\eta\nu} &= \tau \big[M_{\eta\nu}^{(\rho)} \Lambda + \alpha^{(\rho)} \delta_{\eta\nu} \big] + M_{\eta\nu} \Omega + \alpha^{(\rho)} \delta_{\eta\nu} \sum_{s=1}^{3} b_s \varphi^{(s)} \\
m_{3\alpha} &= x_3 \sum_{s=1}^{3} b_s m_{3\alpha}^{(s)} + \gamma^{(\rho)} \varepsilon_{\alpha\beta} b_\beta x_3 \\
m_{\alpha 3} &= \beta^{(\rho)} \varepsilon_{\alpha\beta} b_\beta x_3 + x_3 \sum_{s=1}^{3} b_s m_{\alpha 3}^{(s)} \\
m_{33} &= (\alpha^{(\rho)} + \beta^{(\rho)} + \gamma^{(\rho)}) \left(\tau + \sum_{s=1}^{3} b_s \varphi_3^{(s)} \right) + \alpha^{(\rho)} (\tau \Phi_{\alpha,\alpha} + \Psi_{\alpha,\alpha})
\end{aligned}
\tag{6.5.32}
$$

where $\Omega = (\Psi, \Psi_1, \Psi_2)$. On the basis of the relations 6.5.10 and 6.5.15, we conclude that the equilibrium equations 6.2.1 and the conditions 6.2.2 and 6.5.2 reduce to

$$
\begin{aligned}
\mathcal{L}_\nu^{(\rho)} \Omega &= -\gamma^{(\rho)} \varepsilon_{\nu\beta} b_\beta - \sum_{s=1}^{3} b_s \big[\big(\alpha^{(\rho)} \varphi_3^{(s)} \big)_{,\nu} + m_{3\nu}^{(s)} - \varepsilon_{\nu\beta} \kappa^{(\rho)} u_\beta^{(s)} \big] \\
\mathcal{L}_3^{(\rho)} \Omega &= -\sum_{s=1}^{3} b_s \big[\lambda^{(\rho)} e_{\alpha\alpha}^{(s)} + (\mu^{(\rho)} u_\alpha^{(s)})_{,\alpha} \big] \\
&\quad - (\lambda^{(\rho)} + 2\mu^{(\rho)} + \kappa^{(\rho)})(b_1 x_1 + b_2 x_2 + b_3) \text{ on } A_\rho
\end{aligned}
\tag{6.5.33}
$$

$$
\begin{aligned}
&[\Psi]_1 = [\Psi]_2, \qquad [\Psi_\alpha]_1 = [\Psi_\alpha]_2 \\
&\big[\mathcal{N}_\eta^{(1)} \Omega \big](n^0) - \big[\mathcal{N}_\eta^{(2)} \Omega \big](n^0) = (\alpha^{(2)} - \alpha^{(1)}) \nu_\eta \sum_{s=1}^{3} b_s \varphi_3^{(s)}
\end{aligned}
\tag{6.5.34}
$$

$$[\mathscr{N}_3^{(1)}\Omega](n^0) - [\mathscr{N}_3^{(2)}\Omega](n^0) = (\mu^{(2)} - \mu^{(1)})\nu_\alpha \sum_{s=1}^{3} b_s u_\alpha^{(s)} \text{ on } \Gamma_0$$

$$\mathscr{N}_\nu^{(\rho)}\Omega = -\alpha^{(\rho)} n_\nu \sum_{s=1}^{3} b_s \varphi^{(s)}, \qquad \mathscr{N}_3^{(\rho)}\Omega = -\mu^{(\rho)} n_\alpha \sum_{s=1}^{3} b_s u_\alpha^{(s)} \text{ on } \Gamma_\rho \tag{6.5.35}$$

The necessary and sufficient condition for the existence of the solution of this boundary-value problem becomes

$$Y_{3s} b_s = 0 \tag{6.5.36}$$

where Y_{3s} are given by 6.5.19. Let us impose the conditions 6.2.26. As in Section 6.2, we can prove that the relations 6.2.34 and 6.2.35 hold. Thus, taking into account 6.5.32, the conditions 6.2.26 reduce to

$$Y_{\alpha s} b_s = -F_\alpha \tag{6.5.37}$$

The system 6.5.36 and 6.5.37 determines the constants b_s. Since the conditions 6.5.36 are satisfied, we shall consider that the functions Ψ and Ψ_α are known.

From Equations 6.2.28 and 6.5.32, we obtain

$$\tau D' = -\sum_{\rho=1}^{2} \int_{A_\rho} \left\{ \varepsilon_{\alpha\beta} x_\alpha \left[\mu^{(\rho)} \Psi_{,\beta} + \varepsilon_{\nu\beta} \kappa^{(\rho)} \Psi_\nu + (\mu^{(\rho)} + \kappa^{(\rho)}) \sum_{s=1}^{3} b_s u_\beta^{(s)} \right] \right.$$
$$\left. + (\alpha^{(\rho)} + \beta^{(\rho)} + \gamma^{(\rho)}) \sum_{s=1}^{3} b_s \varphi_3^{(s)} + \alpha^{(\rho)} \Psi_{\alpha,\alpha} \right\} da$$

where D' is given by Equation 6.5.38. The above relation determines the constant τ. The conditions 6.2.27 are identically satisfied on the basis of the relations 6.5.32. Thus, the flexure problem is solved.

6.5.5 Problem of Loaded Cylinders

In order to solve the Almansi problem, we first investigate the problem of uniformly loaded cylinders. We assume that the body loads have the form

$$f_i = R_i^{(\rho)}(x_1, x_2), \qquad g_i = L_i^{(\rho)}(x_1, x_2) \text{ on } A_\rho \tag{6.5.38}$$

and consider the boundary conditions

$$[t_{\alpha i} n_\alpha]_\rho = p_i^{(\rho)}(x_1, x_2), \qquad [m_{\alpha i} n_\alpha]_\rho = q_i^{(\rho)}(x_1, x_2) \text{ on } \Pi_\rho \tag{6.5.39}$$

Let us establish a solution of the Equations 5.1.11, 5.1.17, and 5.1.19 on B_ρ which satisfies the conditions on the end Σ_1, the conditions 6.5.39 on Π_ρ and the conditions 6.5.2 on Π_0, when the body loads are given by Equation 6.5.38. On the basis of Theorem 5.6.2, we try to solve the problem assuming that

$$
\begin{aligned}
u_\alpha &= -\frac{1}{2}a_\alpha x_3^2 - \frac{1}{6}b_\alpha x_3^3 - \frac{1}{24}c_\alpha x_3^4 + \varepsilon_{\beta\alpha}\left(\tau_1 x_3 + \frac{1}{2}\tau_2 x_3^2\right)x_\beta \\
&\quad + \sum_{s=1}^{3}\left(a_s + b_s x_3 + \frac{1}{2}c_s x_3^2\right)u_\alpha^{(s)} + v_\alpha(x_1, x_2) \\
u_3 &= (a_1 x_1 + a_2 x_2 + a_3)x_3 + \frac{1}{2}(b_1 x_1 + b_2 x_2 + b_3)x_3^2 \\
&\quad + \frac{1}{6}(c_1 x_1 + c_2 x_2 + c_3)x_3^3 + (\tau_1 + x_3\tau_2)\Phi + \Psi(x_1, x_2) + x_3\chi(x_1, x_2) \\
\varphi_\alpha &= \varepsilon_{\alpha\beta}\left(a_\beta x_3 + \frac{1}{2}b_\beta x_3^2 + \frac{1}{6}c_\beta x_3^3\right) \\
&\quad + (\tau_1 + \tau_2 x_3)\Phi_\alpha + \Psi_\alpha(x_1, x_2) + x_3\chi_\alpha(x_1, x_2) \\
\varphi_3 &= \sum_{s=1}^{3}\left(a_s + b_s x_3 + \frac{1}{2}c_s x_3^2\right)\varphi_3^{(s)} + \tau_1 x_3 + \frac{1}{2}\tau_2 x_3^2 + w(x_1, x_2)
\end{aligned}
\tag{6.5.40}
$$

where $u_\alpha^{(s)}$ and $\varphi_3^{(s)}$ are the solutions of the problems $\mathcal{E}^{(s)}$; Φ, Φ_1, and Φ_2 are the torsion functions considered in Section 6.5.3; Ψ, Ψ_α, χ, χ_α, v_α, and w are unknown functions, and a_s, b_s, c_s, τ_1, and τ_2 are unknown constants. In view of Equations 5.1.11, 5.1.17, and 6.5.40, we find that

$$
\begin{aligned}
t_{\alpha\beta} &= \lambda^{(\rho)}\Big[a_1 x_1 + a_2 x_2 + a_3 + (b_1 x_1 + b_2 x_2 + b_3)x_3 \\
&\quad + \frac{1}{2}(c_1 x_1 + c_2 x_2 + c_3)x_3^2\Big]\delta_{\alpha\beta} + \lambda^{(\rho)}(\chi + \tau_2\Phi)\delta_{\alpha\beta} \\
&\quad + \sum_{s=1}^{3}\left(a_s + b_s x_3 + \tfrac{1}{2}c_s x_3^2\right)t_{\alpha\beta}^{(s)} + \sigma_{\alpha\beta} \\
t_{33} &= (\lambda^{(\rho)} + 2\mu^{(\rho)} + \kappa^{(\rho)})\Big[a_1 x_1 + a_2 x_2 + a_3 + (b_1 x_1 + b_2 x_2 + b_3)x_3 \\
&\quad + \frac{1}{2}(c_1 x_1 + c_2 x_2 + c_3)x_3^2\Big] + (\lambda^{(\rho)} + 2\mu^{(\rho)} + k^{(\rho)})(\chi + \tau_2\Phi) \\
&\quad + \lambda^{(\rho)}\sum_{s=1}^{3}\left(a_s + b_s x_3 + \frac{1}{2}c_s x_3^2\right)e_{\alpha\alpha}^{(s)} + \lambda^{(\rho)}\gamma_{\alpha\alpha} \\
t_{\alpha 3} &= T^{(\rho)}\Omega + x_3 T_\alpha^{(\rho)}\widetilde{V} + (\tau_1 + \tau_2 x_3)(T_\alpha^{(\rho)}\Lambda + \mu^{(\rho)}\varepsilon_{\beta\alpha}x_\beta) \\
&\quad + \mu^{(\rho)}\sum_{s=1}^{3}(b_s + c_s x_3)u_\alpha^{(s)} \\
t_{3\alpha} &= S_\alpha^{(\rho)}\Omega + x_3 S_\alpha^{(\rho)}\widetilde{V} + (\tau_1 + \tau_2 x_3)\left[S_\alpha^{(\rho)}\Lambda + (\mu^{(\rho)} + \kappa^{(\rho)})\varepsilon_{\beta\alpha}x_\beta\right] \\
&\quad + (\mu^{(\rho)} + \kappa^{(\rho)})\sum_{s=1}^{3}(b_s + c_s x_3)u_\alpha^{(s)}
\end{aligned}
$$

$$\begin{aligned}
m_{\lambda\nu} &= M_{\lambda\nu}^{(\rho)}\Omega + x_3 M_{\lambda\nu}^{(\rho)}\widetilde{V} + (\tau_1+\tau_2 x_3)\big(M_{\lambda\nu}^{(\rho)}\Lambda + \alpha^{(\rho)}\delta_{\lambda\nu}\big)\\
&\quad + \alpha^{(\rho)}\delta_{\lambda\nu}\sum_{s=1}^{3}(b_s + c_s x_3)\varphi_3^{(s)}\\
m_{\alpha 3} &= \beta^{(\rho)}\varepsilon_{\alpha\nu}\left(a_\nu + b_\nu x_3 + \frac{1}{2}c_\nu x_3^2\right) + \beta^{(\rho)}(\chi_\alpha + \tau_2\Phi_\alpha)\\
&\quad + \sum_{s=1}^{3}\left(a_s + b_s x_3 + \frac{1}{2}c_s x_3^2\right)m_{\alpha 3}^{(s)} + \mu_{\alpha 3}\\
m_{3\alpha} &= \gamma^{(\rho)}\varepsilon_{\alpha\nu}\left(a_\nu + b_\nu x_3 + \frac{1}{2}c_\nu x_3^2\right) + \gamma^{(\rho)}(\chi_\alpha + \tau_2\Phi_\alpha)\\
&\quad + \sum_{s=1}^{3}\left(a_s + b_s x_3 + \frac{1}{2}c_s x_3^2\right)m_{3\alpha}^{(s)} + \mu_{3\alpha}\\
m_{33} &= \big(\alpha^{(\rho)} + \beta^{(\rho)} + \gamma^{(\rho)}\big)\left[\tau_1 + \tau_2 x_3 + \sum_{s=1}^{3}(b_s + c_s x_3)\varphi^{(s)}\right]\\
&\quad + \alpha^{(\rho)}(\tau_1+\tau_2 x_3)\Phi_{\lambda,\lambda} + \alpha^{(\rho)}(\Psi_{\lambda,\lambda} + x_3\chi_{\lambda,\lambda})
\end{aligned} \tag{6.5.41}$$

where $\Omega = (\Psi, \Psi_1, \Psi_2)$, $\widetilde{V} = (\chi, \chi_1, \chi_2)$ and

$$\begin{aligned}
&\sigma_{\alpha\beta} = \lambda^{(\rho)}\gamma_{\nu\nu}\delta_{\alpha\beta} + (\mu^{(\rho)} + \kappa^{(\rho)})\gamma_{\alpha\beta} + \mu^{(\rho)}\gamma_{\beta\alpha}\\
&\mu_{\alpha 3} = \gamma^{(\rho)} w_{,\alpha}, \qquad \mu_{3\alpha} = \beta^{(\rho)} w_{,\alpha}, \qquad \gamma_{\alpha\beta} = v_{\beta,\alpha} + \varepsilon_{\beta\alpha} w
\end{aligned} \tag{6.5.42}$$

Taking into account Equations 6.5.10 and 6.5.41, the equations of equilibrium lead to the following equations

$$\sigma_{\beta\alpha,\beta} + H_\alpha^{(\rho)} = 0, \qquad \mu_{\alpha 3,\alpha} + \varepsilon_{\alpha\beta}\sigma_{\alpha\beta} + H^{(\rho)} = 0 \text{ on } A_\rho \tag{6.5.43}$$

$$\mathcal{L}_i^{(\rho)}\Omega = G_i^{(\rho)} \text{ on } A_\rho \tag{6.5.44}$$

$$\mathcal{L}_i^{(\rho)}\widetilde{V} = K_i^{(\rho)} \text{ on } A_\rho \tag{6.5.45}$$

where we have used the notations

$$\begin{aligned}
H_\alpha^{(\rho)} &= [\lambda^{(\rho)}(\chi + \tau_2\Phi)]_{,\alpha} + S_\alpha^{(\rho)}\widetilde{V} + \tau_2\big[S_\alpha^{(\rho)}\Lambda\\
&\quad + (\mu^{(\rho)} + \kappa^{(\rho)})\varepsilon_{\beta\alpha}x_\beta\big] + (\mu^{(\rho)} + \kappa^{(\rho)})\sum_{s=1}^{3}c_s u_\alpha^{(s)} + R_\alpha^{(\rho)}\\
H^{(\rho)} &= [\beta^{(\rho)}(\chi_\alpha + \tau_2\Phi_\alpha)]_{,\alpha} + (\alpha^{(\rho)} + \beta^{(\rho)} + \gamma^{(\rho)})\left(\tau_2 + \sum_{s=1}^{3}c_s\varphi^{(s)}\right)\\
&\quad + \alpha^{(\rho)}(\chi_\nu + \tau_2\Phi_\nu)_{,\nu} + L_3^{(\rho)}
\end{aligned}$$

$$G_3^{(\rho)} = -\sum_{s=1}^{3} b_s\big[\lambda^{(\rho)} e_{\alpha\alpha}^{(s)} + \big(\mu^{(\rho)} u_\alpha^{(s)}\big)_{,\alpha}\big]$$
$$- (\lambda^{(\rho)} + 2\mu^{(\rho)} + \kappa^{(\rho)})(b_1x_1 + b_2x_2 + b_3) - R_3^{(\rho)}$$
$$G_\nu^{(\rho)} = -\sum_{s=1}^{3} b_s\big[\big(\alpha^{(\rho)}\varphi^{(s)}\big)_{,\nu} + m_{3\nu}^{(s)} - \varepsilon_{\nu\beta}\kappa^{(\rho)} u_\beta^{(s)}\big]$$
$$- \gamma^{(\rho)}\varepsilon_{\nu\beta}b_\beta - L_\nu^{(\rho)}$$
$$K_3^{(\rho)} = -\sum_{s=1}^{3} c_s\big[\lambda^{(\rho)} e_{\alpha\alpha}^{(s)} + \big(\mu^{(\rho)} u_\alpha^{(s)}\big)_{,\alpha}\big]$$
$$- (\lambda^{(\rho)} + 2\mu^{(\rho)} + \kappa^{(\rho)})(c_1x_1 + c_2x_2 + x_3)$$
$$K_\nu^{(\rho)} = -\sum_{s=1}^{3} c_s\big[(\alpha^{(\rho)}\varphi^{(s)})_{,\nu} + m_{3\nu}^{(s)} - \varepsilon_{\nu\beta}\kappa^{(\rho)} u_\beta^{(s)}\big] - \gamma^{(\rho)}\varepsilon_{\nu\beta}c_\beta \tag{6.5.46}$$

The boundary conditions 6.5.39 are satisfied if we have

$$\sigma_{\alpha\beta}n_\alpha = s_\beta^{(\rho)}, \qquad \mu_{\alpha 3}n_\alpha = \eta^{(\rho)} \text{ on } \Gamma_\rho \tag{6.5.47}$$

$$\mathscr{N}_i^{(\rho)}\Omega = N_i^{(\rho)} \text{ on } \Gamma_\rho \tag{6.5.48}$$

$$\mathscr{N}_i^{(\rho)}\widetilde{V} = Q_i^{(\rho)} \text{ on } \Gamma_\rho \tag{6.5.49}$$

where

$$s_\beta^{(\rho)} = p_\beta^{(\rho)} - \lambda^{(\rho)}(\chi + \tau_2\Phi)n_\beta, \qquad \eta^{(\rho)} = q_3^{(\rho)} - \beta^{(\rho)}(\chi_\alpha + \tau_2\Phi_\alpha)n_\alpha \tag{6.5.50}$$

$$N_3^{(\rho)} = p_3^{(\rho)} - \mu^{(\rho)}\sum_{s=1}^{3} b_s u_\alpha^{(s)} n_\alpha, \qquad N_\nu^{(\rho)} = q_\nu^{(\rho)} - \alpha^{(\rho)} n_\nu \sum_{s=1}^{3} b_s \varphi_3^{(s)} \tag{6.5.51}$$

$$Q_3^{(\rho)} = -\mu^{(\rho)}\sum_{s=1}^{3} c_s u_\alpha^{(s)} n_\alpha, \qquad Q_\nu^{(\rho)} = -\alpha^{(\rho)} n_\nu \sum_{s=1}^{3} c_s \varphi_3^{(s)} \tag{6.5.52}$$

The conditions 6.5.2 reduce to the following conditions on Γ_0

$$\begin{aligned} [v_\alpha]_1 &= [v_\alpha]_2, & [w]_1 &= [w]_2 \\ [\sigma_{\alpha\beta}]_1 n_\alpha^0 - [\sigma_{\alpha\beta}]_2 n_\alpha^0 &= Z_\beta, & [\mu_{\alpha 3}]_1 n_\alpha^0 - [\mu_{\alpha 3}]_2 n_\alpha^0 &= Z \end{aligned} \tag{6.5.53}$$

$$\begin{aligned} [\Psi]_1 = [\Psi]_2, \qquad [\Psi_\alpha]_1 = [\Psi_\alpha]_2 \\ \big[\mathscr{N}_i^{(1)}\Omega\big](n^0) - \big[\mathscr{N}_i^{(2)}\Omega\big](n^0) = X_i \end{aligned} \tag{6.5.54}$$

$$\begin{aligned} [\chi]_1 = [\chi]_2, \qquad [\chi_\alpha]_1 = [\chi_\alpha]_2 \\ \big[\mathscr{N}_i^{(1)}\widetilde{V}\big](n^0) - \big[\mathscr{N}_i^{(2)}\widetilde{V}\big](n^0) = Y_i \end{aligned} \tag{6.5.55}$$

where

$$Z_\beta = (\lambda^{(2)} - \lambda^{(1)})(\chi + \tau_2\Phi)n_\beta^0, \qquad Z = (\beta^{(2)} - \beta^{(1)})(\chi_\alpha + \tau_2\Phi_\alpha)n_\alpha^0$$
$$X_\beta = (\alpha^{(2)} - \alpha^{(1)})n_\beta^0 \sum_{s=1}^{3} b_s\varphi_3^{(s)}, \qquad X_3 = (\mu^{(2)} - \mu^{(1)}) \sum_{s=1}^{3} b_s u_\alpha^{(s)} n_\alpha^0$$
$$Y_\beta = (\alpha^{(2)} - \alpha^{(1)})n_\beta^0 \sum_{s=1}^{3} c_s\varphi_3^{(s)}, \qquad Y_3 = (\mu^{(2)} - \mu^{(1)}) \sum_{s=1}^{3} c_s u_\alpha^{(s)} n_\alpha^0 \tag{6.5.56}$$

From Equations 6.5.42, 6.5.43, 6.5.47, and 6.5.53 it follows that the functions v_α and w satisfy the equations and the boundary conditions in a plane strain problem. The necessary and sufficient conditions to solve this problem are

$$\sum_{\rho=1}^{2} \left[\int_{A_\rho} H_\alpha^{(\rho)} da + \int_{\Gamma_\rho} s_\alpha^{(\rho)} ds \right] + \int_{\Gamma_0} Z_\alpha ds = 0$$
$$\sum_{\rho=1}^{2} \left[\int_{A_\rho} (\varepsilon_{\alpha\beta}x_\alpha H_\beta^{(\rho)} + H^{(\rho)}) da + \int_{\Gamma_\rho} (\varepsilon_{\alpha\beta}x_\alpha s_\beta^{(\rho)} + \eta^{(\rho)}) ds \right] \tag{6.5.57}$$
$$+ \int_{\Gamma_0} (\varepsilon_{\alpha\beta}x_\alpha Z_\beta + Z) ds = 0$$

By using Equations 6.5.46, 6.5.50, and 6.5.53 and the divergence theorem, we obtain

$$\sum_{\rho=1}^{2} \left[\int_{A_\rho} H_\alpha^{(\rho)} da + \int_{\Gamma_\rho} s_\alpha^{(\rho)} ds \right] + \int_{\Gamma_0} Z_\alpha ds$$
$$= \sum_{\rho=1}^{2} \left[\int_{A_\rho} R_\alpha^{(\rho)} da + \int_{\Gamma_\rho} p_\alpha^{(\rho)} ds \right] + \int_{\Sigma_1} t_{3\alpha,3} da \tag{6.5.58}$$

In a similar way, the last condition from Equation 6.5.57 becomes

$$\sum_{\rho=1}^{2} \left[\int_{A_\rho} (\varepsilon_{\alpha\beta}x_\alpha R_\beta^{(\rho)} + L_3^{(\rho)}) da + \int_{\Gamma_\rho} (\varepsilon_{\alpha\beta}x_\alpha p_\beta^{(\rho)} + q_3^{(\rho)}) ds \right]$$
$$+ \int_{\Sigma_1} (\varepsilon_{\alpha\beta}x_\alpha t_{3\beta,3} + m_{33,3}) da = 0 \tag{6.5.59}$$

With the help of Equation 6.5.40, from Equation 6.3.16 we find that

$$\int_{\Sigma_1} t_{3\alpha,3} da = Y_{\alpha i} c_i \tag{6.5.60}$$

From Equations 6.5.58 and 6.5.60 it follows that the conditions $6.5.57_1$ reduce to

$$Y_{\alpha i}c_i = -\sum_{\rho=1}^{2}\left[\int_{A_\rho} R_\alpha^{(\rho)} da + \int_{\Gamma_\rho} p_\alpha^{(\rho)} ds\right] \tag{6.5.61}$$

Let us consider the boundary-value problem 6.5.45, 6.5.49, and 6.5.55. The necessary and sufficient condition for the existence of the solution of this problem is

$$\sum_{\rho=1}^{2}\left[\int_{A_\rho} K_3^{(\rho)} da - \int_{\Gamma_\rho} Q_3^{(\rho)} ds\right] - \int_{\Gamma_0} Y_3 ds = 0$$

Taking into account the relations 6.5.46, 6.5.52, and 6.5.56, the above condition becomes

$$Y_{3i}c_i = 0 \tag{6.5.62}$$

Equations 6.5.61 and 6.5.62 uniquely determine the constants c_k. Let us study now the boundary-value problem 6.5.14, 6.5.48, and 6.5.54. The necessary and sufficient condition for the existence of the solution of this problem is

$$\sum_{\rho=1}^{2}\left[\int_{A_\rho} G_3^{(\rho)} da - \int_{\Gamma_3} N_3^{(\rho)} ds\right] - \int_{\Gamma_0} X_3 ds = 0$$

By using Equations 6.5.46, 6.5.51, and 6.5.56, this condition reduces to

$$Y_{3s}b_s = -\sum_{\rho=1}^{2}\left[\int_{A_\rho} R_3^{(\rho)} da + \int_{\Gamma_\rho} p_3^{(\rho)} ds\right] \tag{6.5.63}$$

Let us investigate the conditions 6.2.3. We can write

$$\int_{\Sigma_1} t_{3\alpha} da = \sum_{\rho=1}^{2}\left[\int_{A_\rho} (x_\alpha R_3^{(\rho)} + \varepsilon_{\beta\alpha} L_\beta^{(\rho)}) da + \int_{\Gamma_\rho} (x_\alpha p_3^{(\rho)} + \varepsilon_{\beta\alpha} q_\beta^{(\rho)}) ds\right] + \int_{\Sigma_1} (x_\alpha t_{33,3} + \varepsilon_{\beta\alpha} m_{3\beta,3}) da \tag{6.5.64}$$

With the help of Equations 6.5.40 and 6.5.64, the conditions 6.2.3 become

$$Y_{\alpha i}b_i = -F_\alpha - \sum_{\rho=1}^{2}\left[\int_{A_\rho} (x_\alpha R_3^{(\rho)} + \varepsilon_{\beta\alpha} L_\beta^{(\rho)}) da + \int_{\Gamma_\rho} (x_\alpha p_3^{(\rho)} + \varepsilon_{\beta\alpha} q_\beta^{(\rho)}) ds\right] \tag{6.5.65}$$

The Equations 6.5.63 and 6.5.65 determine the constants b_k. In what follows, we assume that the constants c_s and b_s, and the functions χ, χ_α, Ψ, and Ψ_α are known.

Let us consider now the condition $6.5.57_3$. Taking into account Equations 6.5.41 and 6.5.59, the last condition of 6.5.57 reduces to

$$\begin{aligned}\tau_2 D' = &-\sum_{\rho=1}^{2}\left[\int_{A_\rho}(\varepsilon_{\alpha\beta}x_\alpha R_\beta^{(\rho)} + L_3^{(\rho)})da + \int_{\Gamma_\rho}(\varepsilon_{\alpha\beta}x_\alpha p_\beta^{(\rho)} + q_3^{(\rho)})ds\right] \\ &- \sum_{\rho=1}^{2}\int_{A_\rho}\left\{\varepsilon_{\alpha\beta}x_\alpha\left[\mu^{(\rho)}\chi_{,\beta} + \varepsilon_{\nu\beta}\kappa^{(\rho)}\chi_\nu + (\mu^{(\rho)} + \kappa^{(\rho)})\sum_{s=1}^{3}c_s u_\beta^{(s)}\right]\right. \\ &\left. + (\alpha^{(\rho)} + \beta^{(\rho)} + \gamma^{(\rho)})\sum_{s=1}^{3}c_s\varphi_3^{(s)} + \alpha^{(\rho)}\chi_{\nu,\nu}\right\}da \end{aligned} \tag{6.5.66}$$

where D' is given by Equation 6.5.30. The relation 6.5.66 determines the constant τ_2. By Equations 6.2.4, 6.2.5, and 6.5.41, we get

$$Y_{ij}a_j = r_i \tag{6.5.67}$$

where

$$\begin{aligned} r_\alpha = \varepsilon_{\alpha\beta}M_\beta &- \sum_{\rho=1}^{2}\int_{A_\rho}\{x_\alpha[\lambda^{(\rho)}\gamma_{\nu\nu} + (\lambda^{(\rho)} + 2\mu^{(\rho)} + \kappa^{(\rho)})(\chi + \tau_2\Phi) \\ &- \varepsilon_{\alpha\beta}[\gamma^{(\rho)}(\chi_\beta + \tau_2\Phi_\beta) + \mu_{3\beta}]\}da \\ r_3 = -F_3 &- \sum_{\rho=1}^{2}\int_{A_\rho}[(\lambda^{(\rho)} + 2\mu^{(\rho)} + \kappa^{(\rho)})(\chi + \tau_2\Phi) + \lambda^{(\rho)}\gamma_{\alpha\alpha}]da \end{aligned}$$

Equations 6.5.67 uniquely determine the constants a_k. From Equations 6.2.6 and 6.5.41, we obtain

$$\begin{aligned}\tau_1 D' = -M_3 &- \sum_{\rho=1}^{2}\int_{A_\rho}\left\{\varepsilon_{\alpha\beta}x_\alpha\left[\mu^{(\rho)}\Psi_{,\beta} + \varepsilon_{\nu\beta}\kappa^{(\rho)}\Psi_\nu\right.\right. \\ &\left. + (\mu^{(\rho)} + \lambda^{(\rho)})\sum_{s=1}^{3}b_s u_\beta^{(s)}\right] + (\alpha^{(\rho)} + \beta^{(\rho)} \\ &\left. + \gamma^{(\rho)})\sum_{s=1}^{3}b_s\varphi^{(s)} + \alpha^{(\rho)}\Psi_{\lambda,\lambda}\right\}da\end{aligned}$$

so that we can determine the constant τ_1. The problem is therefore solved.

On the basis of the method presented in Sections 2.1 and 5.6, we have to study now the recurrence process. Let us determine the functions u_i and φ_i

that satisfy the equations

$$
\begin{aligned}
&t_{ji,j} + F_i^{(\rho)}(x_1, x_2)x_3^{n+1} = 0, \qquad m_{ji,j} + \varepsilon_{irs}t_{rs} + L_i^{(\rho)}(x_1, x_2)x_3^{n+1} = 0 \\
&t_{ij} = \lambda^{(\rho)}e_{rr}\delta_{ij} + (\mu^{(\rho)} + \kappa^{(\rho)})e_{ij} + \mu^{(\rho)}e_{ji} \\
&m_{ij} = \alpha^{(\rho)}\varphi_{r,r}\delta_{ij} + \beta^{(\rho)}\varphi_{i,j} + \gamma^{(\rho)}\varphi_{j,i} \\
&e_{ij} = u_{j,i} + \varepsilon_{jir}\varphi_r \text{ on } B_\rho
\end{aligned}
\tag{6.5.68}
$$

subjected to the conditions

$$
\begin{aligned}
&[u_i]_1 = [u_i]_2, \qquad [\varphi_i]_1 = [\varphi_i]_2 \\
&[t_{\beta i}]_1 n_\beta^0 = [t_{\beta i}]_2 n_\beta^0, \qquad [m_{\beta i}]_1 n_\beta^0 = [m_{\beta i}]_2 n_\beta^0 \text{ on } \Pi_0 \\
&\int_{\Sigma_1} t_{3i} da = 0, \qquad \int_{\Sigma_1} (\varepsilon_{irs}x_r t_{3s} + m_{3i}) da = 0 \\
&[t_{\alpha i}n_\alpha]_\rho = p_i^{(\rho)}(x_1, x_2)x_3^{n+1}, \qquad [m_{\alpha i}n_\alpha]_\rho = q_i^{(\rho)}(x_1, x_2)x_3^{n+1} \text{ on } \Pi_\rho
\end{aligned}
\tag{6.5.69}
$$

when the solution of the equations

$$
\begin{aligned}
&t^*_{ji,j} + F_i^{(\rho)}(x_1, x_2)x_3^n = 0, \qquad m^*_{ji,j} + \varepsilon_{irs}t^*_{rs} + L_i^{(\rho)}(x_1, x_2)x_3^n = 0 \\
&t^*_{ij} = \lambda^{(\rho)}e^*_{rr}\delta_{ij} + (\mu^{(\rho)} + \kappa^{(\rho)})e^*_{ij} + \mu^{(\rho)}e^*_{ji} \\
&m^*_{ij} = \alpha^{(\rho)}\varphi^*_{r,r}\delta_{ij} + \beta^{(\rho)}\varphi^*_{i,j} + \gamma^{(\rho)}\varphi^*_{j,i}, \qquad e^*_{ij} = u^*_{j,i} + \varepsilon_{jir}\varphi^*_r
\end{aligned}
\tag{6.5.70}
$$

with the conditions

$$
\begin{aligned}
&[u^*_i]_1 = [u^*_i]_2, \qquad [\varphi^*_i]_1 = [\varphi^*_i]_2 \\
&[t^*_{\beta i}]_1 n_\beta^0 = [t^*_{\beta i}]_2 n_\beta^0, \qquad [m^*_{\beta i}]_1 n_\beta^0 = [m^*_{\beta i}]_2 n_\beta^0 \text{ on } \Pi_0 \\
&\int_{\Sigma_1} t^*_{3i} da = 0, \qquad \int_{\Sigma_1} (\varepsilon_{ijk}x_j t^*_{3k} + m^*_{3i}) da = 0 \\
&[t^*_{\alpha i}n_\alpha]_\rho = p_i^{(\rho)}(x_1, x_2)x_3^n, \qquad [m^*_{\alpha i}n_\alpha]_\rho = q_i^{(\rho)}(x_1, x_2)x_3^n \text{ on } \Pi_\rho
\end{aligned}
\tag{6.5.71}
$$

is known. In the above relations $F_i^{(\rho)}, L_i^{(\rho)}, p_i^{(\rho)}$ and $q_i^{(\rho)}$ are prescribed functions which belong to C^∞. We seek the solution of the problem 6.5.68 and 6.5.69 in the form 6.3.33, where v_i and ψ_i are unknown functions. From Equations 6.5.68, 6.5.69, 6.5.70, and 6.3.33, we obtain

$$
t_{ij} = (n+1)\left[\int_0^{x_3} t^*_{ij}dx_3 + \tau_{ij} + k_{ij}^{(\rho)}\right], \qquad m_{ij} = (n+1)\left[\int_0^{x_3} m^*_{ij}dx_3 + \mu_{ij} + h_{ij}^{(\rho)}\right]
\tag{6.5.72}
$$

where we have used the notations

$$
\begin{aligned}
&\tau_{ij} = \lambda^{(\rho)}\gamma_{rr}\delta_{ij} + (\mu^{(\rho)} + \kappa^{(\rho)})\gamma_{ij} + \mu^{(\rho)}\gamma_{ji} \\
&\mu_{ij} = \alpha^{(\rho)}\psi_{r,r}\delta_{ij} + \beta^{(\rho)}\psi_{i,j} + \gamma^{(\rho)}\psi_{j,i}, \qquad \gamma_{ij} = v_{j,i} + \varepsilon_{jik}\psi_k
\end{aligned}
\tag{6.5.73}
$$

and

$$\begin{aligned}
&k^{(\rho)}_{\alpha\beta} = \lambda^{(\rho)}\delta_{\alpha\beta}u^*_3(x_1,x_2,0), \qquad k^{(\rho)}_{33} = (\lambda^{(\rho)} + 2\mu^{(\rho)} + \kappa^{(\rho)})u^*_3(x_1,x_2,0)\\
&k^{(\rho)}_{\alpha 3} = \mu^{(\rho)}u^*_\alpha(x_1,x_2,0), \qquad k^{(\rho)}_{3\alpha} = (\mu^{(\rho)} + \kappa^{(\rho)})u^*_\alpha(x_1,x_2,0)\\
&h^{(\rho)}_{\lambda\nu} = \alpha^{(\rho)}\delta_{\lambda\nu}\varphi^*_3(x_1,x_2,0), \qquad h^{(\rho)}_{33} = (\alpha^{(\rho)} + \beta^{(\rho)} + \gamma^{(\rho)})\varphi^*_3(x_1,x_2,0)\\
&h^{(\rho)}_{\alpha 3} = \beta^{(\rho)}\varphi^*_\alpha(x_1,x_2,0), \qquad h^{(\rho)}_{3\alpha} = \gamma^{(\rho)}\varphi^*_\alpha(x_1,x_2,0)
\end{aligned} \tag{6.5.74}$$

By using Equations $6.5.70_1$, the equations of equilibrium reduce to

$$\tau_{ji,j} + G^{(\rho)}_i = 0, \qquad \mu_{ji,j} + \varepsilon_{irs}\tau_{rs} + H^{(\rho)}_i = 0 \text{ on } B_\rho \tag{6.5.75}$$

where

$$G^{(\rho)}_i = k^{(\rho)}_{\alpha i,\alpha} + t^*_{3i}(x_1,x_2,0), \qquad H^{(\rho)}_i = h^{(\rho)}_{\alpha i,\alpha} + m^*_{3i}(x_1,x_2,0)$$

The conditions on the surface of separation and the conditions 6.5.39 lead to the following conditions

$$\begin{aligned}
&[v_i]_1 = [v_i]_2, \qquad [\psi_i]_1 = [\psi_i]_2\\
&[\tau_{\beta i}]_1 n^0_\beta = [\tau_{\beta i}]_2 n^0_\beta + s_i, \qquad [\mu_{\beta i}]_1 n^0_\beta = [\mu_{\beta i}]_2 n^0_\beta + r_i \text{ on } \Pi_0\\
&[\tau_{\beta i} n_\beta]_\rho = \widetilde{t}^{(\rho)}_i, \qquad [\mu_{\beta i} n_\beta]_\rho = \widetilde{m}^{(\rho)}_i \text{ on } \Pi_\rho
\end{aligned} \tag{6.5.76}$$

where

$$\begin{aligned}
s_i &= (k^{(2)}_{\alpha i} - k^{(1)}_{\alpha i})n^0_\alpha, \qquad r_i = (h^{(2)}_{\alpha i} - h^{(1)}_{\alpha i})n^0_\alpha,\\
\widetilde{t}^{(\rho)}_i &= -k^{(\rho)}_{\alpha i} n_\alpha, \qquad \widetilde{m}^{(\rho)}_i = -h^{(\rho)}_{\alpha i} n_\alpha
\end{aligned}$$

The conditions on the end Σ_1 reduce to

$$\int_{\Sigma_1} \tau_{3i}\,da = -T_i, \qquad \int_{\Sigma_1} (\varepsilon_{ijk}x_j\tau_{3k} + \mu_{3i})\,da = -N_i \tag{6.5.77}$$

where

$$T_i = \sum_{\rho=1}^{2}\int_{A_\rho} k^{(\rho)}_{3i}, \qquad N_i = \sum_{\rho=1}^{2}\int_{A_\rho} (\varepsilon_{irs}x_r k^{(\rho)}_{3s} + h^{(\rho)}_{3i})\,da$$

We note that the functions $G^{(\rho)}_i, H^{(\rho)}_i, \widetilde{t}^{(\rho)}_i, \widetilde{m}^{(\rho)}_i, s_i$, and r_i are independent of the axial coordinate. We conclude that the functions v_k and ψ_k are characterized by a problem of Almansi-Michell type. If $s_i = r_i = 0$, then the solution of this problem can be taken as in Equation 6.5.40. However, it is easy to see that for $s_i \neq 0, r_i \neq 0$, the solution has the same form. Thus, the Almansi problem is solved.

It is easy to extend the solution to the case when B consists of n elastic bodies with different elasticities.

6.6 Exercises

6.6.1 A continuum body occupies the domain $B^* = \{x : (x_1, x_2) \in \Sigma_1, 0 < x_3 < h\}$, where the cross section Σ_1 is the assembly of the regions $A_1^* = \{x : r_2^2 < x_1^2 + x_2^2 < r_1^2, x_3 = 0\}$ and $A_2^* = \{x : x_1^2 + x_2^2 < r_2^2, x_3 = 0\}$, $(r_1 > r_2 > 0)$. The domain A_1^* is bounded by the circles L and Γ^*, of radius r_1 and r_2, respectively. Study the torsion of the cylinder B^* if the domains $B_\rho^* = \{x : (x_1, x_2) \in A_\rho^*, 0 < x_3 < h\}, (\rho = 1, 2)$, are occupied by different homogeneous and isotropic Cosserat elastic materials.

6.6.2 Investigate the extension and bending of the nonhomogeneous cylinder B^* defined in the preceding exercise.

6.6.3 Study the deformation of a heterogeneous circular cylinder subjected to a constant temperature variation.

6.6.4 Investigate the Saint-Venant problem for heterogeneous anisotropic Cosserat elastic cylinders.

6.6.5 A nonhomogeneous and isotropic Cosserat elastic material occupies the domain $B = \{x : x_1^2 + x_2^2 < a^2, 0 < x_3 < h\}, (a > 0)$. The constitutive coefficients are given by

$$\lambda = \lambda_0 e^{-\xi r}, \quad \mu = \mu_0 e^{-\xi r}, \quad \kappa = \kappa_0 e^{-\xi r}$$
$$\alpha = \alpha_0 e^{-\xi r}, \quad \beta = \beta_0 e^{-\xi r}, \quad \gamma = \gamma_0 e^{-\xi r}, \quad \xi > 0$$

where $\lambda_0, \mu_0, \kappa_0, \alpha_0, \beta_0, \gamma_0$, and ξ are prescribed constants. Study the extension problem.

6.6.6 Investigate the Almansi–Michell problem for inhomogeneous and hemitropic Cosserat elastic cylinders.

6.6.7 Study the problem of uniformly loaded cylinders composed of different inhomogeneous and anisotropic Cosserat elastic continua.

Chapter 7

Porous Elastic Bodies

7.1 Basic Equations

In recent years, the mechanics of porous media has been a subject of intensive study. In Ref. 106, Goodman and Cowin established a continuum theory of granular materials with interstitial voids. The basic concept underlying this theory is that of a material for which the bulk density is written as the product of two fields, the density field of the matrix material and the volume fraction field. This representation was employed by Nunziato and Cowin [258] to develop a nonlinear theory of elastic materials with voids. The intended applications of the theory are to elastic bodies with small voids or vacuous pores which are distributed throughout the material. The linear theory of elastic materials with voids has been established by Cowin and Nunziato [57]. The theory of porous elastic materials was explored in various papers (see, e.g., Refs. 25 and 304, and the literature cited therein).

In the framework of the linear theory of isotropic elastic materials with voids, the deformation of right cylinders has been the subject of various investigations. In Ref. 57, Cowin and Nunziato have established the solution of the pure bending of a cylinder made of a homogeneous and isotropic elastic material with voids. Saint-Venant's principle in the theory of elastic materials with voids has been studied by Batra and Yang [16]. The problem of extension and bending for nonhomogeneous and isotropic porous elastic materials has been investigated in Ref. 50. A general treatment of Saint-Venant's problem for homogeneous and isotropic porous elastic cylinders has been presented by Dell'Isola and Batra [64]. The plane strain problem for elastic materials with voids has been investigated in various papers [50,59,214].

The concept of a distributed body asserts that the mass density ρ in the deformed configuration has the decomposition $\rho = \nu\widehat{\rho}$, where $\widehat{\rho}$ is the density of the matrix material and ν is the volume fraction field. This representation introduces an additional degree of kinematic freedom. In the undeformed state, we have $\rho_0 = \nu_0\widehat{\rho}_0$, where ρ_0 is the mass density in the reference configuration, ν_0 is the volume fraction field in the reference configuration, and $\widehat{\rho}_0$ is the density of the matrix material. We introduce the notation $\varphi = \nu - \nu_0$. In what follows we assume that ν_0 is a given constant.

If we know the displacement vector field $\mathbf{u}$ and the volume distribution function φ, then we know the behavior of a material with voids. In what follows we consider the linear theory of equilibrium. We assume that B is occupied by an elastic material with voids. The linear strain measure is defined by

$$2e_{ij} = u_{i,j} + u_{j,i} \tag{7.1.1}$$

Let t_{ij} be the stress tensor and let h_i be the equilibrated stress vector [106,258]. Then, the surface force and the equilibrated surface force at a regular point of ∂B are given by

$$s_i = t_{ji}n_j, \qquad \sigma = h_i n_i \tag{7.1.2}$$

respectively. The equilibrium equations of a porous continuum are

$$\begin{aligned} t_{ji,j} + f_i = 0 \\ h_{i,i} + g + q = 0 \end{aligned} \tag{7.1.3}$$

where $\mathbf{f}$ is the body force, q is the extrinsic equilibrated body force, and g is the intrinsic equilibrated body force. The constitutive equations of a porous elastic material are given by

$$\begin{aligned} t_{ij} &= C_{ijrs}e_{rs} + B_{ij}\varphi + D_{ijk}\varphi_{,k} \\ h_i &= D_{rsi}e_{rs} + d_i\varphi + A_{ij}\varphi_{,j} \\ g &= -B_{ij}e_{ij} - \xi\varphi - d_i\varphi_{,i} \end{aligned} \tag{7.1.4}$$

where $C_{ijrs}, B_{ij}, D_{ijk}, d_i, A_{ij}$, and ξ are constitutive coefficients which satisfy the relations

$$C_{ijrs} = C_{rsij} = C_{jirs}, \qquad B_{ij} = B_{ji}, \qquad D_{ijk} = D_{jik}, \qquad A_{ij} = A_{ji} \tag{7.1.5}$$

In the case of an isotropic material, the constitutive equations are

$$\begin{aligned} t_{ij} &= \lambda e_{rr}\delta_{ij} + 2\mu e_{ij} + \beta\varphi\delta_{ij} \\ h_i &= \alpha\varphi_{,i} \\ g &= -\beta e_{rr} - \xi\varphi \end{aligned} \tag{7.1.6}$$

where $\lambda, \mu, \alpha, \beta$, and ξ are constitutive coefficients.

The *first boundary-value problem* is characterized by the boundary conditions

$$u_i = \widetilde{u}_i, \qquad \varphi = \widetilde{\varphi} \text{ on } \partial B \tag{7.1.7}$$

where $\widetilde{u}_i$ and $\widetilde{\varphi}$ are prescribed functions. In *the second boundary-value problem* (traction problem), the boundary conditions are

$$t_{ji}n_j = \widetilde{t}_i, \qquad h_j n_j = \widetilde{h} \text{ on } \partial B \tag{7.1.8}$$

where $\widetilde{t}_i$ and $\widetilde{h}$ are given functions.

We assume that (i) B is a bounded regular region; (ii) f_i and q are continuous on $\overline{B}$; (iii) $C_{ijrs}, B_{ij}, D_{ijk}, A_{ij}, d_i$, and ξ are smooth on $\overline{B}$ and satisfy Equation 7.1.5; (iv) $\widetilde{t}_i$ and $\widetilde{h}$ are piecewise regular on ∂B; and (v) $\widetilde{u}_i$ and $\widetilde{\varphi}$ are continuous on ∂B.

The elastic potential associated with the porous material is

$$\begin{aligned}\widetilde{W} = & \frac{1}{2}C_{ijrs}e_{ij}e_{rs} + \frac{1}{2}\xi\varphi^2 + \frac{1}{2}A_{ij}\varphi_{,i}\varphi_{,j} \\ & + B_{ij}e_{ij}\varphi + D_{ijk}e_{ij}\varphi_{,k} + d_i\varphi\varphi_{,i}\end{aligned} \tag{7.1.9}$$

Throughout this chapter, we assume that the elastic potential is a positive definite quadratic form in the variables e_{ij}, $\varphi_{,k}$, and φ. It follows that the constitutive coefficients for isotropic bodies satisfy the following inequalities [57]

$$\mu > 0, \quad \alpha > 0, \quad \xi > 0, \quad 3\lambda + 2\mu > 0, \quad (3\lambda + 2\mu)\xi > 3\beta^2 \tag{7.1.10}$$

The traction problem consists in finding the functions $u_i, \varphi \in C^2(B) \cap C^1(\overline{B})$ that satisfy Equations 7.1.1, 7.1.3, and 7.1.4 on B, and the boundary conditions 7.1.8. We recall the following result (cf. [59,288]).

Theorem 7.1.1 *Assume that $\widetilde{W}$ is positive definite and that hypotheses (i)–(iv) hold. Then the second boundary-value problem has solution if and only if*

$$\int_B f_i dv + \int_{\partial B} \widetilde{t}_i da = 0, \qquad \int_B \varepsilon_{ijr} x_j f_r dv + \int_{\partial B} \varepsilon_{ijr} x_j \widetilde{t}_r da = 0 \tag{7.1.11}$$

We note that from Equations 7.1.1, 7.1.3, and 7.1.4, we obtain the basic field equations in terms of displacement and volume fraction fields

$$\begin{aligned}&(C_{ijrs}u_{r,s} + B_{ij}\varphi + D_{ijk}\varphi_{,k})_{,j} + f_i = 0 \\ &(A_{ij}\varphi_{,j} + D_{rsi}u_{r,s} + d_i\varphi)_{,i} - B_{ij}u_{i,j} - \xi\varphi - d_i\varphi_{,i} + q = 0 \text{ on } B\end{aligned} \tag{7.1.12}$$

In the case of homogeneous and isotropic bodies Equations 7.1.12 reduce to

$$\begin{aligned}&\mu\Delta u_i + (\lambda + \mu)u_{j,ji} + \beta\varphi_{,i} + f_i = 0 \\ &\alpha\Delta\varphi - \beta u_{j,j} - \xi\varphi + q = 0\end{aligned} \tag{7.1.13}$$

7.2 Plane Strain

In this section, we present some results concerning the plane strain problem of isotropic materials with voids. In what follows the region B refers to the right cylinder considered in Section 1.2.

7.2.1 Nonhomogeneous Bodies

We assume that the constitutive coefficients are independent of the axial coordinate,

$$\lambda = \lambda(x_1, x_2), \qquad \mu = \mu(x_1, x_2), \qquad \alpha = \alpha(x_1, x_2), \qquad \beta = \beta(x_1, x_2),$$
$$\xi = \xi(x_1, x_2), \quad (x_1, x_2) \in \Sigma_1 \tag{7.2.1}$$

We suppose that the domain Σ_1 is C^∞-smooth and that the functions $\lambda, \mu, \alpha, \beta$, and ξ belong to C^∞ and satisfy the conditions 7.1.10. Moreover, we assume that the functions $\mathbf{f}, g, \widetilde{\mathbf{u}}, \widetilde{\varphi}, \widetilde{\mathbf{t}}$, and $\widetilde{h}$ are independent of x_3 and belong to C^∞, and that $\mathbf{f}, \widetilde{\mathbf{u}}$, and $\widetilde{\mathbf{t}}$ are parallel to the x_1, x_2-plane. The state of plane strain of the cylinder B is characterized by

$$u_\alpha = u_\alpha(x_1, x_2), \qquad u_3 = 0, \qquad \varphi = \varphi(x_1, x_2), \quad (x_1, x_2) \in \Sigma_1 \tag{7.2.2}$$

The restrictions 7.2.2, in conjunction with the relations 7.1.1 and 7.1.6, imply that e_{ij}, t_{ij}, h_i, and g are all independent of x_3. The nonzero components of the strain tensor are given by

$$e_{\alpha\beta} = \frac{1}{2}(u_{\alpha,\beta} + u_{\beta,\alpha}) \tag{7.2.3}$$

The constitutive equations 7.1.6 imply that

$$\begin{aligned} t_{\alpha\nu} &= \lambda e_{\rho\rho}\delta_{\alpha\nu} + 2\mu e_{\alpha\nu} + \beta\varphi\delta_{\alpha\nu} \\ h_\rho &= \alpha\varphi_{,\rho}, \qquad g = -be_{\alpha\alpha} - \xi\varphi \end{aligned} \tag{7.2.4}$$

The equilibrium equations 7.1.3 become

$$t_{\beta\alpha,\beta} + f_\alpha = 0, \qquad h_{\alpha,\alpha} + g + q = 0 \tag{7.2.5}$$

on Σ_1. The first boundary-value problem is characterized by the boundary conditions

$$u_\alpha = \widetilde{u}_\alpha, \qquad \varphi = \widetilde{\varphi} \text{ on } \Gamma \tag{7.2.6}$$

In the case of the second boundary-value problem, the boundary conditions are

$$t_{\beta\alpha}n_\beta = \widetilde{t}_\alpha, \qquad h_\alpha n_\alpha = \widetilde{h} \text{ on } \Gamma \tag{7.2.7}$$

The elastic potential in the plane strain of isotropic bodies is

$$W^* = \frac{1}{2}\lambda e_{\rho\rho}e_{\gamma\gamma} + \mu e_{\alpha\beta}e_{\alpha\beta} + \frac{1}{2}\xi\varphi^2 + \frac{1}{2}\alpha\varphi_{,\rho}\varphi_{,\rho} + \beta e_{\rho\rho}\varphi \tag{7.2.8}$$

Theorem 7.2.1 *If W^* is a positive definite quadratic form, then*

(i) *the first boundary-value problem has at most one solution;*

(ii) *any two solutions of the second boundary-value problem are equal modulo a plane rigid displacement.*

Proof. By Equations 7.2.4 and 7.2.8,

$$t_{\alpha\beta}e_{\alpha\beta} + h_\alpha\varphi_{,\alpha} - g\varphi = 2W^* \tag{7.2.9}$$

In view of Equations 7.2.3 and 7.2.5, we get

$$t_{\alpha\beta}e_{\alpha\beta} + h_\alpha\varphi_{,\alpha} - g\varphi = (u_\alpha t_{\beta\alpha} + \varphi h_\beta)_{,\beta} + f_\alpha u_\alpha + q\varphi \tag{7.2.10}$$

From Equations 7.2.9 and 7.2.10, we obtain

$$2W^* = (u_\alpha t_{\beta\alpha} + \varphi h_\beta)_{,\beta} + f_\alpha u_\alpha + q\varphi$$

This relation implies that

$$2\int_{\Sigma_1} W^* da = \int_\Gamma (u_\alpha t_{\beta\alpha} n_\beta + \varphi h_\beta n_\beta) ds + \int_{\Sigma_1} (f_\alpha u_\alpha + q\varphi) da \tag{7.2.11}$$

Suppose that there are two solutions $(u_\alpha^{(1)}, \varphi^{(1)})$ and $(u_\alpha^{(2)}, \varphi^{(2)})$. We denote $u_\alpha^0 = u_\alpha^{(1)} - u_\alpha^{(2)}$, $\varphi^0 = \varphi^{(1)} - \varphi^{(2)}$, $e_{\alpha\beta}^0 = e_{\alpha\beta}^{(1)} - e_{\alpha\beta}^{(2)}$. Since the theory is linear, the functions u_α^0 and φ^0 satisfy a plane strain problem corresponding to null external data. From Equation 7.2.11, we find that

$$\int_{\Sigma_1} W_0 da = 0 \tag{7.2.12}$$

where

$$W_0 = \frac{1}{2}\lambda e_{\rho\rho}^0 e_{\alpha\alpha}^0 + \mu e_{\alpha\beta}^0 e_{\alpha\beta}^0 + \frac{1}{2}\xi(\varphi^0)^2 + \alpha\varphi_{,\rho}^0\varphi_{,\rho}^0 + \beta e_{\alpha\alpha}^0\varphi^0$$

Since the elastic potential is positive definite, from Equation 7.2.12, we get

$$u_1^0 = a_1 - bx_2, \qquad u_2^0 = a_2 + bx_1, \qquad \varphi^0 = 0 \tag{7.2.13}$$

where a_α and b are arbitrary constants. In the case of the first boundary-value problem, we obtain $a_1 = a_2 = b = 0$. □

We now state, without proof, the following result [59,88].

Theorem 7.2.2 *Assume that W^* is positive definite, then*

(i) *the first boundary-value problem has solution;*

(ii) *the second boundary-value problem has solution if and only if f_α and $\widetilde{t}_\alpha$ satisfy the conditions*

$$\int_{\Sigma_1} f_\alpha da + \int_\Gamma \widetilde{t}_\alpha ds = 0, \qquad \int_{\Sigma_1} \varepsilon_{\alpha\beta} x_\alpha f_\beta da + \int_\Gamma \varepsilon_{\alpha\beta} x_\alpha \widetilde{t}_\beta ds = 0 \tag{7.2.14}$$

Our assumptions of regularity imply that the solution belong to $C^\infty(\overline{\Sigma})$ (cf. [88]).

We consider three plane strain problems characterized by the constitutive equations

$$\begin{aligned} t_{\alpha\nu}^{(k)} &= \lambda e_{\rho\rho}^{(k)}\delta_{\alpha\nu} + 2\mu e_{\alpha\nu}^{(k)} + \beta\varphi^{(k)}\delta_{\alpha\nu} \\ h_\beta^{(k)} &= \alpha\varphi_{,\beta}^{(k)}, \qquad g^{(k)} = -\beta e_{\rho\rho}^{(k)} - \xi\varphi^{(k)}, \quad (k = 1, 2, 3) \end{aligned} \tag{7.2.15}$$

the geometrical equations

$$2e_{\alpha\beta}^{(k)} = u_{\alpha,\beta}^{(k)} + u_{\beta,\alpha}^{(k)} \tag{7.2.16}$$

the equilibrium equations

$$\begin{aligned} &t_{\beta\alpha,\beta}^{(\nu)} + (\lambda x_\nu)_{,\alpha} = 0, \qquad h_{\rho,\rho}^{(\nu)} + g^{(\nu)} - \beta x_\nu = 0 \\ &t_{\beta\alpha,\beta}^{(3)} + \lambda_{,\alpha} = 0, \qquad h_{\rho,\rho}^{(3)} + g^{(3)} - \beta = 0 \text{ on } \Sigma_1 \end{aligned} \tag{7.2.17}$$

and the boundary conditions

$$t_{\beta\alpha}^{(\nu)} n_\beta = -\lambda x_\nu n_\alpha, \qquad t_{\beta\alpha}^{(3)} n_\beta = -\lambda n_\alpha, \qquad h_\alpha^{(k)} n_\alpha = 0 \text{ on } \Gamma \tag{7.2.18}$$

It is easy to see that the necessary and sufficient conditions 7.2.14 for the existence of the solution are satisfied for each boundary-value problem. In what follows we assume that the functions $u_\alpha^{(k)}, \varphi^{(k)}, e_{\alpha\beta}^{(k)}, t_{\alpha\beta}^{(k)}, h_\alpha^{(k)}$, and $g^{(k)}$ are known. These problems have an important role in solving Saint-Venant's problem.

7.2.2 Homogeneous and Isotropic Materials

The plane strain problems for homogeneous and isotropic porous materials can be studied with the aid of the method of functions of complex variables, as in Section 1.5 (cf. [214]). In what follows we present some results established by this method. In the absence of body loads, we obtain the field equations in terms of displacement and the volume distribution function,

$$\begin{aligned} &\mu\Delta u_\alpha + (\lambda + \mu)u_{\beta,\beta\alpha} + \beta\varphi_{,\alpha} = 0 \\ &\alpha\Delta\varphi - \beta u_{\rho,\rho} - \xi\varphi = 0 \end{aligned} \tag{7.2.19}$$

on Σ_1. We introduce the complex coordinates z and $\overline{z}$ on Σ_1 by Equations 1.5.33 and the complex displacement w by Equation 1.5.34. With the aid of relations 1.5.33, 1.5.34, and 1.5.38, the system of equations 7.2.19 can be expressed as

$$\begin{aligned} 2\mu\frac{\partial^2 w}{\partial z\partial\overline{z}} + (\lambda+\mu)\frac{\partial}{\partial\overline{z}}\left(\frac{\partial w}{\partial z} + \frac{\partial\overline{w}}{\partial\overline{z}}\right) + \beta\frac{\partial\varphi}{\partial\overline{z}} = 0 \\ 4\alpha\frac{\partial^2\varphi}{\partial z\partial\overline{z}} - \beta\left(\frac{\partial w}{\partial z} + \frac{\partial\overline{w}}{\partial\overline{z}}\right) - \xi\varphi = 0 \end{aligned} \tag{7.2.20}$$

The first equation of 7.2.20 may be integrated to give the following result

$$2\mu\frac{\partial w}{\partial z} + (\lambda+\mu)\left(\frac{\partial w}{\partial z} + \frac{\partial \overline{w}}{\partial \overline{z}}\right) + \beta\varphi = \Omega'(z) \tag{7.2.21}$$

where Ω is an arbitrary analytic complex function on z. By Equation 7.2.21,

$$2\mu\frac{\partial \overline{w}}{\partial \overline{z}} + (\lambda+\mu)\left(\frac{\partial w}{\partial z} + \frac{\partial \overline{w}}{\partial \overline{z}}\right) + \beta\varphi = \overline{\Omega}'(\overline{z}) \tag{7.2.22}$$

From Equations 7.2.21 and 7.2.22, we get

$$\frac{\partial w}{\partial z} + \frac{\partial \overline{w}}{\partial \overline{z}} = \frac{1}{2(\lambda+2\mu)}[\Omega'(z) + \overline{\Omega}'(z) - 2\beta\varphi(z,\overline{z})] \tag{7.2.23}$$

We may now eliminate w and $\overline{w}$ from Equation 7.2.23 and the second equation of 7.2.20 to give

$$\left(4\frac{\partial^2}{\partial z\partial \overline{z}} - p^2\right)\varphi = \frac{\beta}{2\alpha(\lambda+2\mu)}[\Omega'(z) + \overline{\Omega}'(\overline{z})] \tag{7.2.24}$$

where

$$p^2 = \frac{\xi(\lambda+2\mu) - \beta^2}{\alpha(\lambda+2\mu)} \tag{7.2.25}$$

The relations 7.1.10 show that $p^2 > 0$. From Equation 7.2.24, we conclude that the function φ can be expressed in the form

$$\varphi = \Phi - \frac{\beta}{2\alpha p^2(\lambda+2\mu)}[\Omega'(z) + \overline{\Omega}'(z)] \tag{7.2.26}$$

where Φ is a real function which satisfies the equation

$$\left(4\frac{\partial^2}{\partial z\partial \overline{z}} - p^2\right)\Phi = 0 \tag{7.2.27}$$

In view of Equations 7.2.23 and 7.2.26, we get

$$\frac{\partial w}{\partial z} + \frac{\partial \overline{w}}{\partial \overline{z}} = C[\Omega'(z) + \overline{\Omega}'(\overline{z})] - \frac{\beta}{\lambda+2\mu}\Phi \tag{7.2.28}$$

where

$$C = \frac{\beta^2 + \alpha p^2(\lambda+2\mu)}{2\alpha p^2(\lambda+2\mu)^2} \tag{7.2.29}$$

By Equations 7.2.21, 7.2.26, 7.2.27, and 7.2.28, we find that

$$\frac{\partial w}{\partial z} = \eta_1\Omega'(z) - \eta_2\overline{\Omega}'(z) - 4\eta_3\frac{\partial^2\Phi}{\partial z\partial \overline{z}} \tag{7.2.30}$$

where

$$\eta_1 = \frac{\lambda + 3\mu}{4(\lambda + 2\mu)\mu} + \frac{\beta^2}{4\alpha p^2(\lambda + 2\mu)^2}$$
$$\eta_2 = \frac{\lambda + \mu}{4\mu(\lambda + 2\mu)} - \frac{\beta^2}{4\alpha p^2(\lambda + 2\mu)^2}, \qquad \eta_3 = \frac{\beta}{2p^2(\lambda + 2\mu)} \tag{7.2.31}$$

Equation 7.2.30 may be integrated to give

$$w = \eta_1 \Omega(z) - \eta_2 z \overline{\Omega}'(\overline{z}) - \overline{\omega}(\overline{z}) - 4\eta_3 \frac{\partial \Phi}{\partial \overline{z}} \tag{7.2.32}$$

where ω is an arbitrary analytic complex function on z.

The relations 7.2.26 and 7.2.32 give a representation of φ and w in terms of the complex analytic functions Ω and ω, and the real function Φ which satisfies a homogeneous Helmholtz equation.

A simple calculation shows that the constitutive equations 7.2.4 may be written as

$$t_{11} + t_{22} = 2(\lambda + \mu)\left(\frac{\partial w}{\partial z} + \frac{\partial \overline{w}}{\partial \overline{z}}\right) + 2\beta\varphi$$
$$t_{11} - t_{22} + 2it_{12} = 4\mu \frac{\partial w}{\partial \overline{z}} \tag{7.2.33}$$
$$h_1 - ih_2 = 2\alpha \frac{\partial \varphi}{\partial z}, \qquad g = -\beta\left(\frac{\partial w}{\partial z} + \frac{\partial \overline{w}}{\partial \overline{z}}\right) - \xi\varphi$$

In view of Equations 7.2.26 and 7.2.32, the relations 7.2.33 become

$$t_{11} + t_{22} = \frac{\xi(\lambda + \mu) - \beta^2}{\alpha p^2(\lambda + 2\mu)}[\Omega'(z) + \overline{\Omega}'(\overline{z})] + \frac{2\mu\beta}{\lambda + 2\mu}\Phi$$
$$t_{11} - t_{22} + 2it_{12} = -4\mu\left[\eta_2 z \overline{\Omega}''(\overline{z}) + \overline{\omega}'(\overline{z}) + 4\eta_3 \frac{\partial^2 \Phi}{\partial \overline{z}^2}\right] \tag{7.2.34}$$
$$h_1 - ih_2 = 2\alpha \frac{\partial \Phi}{\partial z} - 2\eta_3 \Omega''(z), \qquad g = -\alpha p^2 \Phi$$

In the case of the plane strain, the relations 7.1.2 become

$$s_\alpha = t_{\beta\alpha} n_\beta, \qquad h = h_\alpha n_\alpha \tag{7.2.35}$$

On the basis of Equations 1.3.39 and 1.5.48, from Equations 7.2.35, we find that

$$2(s_1 + is_2) = -i(t_{11} + t_{22})\frac{dz}{ds} + i(t_{11} - t_{22} + 2it_{12})\frac{d\overline{z}}{ds}$$
$$\sigma = \Im\text{m}\left\{(h_1 - ih_2)\frac{dz}{ds}\right\} \tag{7.2.36}$$

We note that

$$t_{11} + t_{22} = 4\mu\{\eta_2[\Omega'(z) + \overline{\Omega}'(\overline{z})] + \eta_3 p^2 \Phi\} \tag{7.2.37}$$

With the help of Equations 7.2.33, 7.2.34, 7.2.36, and 7.2.37, we find that

$$
\begin{aligned}
s_1 + is_2 &= -2\mu i \frac{d}{ds}\left\{\eta_2[\Omega(z) + z\overline{\Omega}'(\overline{z})] + \overline{\omega}(\overline{z}) + 4\eta_3 \frac{\partial \Phi}{\partial \overline{z}}\right\} \\
\sigma &= \Im\mathrm{m}\left\{\left[2\alpha\frac{\partial \Phi}{\partial z} - 2\eta_3\Omega''(z)\right]\frac{dz}{ds}\right\}
\end{aligned}
\tag{7.2.38}
$$

Thus, we obtain

$$
\begin{aligned}
R_1 + iR_2 &= \int_\Gamma (s_1 + is_2)ds \\
&= -2\mu i\left\{\eta_2[\Omega(z) + z\overline{\Omega}'(\overline{z})] + \overline{\omega}(\overline{z}) + 4\eta_3\frac{\partial \Phi}{\partial \overline{z}}\right\}_A^A
\end{aligned}
\tag{7.2.39}
$$

The boundary conditions 7.2.6 can be written in the form

$$
\begin{aligned}
&\eta_1\Omega(z) - \eta_2 z\overline{\Omega}'(\overline{z}) - \overline{\omega}(\overline{z}) - 4\eta_3\frac{\partial \Phi}{\partial \overline{z}} = U(s) \\
&\Phi(z,\overline{z}) - \eta_3\left[\Omega'(z) + \overline{\Omega}'(\overline{z})\right] = \widetilde{\varphi}(s), \quad z \in \Gamma
\end{aligned}
\tag{7.2.40}
$$

where $U = \widetilde{u}_1 + i\widetilde{u}_2$. In view of Equation 7.2.38, the boundary conditions 7.2.7 become

$$
\begin{aligned}
&2\mu\frac{d}{ds}\left\{\eta_2[\Omega(z) + z\overline{\Omega}'(\overline{z})] + \overline{\omega}(\overline{z}) + 4\eta_3\frac{\partial \Phi}{\partial \overline{z}}\right\} = T(s) \\
&\Im\mathrm{m}\left\{2\left[\alpha\frac{\partial \Phi}{\partial z} - \eta_3\Omega''(z)\right]\frac{dz}{ds}\right\} = \widetilde{h}(s), \quad z \in \Gamma
\end{aligned}
\tag{7.2.41}
$$

where $T = i(\widetilde{t}_1 + i\widetilde{t}_2)$.

The arbitrariness and the structure of complex potentials can be investigated as in Section 1.5.

Let us suppose that the domain Σ_1 is unbounded, with certain contours $\Gamma_1, \Gamma_2, \ldots, \Gamma_m$ as internal boundaries. We assume that the origin is taken outside Σ_1, and suppose that the functions $t_{\alpha\beta}, h_\alpha$, and g are bounded in the neighborhood of the point at infinity. We can show that for sufficiently large $|z|$, we have

$$
\begin{aligned}
\Omega(z) &= -\frac{1}{2\pi}(X + iY)\log z + (c_1 + ic_2)z + \Omega_0(z) \\
\omega(z) &= \frac{1}{2\pi}\eta_1(X - iY)\log z + (d_1 + id_2)z + \omega_0(z) \\
\Phi(z,\overline{z}) &= \Sigma(\Phi_n e^{in\theta} + \overline{\Phi}_n e^{-in\theta})K_n(pr)
\end{aligned}
\tag{7.2.42}
$$

Here, c_α and d_α are real constants, Φ_n are complex constants, Ω_0 and ω_0 are single-valued analytic functions on Σ_1 including the point at infinity, K_n are modified Bessel functions of order n, X and Y are defined by Equation 1.5.66

and $z = re^{i\theta}$. For sufficiently large $|z|$ the functions Ω_0 and ω_0 can be represented in the form

$$\Omega_0(z) = \sum_{n=0}^{\infty} U_n z^{-n}, \qquad \omega_0(z) = \sum_{n=0}^{\infty} V_n z^{-n} \tag{7.2.43}$$

It follows from Equations 7.2.34, 7.2.37, and 7.2.42 that

$$t_{11}^{(\infty)} = 2\mu(2\eta_2 c_1 - d_1), \qquad t_{22}^{(\infty)} = 2\mu(2\eta_2 c_1 + d_1), \qquad t_{12}^{(\infty)} = 2\mu d_2 \tag{7.2.44}$$

7.2.3 Stresses around a Hole

We use the method of functions of complex variables to solve the plane strain problem for an unbounded region with a circular hole of radius c whose center is at the origin of coordinates. This problem has been studied by Cowin [58] by using a different method. In this case, we have $\Sigma_1 = \{(x_1, x_2) \in \mathbb{R}^2, x_1^2 + x_2^2 > c^2\}$. We suppose that the hole is free of loads and that the body is subjected to uniform axial tensions at infinity acting in the x_1-direction. The conditions at infinity are

$$t_{11}^{(\infty)} = P, \qquad t_{12}^{(\infty)} = t_{22}^{(\infty)} = 0, \qquad h_\alpha^{(\infty)} = 0, \qquad g^{(\infty)} = 0 \tag{7.2.45}$$

where P is a given constant. If we take into account the arbitrariness in choosing Ω and ω, we find that on the boundary of the hole we have the conditions

$$\begin{aligned} &\eta_2[\Omega(z) + z\overline{\Omega}'(z)] + \overline{\omega}(\overline{z}) + 4\eta_3 \frac{\partial \Phi}{\partial \overline{z}} = 0 \\ &\Im\mathrm{m}\left\{\left[\alpha\frac{\partial \Phi}{\partial z} - \eta_3 \Omega''(z)\right]\frac{dz}{ds}\right\} = 0, \quad \text{for } |z| = c \end{aligned} \tag{7.2.46}$$

We seek the solution in the form 7.2.42 and 7.2.43. We note that $X = Y = 0$. In the analysis of stresses, the constants c_2, U_0, and V_0 can be set equal to zero. From Equations 7.2.44 and 7.2.45, we find that

$$c_1 = \frac{1}{8\mu\eta_2}P, \qquad d_1 = -\frac{1}{4\mu}P, \qquad d_2 = 0 \tag{7.2.47}$$

By using Equations 7.2.42 and 7.2.43 in Equation 7.2.46, we obtain

$$\begin{aligned} &\Omega(z) = c_1 z + \frac{1}{z}U_1, \qquad \omega(z) = d_1 z + \frac{1}{z}V_1 + \frac{1}{z^3}V_3 \\ &\Phi(z, \overline{z}) = H\left(\frac{z}{\overline{z}} + \frac{\overline{z}}{z}\right) K_2(pr), \qquad r = (z\overline{z})^{1/2} \end{aligned} \tag{7.2.48}$$

where

$$\begin{aligned} &U_1 = \frac{Pc^2}{4\mu[\eta_2 + 4\eta_3 S K_2(pc)]}, \qquad V_1 = -\frac{Pc^2}{4\mu} \\ &H = U_1 S, \qquad S = \frac{4\eta_3}{\alpha p c^3[K_1(pc) + K_3(pc)]} \\ &V_3 = c^2 U_1 \Lambda, \qquad \Lambda = \eta_2 + 4S\eta_3 K_2(pc) + S\eta_3 pc[K_1(pc) + K_3(pc)] \end{aligned}$$

It is a simple matter to determine the functions u_α and φ and the corresponding stresses.

7.3 Extension, Bending, and Torsion of Porous Elastic Cylinders

In Ref. 57, Cowin and Nunziato established the solution of the pure bending of a cylinder composed of a homogeneous and isotropic elastic material with voids. In this section, an adaptation of the method from Section 3.3 is used to study the extension, bending, and torsion of inhomogeneous and isotropic elastic materials with voids. In what follows the region B refers to the cylinder considered in Section 1.2 and suppose that Σ_1 is C^∞-smooth. We consider an inhomogeneous material with constitutive coefficients independent of the axial coordinate, given by Equations 7.2.1. We assume that the constitutive coefficients belong to C^∞ and that the elastic potential is positive definite.

The equilibrium equations in absence of body loads are given by

$$t_{ji,j} = 0, \qquad h_{j,j} + g = 0 \tag{7.3.1}$$

The cylinder is supposed to be free from lateral surface forces. Thus, on the lateral surface of the cylinder, we have the conditions

$$t_{\alpha i} n_\alpha = 0, \qquad h_\alpha n_\alpha = 0 \text{ on } \Pi \tag{7.3.2}$$

Following Cowin and Nunziato [57], we assume that the equilibrated stress vanishes on the ends of the cylinder. Thus, we have the conditions

$$h_3 = 0 \text{ on } \Sigma_\alpha, \quad (\alpha = 1, 2) \tag{7.3.3}$$

We suppose that the resultant force and the resultant moment about O of the tractions acting on Σ_1 are given by $\mathbf{F} = F_3 \mathbf{e}_3$ and $\mathbf{M} = M_i \mathbf{e}_i$, respectively. Thus, for $x_3 = 0$, we have the following conditions

$$\int_{\Sigma_1} t_{\alpha 3} da = 0 \tag{7.3.4}$$

$$\int_{\Sigma_1} t_{33} da = -F_3, \qquad \int_{\Sigma_1} x_\alpha t_{33} da = \varepsilon_{\alpha\beta} M_\beta \tag{7.3.5}$$

$$\int_{\Sigma_1} \varepsilon_{\alpha\beta} x_\alpha t_{\beta 3} da = -M_3 \tag{7.3.6}$$

On Σ_2 there are tractions applied so as to satisfy the equilibrium conditions of the body.

The problem consists in finding of the functions u_i and φ which satisfy Equations 7.1.1, 7.1.6, and 7.3.1 on B, and the boundary conditions 7.3.2 through 7.3.6. We seek the solution in the form

$$u_\alpha = -\frac{1}{2}a_\alpha x_3^2 + \sum_{k=1}^{3} a_k u_\alpha^{(k)} + \varepsilon_{\beta\alpha}\tau x_\beta x_3$$
$$u_3 = (a_1x_1 + a_2x_2 + a_3)x_3 + \tau\Phi(x_1, x_2), \qquad \varphi = \sum_{k=1}^{3} a_k\varphi^{(k)} \tag{7.3.7}$$

where $u_\alpha^{(k)}$ and $\varphi^{(k)}$ are the solutions of the plane strain problems 7.2.17 and 7.2.18, Φ is an unknown function, and a_k and τ are unknown constants. From Equations 7.1.1, 7.2.6, and 7.3.7, we obtain

$$e_{\alpha\beta} = \sum_{k=1}^{3} a_k e_{\alpha\beta}^{(k)}, \qquad e_{\alpha 3} = \mu\tau(\Phi_{,\alpha} + \varepsilon_{\beta\alpha}x_\beta), \qquad e_{33} = a_1x_1 + a_2x_2 + a_3 \tag{7.3.8}$$

By Equations 7.2.4, 7.3.7, 7.3.8, and 7.2.15, we find that

$$t_{\alpha\beta} = \lambda(a_1x_1 + a_2x_2 + a_3)\delta_{\alpha\beta} + \sum_{k=1}^{3} a_k t_{\alpha\beta}^{(k)}, \ t_{\alpha 3} = \mu\tau(\Phi_{,\alpha} + \varepsilon_{\beta\alpha}x_\beta)$$
$$t_{33} = (\lambda + 2\mu)(a_1x_1 + a_2x_2 + a_3) + \sum_{k=1}^{3} a_k\left(\lambda\mu_{\rho,\rho}^{(k)} + \beta\varphi^{(k)}\right) \tag{7.3.9}$$
$$h_\alpha = \sum_{k=1}^{3} a_k h_\alpha^{(k)}, \qquad h_3 = 0, \qquad g = -\beta(a_1x_1 + a_2x_2 + a_3) + \sum_{k=1}^{3} a_k g^{(k)}$$

In view of Equations 7.3.9, the equilibrium equations 7.3.1 reduce to

$$t_{\beta\alpha,\beta} = 0, \qquad h_{\alpha,\alpha} + g = 0 \text{ on } \Sigma_1 \tag{7.3.10}$$

and

$$(\mu\Phi_{,\alpha})_{,\alpha} = \varepsilon_{\alpha\beta}(\mu x_\beta)_{,\alpha} \text{ on } \Sigma_1 \tag{7.3.11}$$

The boundary conditions 7.3.2 become

$$t_{\beta\alpha}n_\beta = 0, \qquad h_\alpha n_\alpha = 0 \text{ on } \Gamma \tag{7.3.12}$$

and

$$\Phi_{,\alpha}n_\alpha = \varepsilon_{\alpha\beta}x_\beta n_\alpha \text{ on } \Gamma \tag{7.3.13}$$

Equations 7.3.10 and the boundary conditions 7.3.12 are satisfied on the basis of the relations 7.2.17 and 7.2.18. We conclude that Φ is the solution of the boundary-value problem 7.3.11 and 7.3.13. Clearly, the function Φ exists. In what follows we shall treat Φ as a known function.

The conditions 7.3.3 are identically satisfied on the basis of the relations 7.3.9. With the aid of equilibrium equations 7.3.10 and the boundary conditions 7.3.12 and 7.3.9, we can write

$$\int_{\Sigma_1} t_{\alpha 3} da = \int_{\Sigma_1} (t_{\alpha 3} + x_\alpha t_{ji,j}) da = \int_\Gamma x_\alpha t_{\beta 3} n_\beta ds + \int_{\Sigma_1} x_\alpha t_{33,3} da = 0$$

so that the conditions 7.3.4 are satisfied. From Equations 7.3.5 and 7.3.9, we obtain the following system for the unknown constants a_1, a_2, and a_3

$$B_{\alpha i} a_i = \varepsilon_{\alpha\beta} M_\beta, \qquad B_{3i} a_i = -F_3 \tag{7.3.14}$$

where

$$\begin{aligned}
B_{\alpha\rho} &= \int_{\Sigma_1} x_\alpha \left[(\lambda + 2\mu) x_\rho + \lambda e_{\nu\nu}^{(\rho)} + \beta \varphi^{(\rho)}\right] da \\
B_{\alpha 3} &= \int_{\Sigma_1} x_\alpha \left(\lambda + 2\mu + \lambda e_{\rho\rho}^{(3)} + \beta \varphi^{(3)}\right) da \\
B_{3\alpha} &= \int_{\Sigma_1} \left[(\lambda + 2\mu) x_\alpha + \lambda e_{\rho\rho}^{(\alpha)} + \beta \varphi^{(\alpha)}\right] da \\
B_{33} &= \int_{\Sigma_1} \left(\lambda + 2\mu + \lambda e_{\rho\rho}^{(3)} + \beta \varphi^{(3)}\right) da
\end{aligned} \tag{7.3.15}$$

By using the reciprocity relation and the positive definiteness of the elastic potential, as in Section 3.2, we can show that

$$B_{ij} = B_{ji}, \qquad \det(B_{ij}) \neq 0 \tag{7.3.16}$$

The system 7.3.14 uniquely determines the constants a_k. In view of Equation 7.3.9, the condition 7.3.6 reduces to

$$\tau D_* = -M_3 \tag{7.3.17}$$

where D_* is the torsional rigidity,

$$D_* = \int_{\Sigma_1} \mu \varepsilon_{\alpha\beta} x_\alpha (\Phi_{,\beta} + \varepsilon_{\eta\beta} x_\eta) da$$

It is known from Section 3.3 that $D_* \neq 0$ so that Equation 7.3.17 determines the constant τ. We conclude that Equation 7.3.7 is a solution of the problem.

In the case of a homogeneous body, we can find that

$$u_\alpha^{(3)} = \frac{\lambda\xi - \beta^2}{2\beta^2 - 2\xi(\lambda + \mu)} x_\alpha, \qquad \varphi^{(3)} = \frac{\beta\mu}{\beta^2 - \xi(\lambda + \mu)}$$

If the rectangular cartesian coordinate frame is supposed to be chosen in such a way that the x_3-axis coincides with the line of centroids of cross-sections, then we obtain $B_{\alpha 3} = 0$. Thus, for the extension problem, we get

$$a_\alpha = 0, \qquad a_3 = \frac{[\beta^2 - \xi(\lambda + \mu)] F_3}{[\xi(\lambda + \mu)(\lambda + 2\mu) - 3\mu\beta^2 - \lambda^2\xi] A}, \qquad \tau = 0$$

where A is the area of the cross-section. The solution of the extension problem is given by

$$u_\alpha = a_3 u_\alpha^{(3)}, \qquad u_3 = a_3 x_3, \qquad \varphi = a_3 \varphi^{(3)}$$

7.4 Cylinders Composed of Different Porous Materials

We study now the deformation of composed cylinders introduced in Section 3.1. We suppose that B_ρ is occupied by an elastic material with voids with the constitutive coefficients $\lambda^{(\rho)}, \mu^{(\rho)}, \alpha^{(\rho)}, \beta^{(\rho)}$, and $\xi^{(\rho)}$. We assume that

$$\begin{aligned} &\lambda^{(\rho)} = \lambda^{(\rho)}(x_1, x_2), \qquad \mu^{(\rho)} = \mu^{(\rho)}(x_1, x_2), \qquad \alpha^{(\rho)} = \alpha^{(\rho)}(x_1, x_2), \\ &\beta^{(\rho)} = \beta^{(\rho)}(x_1, x_2), \qquad \xi^{(\rho)} = \xi^{(\rho)}(x_1, x_2), \quad (x_1, x_2) \in A_\rho \end{aligned} \tag{7.4.1}$$

The constitutive coefficients are supposed to belong to C^∞ and the elastic potential corresponding to the material that occupies B_ρ is assumed to be positive definite. We consider that the cylinder B is composed of two materials which are welded together along Π_0.

The displacement vector, the volume distribution function, and the surface forces must be continuous in passing from one medium to another, so that we have the conditions

$$\begin{aligned} &[u_i]_1 = [u_i]_2, \qquad [\varphi]_1 = [\varphi]_2 \\ &[t_{\alpha i}]_1 n_\alpha^0 = [t_{\alpha i}]_2 n_\alpha^0, \qquad [h_\alpha]_1 n_\alpha^0 = [h_\alpha]_2 n_\alpha^0 \text{ on } \Pi_0 \end{aligned} \tag{7.4.2}$$

We study the problem of extension and bending of heterogeneous porous cylinders. The problem consists in finding of the functions u_i and φ which satisfy Equations 7.1.1, 7.1.6, and 7.3.1 on B_ρ, the conditions 7.3.2 on Π, the conditions 7.3.3 on the ends, the conditions 3.3.1 and 3.3.2 for $x_3 = 0$, and the conditions 7.4.2 on the surface of separation Π_0.

7.4.1 Auxiliary Plane Strain Problems

Let us consider the plane strain of the composed cylinder B. In this case the displacement vector field and the volume distribution function have the form 7.2.2. We assume that the elastic coefficients are given by Equations 7.4.1. The equations of equilibrium, with the body forces $f_\alpha^{(\rho)}$ and extrinsic equilibrated body force $q^{(\rho)}$, are given by

$$t_{\beta\alpha,\beta} + f_\alpha^{(\rho)} = 0, \qquad h_{\alpha,\alpha} + g + q^{(\rho)} = 0 \text{ on } A_\rho \tag{7.4.3}$$

In the plane strain, the constitutive equations are

$$\begin{aligned} &t_{\alpha\beta} = \lambda^{(\rho)} e_{\eta\eta} \delta_{\alpha\beta} + 2\mu^{(\rho)} e_{\alpha\beta} + \beta^{(\rho)} \varphi \delta_{\alpha\beta} \\ &h_\nu = \alpha \varphi_{,\nu}^{(\rho)}, \qquad g = -\beta^{(\rho)} e_{\alpha\alpha} - \xi^{(\rho)} \varphi \text{ on } A_\rho \end{aligned} \tag{7.4.4}$$

where

$$e_{\alpha\beta} = \frac{1}{2}(u_{\alpha,\beta} + u_{\beta,\alpha}) \tag{7.4.5}$$

The conditions on the surface of separation Π_0 reduce to

$$\begin{aligned}
&[u_\alpha]_1 = [u_\alpha]_2, \qquad [\varphi]_1 = [\varphi]_2 \\
&[t_{\alpha\beta}]_1 n^0_\alpha = [t_{\alpha\beta}]_2 n^0_\alpha, \qquad [h_\alpha]_1 n^0_\alpha = [h_\alpha]_2 n^0_\alpha \text{ on } \Gamma_0
\end{aligned} \tag{7.4.6}$$

We consider the second boundary-value problem characterized by the boundary conditions

$$[t_{\beta\alpha} n_\beta]_\rho = \widetilde{t}^{(\rho)}_\alpha, \qquad [h_\alpha n_\alpha]_\rho = \widetilde{h}^{(\rho)} \text{ on } \Gamma_\rho \tag{7.4.7}$$

We assume that $f^{(\rho)}_\alpha, q^{(\rho)}, \widetilde{t}^{(\rho)}_\alpha$, and $\widetilde{h}^{(\rho)}$ are C^∞ functions.

If the domains Σ_1 and A_ρ satisfy some conditions of regularity ([88], p. 369), then the second boundary-value problem for a composed cylinder has a solution $u_\alpha, \varphi \in C^\infty(A_1 \cup \Gamma_1) \cup C^\infty(A_2 \cup \Gamma_2) \cap C^0(\Sigma_1)$ if and only if the conditions 3.6.7 hold. In what follows we assume that the domains Σ_1 and A_ρ satisfy the conditions that ensure this existence result. If the conditions 7.4.6 are replaced by

$$\begin{aligned}
&[u_\alpha]_1 = [u_\alpha]_2, \qquad [\varphi]_1 = [\varphi]_2 \\
&[t_{\alpha\beta}]_1 n^0_\alpha = [t_{\alpha\beta}]_2 n^0_\alpha + g_\beta, \qquad [h_\alpha]_1 n^0_\alpha = [h_\alpha] n^0_\alpha \text{ on } \Gamma_0
\end{aligned} \tag{7.4.8}$$

where g_α are C^∞ functions, then the conditions 3.6.7 are replaced by conditions 3.6.9. We shall have occasion to use three special problems $\Lambda^{(s)}$, $(s = 1, 2, 3)$, of plane strain. We denote by $u^{(s)}_\alpha, \varphi^{(s)}, e^{(s)}_{ij}, t^{(s)}_{ij}, g^{(s)}$, and $h^{(s)}_i$ the functions $u_\alpha, \varphi, e_{ij}, t_{ij}, g$, and h_i for the problem $\Lambda^{(s)}$. The problems $\Lambda^{(s)}$ are characterized by the equations of equilibrium

$$\begin{aligned}
&t^{(\eta)}_{\beta\alpha,\beta} + (\lambda^{(\rho)} x_\eta)_{,\alpha} = 0, \qquad t^{(3)}_{\beta\alpha,\beta} + \lambda^{(\rho)}_{,\alpha} = 0, \quad (\eta = 1, 2) \\
&h^{(\eta)}_{\alpha,\alpha} + g^{(\eta)} - \beta^{(\rho)} x_\eta = 0, \qquad h^{(3)}_{\alpha,\alpha} + g^{(3)} - \beta^{(\rho)} = 0 \text{ on } A_\rho
\end{aligned} \tag{7.4.9}$$

the constitutive equations

$$\begin{aligned}
&t^{(s)}_{\alpha\beta} = \lambda^{(\rho)} e^{(s)}_{\nu\nu} \delta_{\alpha\beta} + 2\mu^{(\rho)} \varepsilon^{(s)}_{\alpha\beta} + \beta^{(\rho)} \varphi^{(s)} \delta_{\alpha\beta} \\
&h^{(s)}_\nu = \alpha \varphi^{(s)}_{,\nu}, \qquad g^{(s)} = -\beta^{(\rho)} e^{(s)}_{\alpha\alpha} - \xi^{(\rho)} \varphi^{(s)} \text{ on } A_\rho
\end{aligned} \tag{7.4.10}$$

and the geometrical equations

$$2e^{(s)}_{\alpha\beta} = u^{(s)}_{\alpha,\beta} + u^{(s)}_{\beta,\alpha} \tag{7.4.11}$$

To these equations, we adjoin the conditions

$$t^{(\eta)}_{\alpha\beta} n_\beta = -\lambda^{(\rho)} x_\eta n_\alpha, \qquad t^{(3)}_{\alpha\beta} n_\beta = -\lambda^{(\rho)} n_\alpha, \qquad h^{(s)}_\alpha n_\alpha = 0 \text{ on } \Gamma_\rho \tag{7.4.12}$$

and the conditions

$$\begin{aligned} &[u_\alpha^{(s)}]_1 = [u_\alpha^{(s)}]_2, \qquad [\varphi^{(s)}]_1 = [\varphi^{(s)}]_2 \\ &[t_{\alpha\beta}^{(s)}]_1 n_\beta^0 = [t_{\alpha\beta}^{(s)}]_2 n_\beta^0 + p_\alpha^{(s)}, \qquad [h_\alpha^{(s)}]_1 n_\alpha^0 = [h_\alpha^{(s)}]_2 n_\alpha^0 \text{ on } \Gamma_0 \end{aligned} \tag{7.4.13}$$

where

$$p_\alpha^{(\eta)} = (\lambda^{(2)} - \lambda^{(1)}) x_\eta n_\alpha^0, \qquad p_\alpha^{(3)} = (\lambda^{(2)} - \lambda^{(1)}) n_\alpha^0 \tag{7.4.14}$$

The necessary and sufficient conditions for the existence of the solution are satisfied for each boundary-value problem $\Lambda^{(s)}$, $(s = 1, 2, 3)$. In what follows we shall consider that the functions $u_\alpha^{(s)}$ and $\varphi^{(s)}$ have been determined.

7.4.2 Solution of the Problem

We seek the solution of the problem of extension and bending in the form

$$\begin{aligned} &u_\alpha = -\frac{1}{2} d_\alpha x_3^2 + \sum_{s=1}^{3} d_a u_\alpha^{(s)}, \qquad \varphi = \sum_{s=1}^{3} d_s \varphi^{(s)} \\ &u_3 = (d_1 x_1 + d_2 x_2 + d_3) x_3 \end{aligned} \tag{7.4.15}$$

where $\{u_\alpha^{(s)}, \varphi^{(s)}\}$ are the solutions of the problems $\Lambda^{(s)}$, $(s = 1, 2, 3, 4)$, and d_k are unknown constants. It follows from Equations 7.1.1, 7.1.6, 7.4.10, 7.4.11, and 7.4.15 that

$$\begin{aligned} &t_{\alpha\beta} = \lambda^{(\rho)} (d_1 x_1 + d_2 x_2 + d_3) \delta_{\alpha\beta} + \sum_{s=1}^{3} d_s t_{\alpha\beta}^{(s)} \\ &t_{33} = (\lambda^{(\rho)} + 2\mu^{(\rho)})(d_1 x_1 + d_2 x_2 + d_3) + \sum_{s=1}^{3} d_s \left(\lambda^{(\rho)} e_{\alpha\alpha}^{(s)} + \beta^{(\rho)} \varphi^{(s)}\right) \\ &t_{\alpha 3} = 0, \qquad h_\alpha = \sum_{s=1}^{3} d_s h_\alpha^{(s)}, \quad h_3 = 0 \\ &g = \sum_{s=1}^{3} d_s g^{(s)} - \beta^{(\rho)} (d_1 x_1 + d_2 x_2 + d_3) \end{aligned} \tag{7.4.16}$$

It is a simple matter to see that the equations of equilibrium and the boundary conditions 7.3.2 are satisfied on the basis of the relations 7.4.9, 7.4.12, and 7.4.16. The conditions 7.3.3 and 3.3.1 are identically satisfied. It follows from Equations 7.4.13, 7.4.15, and 7.4.16 that the conditions 7.4.2 on the surface of separation are satisfied. With the help of Equation 7.4.16, the conditions 3.3.2 lead to the following system for the unknown constants d_s,

$$D_{\alpha i} d_i = \varepsilon_{\alpha\beta} M_\beta, \qquad D_{3i} d_i = -F_3 \tag{7.4.17}$$

where

$$
\begin{aligned}
D_{\alpha\eta} &= \sum_{\rho=1}^{2} \int_{A_\rho} x_\alpha \big[(\lambda^{(\rho)} + 2\mu^{(\rho)})x_\eta + \lambda^{(\rho)} e_{\nu\nu}^{(\eta)} + \beta^{(\rho)}\varphi^{(\eta)}\big] da \\
D_{\alpha 3} &= \sum_{\rho=1}^{2} \int_{A_\rho} x_\alpha \big[\lambda^{(\rho)} + 2\mu^{(\rho)} + \lambda^{(\rho)} e_{\nu\nu}^{(3)} + \beta^{(\rho)}\varphi^{(3)}\big] da \\
D_{3\alpha} &= \sum_{\rho=1}^{2} \int_{A_\rho} [(\lambda^{(\rho)} + 2\mu^{(\rho)})x_\alpha + \lambda^{(\rho)} e_{\nu\nu}^{(\alpha)} + \beta^{(\rho)}\varphi^{(\alpha)}] da \\
D_{33} &= \sum_{\rho=1}^{2} \int_{A_\rho} (\lambda^{(\rho)} + 2\mu^{(\rho)} + \lambda^{(\rho)} e_{\nu\nu}^{(3)} + \beta^{(\rho)}\varphi^{(3)}) da
\end{aligned}
\tag{7.4.18}
$$

As in Section 3.3 we can prove that the system 7.4.17 can always be solved for d_1, d_2, and d_3. Thus, the functions 7.4.1 represent a solution of the problem.

7.5 Applications

7.5.1 Deformation of Porous Elastic Tube

We consider a homogeneous and isotropic porous elastic material which occupies the right cylinder B with the cross-section $\Sigma_1 = \{x : R_1^2 < x_1^2 + x_2^2 < R_2^2, x_3 = 0\}$. The body is in equilibrium in the absence of body loads. Let us study the plane strain problem of the tube, when the lateral boundaries are subjected to constant pressures. We seek the solution in the form

$$u_\alpha = x_\alpha \chi(r), \qquad u_3 = 0, \qquad \varphi = \varphi(r), \tag{7.5.1}$$

where $r = (x_1^2 + x_2^2)^{1/2}$, and χ and φ are unknown functions. With the aid of the relations 1.11.22 we find that the equations of equilibrium 7.1.13, with $f_i = q = 0$, reduce to

$$
\begin{aligned}
(\lambda + 2\mu)(u_{\rho,\rho})_{,\alpha} + \beta\varphi_{,\alpha} &= 0 \\
\alpha\Delta\varphi - \beta u_{\rho,\rho} - \xi\varphi &= 0
\end{aligned}
\tag{7.5.2}
$$

From the first equation of 7.5.2 we get

$$u_{\rho,\rho} = -\frac{\beta}{\lambda + 2\mu}\varphi + C_1 \tag{7.5.3}$$

where C_1 is an arbitrary constant. Substituting Equation 7.5.3 into the second Equation 7.5.5 we obtain

$$\Delta\varphi - p^2\varphi = \frac{\beta}{\alpha}C_1 \tag{7.5.4}$$

where p^2 is given by Equation 7.2.25. We note that the relations 7.1.10 imply that p^2 is strictly positive. It follows from Equations 7.5.1 that Equations 7.5.3 can be written in the form

$$(r^2\chi)' + \frac{\beta}{\lambda + 2\mu} r\varphi = rC_1 \tag{7.5.5}$$

where prime denotes the derivative with respect to r. From Equation 7.5.4 we obtain

$$\varphi = \varphi_0(r) - \frac{1}{\alpha p^2}\beta C_1 \tag{7.5.6}$$

where

$$\varphi_0(r) = E_1 I_0(pr) + E_2 K_0(pr) \tag{7.5.7}$$

E_α are arbitrary constants, and I_n, K_n are the modified Bessel functions of order n. If we take into account that

$$\Delta\varphi = \frac{1}{r}\frac{d}{dr}\left(r\frac{d\varphi}{dr}\right)$$

from Equation 7.5.4 we find that

$$r\varphi = \frac{1}{p^2}\frac{d}{dr}(r\varphi') - \frac{\beta r}{\alpha p^2}C_1$$

Thus, Equation 7.5.5 becomes

$$r^2\chi + \frac{\beta}{(\lambda + 2\mu)p^2} r\varphi' = \frac{\xi}{2\alpha p^2} r^2 C_1 + C_2 \tag{7.5.8}$$

where C_2 is an arbitrary constant. In view of Equation 7.5.6, from Equation 7.5.8 we get

$$\chi = \frac{\xi}{2\alpha p^2}C_1 + \frac{1}{r^2}C_2 + F(r) \tag{7.5.9}$$

where

$$F(r) = -\frac{\beta}{(\lambda + 2\mu)p^2 r}\varphi_0'(r) \tag{7.5.10}$$

We have the following boundary conditions

$$\begin{aligned} t_{\beta\alpha}n_\beta &= -p_1 n_\alpha, \qquad h_\alpha n_\alpha = 0 \text{ for } r = R_1 \\ t_{\beta\alpha}n_\beta &= -p_2 n_\alpha, \qquad h_\alpha n_\alpha = 0 \text{ for } r = R_2 \end{aligned} \tag{7.5.11}$$

where p_α are prescribed constants. From the constitutive Equations 7.2.4 and 7.5.1 we obtain

$$\begin{aligned} t_{\alpha\beta} &= 2(\lambda + \mu)\chi\delta_{\alpha\beta} + (\lambda r\delta_{\alpha\beta} + 2\mu x_\alpha x_\beta r^{-1})\chi' + \beta\varphi\delta_{\alpha\beta} \\ h_\nu &= \alpha x_\nu r^{-1}\varphi'(r) \end{aligned} \tag{7.5.12}$$

Since $n_\beta = -x_\beta/R_1$ on $r = R_1$ and $n_\beta = x_\beta/R_2$ on $r = R_2$, the conditions 7.5.11 reduce to

$$\begin{aligned} t_{\beta\alpha}x_\beta &= -p_1 x_\alpha, & \varphi'(r) &= 0 \text{ for } r = R_1 \\ t_{\beta\alpha}x_\beta &= -p_2 x_\alpha, & \varphi'(r) &= 0 \text{ for } r = R_2 \end{aligned} \tag{7.5.13}$$

The conditions $\varphi'(R_\alpha) = 0, (\alpha = 1, 2)$, imply that $E_1 = E_2 = 0$, so that $\varphi_0(r) = 0$. Thus, we obtain

$$\varphi = -\frac{1}{\alpha p^2}\beta C_1 \tag{7.5.14}$$

Since $\varphi_0 = 0$, we have

$$\chi = \frac{\xi}{2\alpha p^2}C_1 + C_2 r^{-2} \tag{7.5.15}$$

It follows from Equations 7.5.12, 7.5.14, and 7.5.15 that

$$t_{\beta\alpha}x_\beta = 2(\zeta C_1 - \mu r^{-2}C_2)x_\alpha \tag{7.5.16}$$

where

$$\zeta = \frac{(\lambda + \mu)\xi - \beta^2}{2\alpha p^2}$$

In view of Equation 7.1.10, we have $\zeta \neq 0$. By using Equation 7.5.16 and the conditions for $t_{\beta\alpha}x_\beta$, from Equations 7.5.13, we obtain

$$\begin{aligned} 2(\zeta C_1 - \mu R_1^{-2}C_2) &= -p_1 \\ 2(\zeta C_1 - \mu R_2^{-2}C_2) &= -p_2 \end{aligned}$$

We find that

$$C_1 = \frac{p_2R_2^2 - p_1R_1^2}{2\zeta(R_1^2 - R_2^2)}, \qquad C_2 = \frac{R_1^2R_2^2(p_2 - p_1)}{2\mu(R_1^2 - R_2^2)}$$

Thus, from Equation 7.5.14, we get

$$\varphi = \frac{(p_2R_2^2 - p_1R_1^2)\beta}{2\zeta(R_2^2 - R_1^2)\alpha p^2}$$

7.5.2 Extension and Bending of Porous Circular Cylinder

In what follows, we use the method given in Section 7.4 in order to study the extension and bending of a composed circular cylinder. We assume that the domain A_1 is bounded by two concentric circles, L and Γ^*, of radius r_1 and r_2, respectively, where $r_2 < r_1$. The domain A_2 is bounded by the circle Γ^*. We consider the case when the domains A_1 and A_2 are occupied by two different homogeneous and isotropic elastic materials. In what follows, we shall investigate the plane strain problem by using the polar coordinates (r, θ). We denote by u and v the physical components of the displacement vector in polar coordinates,

$$u + iv = (u_1 + iu_2)e^{-i\theta}$$

The geometrical Equations 7.2.3 become

$$\varepsilon_{rr} = \frac{\partial u}{\partial r}, \qquad \varepsilon_{\theta\theta} = \frac{1}{r}\left(\frac{\partial v}{\partial \theta} + u\right), \qquad \varepsilon_{r\theta} = \frac{1}{2}\left(\frac{1}{r}\frac{\partial u}{\partial \theta} + \frac{\partial v}{\partial r} - \frac{1}{r}v\right) \tag{7.5.17}$$

where $\varepsilon_{rr}, \varepsilon_{\theta\theta}$, and $\varepsilon_{r\theta}$ are the physical components of the strain tensor. The constitutive equations for the plane strain problem can be written in the form

$$\begin{aligned} &t_{rr} = (\lambda + 2\mu)\varepsilon_{rr} + \lambda\varepsilon_{\theta\theta} + \beta\varphi, \qquad t_{\theta\theta} = \lambda\varepsilon_{rr} + (\lambda + 2\mu)\varepsilon_{\theta\theta} + \beta\varphi \\ &t_{r\theta} = 2\mu\varepsilon_{r\theta}, \qquad h_r = \alpha\frac{\partial \varphi}{\partial r}, \qquad h_\theta = \frac{1}{r}\alpha\frac{\partial \varphi}{\partial \theta} \\ &g = -\beta\left[\frac{1}{r}\frac{\partial}{\partial r}(ru) + \frac{1}{r}\frac{\partial v}{\partial \theta}\right] - \xi\varphi \end{aligned} \tag{7.5.18}$$

where $t_{rr}, t_{\theta\theta}$, and $t_{r\theta}$ are the physical components of the stress tensor, and h_r and h_θ are the physical components of the equilibrated stress vector.

The equilibrium Equations 7.2.5 reduce to

$$\begin{aligned} &\frac{\partial t_{rr}}{\partial r} + \frac{1}{r}\frac{\partial t_{r\theta}}{\partial \theta} + \frac{1}{r}(t_{rr} - t_{\theta\theta}) + f_r = 0 \\ &\frac{\partial t_{r\theta}}{\partial r} + \frac{1}{r}\frac{\partial t_{\theta\theta}}{\partial \theta} + \frac{2}{r}t_{r\theta} + f_\theta = 0 \\ &\frac{1}{r}\frac{\partial}{\partial r}(rh_r) + \frac{1}{r}\frac{\partial h_\theta}{\partial \theta} + g + q = 0 \end{aligned}$$

where f_r and f_θ are the physical components of the body force.

First, we determine the solutions of the auxiliary plane strain problems $\Lambda^{(s)}, (s = 1, 2, 3)$, introduced in Section 7.4. We denote by $u^{(1)}$ and $v^{(1)}$ the physical components of the displacement vector in polar coordinates in the problem $\Lambda^{(1)}$. We seek the solution of the problem $\Lambda^{(1)}$ in the form

$$u^{(1)} = U(r)\cos\theta, \qquad v^{(1)} = V(r)\sin\theta, \qquad \varphi^{(1)} = W(r)\cos\theta \tag{7.5.19}$$

where U, V, and W are functions only on r. It follows from Equations 7.5.17, 7.5.18, and 7.5.19 that

$$
\begin{aligned}
t_{rr}^{(1)} &= \left[(\lambda^{(\rho)} + 2\mu^{(\rho)})\frac{dU}{dr} + \frac{1}{r}\lambda^{(\rho)}(U+V) + \beta^{(\rho)}W\right]\cos\theta \\
t_{\theta\theta}^{(1)} &= \left[\lambda^{(\rho)}\frac{dU}{dr} + \frac{1}{r}(\lambda^{(\rho)} + 2\mu^{(\rho)})(U+V) + \beta^{(\rho)}W\right]\cos\theta \\
t_{r\theta}^{(1)} &= \mu^{(\rho)}\left[\frac{dV}{dr} - \frac{1}{r}(U+V)\right]\sin\theta \qquad (7.5.20) \\
h_r^{(1)} &= \alpha^{(\rho)}\frac{dW}{dr}\cos\theta, \qquad h_\theta^{(1)} = -\frac{1}{r}\alpha^{(\rho)}\frac{dW}{dr}\sin\theta \\
g &= -\left\{\beta^{(\rho)}\left[\frac{1}{r}\frac{d}{dr}(rU) + \frac{1}{r}V\right] + \xi^{(\rho)}W\right\}\cos\theta \text{ on } A_\rho
\end{aligned}
$$

The equilibrium equations for the problem $\Lambda^{(1)}$ reduce to

$$
\begin{aligned}
&r^2\frac{d^2U}{dr^2} + r\frac{dU}{dr} - (1+\kappa_1^{(\rho)})U + r(1-\kappa_1^{(\rho)})\frac{dV}{dr} - (1+\kappa_1^{(\rho)})V \\
&\qquad + \kappa_2^{(\rho)}r^2\frac{dW}{dr} = -(1-2\kappa_1^{(\rho)})r^2 \\
&\kappa_1^{(\rho)}\left(r^2\frac{d^2V}{dr^2} + r\frac{dV}{dr}\right) - (1+\kappa_1^{(\rho)})V - r(1-\kappa_1^{(\rho)})\frac{dU}{dr} - (1+\kappa_1^{(\rho)})U \qquad (7.5.21) \\
&\qquad - \kappa_2^{(\rho)}rW = (1-2\kappa_1^{(\rho)})r^2 \\
&\alpha^{(\rho)}\left[\frac{1}{r}\frac{d}{dr}\left(r\frac{dW}{dr}\right) - \frac{1}{r^2}W\right] - \frac{1}{r}\beta^{(\rho)}\left[\frac{d}{dr}(rU) + V\right] - \xi^{(\rho)}W \\
&\qquad - \beta^{(\rho)}r = 0
\end{aligned}
$$

on A_ρ, where

$$
\kappa_1^{(\rho)} = \frac{\mu^{(\rho)}}{\lambda^{(\rho)} + 2\mu^{(\rho)}}, \qquad \kappa_2^{(\rho)} = \frac{\beta^{(\rho)}}{\lambda^{(\rho)} + 2\mu^{(\rho)}} \qquad (7.5.22)
$$

A particular solution of Equation 7.5.21 is given by

$$
\overline{U} = 0, \qquad \overline{V} = a^{(\rho)}r^2, \qquad \overline{W} = b^{(\rho)}r \text{ on } A_\rho
$$

where

$$
\begin{aligned}
a^{(\rho)} &= [\beta^{(\rho)}\kappa_2^{(\rho)} + 2\xi^{(\rho)}\kappa_1^{(\rho)} - \xi^{(\rho)}]/[\xi^{(\rho)} - \beta^{(\rho)}\kappa_2^{(\rho)} - 3\kappa_1^{(\rho)}\xi^{(\rho)}] \\
b^{(\rho)} &= \beta^{(\rho)}\kappa_1^{(\rho)}/[\xi^{(\rho)} - \beta^{(\rho)}\kappa_2^{(\rho)} - 3\kappa_1^{(\rho)}\xi^{(\rho)}]
\end{aligned} \qquad (7.5.23)
$$

We denote by (U_0, V_0, W_0) the general solution of the homogeneous system 7.5.21. We now introduce the independent variable t through the relation

$$
t = \ln r \qquad (7.5.24)
$$

and denote $D = d/dt$. The first two equations of the homogeneous system 7.5.21 become

$$\begin{gathered}\left[D^2-(1+\kappa_1^{(\rho)})\right]U_0+\left[(1-\kappa_1^{(\rho)})D-(1+\kappa_1^{(\rho)})\right]V_0=e^{-t}\kappa_2^{(\rho)}DW_0\\ \left[(\kappa_1^{(\rho)}-1)D-(1+\kappa_1^{(\rho)})\right]U_0+\left[\kappa_1^{(\rho)}D^2-(1+\kappa_1^{(\rho)})\right]V_0=e^{t}\kappa_2^{(\rho)}W_0\end{gathered} \tag{7.5.25}$$

The general solution of the homogeneous system 7.5.25 is given by

$$U^*=N_0^{(\rho)}+N_1^{(\rho)}t+N_2^{(\rho)}e^{2t}+N_3^{(\rho)}e^{-2t} \tag{7.5.26}$$

$$V^*=-N_0^{(\rho)}-\frac{1-\kappa_1^{(\rho)}}{1+\kappa_1^{(\rho)}}N_1^{(\rho)}-N_1^{(\rho)}t-\frac{3-\kappa_1^{(\rho)}}{1-3\kappa_1^{(\rho)}}N_2^{(\rho)}e^{2t}+N_3^{(\rho)}e^{-2t}$$

where $N_0^{(\rho)}$ and $N_i^{(\rho)}$ are arbitrary constants. Particular solution of the system 7.5.25 can be seen to be

$$\widetilde{U}=-\frac{1}{2}\kappa_2^{(\rho)}(J_1+e^{-2t}J_2),\qquad \widetilde{V}=\frac{1}{2}\kappa_2^{(\rho)}(J_1-e^{-2t}J_2)$$

where

$$J_1(t)=\int^t e^{s}W_0(s)ds,\qquad J_2(t)=\int^t e^{3s}W_0(s)ds$$

Thus, we obtain

$$\begin{aligned}U^0&=N_0^{(\rho)}+N_1^{(\rho)}\ln r+N_2^{(\rho)}r^2+N_3^{(\rho)}r^{-2}-\frac{1}{2}\kappa_2^{(\rho)}\left(\int^r W_0(x)dx\right.\\&\quad\left.+\,r^{-2}\int^r x^2W_0(x)dx\right)\\ V^0&=-N_0^{(\rho)}-\frac{1-\kappa_1^{(\rho)}}{1+\kappa_1^{(\rho)}}N_1^{(\rho)}+N_1^{(\rho)}\ln r-\frac{3-\kappa_1^{(\rho)}}{1-3\kappa_1^{(\rho)}}N_2^{(\rho)}r^2+N_3^{(\rho)}r^{-2}\\&\quad+\frac{1}{2}\kappa_2^{(\rho)}\left(\int^r W_0(x)dx-r^{-2}\int^r x^2W_0(x)dx\right)\end{aligned} \tag{7.5.27}$$

If we substitute U_0 and V_0 from Equations 7.5.27 into the third equation of the homogeneous system 7.5.21, then we obtain the following equation

$$\begin{aligned}&r^2\frac{d^2W_0}{dr^2}+r\frac{dW_0}{dr}-(1+p_{(\rho)}^2r^2)W_0\\&\quad=\frac{2\beta^{(\rho)}\kappa_1^{(\rho)}r}{\alpha^{(\rho)}}\left(\frac{1}{1+\kappa_1^{(\rho)}}N_1^{(\rho)}-\frac{4r^2}{1-3\kappa_1^{(\rho)}}N_2^{(\rho)}\right)\end{aligned} \tag{7.5.28}$$

where

$$p_{(\rho)}=\{[\xi^{(\rho)}(\lambda^{(\rho)}+2\mu^{(\rho)})-(\beta^{(\rho)})^2]/[\alpha^{(\rho)}(\lambda^{(\rho)}+2\mu^{(\rho)})]\}^{1/2} \tag{7.5.29}$$

We note that the positive definiteness of the elastic potential implies that $p^2_{(\rho)}$ is strictly positive. Thus, we find that

$$W_0 = N_4^{(\rho)} K_1(p_{(\rho)}r) + N_5^{(\rho)} I_1(p_{(\rho)}r) - R^{(\rho)}\left(\frac{1}{1+\kappa_1^{(\rho)}} N_1^{(\rho)} r^{-1} - \frac{4r}{1-3\kappa_1^{(\rho)}} N_2^{(\rho)}\right) \quad (7.5.30)$$

where I_n and K_n are the modified Bessel functions of order n, $N_4^{(\rho)}$ and $N_5^{(\rho)}$ are arbitrary constants, and

$$R^{(\rho)} = 2\beta^{(\rho)}\kappa_1^{(\rho)}/\alpha^{(\rho)}p^2_{(\rho)} \text{ on } A_\rho$$

The solution of the system 7.5.21 can be expressed in the form

$$\begin{aligned} U &= N_0^{(\rho)} + (B_1^{(\rho)} + B_2^{(\rho)} \ln r)N_1^{(\rho)} + B_3^{(\rho)} r^2 N_2^{(\rho)} + N_3^{(\rho)} r^{-2} \\ &\quad - B_4^{(\rho)}\{[I_2(p_{(\rho)}r) + I_0(p_{(\rho)}r)]N_5^{(\rho)} - [K_2(p_{(\rho)}r) + K_0(p_{(\rho)}r)]N_4^{(\rho)}\} \\ V &= -N_0^{(\rho)} + (C_1^{(\rho)} + C_2^{(\rho)} \ln r)N_1^{(\rho)} + C_3^{(\rho)} r^2 N_2^{(\rho)} + N_3^{(\rho)} r^{-2} \\ &\quad + B_4^{(\rho)}\{[K_2(p_{(\rho)}r) - K_0(p_{(\rho)}r)]N_4^{(\rho)} - [I_2(p_{(\rho)}r) \\ &\quad - I_0(p_{(\rho)}r)]N_5^{(\rho)}\} + a^{(\rho)} r^2 \\ W &= b^{(\rho)} r + E_1^{(\rho)} r^{-1} N_1^{(\rho)} + E_2^{(\rho)} r N_2^{(\rho)} + N_4^{(\rho)} K_1(p_{(\rho)}r) + N_5^{(\rho)} I_1(p_{(\rho)}r) \end{aligned} \quad (7.5.31)$$

on A_ρ, where we have used the notations

$$\begin{aligned} B_1^{(\rho)} &= \frac{1}{4(1+\kappa_1^{(\rho)})} R^{(\rho)}\kappa_2^{(\rho)}, \qquad B_2^{(\rho)} = 1 + \frac{1}{2(1+\kappa_1^{(\rho)})} R^{(\rho)}\kappa_2^{(\rho)} \\ B_3^{(\rho)} &= 1 - \frac{3}{2(1-3\kappa_1^{(\rho)})} R^{(\rho)}\kappa_2^{(\rho)}, \qquad B_4^{(\rho)} = \frac{1}{2p_{(\rho)}}\kappa_2^{(\rho)} \\ C_1^{(\rho)} &= \frac{1}{4(1+\kappa_1^{(\rho)})} R^{(\rho)}\kappa_2^{(\rho)} - \frac{1-\kappa_1^{(\rho)}}{1+\kappa_1^{(\rho)}}, \qquad C_2^{(\rho)} = 1 - \frac{1}{2(1+\kappa_1^{(\rho)})} R^{(\rho)}\kappa_2^{(\rho)} \\ E_1^{(\rho)} &= -\frac{1}{1+\kappa_1^{(\rho)}} R^{(\rho)}, \qquad E_2^{(\rho)} = \frac{4}{1-3\kappa_1^{(\rho)}} R^{(\rho)} \text{ on } A_\rho \end{aligned} \quad (7.5.32)$$

Since U, V, and W must be bounded for $r = 0$, we conclude from Equations 7.5.31 that

$$N_1^{(2)} = 0, \qquad N_3^{(2)} = 0, \qquad N_4^{(2)} = 0 \quad (7.5.33)$$

The boundary conditions on L become

$$[t_{rr}^{(1)}]_1 = -\lambda^{(1)} r_1 \cos\theta, \qquad [t_{r\theta}^{(1)}]_1 = 0, \qquad [h_r^{(1)}]_1 = 0 \text{ on } L \quad (7.5.34)$$

The conditions on the curve Γ^* can be written in the form

$$\begin{aligned} &[U]_1 = [U]_2, \qquad [V]_1 = [V]_2, \qquad [W]_1 = [W]_2 \\ &\left[t_{rr}^{(1)}\right]_1 - \left[t_{rr}^{(1)}\right]_2 = \zeta r_2 \cos\theta \\ &\left[t_{r\theta}^{(1)}\right]_1 = \left[t_{r\theta}^{(1)}\right]_2, \qquad \left[h_r^{(1)}\right]_1 = \left[h_r^{(1)}\right]_2 \text{ on } \Gamma^* \end{aligned} \tag{7.5.35}$$

where $\zeta = \lambda^{(2)} - \lambda^{(1)}$. It follows from Equations 7.5.20 and 7.5.31 that

$$\begin{aligned} t_{rr}^{(1)} &= \left[\sigma^{(\rho)}(r) + \sum_{k=1}^{5} \Lambda_{1k}^{(\rho)}(r) N_k^{(\rho)}\right] \cos\theta \\ t_{r\theta}^{(1)} &= \left[s^{(\rho)}(r) + \sum_{k=1}^{5} \Lambda_{2k}^{(\rho)}(r) N_k^{(\rho)}\right] \sin\theta \\ h_r^{(1)} &= \left[\gamma^{(\rho)}(r) + \sum_{k=1}^{5} \Lambda_{3k}^{(\rho)}(r) N_k^{(\rho)}\right] \cos\theta \text{ on } A_\rho \end{aligned} \tag{7.5.36}$$

where

$$\begin{aligned} &\sigma^{(\rho)}(r) = (\lambda^{(\rho)} a^{(\rho)} + b^{(\rho)} \beta^{(\rho)}) r, \qquad s^{(\rho)}(r) = \mu^{(\rho)} a^{(\rho)} r, \qquad \gamma^{(\rho)}(r) = \alpha^{(\rho)} b^{(\rho)} \\ &\Lambda_{11}^{(\rho)}(r) = [(\lambda^{(\rho)} + 2\mu^{(\rho)}) B_2^{(\rho)} + \lambda^{(\rho)} D_1^{(\rho)} + \beta^{(\rho)} E_1^{(\rho)} + \lambda^{(\rho)} D_2^{(\rho)} \ln r] r^{-1} \\ &\Lambda_{12}^{(\rho)}(r) = [2(\lambda^{(\rho)} + 2\mu^{(\rho)}) B_3^{(\rho)} + \lambda^{(\rho)} D_3^{(\rho)} + \beta^{(\rho)} E_2^{(\rho)}] r \\ &\Lambda_{13}^{(\rho)}(r) = V_3^{(\rho)} = -4\mu^{(\rho)} r^{-3}, \qquad D_i^{(\rho)} = B_i^{(\rho)} + C_i^{(\rho)} \\ &\Lambda_{14}^{(\rho)}(r) = -2B_4^{(\rho)} [(\lambda^{(\rho)} + 2\mu^{(\rho)}) p_{(\rho)} K_1(p_{(\rho)} r) - \mu^{(\rho)} r^{-1} K_2(p_{(\rho)} r)] \\ &\qquad + \beta^{(\rho)} K_1(p_{(\rho)} r) \\ &\Lambda_{15}^{(\rho)}(r) = -\frac{1}{2}\{4\lambda^{(\rho)} r^{-1} I_2(p_{(\rho)} r) + (\lambda^{(\rho)} + 2\mu^{(\rho)})[3I_1(p_{(\rho)} r) \\ &\qquad + I_3(p_{(\rho)} r)]\} B_4^{(\rho)} \\ &\Lambda_{21}^{(\rho)}(r) = \mu^{(\rho)} (C_2^{(\rho)} - D_1^{(\rho)} - D_2^{(\rho)} \ln r) r^{-1} \\ &\Lambda_{22}^{(\rho)}(r) = (2C_3^{(\rho)} - D_3^{(\rho)}) \mu^{(\rho)} r, \qquad \Lambda_{24}^{(\rho)}(r) = -4B_4^{(\rho)} \mu^{(\rho)} r^{-1} K_2(p_{(\rho)} r) \\ &\Lambda_{25}^{(\rho)}(r) = 3B_4^{(\rho)} \mu^{(\rho)} r^{-1} I_2(p_{(\rho)} r), \qquad \Lambda_{31}^{(\rho)}(r) = -\alpha^{(\rho)} E_1^{(\rho)} r^{-2} \\ &\Lambda_{32}^{(\rho)}(r) = \alpha^{(\rho)} E_2^{(\rho)}, \qquad \Lambda_{34}^{(\rho)}(r) = -\alpha^{(\rho)} [p_{(\rho)} K_0(p_{(\rho)} r) \\ &\qquad + r^{-1} K_1(p_{(\rho)} r)], \qquad \Lambda_{35}^{(\rho)}(r) = \frac{1}{2}\alpha^{(\rho)} p_{(\rho)} [I_0(p_{(\rho)} r) + I_2(p_{(\rho)} r)] \end{aligned} \tag{7.5.37}$$

The conditions 7.5.34 and 7.5.35 reduce to the following system of equations for the constants $N_0^{(\rho)}, N_1^{(1)}, N_2^{(\rho)}, N_3^{(1)}, N_4^{(1)}$, and $N_5^{(\rho)}, (\rho = 1, 2)$,

$$
\begin{aligned}
&N_0^{(1)} = N_0^{(2)} + B_3^{(2)} r_2^2 N_2^{(2)} + Z_5^{(2)}(r_2) N_5^{(2)} - \sum_{k=1}^{5} Z_k^{(1)}(r_2) N_k^{(1)} \\
&\sum_{k=1}^{5} \Lambda_{1k}^{(1)}(r_1) N_k^{(1)} = -\lambda^{(1)} r_1 - \sigma^{(1)}(r_1), \qquad \sum_{k=1}^{5} \Lambda_{2k}^{(1)}(r_1) N_k^{(1)} = -s^{(1)}(r_1) \\
&\sum_{k=1}^{5} \Lambda_{3k}^{(1)}(r_1) N_k^{(1)} = -\gamma^{(1)}(r_1) \\
&\sum_{k=1}^{5} \Lambda_{1k}^{(1)}(r_2) N_k^{(1)} - \Lambda_{12}^{(2)}(r_2) N_2^{(2)} - \Lambda_{15}^{(2)} N_5^{(2)} = \zeta r_2 + \sigma^{(2)}(r_2) - \sigma^{(1)}(r_2) \\
&\sum_{k=1}^{5} \Lambda_{2k}^{(1)}(r_2) N_k^{(1)} - \Lambda_{22}^{(2)}(r_2) N_2^{(2)} - \Lambda_{25}^{(2)} N_5^{(2)} = s^{(2)}(r_2) - s^{(1)}(r_2) \qquad (7.5.38) \\
&\sum_{k=1}^{5} \Lambda_{3k}^{(1)}(r_2) N_k^{(1)} - \Lambda_{32}^{(2)}(r_2) N_2^{(2)} - \Lambda_{35}^{(2)}(r_2) N_5^{(2)} = \gamma^{(2)}(r_2) - \gamma^{(1)}(r_2) \\
&\sum_{k=1}^{5} \Gamma_{4k} N_k^{(1)} - D_3^{(2)} r_2^{(2)} N_2^{(2)} + \omega N_5^{(2)} = (a^{(2)} - a^{(1)}) r_2^2 \\
&\sum_{k=1}^{5} \Gamma_{5k} N_k^{(1)} - E_2^{(2)} r_2 N_2^{(2)} - I_1(p_{(2)} r_2) N_5^{(2)} = (b^{(2)} - b^{(1)}) r_2
\end{aligned}
$$

where

$$
\begin{aligned}
&Z_1^{(1)}(r_2) = B_1^{(1)} + B_2^{(1)} \ln r_2, Z_2^{(1)}(r_2) = B_3^{(1)} r_2^2, Z_3^{(1)}(r) = r^{-2} \\
&Z_4^{(\rho)}(r) = B_4^{(\rho)} [K_2(p_{(\rho)} r) + K_0(p_{(\rho)} r)] \\
&Z_5^{(\rho)}(r) = -B_4^{(\rho)} [I_2(p_{(\rho)} r) + I_0(p_{(\rho)} r)] \\
&\Gamma_{41} = D_1^{(1)} + D_2^{(1)} \ln r_2, \qquad \Gamma_{42} = D_3^{(1)} r_2^2, \Gamma_{43} = 2 r_2^{-2} \\
&\Gamma_{44} = Z_4^{(1)}(r_2) + Y_4^{(1)}(r_2), \qquad \Gamma_{45} = Z_5^{(1)}(r_2) + Y_5^{(1)}(r_2) \qquad (7.5.39) \\
&Y_4^{(1)}(r_2) = B_4^{(1)} [K_2(p_{(1)} r_2) - K_0(p_{(1)} r_2)] \\
&Y_5^{(\rho)}(r_2) = -B_4^{(\rho)} [I_2(p_{(\rho)} r_2) - I_0(p_{(\rho)} r_2)], \qquad \omega = Z_5^{(2)}(r_2) + Y_5^{(2)}(r_2) \\
&\Gamma_{51} = E_1^{(1)} r_2^{-1}, \qquad \Gamma_{52} = E_2^{(1)} r_2, \qquad \Gamma_{53} = 0, \qquad \Gamma_{54} = K_1(p_{(1)} r_2), \\
&\Gamma_{55} = I_1(p_{(1)} r_2)
\end{aligned}
$$

We note that the constant $N_0^{(2)}$ characterizes a rigid translation. Thus, the solution of the problem $\Lambda^{(1)}$ has the form 7.5.19 where U, V, and W are given by the relations 7.5.31. Similarly, we can prove that the solution of the problem

$\Lambda^{(2)}$ has the form

$$\begin{aligned} &u_1^{(2)} + iu_2^{(2)} = (u^{(2)} + iv^{(2)})e^{i\theta} \\ &u^{(2)} = U(r)\sin\theta, \qquad v^{(2)} = -V(r)\cos\theta, \qquad \varphi^{(2)} = W(r)\sin\theta \end{aligned} \tag{7.5.40}$$

where U, V, and W have the expressions 7.5.31.

We seek the solution of the problem $\Lambda^{(3)}$ in the form

$$u^{(3)} = rF(r), \qquad v^{(3)} = 0, \qquad \varphi^{(3)} = H(r) \tag{7.5.41}$$

where F and H are unknown functions. In this case, we obtain

$$\begin{aligned} t_{rr}^{(3)} &= (\lambda^{(\rho)} + 2\mu^{(\rho)})r\frac{dF}{dr} + 2(\lambda^{(\rho)} + \mu^{(\rho)})F + \beta^{(\rho)}H \\ t_{\theta\theta}^{(3)} &= \lambda^{(\rho)}r\frac{dF}{dr} + 2(\lambda^{(\rho)} + \mu^{(\rho)})F + \beta^{(\rho)}H \\ t_{r\theta}^{(3)} &= 0, \qquad h_r^{(3)} = \alpha^{(\rho)}\frac{dH}{dr}, \qquad h_\theta^{(3)} = 0 \\ g^{(3)} &= -\beta^{(\rho)}\left(2F + r\frac{dF}{dr}\right) - \xi^{(\rho)}H \end{aligned} \tag{7.5.42}$$

The equilibrium equations become

$$\begin{aligned} &\frac{d}{dr}\left[r^{-1}\frac{d}{dr}(r^2F)\right] = -\kappa_2^{(\rho)}\frac{dH}{dr} \\ &\alpha^{(\rho)}\frac{d}{dr}\left(r\frac{dH}{dr}\right) - \xi^{(\rho)}rH = \beta^{(\rho)}\frac{d}{dr}(r^2F) + r\beta^{(\rho)} \end{aligned} \tag{7.5.43}$$

From Equation $7.5.43_1$, we get

$$\frac{d}{dr}(r^2F) = -\kappa_2^{(\rho)}rH + d_1^{(\rho)}r \text{ on } A_\rho \tag{7.5.44}$$

where $d_1^{(\rho)}$ are arbitrary constants. Equation $7.5.43_2$ becomes

$$\frac{d^2H}{dr^2} + \frac{1}{r}\frac{dH}{dr} - p_{(\rho)}^2H = \tau^{(\rho)}(1 + d_1^{(\rho)}) \tag{7.5.45}$$

Here, we have used the notation $\tau^{(\rho)} = \beta^{(\rho)}/\alpha^{(\rho)}$. The solution of Equation 7.5.45 can be expressed in the form

$$H = Q_{21}^{(\rho)}(r)d_1^{(\rho)} + Q_{23}^{(\rho)}(r)d_3^{(\rho)} + Q_{24}^{(\rho)}(r)d_4^{(\rho)} + \Omega_{(\rho)} \text{ on } A_\rho \tag{7.5.46}$$

where $d_3^{(\rho)}$ and $d_4^{(\rho)}$ are arbitrary constants and

$$\begin{aligned} &Q_{21}^{(\rho)}(r) = -p_{(\rho)}^{-2}\tau^{(\rho)}, \qquad Q_{23}^{(\rho)}(r) = I_0(p_{(\rho)}r) \\ &Q_{24}^{(\rho)}(r) = K_0(p_{(\rho)}r), \qquad \Omega_{(\rho)} = -\frac{\tau^{(\rho)}}{p_{(\rho)}^2} \end{aligned} \tag{7.5.47}$$

From Equation 7.5.44, we obtain

$$F = \sum_{s=1}^{4} Q_{1s}^{(\rho)}(r) d_s^{(\rho)} \text{ on } A_\rho \tag{7.5.48}$$

where $d_2^{(\rho)}$ are arbitrary constants and we have denoted

$$\begin{aligned} &Q_{11}^{(\rho)}(r) = \frac{1}{2}\xi^{(\rho)}/(\alpha^{(\rho)} p_{(\rho)}^2), \qquad Q_{12}^{(\rho)}(r) = r^{-2} \\ &Q_{13}^{(\rho)}(r) = -\frac{1}{p_{(\rho)} r}\kappa_2^{(\rho)} I_1(p_{(\rho)} r), \qquad Q_{14}^{(\rho)}(r) = \frac{1}{p_{(\rho)} r}\kappa_2^{(\rho)} K_1(p_{(\rho)} r) \end{aligned} \tag{7.5.49}$$

With the help of Equation 7.5.42 we find the expressions

$$\begin{aligned} &t_{rr}^{(3)} = \sum_{s=1}^{4} Q_{3s}^{(\rho)}(r) d_s^{(\rho)} + \mathscr{T}^{(\rho)}, \qquad h_r^{(3)} = Q_{43}^{(\rho)}(r) d_3^{(\rho)} + Q_{44}^{(\rho)}(r) d_4^{(\rho)} \\ &t_{r\theta}^{(3)} = 0, \qquad h_\theta^{(3)} = 0 \end{aligned} \tag{7.5.50}$$

on A_ρ, in which we have used the notations

$$\begin{aligned} &Q_{31}^{(\rho)}(r) = [(\lambda^{(\rho)} + \mu^{(\rho)})\xi^{(\rho)} - (\beta^{(\rho)})^2]/(\alpha^{(\rho)} p_{(\rho)}^2) \\ &Q_{32}^{(\rho)}(r) = -2\mu^{(\rho)} r, \qquad Q_{33}^{(\rho)}(r) = \frac{2}{r p_{(\rho)}}\mu^{(\rho)}\kappa_2^{(\rho)} I_1(p_{(\rho)} r) \\ &Q_{34}^{(\rho)}(r) = -\frac{2}{r p_{(\rho)}}\mu^{(\rho)}\kappa_2^{(\rho)} K_1(p_{(\rho)} r) \\ &Q_{43}^{(\rho)}(r) = \alpha^{(\rho)} p_{(\rho)} I_1(p_{(\rho)} r), \qquad Q_{44}^{(\rho)}(r) = -\alpha^{(\rho)} p_{(\rho)} K_1(p_{(\rho)} r) \\ &\mathscr{T}^{(\rho)} = -\tau^{(\rho)}\beta^{(\rho)}/p_{(\rho)}^2 \end{aligned} \tag{7.5.51}$$

Since $u_\alpha^{(3)}$ and $\varphi^{(3)}$ must be bounded for $r = 0$, we have

$$d_2^{(2)} = 0, \qquad d_4^{(2)} = 0 \tag{7.5.52}$$

The boundary conditions on L reduce to

$$[t_{rr}^{(3)}]_1 = -\lambda^{(1)}, \qquad [h_r^{(3)}]_1 = 0 \tag{7.5.53}$$

The conditions 7.4.13 are satisfied if we have

$$\begin{aligned} &[F]_1 = [F]_2, \qquad [H]_1 = [H]_2 \\ &[t_{rr}^{(3)}]_1 = [t_{rr}^{(3)}]_2 + \zeta, \qquad [h_r^{(3)}]_1 = [h_r^{(3)}]_2 \end{aligned} \tag{7.5.54}$$

We introduce the notations

$$\begin{aligned} &\mathscr{H}_1 = -Q_{44}^{(1)}(r_1)/Q_{43}^{(1)}(r_1) \\ &\mathscr{H}_2 = [Q_{43}^{(1)}(r_1)Q_{44}^{(1)}(r_2) - Q_{43}^{(1)}(r_2)Q_{44}^{(1)}(r_1)]/Q_{43}^{(1)}(r_1)Q_{43}^{(2)}(r_2) \\ &\mathscr{H}_3 = Q_{13}^{(1)}(r_2)\mathscr{H}_1 + Q_{14}^{(1)}(r_2) - Q_{13}^{(2)}(r_2)\mathscr{H}_2 \end{aligned} \tag{7.5.55}$$

With the help of Equations 7.5.46, 7.5.48, and 7.5.50 the conditions 7.5.53 and 7.5.54 reduce to a system for the constants $d_s^{(1)}, (s = 1, 2, 3, 4), d_1^{(2)}$ and $d_4^{(2)}$. Thus, we obtain

$$\begin{aligned} &d_1^{(1)} = e_1, \qquad d_2^{(1)} = e_2, \qquad d_3^{(1)} = \mathscr{H}_1 e_3, \qquad d_4^{(1)} = e_3 \\ &d_1^{(2)} = \left[Q_{11}^{(1)}(r_2)e_1 + Q_{12}^{(1)}(r_2)e_2 + \mathscr{H}_3 e_3\right], \qquad d_3^{(2)} = \mathscr{H}_2 e_3 \end{aligned} \tag{7.5.56}$$

The constants e_j are given by the following equations

$$R_{ij}e_j = q_i \tag{7.5.57}$$

where

$$\begin{aligned} &R_{11} = Q_{21}^{(1)}(r_2) - Q_{21}^{(2)}(r_2)Q_{11}^{(1)}(r_1), \qquad R_{12} = -Q_{21}^{(2)}(r_2)Q_{12}^{(1)}(r_2) \\ &R_{13} = Q_{23}^{(1)}(r_2)\mathscr{H}_1 + Q_{24}^{(1)}(r_2) - Q_{21}^{(2)}(r_1)\mathscr{H}_3 - Q_{23}^{(2)}(r_2)\mathscr{H}_2 \\ &R_{21} = Q_{31}^{(1)}(r_2) - Q_{31}^{(2)}(r_2)Q_{11}^{(1)}(r_2), \qquad R_{22} = Q_{32}^{(1)}(r_2) - Q_{31}^{(2)}(r_2)Q_{12}^{(1)}(r_2) \\ &R_{23} = Q_{34}^{(1)}(r_2) + Q_{33}^{(1)}\mathscr{H}_1 - Q_{31}^{(2)}(r_2)\mathscr{H}_3 - Q_{33}^{(2)}\mathscr{H}_2 \\ &R_{31} = Q_{31}^{(1)}(r_1), \qquad R_{32} = Q_{32}^{(1)}(r_1), \qquad R_{33} = Q_{34}^{(1)}(r_1) + Q_{33}^{(1)}(r_1)\mathscr{H}_1 \\ &q_1 = \Omega_{(2)} - \Omega_{(1)}, \qquad q_2 = \mathscr{T}^{(2)} - \mathscr{T}^{(1)}, \qquad q_3 = -\lambda^{(1)} - \mathscr{T}^{(1)} \end{aligned} \tag{7.5.58}$$

Thus, the solution of the problem $\Lambda^{(3)}$ has the form 7.5.41 where F and H are given by Equations 7.5.46 and 7.5.48, respectively. With the help of relations 7.5.17, 7.5.31, 7.5.40, and 7.5.41, from Equations 7.4.18 we obtain

$$D_{11} = D_{22} = \mathscr{D}_1, \qquad D_{33} = \mathscr{D}_2, \qquad D_{12} = D_{\alpha 3} = 0 \tag{7.5.59}$$

where

$$\begin{aligned} \mathscr{D}_1 &= \frac{1}{4}\pi[(\lambda^{(2)} + 2\mu^{(2)})r_2^4 + (\lambda^{(1)} + 2\mu^{(1)})(r_1^4 - r_2^4)] + 2\pi\lambda^{(2)}\{\mathscr{T}_1(r_2) \\ &\quad + \frac{1}{2}B_2^{(2)}N_2^{(2)} - \mathscr{T}_1(0) - 2p_{(2)}[S_1(r_2) - S_1(0)]\} + 2\pi\lambda^{(1)}[S_2(r_1) \\ &\quad + \mathscr{T}_2(r_1) - S_2(r_2) - \mathscr{T}_2(r_2)] + \frac{1}{2}\pi\beta^{(2)}\{(E_2^{(2)}N_2^{(2)} + b^{(2)})r_2^4 \\ &\quad + 4N_5^{(2)}[\mathscr{L}_1(r_2) - \mathscr{L}_1(0)]\} + 2\pi\beta^{(1)}[\mathscr{L}_2(r_1) - \mathscr{L}_2(r_2)] \\ S_1(r) &= p_{(2)}r^2 I_1(p_{(1)}r) - 2rI_1(p_{(1)}r) + 2p_{(1)}^{-1}I_0(p_{(1)}r) \\ S_2(r) &= \frac{1}{2}B_2^{(1)}N_1^{(1)}r^2 + \frac{1}{2}B_3^{(1)}N_2^{(1)}r^4 - 2N_3^{(1)}\ln r \\ &\quad - 2B_4^{(1)}\{[p_{(1)}r^2K_1'(p_{(1)}r) - 2rK_1(p_{(1)}r) - 2p_{(1)}^{-1}K_0(p_{(1)}r)]N_4^{(1)} \\ &\quad - [p_{(1)}r^2I_1'(p_{(1)}r) - 2rI_1(p_{(1)}r) + 2p_{(1)}^{-1}I_0(p_{(1)}r)]N_5^{(1)}\} \end{aligned}$$

$$\mathscr{T}_1(r) = \frac{1}{4}(D_3^{(2)}N_2^{(2)} + a^{(2)})r^4 + 2B_4^{(2)}p_{(2)}^{-2}[2I_0(p_{(2)}r) - rp_{(2)}I_1(p_{(2)}r)]$$

$$\begin{aligned}\mathscr{T}_2(r) &= \frac{1}{2}D_1^{(1)}r^2 + \frac{1}{4}(2\ln r - 1)r^2D_2^{(1)} + \frac{1}{4}(N_2^{(1)}D_3^{(1)} + a^{(1)})r^4 \\ &\quad + 2N_3^{(1)}\ln r - 2B_4^{(1)}\{p_{(1)}^{-2}[p_{(1)}rK_1(p_{(1)}r) + 2K_0(p_{(1)}r)]N_4^{(1)} \\ &\quad + p_1^{-2}[p_{(1)}rI_1(p_{(1)}r) - 2I_0(p_{(1)}r)]N_5^{(1)}\}\end{aligned}$$

$$\mathscr{L}_1(r) = p_{(2)}^{-2}[p_{(2)}r^2I_0(p_{(2)}r) - 2rI_1(p_{(2)}r)]$$

$$\begin{aligned}\mathscr{L}_2(r) &= \frac{1}{2}E_1^{(1)}N_1^{(1)}r^2 + \frac{1}{4}(E_2^{(1)}N_2^{(1)} + b^{(1)})r^4 \\ &\quad - N_4^{(1)}p_{(1)}^{-2}[p_{(1)}r^2K_0(p_{(1)}r) + 2rK_1(p_{(1)}r)] \\ &\quad + N_5^{(1)}p_{(1)}^{-2}[p_{(1)}r^2I_0(p_{(1)}r) - 2rI_1(p_{(1)}r)]\end{aligned}$$

$$\begin{aligned}\mathscr{D}_2 &= \pi r_2^2k_{(2)} + \pi(r_1^2 - r_2^2)k_{(1)} + 2\pi r_2p_{(2)}^{-1}c_{(2)}d_3^{(2)}I_1(p_{(2)}r_2) \\ &\quad + 2\pi p_{(1)}^{-1}c_{(1)}d_3^{(1)}[r_1I_1(p_{(1)}r_1) - r_2I_1(p_{(1)}r_2)] \\ &\quad + 2\pi p_{(1)}^{-1}c_{(1)}d_4^{(1)}[r_2K_1(p_{(1)}r_2) - r_1K_1(p_{(1)}r_2)]\end{aligned}$$

$$c_{(\rho)} = \beta^{(\rho)} - \lambda^{(\rho)}\kappa_2^{(\rho)}$$

$$k_{(\rho)} = \lambda^{(\rho)} + 2\mu^{(\rho)} + (\lambda^{(\rho)} + c_{(\rho)}Q_{21}^{(\rho)})d_1^{(\rho)} + \Omega_{(\rho)}$$

In the case of extension and bending problem, from Equations 7.4.17 we obtain

$$d_1 = \frac{1}{\mathscr{D}_1}M_2, \qquad d_2 = -\frac{1}{\mathscr{D}_1}M_1, \qquad d_3 = -\frac{1}{\mathscr{D}_2}F_3 \tag{7.5.60}$$

The solution of the problem has the form 7.4.15 where $\{u_\alpha^{(s)}, \varphi^{(s)}\}$, $(s = 1, 2, 3)$, are the solutions of the problems $\Lambda^{(s)}$ and d_s are given by Equations 7.5.60.

7.6 Exercises

7.6.1 Investigate the bending of a homogeneous and isotropic porous elastic cylinder with rectangular cross-section.

7.6.2 Study the extension of a homogeneous and isotropic porous cylinder with elliptical cross-section.

7.6.3 Investigate the elastic deformation of a homogeneous and isotropic porous cylinder subjected to a constant temperature variation.

7.6.4 An inhomogeneous and isotropic porous elastic cylinder occupies the domain $B = \{x : x_1^2 + x_2^2 < a^2, 0 < x_3 < h\}, (a > 0)$. Study the extension problem when the constitutive coefficients are given by

$$\lambda = \lambda_0 e^{-\kappa r}, \quad \mu = \mu_0 e^{-\kappa r}, \quad \alpha = \alpha_0 e^{-\kappa r}, \quad \beta = \beta_0 e^{-\kappa r},$$
$$\xi = \xi_0 e^{-\kappa r}, \quad \kappa > 0$$

where λ_0, μ_0, α_0, β_0, ξ_0, and κ are prescribed constants, and $r = (x_1^2 + x_2^2)^{1/2}$.

7.6.5 Study extension and bending of a porous elastic right cylinder made from a homogeneous and orthotropic material.

7.6.6 Investigate the deformation of an orthotropic porous elastic cylinder subjected to a temperature distribution which is independent of the axial coordinate.

7.6.7 Study the flexure of a nonhomogeneous and isotropic porous elastic cylinder when the constitutive coefficients are independent of the axial coordinate.

Answers to Selected Problems

1.11.1 In the boundary-value problem 1.3.43, and 1.3.44, the curve Γ is defined by the equation

$$\frac{x_1^2}{a^2} + \frac{x_2^2}{b^2} = 1 \tag{A.1}$$

If we take the stress function of Prandtl in the form

$$\Psi = C\left(\frac{x_1^2}{a^2} + \frac{x_2^2}{b^2} - 1\right), \quad (x_1, x_2) \in \Sigma_1 \tag{A.2}$$

where C is an unknown constant, then the function Ψ satisfies the condition 1.3.44. The stress function satisfies Equation 1.3.43 if

$$C = -\frac{a^2 b^2}{a^2 + b^2} \tag{A.3}$$

By using the relations

$$\int_{\Sigma_1} x_1^2 da = \frac{1}{4}\pi a^3 b, \qquad \int_{\Sigma_1} x_2^2 da = \frac{1}{4}\pi a b^3, \qquad \int_{\Sigma_1} da = \pi ab$$

from Equations 1.3.46 and A.2, we obtain the torsional rigidity,

$$D = \frac{\pi a^3 b^3 \mu}{a^2 + b^2} \tag{A.4}$$

It follows from Equations 1.3.31 and A.4 that

$$\tau = -\frac{M_3(a^2 + b^2)}{\pi \mu a^3 b^3} \tag{A.5}$$

In view of Equations 1.3.36 and 1.3.42,

$$\varphi_{,1} = \Psi_{,2} + x_2 = \left(\frac{2C}{b^2} + 1\right) x_2 = \frac{b^2 - a^2}{a^2 + b^2} x_2$$

$$\varphi_{,2} = -\Psi_{,1} - x_1 = -\left(\frac{2C}{a^2} + 1\right) x_1 = \frac{b^2 - a^2}{a^2 + b^2} x_1$$

so that

$$\varphi = \frac{b^2 - a^2}{a^2 + b^2} x_1 x_2, \quad (x_1, x_2) \in \Sigma_1 \tag{A.6}$$

Thus, the solution of the problem has the form 1.3.23 where φ is defined in A.6 and the constant τ has the value A.5. From Equations A.2 and 1.3.45, we get

$$t_{13} = 2\mu\tau C x_2 b^{-2}, \qquad t_{23} = -2\mu\tau C x_1 a^{-2} \tag{A.7}$$

The stress vector acting on any cross section is $\mathbf{t}_3 = t_{13}\mathbf{e}_1 + t_{23}\mathbf{e}_2$. The magnitude of the vector $\mathbf{t}_3$ at the point $M(\overline{x}_1, \overline{x}_2)$ on Γ is

$$P = \left(t_{13}^2 + t_{23}^2\right)^{1/2} = 2\mu|\tau C|\left(\frac{\overline{x}_1^2}{a^4} + \frac{\overline{x}_2^2}{b^4}\right)^{1/2} \tag{A.8}$$

The tangent line at the point M on Γ is given by

$$\frac{\overline{x}_1}{a^2}x_1 + \frac{\overline{x}_2}{b^2}x_2 - 1 = 0$$

The distance between origin and this tangent line is

$$d = \left(\frac{\overline{x}_1^2}{a^4} + \frac{\overline{x}_2^2}{b^4}\right)^{-1/2}$$

Thus, by Equation A.8, we get

$$P = \frac{2}{d}\mu|\tau C|$$

The maximum and minimum of P are given by

$$P_{\max} = \frac{2a^2b}{a^2+b^2}\mu|\tau|, \qquad P_{\min} = \frac{2ab^2}{a^2+b^2}\mu|\tau|$$

respectively. The maximum stress occurs at the extremities of the minor axis of the ellipse.

1.11.2 If we introduce the notations $r = (x_1^2 + x_2^2)^{1/2}$, $x_1 = r\cos\theta, x_2 = r\sin\theta$, then the circles C_1 and C_2 can be described by

$$(C_1): r = 2a\sin\theta, \qquad (C_2): r = b$$

We seek the stress function Ψ in the form

$$\Psi = \alpha(r - 2a\sin\theta)\left(r - \frac{b^2}{r}\right) \tag{A.9}$$

where α is an unknown constant. Clearly, the function Ψ satisfies the condition 1.3.44. By using the relation

$$\Delta\Psi = \frac{1}{r}\frac{\partial}{\partial r}\left(r\frac{\partial\Psi}{\partial r}\right) + \frac{1}{r^2}\frac{\partial^2\Psi}{\partial\theta^2}$$

from Equations 1.3.43 and A.9, we obtain

$$\alpha = -\frac{1}{2}$$

We can express Ψ in the form

$$\Psi = -\frac{1}{2(x_1^2 + x_2^2)}\left(x_1^2 + x_2^2 - 2ax_2\right)\left(x_1^2 + x_2^2 - b^2\right)$$

It follows from Equations 1.3.46 and A.9 that

$$D = \mu \int_{\arcsin(b/2a)}^{\pi-\arcsin(b/2a)} \left[\int_b^{2a\sin\theta} (2ar\sin\theta - r^2 + b^2 - 2ab^2r^{-1}\sin\theta)dr \right] d\theta$$

$$= \mu \left\{ \left(a^4 - 2a^2b^2 - \frac{1}{2}b^4\right)\left[\frac{1}{2}\pi - \arcsin(b/2a)\right] + ab\left(\frac{7}{4}b^2 + \frac{1}{2}a^2\right)\left[1 - (b/2a)^2\right]^{1/2} \right\}$$

The torsion function is given by

$$\varphi = a\left(1 + \frac{b^2}{r^2}\right)x_1 \text{ on } \Sigma_1$$

It is not difficult to show that the maximum shearing stress is at the point $(0, b) \in \Gamma$.

1.11.3 We suppose that the loading applied on the end located at $x_3 = 0$ is statically equivalent to the force $\mathbf{F} = F_1\mathbf{e}_1$ and the moment $\mathbf{M} = \mathbf{0}$. In this case, the solution of the flexure problem is given by Equations 1.3.70 where A_1 is given by Equation 1.3.59 and the function Φ satisfies the boundary-value problem 1.3.66 and 1.3.67. We assume that the curve Γ is defined by Equation A.1. In this case, from Equation 1.3.59, we obtain

$$A_1 = -\frac{4}{\pi a^3 bE}F_1 \tag{A.10}$$

Let us study the boundary-value problem 1.3.66 and 1.3.67. We introduce the function Λ on Σ_1 by

$$\Phi = -A_1\left[\Lambda(x_1, x_2) - \frac{1}{3}\left(1 + \frac{1}{2}\nu\right)\left(x_1^3 - 3x_1x_2^2\right) + \frac{1}{3}(1+\nu)x_1^3\right], (x_1, x_2) \in \Sigma_1 \tag{A.11}$$

From Equations 1.3.66, 1.3.67, and A.11, we find that Λ satisfies the equation

$$\Delta\Lambda = 0 \text{ on } \Sigma_1 \tag{A.12}$$

and the boundary condition

$$\frac{\partial \Lambda}{\partial n} = -\left[\frac{1}{2}\nu x_1^2 + \left(1 - \frac{1}{2}\nu\right) x_2^2\right] n_1 - (2+\nu) x_1 x_2 n_2 \text{ on } \Gamma \tag{A.13}$$

We note that in Equation A.13, we have

$$n_1 = \frac{x_1}{a^2} K, \qquad n_2 = \frac{x_2}{b^2} K, \qquad K^{-1} = \left(\frac{x_1^2}{a^4} + \frac{x_2^2}{b^4}\right)^{1/2} \tag{A.14}$$

With the aid of Equation A.14, the condition A.13 reduces to

$$b^2 x_1 \Lambda_{,1} + a^2 x_2 \Lambda_{,2} = -\left[\frac{1}{2}\nu x_1^2 + \left(1 - \frac{1}{2}\nu\right) x_2^2\right] b^2 x_1 - (2+\nu) a^2 x_1 x_2^2 \text{ on } \Gamma \tag{A.15}$$

We seek the solution of the boundary-value problem A.12 and A.15 in the form

$$\Lambda = \gamma_1 x_1 + \gamma_2 (x_1^3 - 3x_1 x_2^2), \quad (x_1, x_2) \in \Sigma_1 \tag{A.16}$$

where γ_1 and γ_2 are arbitrary constants. It is easy to see that Equation A.12 is satisfied. From Equations A.15 and A.16, we obtain the condition

$$\begin{aligned} &\left[\gamma_1 + 3\gamma_2 (x_1^2 - x_2^2)\right] b^2 - 6\gamma_2 a^2 x_2^2 \\ &\quad = -\left[\frac{1}{2}\nu x_1^2 + \left(1 - \frac{1}{2}\nu\right) x_1^2\right] b^2 - (2+\nu) a^2 x_1^2 \text{ on } \Gamma \end{aligned} \tag{A.17}$$

Since on Γ we have

$$x_1^2 = a^2 - \frac{a^2}{b^2} x_2^2$$

the condition A.17 implies that

$$\begin{aligned} \gamma_1 + 3\gamma_2 a^2 &= -\tfrac{1}{2}\nu a^2 \\ 3(3a^2 + b^2)\gamma_2 &= (2 + \tfrac{1}{2}\nu) a^2 + (1 - \tfrac{1}{2}\nu) b^2 \end{aligned}$$

Thus, we find

$$\begin{aligned} \gamma_1 &= -\frac{a^2}{3a^2 + b^2}\left[2(1+\nu)a^2 + b^2\right] \\ \gamma_2 &= \frac{1}{3(3a^2 + b^2)}\left[\left(2 + \frac{1}{2}\nu\right) a^2 + \left(1 - \frac{1}{2}\nu\right) b^2\right] \end{aligned} \tag{A.18}$$

From Equations A.11 and A.16, we obtain

$$\Phi = -A_1\left[\gamma_1 x_1 + \left(\gamma_2 - \frac{1}{3} - \frac{1}{6}\nu\right)(x_1^3 - 3x_1 x_2^2) + \frac{1}{3}(1+\nu) x_1^3\right], \quad (x_1, x_2) \in \Sigma_1 \tag{A.19}$$

With the help of the relations

$$\begin{aligned}\Phi_{,1} &= -A_1\left[\gamma_1 + \left(3\gamma_2 - 1 - \frac{1}{2}\nu\right)(x_1^2 - x_2^2) + (1+\nu)x_1^2\right] \\ \Phi_{,2} &= -A_1(2+\nu-6\gamma_2)x_1x_2 \\ \int_{\Sigma_1} x_1^2 x_2 da &= \int_{\Sigma_1} x_1^3 da = \int_{\Sigma_1} x_1 x_2^2 da = \int_{\Sigma_1} x_2^3 da = 0\end{aligned} \tag{A.20}$$

from Equation 1.3.69, we find that $M^* = 0$. In view of Equation 1.3.68, we get $\tau = 0$. Thus, from Equations 1.3.70, we obtain

$$u_1 = -\frac{1}{2}A_1\left[\frac{1}{3}x_3^2 + \nu(x_1^2 - x_2^2)\right]x_3, \qquad u_2 = -A_1\nu x_1 x_2 x_3$$

$$u_3 = \frac{1}{2}A_1\left[x_3^2 + \nu\left(\frac{1}{3}x_1^2 + x_2^2\right)\right]x_1 + \Phi, \quad (x_1, x_2, x_3) \in B$$

The stress tensor is given by

$$\begin{aligned}t_{\alpha\beta} &= 0, \qquad t_{33} = A_1 E x_1 x_3 \\ t_{23} &= -2\mu A_1(a^{-2}\gamma_1 + 1 + \nu)x_1x_2 \\ t_{31} &= -\mu A_1 \gamma_1 a^{-2}\left[a^2 - x_1^2 + \left(\frac{a^2}{\gamma_1} + 1\right)x_2^2\right]\end{aligned}$$

1.11.4 In this case, we have $f_\alpha = 0$. We seek the solution of Equations 1.5.8 in the form

$$u_\alpha = x_\alpha \varphi(r), \qquad r = \left(x_1^2 + x_2^2\right)^{1/2} \tag{A.21}$$

where φ is an unknown function. By Equation A.21,

$$\begin{aligned}u_{\alpha,\beta} &= \delta_{\alpha\beta}\varphi + x_\alpha x_\beta r^{-1}\varphi', \qquad \varphi' = \frac{d\varphi}{dr} \\ u_{\rho,\rho} &= 2\varphi + r\varphi' = \frac{1}{r}(r^2\varphi)', \qquad u_{\rho,\rho\alpha} = (3r^{-1}\varphi' + \varphi'')x_\alpha \\ u_{\alpha,\beta\gamma} &= (\delta_{\alpha\beta}x_\gamma + \delta_{\alpha\gamma}x_\beta + \delta_{\beta\gamma}x_\alpha)r^{-1}\varphi' \\ &\quad - r^{-3}x_\alpha x_\beta x_\gamma \varphi' + x_\alpha x_\beta x_\gamma r^{-2}\varphi'' \\ \Delta u_\alpha &= (3r^{-1}\varphi' + \varphi'')x_\alpha\end{aligned} \tag{A.22}$$

In view of Equations A.12, the equilibrium equations 1.5.8 reduce to

$$(\lambda + 2\mu)x_\alpha(\varphi'' + 3r^{-1}\varphi') = 0$$

Using the relations 1.1.5, we see that these equations are satisfied if and only if

$$\varphi'' + 3r^{-1}\varphi' = 0 \tag{A.23}$$

Equation A.23 can be written in the form

$$(r^3\varphi')' = 0$$

so that

$$\varphi(r) = C_1 r^{-2} + C_2 \tag{A.24}$$

where C_α are arbitrary constants. From the constitutive equations 1.5.7 and A.22, we obtain

$$t_{\alpha\beta} = 2(\lambda+\mu)\varphi\delta_{\alpha\beta} + (\lambda r\delta_{\alpha\beta} + 2\mu x_\alpha x_\beta r^{-1})\varphi' \tag{A.25}$$

We have the boundary conditions

$$\mathbf{t} = -p_1\mathbf{n} \text{ for } r = R_1, \qquad \mathbf{t} = -p_2\mathbf{n} \text{ for } r = R_2 \tag{A.26}$$

where p_α are prescribed constants. Since

$$n_\beta = -\frac{1}{R_1}x_\beta \text{ on } r = R_1, \qquad n_\beta = \frac{1}{R_2}x_\beta \text{ on } r = R_2$$

the conditions A.26 reduce to

$$\begin{aligned} t_{\beta\alpha}x_\beta &= -p_1 x_\alpha \quad \text{for } r = R_1 \\ t_{\beta\alpha}x_\beta &= -p_2 x_\alpha \quad \text{for } r = R_2 \end{aligned} \tag{A.27}$$

In view of Equations A.24 and A.25, we obtain

$$t_{\beta\alpha}x_\beta = 2x_\alpha[(\lambda+\mu)C_2 - \mu r^{-2}C_1]$$

Thus, the boundary conditions A.27 reduce to

$$\begin{aligned} (\lambda+\mu)C_2 - \mu R_1^{-2}C_1 &= -p_1/2 \\ (\lambda+\mu)C_2 - \mu R_2^{-2}C_1 &= -p_2/2 \end{aligned}$$

We find that

$$C_1 = \frac{R_1^2R_2^2(p_2-p_1)}{2\mu(R_1^2-R_2^2)}, \qquad C_2 = \frac{p_2R_2^2 - p_1R_1^2}{2(\lambda+\mu)(R_1^2-R_2^2)}$$

The components of the stress tensor are given by

$$\begin{aligned} t_{\alpha\beta} &= 2\mu C_1 r^{-2}(\delta_{\alpha\beta} - 2x_\alpha x_\beta r^{-2}) + 2(\lambda+\mu)C_2\delta_{\alpha\beta} \\ t_{33} &= 2\lambda C_2, \qquad t_{\alpha 3} = 0 \end{aligned}$$

1.11.5 Clearly, we have

$$\begin{aligned} \chi_{,11} &= \frac{3}{4a}qx_2 - \frac{1}{4a^3}qx_2^3 + \frac{1}{2}q \\ \chi_{,22} &= \frac{1}{2a^3}qx_2^3 - \frac{3}{4a^3}qx_1^2x_2 - \frac{3}{2a^3}\left(m + \frac{1}{5}qa^2 - \frac{1}{2}qh^2\right)x_2 \\ \chi_{,12} &= \frac{3}{4a^3}qx_1x_2^2 - \frac{3}{4a}qx_1 \end{aligned}$$

$$\chi_{,1111} = 0, \qquad \chi_{,2222} = \frac{3}{a^3}qx_2, \qquad \chi_{,1212} = -\frac{3}{2a^3}qx_2$$

so that $\Delta\Delta\chi = 0$, and χ is a valid Airy stress function. The stresses $t_{\alpha\beta}$ are given by

$$t_{11} = \chi_{,22}, \qquad t_{22} = \chi_{,11}, \qquad t_{12} = -\chi_{,12}$$

The stress vector on the face $x_2 = a$ is $\mathbf{t} = t_1\mathbf{e}_1 + t_2\mathbf{e}_2$ where

$$t_1 = t_{21} = 0, \qquad t_2 = t_{22} = q$$

so that $\mathbf{t} = q\mathbf{e}_2$. The stress vector on the face $x_2 = -a$ is zero. The stress vector on $x_1 = h$ is given by

$$\frac{1}{2a^3}\left(qx_2^3 - 3mx_2 - \frac{3}{5}qa^2x_2\right)\mathbf{e}_1 - \frac{3}{4}qh\left(1 - \frac{x_2^2}{a^2}\right)\mathbf{e}_2$$

The resultant force acting on $x_1 = h$ is $\mathbf{R} = -qh\mathbf{e}_2$. The resultant moment about O of the traction acting on $x_1 = h$ is $\mathbf{M} = (m - qh^2)\mathbf{e}_3$. The stress vector on the face $x_1 = -h$ is

$$-\frac{1}{2a^3}\left(qx_2^3 - 3mx_2 - \frac{3}{5}qa^2x_2\right)\mathbf{e}_1 - \frac{3}{4a}qh\left(1 - \frac{x_2^2}{a^2}\right)\mathbf{e}_2$$

so that the resultant force acting on $x_1 = -h$ is $-qh\mathbf{e}_2$. If $q < 0$, then the stresses are those of a beam which is supported at both sides, and has a uniform distributed load.

1.11.6 We assume that Σ_1 is defined by $\Sigma_1 = \{x : x_1^2 + x_2^2 < a^2, x_3 = 0\}$, where a is a positive constant. We suppose that on the boundary Γ of the domain Σ_1 are imposed the conditions 1.5.6 where $\widetilde{t}_\alpha$ are piecewise regular functions. Since $f_\alpha = 0$, from Equations 1.5.17, we conclude that $\widetilde{t}_\alpha$ must satisfy the relations

$$\int_\Gamma \widetilde{t}_\alpha ds = 0, \qquad \int_\Gamma (x_1\widetilde{t}_2 - x_2\widetilde{t}_1)da = 0 \tag{A.28}$$

First, we assume that on Γ acts a uniform pressure, so that

$$\widetilde{t}_1\mathbf{e}_1 + \widetilde{t}_2\mathbf{e}_2 = -P\mathbf{n}$$

where P is a given constant, and $\mathbf{n}$ is the outward unit normal of the circle Γ. Thus, we can write

$$\widetilde{t}_1 = -\frac{1}{a}Px_1, \qquad \widetilde{t}_2 = -\frac{1}{a}Px_2, \quad (x_1, x_2) \in \Gamma$$

In this case, the representation 1.3.38 of the curve Γ reduces to

$$x_1 = a\cos\frac{s}{a}, \qquad x_2 = a\sin\frac{s}{a}, \quad s \in [0, 2\pi a] \tag{A.29}$$

The functions $\widetilde{t}_\alpha$ can be expressed as

$$\widetilde{t}_1 = -P\cos\frac{s}{a}, \qquad \widetilde{t}_2 = -P\sin\frac{s}{a}, \quad s \in [0, 2\pi a] \tag{A.30}$$

It is easy to verify that the conditions A.28 are satisfied. From Equation 1.5.73, we obtain

$$\sigma = a\left(\cos\frac{s}{a} + i\sin\frac{s}{a}\right), \quad s \in [0, 2\pi a] \tag{A.31}$$

In view of Equations A.30 and A.31, the relation 1.5.75 becomes

$$T(\sigma) = -iP\int_0^s \left(\cos\frac{s}{a} + i\sin\frac{s}{a}\right)ds = -Pa\left(\cos\frac{s}{a} + i\sin\frac{s}{a}\right) = -P\sigma. \tag{A.32}$$

The function 1.5.77 that maps Σ_1 on the region $|\zeta| \leq 1$ is

$$z = \vartheta(\zeta) = a\zeta \tag{A.33}$$

In this case,

$$\frac{\vartheta(\zeta)}{\overline{\vartheta}'(\overline{\zeta})} = \zeta$$

It follows from Equations A.32 and A.33 that

$$N_1(\eta) = T[\vartheta(\eta)] = -Pa\eta \tag{A.34}$$

Thus, the boundary condition $1.5.79_1$ becomes

$$\Omega_1(\eta) + \eta\overline{\Omega}_1'(\overline{\eta}) + \overline{\omega}_1(\overline{\eta}) = -Pa\eta, \quad |\eta| = 1 \tag{A.35}$$

The functions $\Omega_1(\zeta)$ and $\omega_1(\zeta)$ have the representations

$$\Omega_1(\zeta) = \sum_{n=1}^{\infty} a_n\zeta^n, \qquad \omega_1(\zeta) = \sum_{n=0}^{\infty} b_n\zeta^n, \quad |\zeta| \leq 1 \tag{A.36}$$

In view of the arbitrariness of complex potentials discussed in Section 1.5, we have taken $\Omega_1(0) = 0$. Let us impose the condition $\Im m[\Omega_1'(0)/\vartheta'(0)] = 0$. We find that

$$a_1 - \overline{a}_1 = 0 \tag{A.37}$$

The insertion of the functions A.36 in Equation A.35 yields

$$a_1 + \overline{a}_1 = -Pa, \qquad a_n = 0, \quad n \geq 2$$
$$b_n = 0, \quad n = 0, 1, 2, 3, \ldots$$

Thus, we obtain

$$\Omega_1(\zeta) = -\frac{1}{2}Pa\zeta, \qquad \omega_1(\zeta) = 0$$

so that

$$\Omega(z) = -\frac{1}{2}Pz, \qquad \omega(z) = 0 \tag{A.38}$$

By Equations 1.5.45 and A.38, we find that

$$u_\alpha = -\frac{\kappa}{4\mu}Px_\alpha$$

We now consider the general case, when the traction $\widetilde{t}_\alpha \mathbf{e}_\alpha$ is not a uniform pressure. We assume that the function $N_1(\eta)$ can be represented in the form

$$N_1(\eta) = \sum_{-\infty}^{\infty} A_k \eta^k \tag{A.39}$$

where A_k are prescribed complex coefficients. In this case, from Equation 1.5.79$_1$ and A.36, we obtain the following system for the coefficients a_k and b_k,

$$\begin{aligned} &a_1 + \overline{a}_1 = A_1, \qquad a_k = A_k, \quad k \geq 2 \\ &b_k = \overline{A}_{-k} - (k+2)A_{k+2}, \quad k = 0, 1, 2, \ldots \end{aligned} \tag{A.40}$$

By Equations A.37 and A.40, we get

$$a_1 = \frac{1}{2}A_1$$

Thus, we conclude that

$$\Omega_1(\zeta) = \frac{1}{2}A_1\zeta + \sum_{k=2}^{\infty} A_k\zeta^k$$

$$\omega_1(\zeta) = \sum_{k=0}^{\infty} [\overline{A}_{-k} - (k+2)A_{k+2}]\zeta^k$$

Clearly,

$$\Omega(z) = \Omega_1\left(\frac{1}{a}z\right), \qquad \omega(z) = \omega_1\left(\frac{1}{a}z\right)$$

Remark. The first equation from Equations A.40 requires A_1 to be real. Let us show that this fact is the consequence of the vanishing of the resultant moment of forces applied to the boundary,

$$\int_\Gamma (x_1\widetilde{t}_2 - x_2\widetilde{t}_1)ds = 0 \tag{A.41}$$

It follows from Equation 1.5.75 that

$$\widetilde{t}_1 = \frac{dT_2}{ds}, \qquad \widetilde{t}_2 = -\frac{dT_1}{ds}$$

Thus, the condition A.41 becomes

$$\begin{aligned}\int_\Gamma (x_1\widetilde{t}_2 - x_2\widetilde{t}_1)ds &= -\int_\Gamma x_1 dT_1 + x_2 dT_2 = \int_\Gamma T_1 dx_1 + T_2 dx_2 \\ &= a\int_0^{2\pi}(F_2\cos\theta - F_1\sin\theta)d\theta = a\Im m\left\{\int_0^{2\pi} Te^{-i\theta}d\theta\right\} = 0\end{aligned} \tag{A.42}$$

In view of the relations A.33 and A.39, we find that the condition A.42 reduces to $\Im m A_1 = 0$.

1.11.7 We consider the ring

$$R_1 < |z| < R_2$$

formed by a pair of concentric circles L_α of radii $R_\alpha, (\alpha = 1, 2)$. We assume that on the curves L_α are prescribed constant pressures. In this case, the boundary conditions are A.26. From Equations 1.5.75 and A.26, we obtain

$$T(\sigma) = -p_1\sigma \text{ on } L_1, \qquad T(\sigma) = -p_2\sigma \text{ on } L_2$$

Thus, the boundary conditions A.26 can be written in the form

$$\begin{aligned}\Omega(\sigma) + \sigma\overline{\Omega}'(\overline{\sigma}) + \overline{\omega}(\overline{\sigma}) &= -p_1\sigma + d_1 \text{ on } L_1 \\ \Omega(\sigma) + \sigma\overline{\Omega}'(\overline{\sigma}) + \overline{\omega}(\overline{\sigma}) &= -p_2\sigma \text{ on } L_2\end{aligned} \tag{A.43}$$

where d_1 is an arbitrary constant. In the above relation, we have chosen $\omega(0)$ to have no arbitrary constant on L_2. By Equations 1.5.61 and A.20, we obtain

$$X_1 + iY_1 = -p_1\int_{L_1}(n_1 + in_2)ds = 0, \qquad X_2 + iY_2 = 0$$

Thus, from Equations 1.5.64, we find that $\Omega(z) = \Omega_0(z), \omega(z) = \omega_0(z)$, where Ω_0 and ω_0 are analytic and single-valued functions on Σ_1,

$$\Omega(z) = \sum_{-\infty}^{\infty} a_k z^k, \qquad \omega(z) = \sum_{-\infty}^{\infty} b_k z^k, \quad R_1 < |z| < |R_2| \tag{A.44}$$

Clearly, we can take $\Omega(0) = 0$ and $\Im m\Omega'(0) = 0$, so that

$$a_0 = 0, \qquad a_1 - \overline{a}_1 = 0 \tag{A.45}$$

From Equations A.43, A.44, and A.45, we obtain the following system for the unknown coefficients

$$\begin{aligned}
&2R_2^2\bar{a}_2 + \bar{b}_0 = 0\\
&2a_1R_2 + \bar{b}_{-1}R_2^{-1} = -p_2R_2\\
&a_kR_2^k + (2-k)\bar{a}_{2-k}R_2^{2-k} + \bar{b}_{-k}R_2^{-k} = 0\\
&2R_1^2\bar{a}_2 + \bar{b}_0 = d_1\\
&2a_1R_1 + \bar{b}_{-1}R_1^{-1} = -p_1R_1\\
&a_kR_1^k + (2-k)\bar{a}_{2-k}R_1^{2-k} + \bar{b}_{-k}R_1^{-k} = 0, \quad k \neq 0,1
\end{aligned} \tag{A.46}$$

From the above system, we find that the nonvanishing coefficients are

$$a_1 = \frac{1}{2(R_1^2 - R_2^2)}(p_2R_2^2 - p_1R_1^2), \qquad b_{-1} = \frac{1}{R_1^2 - R_2^2}(p_1 - p_2)R_1^2R_2^2$$

We note that from Equation A.46, we obtain $d_1 = 0$. Thus, we have

$$\Omega(z) = a_1 z, \qquad \omega(z) = b_{-1}\frac{1}{z}, \quad R_1 < |z| < R_2$$

The relation 1.5.45 implies that

$$2\mu(u_1 + iu_2) = \left[(\kappa - 1)a_1 - b_{-1}(x_1^2 + x_2^2)\right]z$$

2.7.1 We assume that the temperature field is a polynomial of degree r in the axial coordinate, namely

$$T = \sum_{k=0}^{r} T_k x_3^k$$

where T_k are independent of x_3.

In this case, the problem (Z) considered in Section 2.6 reduces to the Almansi problem. We denote by (Z_n), $(n = 0, 1, 2, \ldots, r)$, the problem (Z) corresponding to the temperature field $T = T_n x_3^n$. Clearly, if we know the solution of the problem (Z_n), for any n, then we can establish a solution of the problem (Z) when the temperature has the form 2.6.19. The solution of the problem (Z_0) has been established previously. We must derive the solution $\mathbf{u}''$ of the problem (Z_{n+1}) when the solution of the problem (Z_n) is known. As the solution of the problem (Z_n) is known for any T_n, it follows that we know the solution $\mathbf{u}^*$ of the problem corresponding to the temperature field $T = T_{n+1}x_3^n$. According to Theorem 2.4.4, the vector field $\mathbf{u}''$ is given by Equation 2.4.16, where $\mathbf{w}'$ is characterized by Equations 2.4.13 and 2.4.14, and $\widehat{a}$ is determined by Equations 2.4.15.

If the temperature field is linear in x_3,

$$T = T_0 + T_1 x_3$$

where T_0 and T_1 are prescribed constants, then a simple calculation shows that

$$u_\alpha = \frac{\beta}{3\lambda + 2\mu} x_\alpha (T_0 + T_1 x_3)$$

$$u_3 = \frac{\beta}{3\lambda + 2\mu} \left[\left(T_0 + \frac{1}{2} T_1 x_3 \right) x_3 - \frac{1}{2} T_1 x_\rho x_\rho \right]$$

3.9.1 It follows from Equations 3.6.41 and 3.6.42 that the torsion function φ satisfies the boundary-value problem

$$\Delta\varphi = 0 \text{ on } A_\rho$$

$$[\varphi]_1 = [\varphi]_2, \quad \mu^{(1)} \left[\frac{\partial \varphi}{\partial n^0} \right]_1 = \mu^{(2)} \left[\frac{\partial \varphi}{\partial n^0} \right]_2 + (\mu^{(1)} - \mu^{(2)}) \varepsilon_{\alpha\beta} x_\beta n^0_\alpha \text{ on } \Gamma_0$$

$$\left[\frac{\partial \varphi}{\partial n} \right]_\rho = \varepsilon_{\alpha\beta} x_\beta n_\alpha \text{ on } \Gamma_\rho$$

If we introduce the functions Λ_1 and Λ_2 by

$$\varphi = \Lambda_1 - x_1 x_2 \text{ on } A_1, \qquad \varphi = \Lambda_2 - x_1 x_2 \text{ on } A_2$$

then we conclude that Λ_1 and Λ_2 satisfy the equations

$$\Delta\Lambda_1 = 0 \text{ on } A_1, \qquad \Delta\Lambda_2 = 0 \text{ on } A_2 \tag{A.47}$$

and the conditions

$$\begin{aligned} &\Lambda_1 = \Lambda_2, \qquad \mu^{(1)} \Lambda_{1,1} - \mu^{(2)} \Lambda_{2,1} = 2(\mu^{(1)} - \mu^{(2)}) x_2, \\ &(x_1 = 0, -\beta \le x_2 \le \beta) \end{aligned} \tag{A.48}$$

$$\begin{aligned} \Lambda_{1,1} &= 2x_2, \qquad (x_1 = -\alpha_1, -\beta \le x_2 \le \beta), \\ \Lambda_{2,1} &= 2x_2, \qquad (x_1 = \alpha_2, -\beta \le x_2 \le \beta) \end{aligned} \tag{A.49}$$

$$\begin{aligned} \Lambda_{1,2} &= 0, \qquad (x_2 = \pm\beta, -\alpha_1 \le x_1 \le 0), \\ \Lambda_{2,2} &= 0, \qquad (x_2 = \pm\beta, \ 0 \le x_1 \le \alpha_2) \end{aligned} \tag{A.50}$$

We seek the functions Λ_1 and Λ_2 in the form of the series

$$\begin{aligned} \Lambda_1 &= \sum_{n=0}^{\infty} \left(A^{(1)}_{2n+1} \operatorname{sh} m x_1 + B_{2n+1} \operatorname{ch} m x_1 \right) \sin m x_2 \\ \Lambda_2 &= \sum_{n=0}^{\infty} \left(A^{(2)}_{2n+1} \operatorname{sh} m x_1 + B_{2n+1} \operatorname{ch} m x_1 \right) \sin m x_2 \end{aligned} \tag{A.51}$$

where

$$m = \frac{1}{2\beta}(2n+1)\pi \tag{A.52}$$

Clearly, each term of Equations A.51 is a harmonic function. In view of Equations A.51 and A.52, we see that the conditions A.50 are satisfied. It is easy to verify that $\Lambda_1 = \Lambda_2$ on Γ_0. Let us study the remaining conditions from Equations A.48 and A.49. The function $f(x_2) = 2x_2$, $x_2 \in (-\beta, \beta)$, can be represented in the form

$$2x_2 = \sum_{n=0}^{\infty} mC_{2n+1} \sin mx_2, \qquad -\beta < x_2 < \beta \tag{A.53}$$

where

$$C_{2n+1} = (-1)^n \frac{32\beta^2}{(2n+1)^3\pi^3} \tag{A.54}$$

mC_{2n+1} are the Fourier coefficients for the function defined on $(-2\beta, 2\beta)$ by $F(x_2) = 2x_2$, $x_2 \in (-\beta, \beta)$; $F(x_2) = 4\beta - 2x_2$, $x_2 \in (\beta, 2\beta)$; $F(x_2) = -4\beta - 2x_2$, $x_2 \in (-\beta, -2\beta)$. In view of Equation A.53, we find that the conditions A.49 reduce to

$$\begin{aligned} A^{(1)}_{2n+1}\mathrm{ch}\, m\alpha_1 - B_{2n+1}\mathrm{sh}\, m\alpha_1 &= C_{2n+1} \\ A^{(2)}_{2n+1}\mathrm{ch}\, m\alpha_2 + B_{2n+1}\mathrm{sh}\, m\alpha_2 &= C_{2n+1} \end{aligned} \tag{A.55}$$

The condition A.48$_2$ is satisfied if we have

$$\mu^{(1)} A^{(1)}_{2n+1} - \mu^{(2)} A^{(2)}_{2n+1} = 2(\mu^{(1)} - \mu^{(2)})C_{2n+1} \tag{A.56}$$

From Equations A.55 and A.56, we can determine the coefficients $A^{(1)}_{2n+1}$, $A^{(2)}_{2n+1}$, and B_{2n+1}. The functions Λ_1 and Λ_2 can be expressed as

$$\begin{aligned} \Lambda_1 &= \sum_{n=0}^{\infty} \frac{1}{d_m} C_{2n+1}\{[\mu^{(2)} + (\mu^{(1)} - \mu^{(2)})\mathrm{ch}\, m\alpha_2]\mathrm{ch}\, m(x_1 + \alpha_1) \\ &\quad + \mu^{(2)}\mathrm{sh}\, m\alpha_2\, \mathrm{sh}\, mx_1 - \mu^{(1)}\mathrm{ch}\, m\alpha_2\, \mathrm{ch}\, mx_1\} \sin mx_2 \\ \Lambda_2 &= \sum_{n=0}^{\infty} \frac{1}{d_m} C_{2n+1}\{[(\mu^{(1)} - \mu^{(2)})\mathrm{ch}\, m\alpha_1 - \mu^{(1)}]\mathrm{ch}\, m(x_1 - \alpha_2) \\ &\quad + \mu^{(1)}\, \mathrm{sh}\, m\alpha_1\, \mathrm{sh}\, mx_1 + \mu^{(2)}\, \mathrm{ch}\, m\alpha_1\, \mathrm{ch}\, mx_1\} \sin mx_2 \end{aligned} \tag{A.57}$$

where

$$d_m = \mu^{(1)}\, \mathrm{ch}\, m\alpha_2\, \mathrm{sh}\, m\alpha_1 + \mu^{(2)}\, \mathrm{ch}\, m\alpha_1\, \mathrm{sh}\, m\alpha_2$$

The above series are absolutely and uniformly convergent, so that the term-by-term differentiation is justified. In view of Equation 3.6.48, we find that the torsional rigidity is given by

$$\begin{aligned} D_0 &= \mu^{(1)} \int_{A_1} (2x_2^2 + x_1\Lambda_{1,2} - x_2\Lambda_{1,1})da \\ &\quad + \mu^{(2)} \int_{A_2} (2x_2^2 + x_1\Lambda_{2,2} - x_2\Lambda_{2,1})da \end{aligned} \tag{A.58}$$

It follows from Equations A.57 and A.58 that

$$D_0 = \frac{8}{3}\left(\mu^{(1)}\alpha_1 + \mu^{(2)}\alpha_2\right)\beta^3 + \left(\frac{4}{\pi}\right)^5 b^4 \sum_{n=0}^{\infty} \frac{1}{(2n+1)^5 d_m} \{(\mu^{(1)})^2 \mathrm{ch}\, m\alpha_2$$
$$+ (\mu^{(2)})^2 \mathrm{ch}\, m\alpha_1 - [(\mu^{(1)})^2 + (\mu^{(2)})^2]\mathrm{ch}\, m\alpha_1 \mathrm{ch}\, m\alpha_2$$
$$- \mu^{(1)}\mu^{(2)}[\mathrm{ch}\, m\alpha_1 + \mathrm{ch}\, m\alpha_2 - \mathrm{ch}\, m(\alpha_1 - \alpha_2) - 1]\}$$

The constant τ is given by

$$\tau = -\frac{1}{D_0} M_3$$

The torsion problem for a homogeneous and isotropic elastic cylinder with rectangular cross section has been solved by Saint-Venant (see, for example, Ref. 211, Sections 221–225).

3.9.2 Let us study the plane strain problems $\mathcal{P}_*^{(k)}$, defined in Section 3.7, when L and Γ are two concentric circles. The results have been established by Muskhelishvili [241]. We assume that the domain A_1^* is bounded by two concentric circles of radius R_1 and R_2, where $R_1 < R_2$. The domain A_2^* is bounded by the circle of radius R_1. We suppose that the domains A_1^* and A_2^* are occupied by two different homogeneous and isotropic elastic materials. Let us study the problem $\mathcal{P}_*^{(1)}$, where the function f has the form 3.7.13. We try to satisfy the relations 3.7.11 and 3.7.12 assuming that

$$\begin{aligned} \Omega(z) &= m_1 z^2, \qquad \omega(z) = 0 \text{ on } A_2^* \\ \Omega(z) &= m_2 z^2, \qquad \omega(z) = m_3 z^{-1} + m_4 \text{ on } A_1^* \end{aligned} \tag{A.59}$$

where m_s, $(s = 1, 2, 3, 4)$, are real constants. We note that m_4 has no influence on the stress tensor. By Equations 3.7.10 and 3.7.12, we find that

$$\begin{aligned} m_1 &= \frac{1}{2}\rho(R_2^4 - R_1^4), \qquad m_2 = -\rho R_1^4, \qquad m_3 = \rho R_1^4 R_2^4 \\ \rho &= \frac{\nu^{(1)} - \nu^{(2)}}{2\left[\beta^{(2)} R_2^4 + \alpha^{(2)} R_1^4 + \alpha^{(1)}(R_2^4 - R_1^4)\right]} \end{aligned} \tag{A.60}$$

It follows from Equations 1.5.38 and A.59 that

$$\gamma_{\eta\eta}^{*(1)} = \frac{4m_1 x_1}{\lambda^{(2)} + \mu^{(2)}} \text{ on } A_2^*, \qquad \gamma_{\eta\eta}^{*(1)} = \frac{4m_2 x_1}{\lambda^{(1)} + \mu^{(1)}} \text{ on } A_1^* \tag{A.61}$$

In the case of the plane strain problem $\mathcal{P}_*^{(2)}$, we have $f = f^{(2)}$, where

$$f^{(2)} = h_1^{(2)} + i h_2^{(2)} = -\frac{1}{2} i(\nu^{(1)} - \nu^{(2)}) z^2 \tag{A.62}$$

We seek the solution of the problem $\mathcal{P}_*^{(2)}$ in the form

$$\begin{aligned} \Omega(z) &= i m_1^* z^2, \qquad \omega(z) = 0 \text{ on } A_2^* \\ \Omega(z) &= i m_2^* z^2, \qquad \omega(z) = i m_3^* z^{-1} + i m_4^* \text{ on } A_1^* \end{aligned} \tag{A.63}$$

where m_k^*, $(k = 1, 2, 3, 4)$, are real constants. From Equations A.63, 3.7.10, and 3.7.12, we obtain

$$m_1^* = -m_1, \qquad m_2^* = -m_2, \qquad m_3^* = m_3 \tag{A.64}$$

where m_k are given in Equation A.60. By Equations 1.5.38 and A.63, we get

$$\gamma_{\eta\eta}^{*(2)} = \frac{4m_1x_2}{\lambda^{(2)} + \mu^{(2)}} \text{ on } A_2^*, \qquad \gamma_{\eta\eta}^{*(2)} = \frac{4m_2x_2}{\lambda^{(1)} + \mu^{(1)}} \text{ on } A_1^* \tag{A.65}$$

In the case of the problem $\mathcal{P}_*^{(3)}$, we take $f = f^{(3)}$, where

$$f^{(3)} = h_1^{(3)} + ih_2^{(3)} = (\nu^{(1)} - \nu^{(2)})z \tag{A.66}$$

We seek the solution in the form

$$\begin{aligned} \Omega(z) &= m_1^0 z, \qquad \omega(z) = 0 \text{ on } A_2^* \\ \Omega(z) &= m_2^0 z, \qquad \omega(z) = m_3^0 z^{-1} \text{ on } A_1^* \end{aligned} \tag{A.67}$$

where m_k^0 are real constants. The conditions 3.7.11 and 3.7.12 are satisfied if

$$\begin{aligned} &2m_2^0 z + m_3^0 \bar{z}^{-1} = 0; \text{on } |z| = R_2, \qquad 2m_1^0 z = 2m_2^0 z + m_3^0 \bar{z}^{-1} \text{ on } |z| = R_1 \\ &(\alpha^{(1)} - \beta^{(1)})m_1^0 z = (\alpha^{(2)} - \beta^{(2)})m_2^0 z - \beta^{(2)} m_3^0 \bar{z}^{-1} \\ &\quad + (\nu^{(1)} - \nu^{(2)})z \text{ on } |z| = R_1 \end{aligned} \tag{A.68}$$

It follows from Equation A.68 that

$$\begin{aligned} &2m_2^0 R_2^2 + m_3^0 = 0, \qquad 2m_1^0 R_1^2 = 2m_2^0 R_1^2 + m_3^0 \\ &(\alpha^{(1)} - \beta^{(1)})m_1^0 R_1^2 = (\alpha^{(2)} - \beta^{(2)})m_2^0 R_1^2 - \beta^{(2)} m_3^0 + (\nu^{(1)} - \nu^{(2)})R_1^2 \end{aligned}$$

The constants m_k^0 are given by

$$\begin{aligned} &m_1^0 = \sigma(R_2^2 - R_1^2), \qquad m_2^0 = -\sigma R_1^2, \qquad m_3^0 = 2\sigma R_1^2 R_2^2 \\ &\sigma = \frac{\nu^{(1)} - \nu^{(2)}}{2\beta^{(2)} R_2^2 + (\alpha^{(2)} - \beta^{(2)})R_1^2 + (\alpha^{(1)} - \beta^{(1)})(R_2^2 - R_1^2)} \end{aligned} \tag{A.69}$$

In view of Equations A.67 and 1.5.38, we obtain

$$\gamma_{\eta\eta}^{*(3)} = \frac{2m_1^0}{(\lambda^{(2)} + \mu^{(2)})} \text{ on } A_2^*, \qquad \gamma_{\eta\eta}^{*(3)} = \frac{2m_2^0}{(\lambda^{(1)} + \mu^{(1)})} \text{ on } A_1^* \tag{A.70}$$

3.9.3 We use the solution 3.7.14 to solve the extension and bending problem for a cylinder composed by two different homogeneous and isotropic elastic materials. We assume that the curves L and Γ are concentric circles of radius

R_1 and R_2, respectively. The solutions of the plane strain problems $\mathcal{P}_*^{(k)}$ associated to the considered cylinder are given by Equations A.59, A.63, and A.67. It follows from Equations 3.7.16, A.61, A.65, and A.70 that

$$\mathscr{I}_{11} = E^{(1)} \int\limits_{A_1^*} x_1^2 da + E^{(2)} \int\limits_{A_2^*} x_1^2 da = \frac{\pi}{4} \left[E^{(1)} (R_2^4 - R_1^4) + E^{(2)} R_1^4 \right]$$

$$\mathscr{I}_{22} = \mathscr{I}_{11}, \qquad \mathscr{I}_{12} = \mathscr{I}_{3\alpha} = \mathscr{I}_{\alpha 3} = 0, \qquad \mathscr{I}_{33} = \pi \left[E^{(1)} (R_2^2 - R_1^2) + E^{(2)} R_1^2 \right]$$

$$\mathscr{K}_{11} = \lambda^{(1)} \int\limits_{A_1^*} x_1 \gamma_{\eta\eta}^{*(1)} da + \lambda^{(2)} \int\limits_{A_2^*} x_1 \gamma_{\eta\eta}^{*(1)} da$$

$$= 2\pi \left[m_2 \nu^{(1)} (R_2^4 - R_1^4) + m_1 \nu^{(2)} R_1^4 \right]$$

$$\mathscr{K}_{22} = \mathscr{K}_{11}, \qquad \mathscr{K}_{12} = \mathscr{K}_{\alpha 3} = \mathscr{K}_{3\alpha} = 0$$

$$\mathscr{K}_{33} = 4\pi m_2^0 \left[\nu^{(1)} (R_2^2 - R_1^2) + \nu^{(2)} R_2^2 \right] \tag{A.71}$$

Thus, with the aid of Equations 3.7.15, we obtain

$$L_{11} = L_{22} = \mathscr{I}_{11} + \mathscr{K}_{11}, \qquad L_{12} = L_{\alpha 3} = 0, \qquad L_{33} = \mathscr{I}_{33} + \mathscr{K}_{33}$$

so that the system 3.6.18 implies that

$$d_1 = \frac{M_2}{\mathscr{I}_{11} + \mathscr{K}_{11}}, \qquad d_2 = -\frac{M_1}{\mathscr{I}_{11} + \mathscr{K}_{11}}, \qquad d_3 = -\frac{F_3}{\mathscr{I}_{33} + \mathscr{K}_{33}} \tag{A.72}$$

where $\mathscr{I}_{11}, \mathscr{I}_{33}, \mathscr{K}_{11}$, and $\mathscr{K}_{33}$ are defined in Equations A.71. The solution of the problem has the form 3.7.14 where the functions $v_\alpha^{*(k)}$ are defined by Equations A.59, A.63, and A.67, and the constants d_k are given by Equations A.72.

3.9.4 Let us consider a continuum body that occupies the region $B = \{x : R_2^2 < x_1^2 + x_2^2 < R_1^2, 0 < x_3 < h\}$, $R_1 > 0, R_2 > 0$. The cross section Σ_1 is the assembly of the regions A_1^* and A_2^*, $\Sigma_1 = A_1^* \cup A_2^*$, where $A_1^* = \{x : R_0^2 < x_1^2 + x_2^2 < R_1^2, x_3 = 0\}$, $A_2^* = \{x : R_2^2 < x_1^2 + x_2^2 < R_0^2, x_3 = 0\}$, $R_0 > 0$. The domains $B_1 = \{x : (x_1, x_2) \in A_1^*, 0 < x_3 < h\}$ and $B_2 = \{x : (x_1, x_2) \in A_2^*, 0 < x_3 < h\}$ are occupied by different homogeneous and isotropic elastic materials. We denote by $\lambda^{(\rho)}$ and $\mu^{(\rho)}$ the Lamé moduli of the material which occupies the cylinder B_ρ. We assume that cylinder B is in equilibrium in the absence of the body forces. Let us investigate the plane strain of B, parallel to the x_1, x_2-plane, when the lateral boundaries are subjected to constant pressures. It follows from Equations 3.6.2 and 3.6.4 that the displacement vector field satisfies the equations

$$\mu^{(\rho)} \Delta u_\alpha + (\lambda^{(\rho)} + \mu^{(\rho)}) u_{\beta,\beta\alpha} = 0 \text{ on } A_\rho^*, \quad (\rho = 1, 2) \tag{A.73}$$

We introduce the notation $r = (x_1^2 + x_2^2)^{1/2}$. The conditions (3.6.5) on the surface of separation reduce to

$$[u_\alpha]_1 = [u_\alpha]_2, \qquad [t_{\alpha\beta}]_1 x_\beta = [t_{\alpha\beta}]_2 x_\beta \text{ on } r = R_0 \tag{A.74}$$

The conditions on the lateral surface become

$$[t_{\beta\alpha}]_1 x_\beta = -p_1 x_\alpha \text{ for } r = R_1, \qquad [t_{\beta\alpha}]_2 x_\beta = -p_2 x_\alpha \text{ for } r = R_2 \tag{A.75}$$

We seek the solution in the form

$$u_\alpha = x_\alpha G^{(\rho)}(r) \text{ on } A_\rho^* \tag{A.76}$$

where $G^{(1)}$ and $G^{(2)}$ are unknown functions of r. With the help of Equations A.22 and A.24, we find that Equations A.73 are satisfied if and only if

$$G^{(1)} = C_1 r^{-2} + C_2 \text{ on } A_1^*, \qquad G^{(2)} = C_3 r^{-2} + C_4 \text{ on } A_2^* \tag{A.77}$$

where C_k, $(k = 1, 2, 3, 4)$, are arbitrary constants. Using the constitutive equations 3.6.3 and A.77, we obtain

$$[t_{\beta\alpha}]_1 x_\beta = 2x_\alpha \left[(\lambda^{(1)} + \mu^{(1)})C_2 - \mu^{(1)} C_1 r^{-2}\right]$$
$$[t_{\beta\alpha}]_2 x_\beta = 2x_\alpha \left[(\lambda^{(2)} + \mu^{(2)})C_4 - \mu^{(2)} C_3 r^{-2}\right]$$

Thus, the conditions A.74 and A.75 reduce to

$$\begin{aligned}
&C_1 R_0^{-2} + C_2 = C_3 R_0^{-2} + C_4\\
&(\lambda^{(1)} + \mu^{(1)})C_2 - \mu^{(1)} C_1 R_0^{-2} = (\lambda^{(2)} + \mu^{(2)})C_4 - \mu^{(2)} C_3 R_0^{-2}\\
&\mu^{(1)} R_1^{-2} C_1 - (\lambda^{(1)} + \mu^{(1)})C_2 = \frac{1}{2}p_1\\
&\mu^{(2)} R_2^{-2} C_3 - (\lambda^{(2)} + \mu^{(2)})C_4 = \frac{1}{2}p_2
\end{aligned} \tag{A.78}$$

The determinant of the system A.78 is

$$\begin{aligned}
\delta_1 = {} & \mu^{(1)}(\lambda^{(1)} + \mu^{(1)})\left(\frac{1}{R_0^2} - \frac{1}{R_1^2}\right)\left(\frac{\lambda^{(2)} + \mu^{(2)}}{R_0^2} + \frac{\mu^{(2)}}{R_2^2}\right)\\
& + \mu^{(2)}(\lambda^{(2)} + \mu^{(2)})\left(\frac{1}{R_2^2} - \frac{1}{R_0^2}\right)\left(\frac{\mu^{(1)}}{R_1^2} + \frac{\lambda^{(1)} + \mu^{(1)}}{R_0^2}\right)
\end{aligned}$$

In view of the relations

$$\mu^{(\rho)} > 0, \qquad \lambda^{(\rho)} + \mu^{(\rho)} > 0, \qquad R_2^{-2} > R_0^{-2} > R_1^{-2}$$

we conclude that δ_1 is different from zero. Thus, the system A.78 uniquely determines the constants C_s, $(s = 1, 2, 3, 4)$.

4.12.1 The solution of the torsion problem is given by

$$u_1 = -\tau x_2 x_3, \qquad u_2 = \tau x_1 x_3, \qquad u_3 = \tau\varphi,$$

where φ is the solution of the boundary-value problem 4.8.16 and 4.8.17, and τ is given by $\tau = -M_3/D_0$. The torsional rigidity D_0 is defined in Equation 4.8.24. The corresponding stress tensor has the components

$$t_{\alpha\beta} = 0, \qquad t_{33} = 0, \qquad t_{23} = A_{44}\tau(\varphi_{,2} + x_1), \qquad t_{13} = A_{55}\tau(\varphi_{,1} - x_2) \tag{A.79}$$

We introduce the function F by

$$A_{44}(\varphi_{,2} + x_1) = -F_{,1}, \qquad A_{55}(\varphi_{,1} - x_2) = F_{,2} \tag{A.80}$$

If F is given, then the integrability condition to determine the function φ is

$$\frac{1}{A_{44}}F_{,11} + \frac{1}{A_{55}}F_{,22} = -2 \text{ on } \Sigma_1 \tag{A.81}$$

The boundary condition 4.8.17 takes the form

$$F_{,2}n_1 - F_{,1}n_2 = 0 \text{ on } \Gamma \tag{A.82}$$

Since Σ_1 is simply-connected, from Equations A.82 and 1.3.39, we obtain the following boundary condition for the function F,

$$F = 0 \text{ on } \Gamma \tag{A.83}$$

Thus, the function F is the solution of the boundary-value problem A.81 and A.83. By Equations A.79 and A.80, we get

$$t_{23} = -\tau F_{,1}, \qquad t_{13} = \tau F_{,2} \tag{A.84}$$

As in Section 1.3, we can prove that the torsional rigidity can be written as

$$D_0 = 2\int_{\Sigma_1} F da \tag{A.85}$$

In our case, the curve Γ is defined by the equation

$$\frac{x_1^2}{a^2} + \frac{x_2^2}{b^2} = 1 \tag{A.86}$$

We seek the function F in the form

$$F = C_1\left(\frac{x_1^2}{a^2} + \frac{x_2^2}{b^2} - 1\right) \tag{A.87}$$

where C_1 is an unknown constant. Clearly, F satisfies the boundary condition A.83. From Equation A.81, we find that

$$C_1 = -\frac{A_{44}A_{55}a^2b^2}{a^2A_{44} + b^2A_{55}} \tag{A.88}$$

It follows from Equations A.85 and A.87 that the torsional rigidity is

$$D_0 = \frac{\pi A_{44} A_{55} a^3 b^3}{a^2 A_{44} + b^2 A_{55}} \tag{A.89}$$

In view of Equations A.80 and A.87, we get

$$\varphi_{,1} = x_2 \left(1 + \frac{2C_1}{b^2 A_{55}}\right) = H x_2, \qquad \varphi_{,2} = H x_1 \tag{A.90}$$

where

$$H = \frac{A_{55} b^2 - A_{44} a^2}{A_{55} b^2 + A_{44} a^2}$$

Thus we find that the torsion function is given by

$$\varphi = H x_1 x_2, \quad (x_1, x_2) \in \Sigma_1$$

We note that for a circular cylinder ($b = a$), we obtain

$$\varphi = \frac{A_{55} - A_{44}}{A_{55} + A_{44}} x_1 x_2, \quad (x_1, x_2) \in \Sigma_1 \tag{A.91}$$

In the case of isotropic circular cylinders, we find that $\varphi = 0$.

4.12.2 The solution of the flexure problem for a homogeneous and orthotropic cylinder has the form 4.8.41, where the constants b_1, b_2, and b_3 are given by Equations 4.2.28, the function φ is the solution of the boundary-value problem 4.8.16 and 4.8.17, the function ψ is characterized by Equations 4.8.35 and 4.8.36, and the constant c_4 is defined by Equation 4.8.39. We assume that $\mathbf{F} = F_1 \mathbf{e}_1$. We suppose that Σ_1 is bounded by the curve Γ, defined by Equation A.86. In this case, from Equations 1.4.9 and 1.7.14, we obtain

$$A = \int_{\Sigma_1} da = \pi ab, \qquad I_{11} = \int_{\Sigma_1} x_1^2 da = \frac{1}{4}\pi a^3 b$$

$$I_{22} = \frac{1}{4}\pi ab^3, \qquad I_{12} = 0, \qquad x_1^0 = x_2^0 = 0$$

so that the system 4.8.28 implies that

$$b_1 = -\frac{4}{\pi a^3 b E_0} F_1, \qquad b_2 = 0, \qquad b_3 = 0 \tag{A.92}$$

Let us study the boundary-value problem 4.8.35 and 4.8.36. In view of Equations A.92, this problem reduces to the equation

$$A_{55} \psi_{,11} + A_{44} \psi_{,22} = q b_1 x_1 \text{ on } \Sigma_1 \tag{A.93}$$

and the boundary condition

$$A_{55}\psi_{,1} n_1 + A_{44}\psi_{,2} n_2 = \frac{1}{2} A_{55} b_1 (\nu_1 x_1^2 - \nu_2 x_2^2) n_1 + A_{44} b_1 \nu_2 x_1 x_2 n_2 \text{ on } \Gamma \tag{A.94}$$

where

$$q = A_{55}\nu_1 + A_{44}\nu_2 - E_0 \tag{A.95}$$

For the curve A.86, the components n_α are given by Equations A.14. Thus, the condition A.94 can be written as

$$b^2 A_{55}\psi_{,1}x_1 + a^2 A_{44}\psi_{,2}x_2 = b_1\left\{\frac{1}{2}A_{55}b^2(\nu_1 x_1^2 - \nu_2 x_2^2) + A_{44}a^2\nu_2 x_2^2\right\}x_1 \text{ on } \Gamma \tag{A.96}$$

We seek the function ψ in the form

$$\psi = b_1(\alpha_1 x_1^3 + \alpha_2 x_1 x_2^2 + \alpha_3 x_1) \text{ on } \Sigma_1 \tag{A.97}$$

where α_1, α_2, and α_3 are unknown constants. Thus, Equation A.93 becomes

$$6A_{55}\alpha_1 + 2A_{44}\alpha_2 = q \tag{A.98}$$

If we take into account Equation A.97 and the relation

$$x_2^2 = b^2 - \frac{b^2}{a^2}x_1^2 \text{ on } \Gamma$$

we find that the boundary condition A.96 reduces to

$$3A_{55}\alpha_1 - \left(2A_{44} + \frac{b^2}{a^2}A_{55}\right)\alpha_2 = \frac{1}{2}A_{55}\left(\nu_1 + \frac{b^2}{a^2}\nu_2\right) - A_{44}\nu_2 \tag{A.99}$$

$$(2a^2 A_{44} + b^2 A_{55})\alpha_2 + A_{55}\alpha_3 = A_{44}\nu_2 a^2 - \frac{1}{2}A_{55}b^2\nu_2 \tag{A.100}$$

Thus, the constants α_1 and α_2 must satisfy Equations A.98 and A.99. The determinant of the system A.98 and A.99 is

$$\delta = -6A_{55}\left(3A_{44} + \frac{b^2}{a^2}A_{55}\right)$$

In view of Equation 4.8.3, we conclude that $\delta \neq 0$ so that the system A.98 and A.99 uniquely determines the constants α_1 and α_2. From Equation A.100, we can obtain the constant α_3. In view of Equations 4.8.9, A.20, and A.97, we find that

$$\int_{\Sigma_1}(A_{55}x_2\psi_{,1} - A_{44}x_1\psi_{,2} + A_{55}x_2 b_1 w_1^{(1)} - A_{44}x_1 b_1 w_2^{(1)})da = 0$$

so that the relation 4.8.39 reduces to $c_4 = 0$. Thus, the solution of the flexure problem for an elliptic cylinder is

$$u_1 = -\frac{1}{2}b_1\left(\frac{1}{3}x_3^2 + \nu_1 x_1^2 - \nu_2 x_2^2\right)x_3, \qquad u_2 = -b_1\nu_2 x_1 x_2 x_3$$

$$u_3 = b_1\left(\frac{1}{2}x_3^2 + \alpha_1 x_1^2 + \alpha_2 x_2^2 + \alpha_3\right)x_1, \qquad (x_1, x_2, x_3) \in B$$

We find that the stress tensor is given by

$$t_{\alpha\beta} = 0, \quad t_{33} = E_0 b_1 x_1 x_3, \quad t_{23} = A_{44} b_1 (2\alpha_2 - \nu_2) x_1 x_2$$
$$t_{13} = A_{55} b_1 \left[\left(3\alpha_1 - \frac{1}{2}\nu_1\right) x_1^2 + \left(\alpha_2 + \frac{1}{2}\nu_2\right) x_2^2 + \alpha_3 \right]$$

4.12.3 We assume that the domain Σ_1 has the form $\Sigma_1 = A_1 \cup A_2$, where $A_1 = \{x : -\alpha_1 < x_1 < 0, \ -\beta < x_2 < \beta, \ x_3 = 0\}$, $A_2 = \{x : 0 < x_1 < \alpha_2, -\beta < x_2 < \beta, x_3 = 0\}$, $(\alpha_1 > 0, \alpha_2 > 0, \beta > 0)$. We define $B_\rho = \{x : (x_1, x_2) \in A_\rho, 0 < x_3 < h\}$ and assume that B_1 and B_2 are occupied by different homogeneous and orthotropic elastic materials.

We assume that the loading applied at the end Σ_1 is equivalent to the force $\mathbf{F} = \mathbf{0}$ and the moment $\mathbf{M} = M_3 \mathbf{e}_3$. In Section 4.11, we have seen that the solution of the torsion problem is given by

$$u_\alpha = \tau \varepsilon_{\beta\alpha} x_\beta x_3, \qquad u_3 = \tau\varphi$$

where the constant τ is defined by

$$D^* \tau = -M_3 \tag{A.101}$$

and the function φ satisfies the boundary-value problem 4.11.27 and 4.11.28. The constant D^* is given by Equation 4.11.34. Let us study the boundary-value problem 4.11.27 and 4.11.28. We introduce the functions G_α by

$$G_1 = \varphi + x_1 x_2 \text{ on } A_1, \qquad G_2 = \varphi + x_1 x_2 \text{ on } A_2 \tag{A.102}$$

From Equations 4.11.27 and 4.11.28, we find that the functions G_α satisfy the equations

$$A_{55}^{(1)} G_{1,11} + A_{44}^{(1)} G_{1,22} = 0 \text{ on } A_1, \qquad A_{55}^{(2)} G_{2,11} + A_{44}^{(2)} G_{2,22} = 0 \text{ on } A_2 \tag{A.103}$$

and the conditions

$$G_1 = G_2, \qquad A_{55}^{(1)} G_{1,1} - A_{55}^{(2)} G_{2,1} = 2\left(A_{55}^{(1)} - A_{55}^{(2)}\right) x_2 \quad (x_1 = 0, -\beta \le x_2 \le \beta) \tag{A.104}$$

$$\begin{aligned} G_{1,1} &= 2x_2, \qquad (x_1 = -\alpha_1, -\beta \le x_2 < \beta) \\ G_{2,1} &= 2x_2, \qquad (x_1 = \alpha_2, -\beta \le x_2 \le \beta) \end{aligned} \tag{A.105}$$

$$\begin{aligned} G_{1,2} &= 0, \qquad (x_2 = \pm\beta, -\alpha_1 \le x_2 < 0) \\ G_{2,2} &= 0, \qquad (x_2 = \pm\beta, 0 \le x_1 \le \alpha_2) \end{aligned} \tag{A.106}$$

We seek the functions G_1 and G_2 in the form

$$G_\alpha = \sum_{n=0}^{\infty} H_{2n+1}^{(\alpha)}(x_1) \sin m x_2 \tag{A.107}$$

where

$$m = \frac{1}{2\beta}(2n+1)\pi \tag{A.108}$$

Clearly, the functions G_α satisfy the conditions A.106. By Equations A.107 and A.103, we obtain

$$A_{55}^{(\alpha)} \frac{d^2}{dx_1^2} H_{2n+1}^{(\alpha)} - A_{44}^{(\alpha)} m^2 H_{2n+1}^{(\alpha)} = 0, \quad (\alpha = 1, 2)$$

so that

$$H_{2n+1}^{(\alpha)} = A_{2n+1}^{(\alpha)} \text{sh}\, \mu_\alpha m x_1 + B_{2n+1}^{(\alpha)} \text{ch}\, \mu_\alpha m x_1$$

where $A_{2n+1}^{(\alpha)}$ and $B_{2n+1}^{(\rho)}$ are arbitrary constants and

$$\mu_\alpha^2 = A_{44}^{(\alpha)} / A_{55}^{(\alpha)}, \quad (\alpha = 1, 2)$$

From the condition A.105$_1$, we obtain

$$B_{2n+1}^{(1)} = B_{2n+1}^{(2)}$$

Thus the functions A.107 have the form

$$\begin{aligned} G_1 &= \sum_{n=1}^{\infty} (A_{2n+1}^{(1)}\ \text{sh}\, \mu_1 m x_1 + B_{2n+1} \text{ch}\, \mu_1 m x_1) \sin m x_2 \\ G_2 &= \sum_{n=1}^{\infty} (A_{2n+1}^{(2)} \text{sh}\, \mu_2 m x_1 + B_{2n+1} \text{ch}\, \mu_2 m x_1) \sin m x_2 \end{aligned} \tag{A.109}$$

We can write

$$2x_2 = \sum_{n=0}^{\infty} m C_{2n+1} \sin m x_2, \qquad -\beta < x_2 < \beta$$

where

$$m C_{2n+1} = (-1)^n \frac{16\beta}{\pi^2 (2n+1)^2}$$

The conditions A.105 reduce to

$$\begin{aligned} A_{2n+1}^{(1)} \text{ch}\, \mu_1 m \alpha_1 - B_{2n+1} \text{sh}\, \mu_1 m \alpha_1 &= \mu_1^{-1} C_{2n+1} \\ A_{2n+1}^{(2)} \text{ch}\, \mu_2 m \alpha_2 + B_{2n+1} \text{sh}\, \mu_2 m \alpha_2 &= \mu_2^{-1} C_{2n+1} \end{aligned} \tag{A.110}$$

The condition A.104$_2$ becomes

$$\left(A_{44}^{(1)} A_{55}^{(1)}\right)^{1/2} A_{2n+1}^{(1)} - \left(A_{44}^{(2)} A_{55}^{(2)}\right)^{1/2} A_{2n+1}^{(2)} = \left(A_{55}^{(1)} - A_{55}^{(2)}\right) C_{2n+1} \tag{A.111}$$

From Equations A.110 and A.111, we obtain

$$A_{2n+1}^{(1)} = \frac{(-1)^n 16\beta^2}{(2n+1)^3\gamma\pi^2}\big[\rho_2(\mu_1 \text{sh}\,\mu_1 m\alpha_1 + \mu_2 \text{sh}\,\mu_2 m\alpha_2)$$
$$+ (A_{55}^{(1)} - A_{55}^{(2)})\mu_1\mu_2 \text{ch}\,\mu_2 m\alpha_2 \text{sh}\mu_1 m\alpha_1\big]$$
$$A_{2n+1}^{(1)} = \frac{(-1)^n 16\beta^2}{(2n+1)^3\gamma\pi^2}\big[\rho_1(\mu_1 \text{sh}\,\mu_1 m\alpha_1 + \mu_2 \text{sh}\,\mu_2 m\alpha_2)$$
$$+(A_{55}^{(1)} - A_{55}^{(2)})\mu_1\mu_2 \text{ch}\,\mu_1 m\alpha_1 \text{sh}\,\mu_2 m\alpha_2\big]$$
$$B_{2n+1} = \frac{(-1)^n 16\beta^2}{(2n+1)^3\gamma\pi^2}\left[\rho_1\rho_2\left(\frac{1}{A_{55}^{(1)}}\text{ch}\,\mu_1 m\alpha_2 - \frac{1}{A_{55}^{(2)}}\text{ch}\,\mu_2 m\alpha_2\right)\right.$$
$$\left. + (A_{55}^{(1)} - A_{55}^{(2)})\mu_1\mu_2 \text{ch}\,\mu_1 m\alpha_1 \text{sh}\,\mu_2 m\alpha_2\right]$$

where

$$\gamma = A_{44}^{(1)}\mu_2 \text{ch}\,\mu_2 m\alpha_2 \text{sh}\,\mu_1 m\alpha_1 + A_{44}^{(2)}\mu_1 \text{ch}\,\mu_1 m\alpha_1 \text{sh}\,\mu_2 m\alpha_2,\ \rho_\alpha = \left(A_{44}^{(\alpha)}A_{55}^{(\alpha)}\right)^{1/2}$$

The series A.109 are absolutely and uniformly convergent.

4.12.4 We denote by $\Pi^{(k)}$, $(k = 1,2,3)$, the plane strain problems characterized by the equations of equilibrium

$$t_{\beta 1,\beta}^{(1)} + (A_{13}x_1)_{,1} = 0, \qquad t_{\beta 2,\beta}^{(1)} + (A_{23}x_1)_{,2} = 0, \qquad t_{\beta 1,\beta}^{(2)} + (A_{13}x_2)_{,1} = 0$$
$$t_{\beta 2,\beta}^{(2)} + (A_{23}x_2)_{,2} = 0, \qquad t_{\beta 1,\beta}^{(3)} + A_{13,1} = 0, \qquad t_{\beta 2,\beta}^{(3)} + A_{23,2} = 0 \tag{A.112}$$

the constitutive equations

$$t_{11}^{(k)} = A_{11}e_{11}^{(k)} + A_{12}e_{22}^{(k)}, \qquad t_{22}^{(k)} = A_{12}e_{11}^{(k)} + A_{22}e_{22}^{(k)}, \qquad t_{12}^{(k)} = 2A_{66}e_{12}^{(k)} \tag{A.113}$$

the geometrical equations

$$2e_{\alpha\beta}^{(k)} = u_{\alpha,\beta}^{(k)} + u_{\beta,\alpha}^{(k)} \tag{A.114}$$

on Σ_1, and the boundary conditions

$$t_{\beta 1}^{(1)}n_\beta = -A_{13}x_1n_1, \qquad t_{\beta 1}^{(1)}n_\beta = -A_{23}x_1n_2, \qquad t_{\beta 1}^{(2)}n_\beta = -A_{13}x_2n_1$$
$$t_{\beta 1}^{(2)}n_\beta = -A_{23}x_2n_2, \qquad t_{\beta 1}^{(3)}n_\beta = -A_{13}n_1, \qquad t_{\beta 1}^{(3)}n_\beta = -A_{23}n_2 \text{ on } \Gamma \tag{A.115}$$

The solution of the extension and bending problem is given by

$$u_\alpha = -\frac{1}{2}a_\alpha x_3^2 + \sum_{k=1}^{3} a_k u_\alpha^{(k)}, \qquad u_3 = (a_1x_1 + a_2x_2 + a_3)x_3 \tag{A.116}$$

where the constants a_k are determined by the following system

$$H_{\alpha j}a_j = \varepsilon_{\alpha\beta}M_\beta, \qquad H_{3j}a_j = -F_3 \tag{A.117}$$

The coefficients H_{ij} are defined by

$$\begin{aligned}
H_{\alpha\beta} &= \int\limits_{\Sigma_1} x_\alpha\big(A_{33}x_\beta + A_{13}e_{11}^{(\beta)} + A_{23}e_{22}^{(\beta)}\big)da \\
H_{\alpha 3} &= \int\limits_{\Sigma_1} x_\alpha\big(A_{33} + A_{13}e_{11}^{(3)} + A_{23}e_{22}^{(3)}\big)da \\
H_{3\alpha} &= \int\limits_{\Sigma_1} \big(A_{33}x_\alpha + A_{13}e_{11}^{(\alpha)} + A_{23}e_{22}^{(\alpha)}\big)da \\
H_{33} &= \int\limits_{\Sigma_1} \big(A_{33} + A_{13}e_{11}^{(3)} + A_{23}e_{22}^{(3)}\big)da
\end{aligned} \tag{A.118}$$

Let us prove that the solution of the problem $\Pi^{(1)}$ is

$$u_1^{(1)} = -\frac{1}{2}\big(\nu_1^* x_1^2 - \nu_2^* x_2^2\big), \qquad u_2^{(1)} = -\nu_2^* x_1 x_2 \tag{A.119}$$

where

$$\begin{gathered}
\nu_1^* = \frac{1}{\delta_1^*}(A_{13}^* A_{22}^* - A_{23}^* A_{12}^*), \qquad \nu_2^* = \frac{1}{\delta_1^*}(A_{23}^* A_{11}^* - A_{13}^* A_{12}^*) \\
\delta_1^* = A_{11}^* A_{22}^* - (A_{12}^*)^2
\end{gathered} \tag{A.120}$$

In view of Equations A.114 and A.119,

$$e_{11}^{(1)} = -\nu_1^* x_1, \qquad e_{22}^{(1)} = -\nu_2^* x_1, \qquad e_{12}^{(1)} = 0 \tag{A.121}$$

By Equations A.113 and A.121, we get

$$\begin{aligned}
t_{11}^{(1)} &= -(A_{11}\nu_1^* + A_{12}\nu_2^*)x_1 = -(A_{11}^*\nu_1^* + A_{12}^*\nu_2^*)x_1 e^{-\alpha r} \\
t_{22}^{(1)} &= -(A_{12}^*\nu_1^* + A_{22}^*\nu_2^*)x_1 e^{-\alpha r}, \qquad t_{12}^{(1)} = 0
\end{aligned}$$

It follows from Equations A.111, A.120, 4.8.10, and 4.8.11 that

$$\begin{gathered}
\nu_1 = \nu_1^*, \qquad \nu_2 = \nu_2^* \\
A_{11}^*\nu_1^* + A_{12}^*\nu_2^* = A_{13}^*, \qquad A_{12}^*\nu_1^* + A_{22}^*\nu_2^* = A_{23}^*
\end{gathered} \tag{A.122}$$

Thus we obtain

$$t_{11}^{(1)} = -A_{13}^* x_1 e^{-\alpha r} = -A_{13}x_1, \qquad t_{22}^{(1)} = -A_{23}x_1, \qquad t_{12}^{(1)} = 0 \tag{A.123}$$

Clearly, the stresses A.123 satisfy the equilibrium equations A.112$_1$ and the boundary conditions A.115$_1$. Similarly, we can prove that the problems $\Pi^{(2)}$ and $\Pi^{(3)}$ have the solutions

$$u_1^{(2)} = -\nu_1^* x_1 x_2, \qquad u_2^{(2)} = \frac{1}{2}\left(\nu_1^* x_1^2 - \nu_2^* x_2^2\right)$$
$$u_1^{(3)} = -\nu_1^* x_1, \qquad u_2^{(3)} = -\nu_2^* x_2 \tag{A.124}$$

From Equation A.124, we obtain

$$e_{11}^{(2)} = -\nu_1^* x_2, \qquad e_{22}^{(2)} = -\nu_2^* x_2, \qquad e_{12}^{(2)} = 0$$
$$e_{11}^{(3)} = -\nu_1^*, \qquad e_{22}^{(3)} = -\nu_2^*, \qquad e_{12}^{(3)} = 0 \tag{A.125}$$

so that

$$t_{11}^{(2)} = -A_{13} x_2, \qquad t_{22}^{(2)} = -A_{23} x_2, \qquad t_{12}^{(2)} = 0$$
$$t_{11}^{(3)} = -A_{13}, \qquad t_{22}^{(3)} = -A_{23}, \qquad t_{12}^{(3)} = 0$$

By Equations 4.8.21, 4.8.22, and A.122, we obtain

$$A_{33} - A_{13}\nu_1^* - A_{23}\nu_2^* = E_0 \tag{A.126}$$

We can write

$$E_0 = E_0^* e^{-\alpha r} \tag{A.127}$$

where

$$E_0^* = A_{33}^* - A_{13}^*\nu_1^* - A_{23}^*\nu_2^* \tag{A.128}$$

We have $\Sigma_1 = \{x : x_1^2 + x_2^2 < a^2, x_3 = 0\}$. In view of Equations A.118, A.121, A.125, and A.126, we find that the constants H_{ij} are given by

$$H_{11} = H_{22} = \frac{\pi}{\alpha^4} E_0^* [6 - (6 + 6a\alpha + 3a^2\alpha^2 + a^3\alpha^3)e^{-a\alpha}]$$
$$H_{3\alpha} = H_{\alpha 3} = H_{12} = 0$$
$$H_{33} = \frac{2\pi}{\alpha^2} E_0^* [1 - (1 + a\alpha)e^{-a\alpha}] \tag{A.129}$$

From Equation A.117, we obtain

$$a_1 = \frac{M_2}{H_{11}}, \qquad a_2 = -\frac{M_1}{H_{11}}, \qquad a_3 = -\frac{F_3}{H_{33}} \tag{A.130}$$

The solution of the extension and bending problem has the form A.116 where $u_\alpha^{(j)}$ are given by Equations A.119 and A.124 and the constants a_k are defined by Equations A.130.

The solution of the torsion problem is $u_1 = -\tau x_2 x_3$, $u_2 = \tau x_1 x_3$, $u_2 = \tau\varphi$, where φ is the solution of the boundary-value problem

$$(A_{55}\varphi_{,1})_1 + (A_{44}\varphi_{,2})_{,2} = (A_{55}x_2)_{,1} - (A_{44}x_1)_{,2} \text{ on } \Sigma_1$$
$$A_{55}\varphi_{,1}n_1 + A_{44}\varphi_{,2}n_2 = A_{55}x_2n_1 - A_{44}x_1n_2 \text{ on } \Gamma \tag{A.131}$$

The constant τ is equal to $-M_3/D_0$ where D_0 is defined in Equation 4.8.24. In this case, we have

$$(A_{55}x_2)_{,1} - (A_{44}x_1)_{,2} = A_{55,1}x_2 - A_{44,2}x_1 = -\alpha e^{-\alpha r}x_1x_2r^{-1}(A_{55}^* - A_{44}^*) \tag{A.132}$$

The condition on boundary can be written as

$$A_{55}\varphi_{,1}x_1 + A_{44}\varphi_{,2}x_2 = (A_{55} - A_{44})x_1x_2 \text{ on } \Gamma \tag{A.133}$$

We seek the function φ in the form

$$\varphi = kx_1x_2 \tag{A.134}$$

where k is a constant. From Equations A.132 and A.133, we get

$$k = \frac{A_{55}^* - A_{44}^*}{A_{55}^* + A_{44}^*}$$

It follows from Equations A.134 and A.4.8.24 that the torsional rigidity is

$$D_0 = \frac{\pi}{\alpha^4}[(1+k)A_{44}^* + (1-k)A_{55}^*][6 - (6 + 6a\alpha + 3\alpha^2a^2 + a^3\alpha^3)e^{-a\alpha}]$$

Thus, the problem of torsion is solved.

5.7.1 We shall use the polar coordinates (r, θ) and the relations 5.2.9 and 5.2.13. The problem consists in the finding of the functions u_r, u_θ, and φ_3 which satisfy Equations 5.2.9, 5.2.10, and 5.2.11 with $f_r = f_\theta = 0$, $g_3 = 0$, and the boundary conditions

$$t_{rr} = 0, \qquad t_{r\theta} = 0, \qquad m_{rz} = q_1\cos\theta + q_2\sin\theta \text{ for } r = a \tag{A.135}$$

We seek the solution of this problem in the form

$$u_r = u^{(1)}(r)\cos\theta + u^{(2)}(r)\sin\theta, \qquad u_\theta = v^{(1)}(r)\cos\theta + v^{(2)}(r)\sin\theta$$
$$\varphi_3 = \psi^{(1)}(r)\cos\theta + \psi^{(2)}(r)\sin\theta \tag{A.136}$$

where $u^{(\alpha)}, v^{(\alpha)}$, and $\psi^{(\alpha)}$ are functions only on r. It follows from Equations 5.3.10, 5.3.11, and A.136 that

$$\begin{aligned} t_{rr} &= t_{rr}^{(1)}\cos\theta + t_{rr}^{(2)}\sin\theta, & t_{\theta\theta} &= t_{\theta\theta}^{(1)}\cos\theta + t_{\theta\theta}^{(2)}\sin\theta \\ t_{r\theta} &= t_{r\theta}^{(1)}\cos\theta + t_{r\theta}^{(2)}\sin\theta, & t_{\theta r} &= t_{\theta r}^{(1)}\cos\theta + t_{\theta r}^{(2)}\sin\theta \\ m_{rz} &= m_{rz}^{(1)}\cos\theta + m_{rz}^{(2)}\sin\theta, & m_{\theta z} &= m_{\theta z}^{(1)}\cos\theta + m_{\theta z}^{(2)}\sin\theta \end{aligned} \tag{A.137}$$

where

$$
\begin{aligned}
t_{rr}^{(1)} &= (\lambda + 2\mu + \kappa)\frac{du^{(1)}}{dr} + \lambda(u^{(1)} + v^{(2)})r^{-1} \\
t_{rr}^{(2)} &= (\lambda + 2\mu + \kappa)\frac{du^{(2)}}{dr} + \lambda(u^{(2)} - v^{(1)})r^{-1} \\
t_{\theta\theta}^{(1)} &= (\lambda + 2\mu + \kappa)(u^{(1)} + v^{(2)})r^{-1} + \lambda\frac{du^{(1)}}{dr} \\
t_{\theta\theta}^{(2)} &= (\lambda + 2\mu + \kappa)(u^{(2)} - v^{(1)})r^{-1} + \lambda\frac{du^{(2)}}{dr} \\
t_{r\theta}^{(1)} &= (\mu + \kappa)\frac{dv^{(1)}}{dr} + \mu(u^{(2)} - v^{(1)})r^{-1} - \kappa\psi^{(1)} \\
t_{r\theta}^{(2)} &= (\mu + \kappa)\frac{dv^{(2)}}{dr} - \mu(u^{(1)} + v^{(2)})r^{-1} - \kappa\psi^{(2)} \\
t_{\theta r}^{(1)} &= \mu\frac{dv^{(1)}}{dr} + (\mu + \kappa)(u^{(2)} - v^{(1)})r^{-1} + \kappa\psi^{(1)} \\
t_{\theta r}^{(2)} &= \mu\frac{dv^{(2)}}{dr} - (\mu + \kappa)(u^{(1)} + v^{(2)})r^{-1} + \kappa\psi^{(2)} \\
m_{rz}^{(1)} &= \gamma\frac{d\psi^{(1)}}{dr}, \qquad m_{rz}^{(2)} = \gamma\frac{d\psi^{(2)}}{dr}, \qquad m_{\theta z}^{(1)} = \gamma r^{-1}\psi^{(2)} \\
m_{\theta z}^{(2)} &= -\gamma r^{-1}\psi^{(1)}
\end{aligned}
\tag{A.138}
$$

The equilibrium equations 5.2.9 reduce to

$$
\begin{aligned}
&\frac{dt_{rr}^{(1)}}{dr} + \left(t_{\theta r}^{(2)} + t_{rr}^{(1)} - t_{\theta\theta}^{(1)}\right)r^{-1} = 0 \\
&\frac{dt_{r\theta}^{(1)}}{dr} + \left(t_{\theta\theta}^{(2)} + t_{r\theta}^{(1)} + t_{\theta r}^{(1)}\right)r^{-1} = 0 \\
&\frac{dm_{rz}^{(1)}}{dr} + \left(m_{\theta z}^{(2)} + m_{rz}^{(1)}\right)r^{-1} + t_{r\theta}^{(1)} - t_{\theta r}^{(1)} = 0 \\
&\frac{dt_{rr}^{(2)}}{dr} + \left(t_{rr}^{(2)} - t_{\theta\theta}^{(2)} - t_{\theta r}^{(1)}\right)r^{-1} = 0 \\
&\frac{dt_{r\theta}^{(2)}}{dr} + \left(t_{r\theta}^{(2)} + t_{\theta r}^{(2)} - t_{\theta\theta}^{(1)}\right)r^{-1} = 0 \\
&\frac{dm_{rz}^{(2)}}{dr} + \left(m_{rz}^{(2)} - m_{\theta z}^{(1)}\right)r^{-1} + t_{r\theta}^{(2)} - t_{\theta r}^{(2)} = 0
\end{aligned}
\tag{A.139}
$$

Substituting the functions A.138 into A.139, we obtain the equations

$$
\begin{aligned}
&r^2\frac{d^2u^{(1)}}{dr^2}+r\frac{du^{(1)}}{dr}-(1+d_1)u^{(1)}+(1-d_1)r\frac{dv^{(2)}}{dr}\\
&\quad-(1+d_1)v^{(2)}=-d_2r\psi^{(2)}\\
&d_1\left(r^2\frac{d^2v^{(2)}}{dr^2}+r\frac{dv^{(2)}}{dr}\right)-(1+d_1)v^{(2)}-(1-d_1)r\frac{du^{(1)}}{dr}\\
&\quad-(1+d_1)u^{(1)}=d_2r^2\frac{d\psi^{(2)}}{dr}\\
&r^2\frac{d^2\psi^{(2)}}{dr^2}+r\frac{d\psi^{(2)}}{dr}-(1+2d_3r^2)\psi^{(2)}=-d_3r(r\frac{dv^{(2)}}{dr}+v^{(2)})-d_3ru^{(1)}
\end{aligned}
\tag{A.140}
$$

and

$$
\begin{aligned}
&d_1\left(r^2\frac{d^2v^{(1)}}{dr^2}+r\frac{dv^{(1)}}{dr}\right)-(1+d_1)v^{(1)}+r(1-d_1)\frac{du^{(2)}}{dr}\\
&\quad+(1+d_1)u^{(2)}=d_2r^2\frac{d\psi^{(1)}}{dr}\\
&r^2\frac{d^2\psi^{(1)}}{dr^2}+r\frac{d\psi^{(1)}}{dr}-(1+2d_3r^2)\psi^{(1)}=-d_3r\left(r\frac{dv^{(1)}}{dr}+v^{(1)}\right)+d_3ru^{(2)}\\
&r^2\frac{d^2u^{(2)}}{dr^2}+r\frac{du^{(2)}}{dr}-(1+d_1)u^{(2)}-(1-d_1)r\frac{dv^{(1)}}{dr}+(1+d_1)v^{(1)}=d_2r\psi^{(1)}
\end{aligned}
\tag{A.141}
$$

where d_j are defined by

$$
d_1=\frac{\mu+\kappa}{\lambda+2\mu+\kappa},\qquad d_2=\frac{\kappa}{\lambda+2\mu+\kappa},\qquad d_3=\frac{\kappa}{\gamma}
\tag{A.142}
$$

Let us study the system A.141. If we introduce the notations

$$
r=e^t,\qquad Y=\frac{d}{dt}
$$

then the first two equations from Equations A.140 become

$$
\begin{aligned}
&[Y^2-(1+d_1)]u^{(1)}+[(1-d_1)Y-(1+d_1)]v^{(2)}=-d_2e^t\psi^{(2)}\\
&\left[d_1Y^2-(1+d_1)\right]v^{(2)}+[(d_1-1)Y-(1+d_1)]u^{(1)}=d_2e^tY\psi^{(2)}
\end{aligned}
\tag{A.143}
$$

The general solution which corresponds to a nonrigid displacement is

$$
\begin{aligned}
u^{(1)}&=A_1t+A_2e^{2t}+A_3e^{-2t}+\frac{1}{2d_1}d_2\int^t(e^{3s-2t}-e^s)\psi^{(2)}(s)ds\\
v^{(2)}&=\frac{d_1-1}{d_1+1}A_1-A_1t-\frac{3-d_1}{1-3d_1}A_2e^{2t}+A_3e^{-2t}\\
&\quad+\frac{1}{2d_1}d_2\int^t(e^{3s-2t}+e^s)\psi^{(2)}(s)ds
\end{aligned}
\tag{A.144}
$$

where A_i are arbitrary constants. The functions $u^{(1)}$ and $v^{(2)}$ must be bounded for $r = 0$ so that from Equation A.144, we obtain

$$\begin{aligned} u^{(1)} &= A_2 r^2 - \frac{1}{2d_1} d_2 \left[\int_0^r \psi^{(2)}(x)dx - r^{-2} \int_0^r x^2 \psi^{(2)}(x)dx \right] \\ v^{(2)} &= \frac{d_1 - 3}{1 - 3d_1} A_2 r^2 + \frac{1}{2d_1} d_2 \left[\int_0^r \psi^{(2)}(x)dx + r^2 \int_0^r x^2 \psi^{(2)}(x)dx \right] \end{aligned} \tag{A.145}$$

If we substitute $u^{(1)}$ and $v^{(2)}$ from Equation A.145 into A.140$_3$, then we find the equation

$$r^2 \frac{d^2\psi^{(2)}}{dr^2} + r\frac{d\psi^{(2)}}{dr} - (1 + k^2 r^2)\psi^{(2)} = \frac{8r^2}{1 - 3d_1} d_3 A_2 \tag{A.146}$$

where k is given by Equation 5.2.15. The solution of Equation A.146, which is bounded for $r = 0$, has the form

$$\psi^{(2)} = A_4 I_1(kr) - \frac{8(\mu + \kappa)r}{(2\mu + \kappa)(1 - 3d_1)} A_2 \tag{A.147}$$

where A_4 is an arbitrary constant. We denote by I_n and K_n the modified Bessel functions of order n. In view of Equation A.147, from A.145, we obtain

$$\begin{aligned} u^{(1)} &= Q_1 A_2 r^2 + \frac{1}{2d_1 k} d_2 A_4 [I_2(kr) - I_0(kd)] \\ v^{(2)} &= -Q_2 A_2 r^2 + \frac{1}{2d_1 k} d_2 A_4 [I_2(kr) + I_0(kr)] \end{aligned} \tag{A.148}$$

where

$$Q_1 = 1 + \frac{\kappa}{(2\mu + \kappa)(1 - 3d_1)}, \qquad Q_2 = \frac{1}{1 - 3d_1}\left(3 - d_1 + \frac{3\kappa}{2\mu + \kappa}\right) \tag{A.149}$$

The solution of the system A.141 can be determined in a similar way. Thus, we get

$$\begin{aligned} u^{(2)} &= Q_1 B_2 r^2 + \frac{1}{2d_1 k} d_2 B_4 [I_0(kr) - I_2(kr)] \\ v^{(1)} &= Q_2 B_2 r^2 + \frac{1}{2d_1 k} d_2 B_4 [I_0(kr) + I_2(kr)] \\ \psi^{(1)} &= B_4 I_1(kr) + \frac{8(\mu + \kappa)r}{(2\mu + \kappa)(1 - 3d_1)} B_2 \end{aligned} \tag{A.150}$$

where B_2 and B_4 are arbitrary constants. It follows from Equations A.139, A.147, A.148, and A.150 that

$$\begin{aligned} t_{rr}^{(1)} &= N_1 A_2 r - k\gamma r^{-1} A_4 I_2(kr), & t_{rr}^{(2)} &= N_1 B_2 r + k\gamma r^{-1} B_4 I_2(kr) \\ t_{r\theta}^{(1)} &= N_2 B_2 r - k\gamma r^{-1} B_4 I_2(kr), & t_{r\theta}^{(2)} &= -N_2 A_2 r - k\gamma r^{-1} A_4 I_2(kr) \\ m_{rz}^{(1)} &= k\gamma B_4 I_1'(kr) + \gamma Q_3 B_2, & m_{rz}^{(2)} &= k\gamma A_4 I_1'(kr) - \gamma Q_3 A_2 \end{aligned} \tag{A.151}$$

where

$$N_1 = (3\lambda + 4\mu + 2\kappa)Q_1 - \lambda Q_2, \qquad N_2 = \mu Q_1 + (\mu + 2\kappa)Q_2 - \kappa Q_3$$
$$Q_3 = 8(\mu + \kappa)[(2\mu + \kappa)(1 - 3d_1)]^{-1} \tag{A.152}$$

It is easy to see that

$$1 - 3d_1 = \frac{\lambda - \mu - 2\kappa}{\lambda + 2\mu + \kappa}, \qquad 3 - d_1 = \frac{3\lambda + 5\mu + 2\kappa}{\lambda + 2\mu + \kappa}$$
$$Q_1 = \frac{1}{Q}[(2\mu + \kappa)(1 - 3d_1) + \kappa], \qquad Q_2 = \frac{1}{Q}[(3 - d_1)(2\mu + \kappa) + 3\kappa]$$
$$Q_3 = \frac{8}{Q}(\mu + \kappa), \qquad Q = (1 - 3d_1)(2\mu + \kappa)$$
$$\begin{aligned} N_1 + N_2 &= (\lambda + 2\mu + \kappa)[(3 - d_1)Q_1 - (3 - d_1)Q_2] - \kappa Q_3 \\ &= \frac{1}{Q}(\lambda + 2\mu + \kappa)\Big\{(3 - d_1)[(2\mu + \kappa)(1 - 3d_1) + \kappa] \\ &\quad - (1 - 3d_1)[(3 - d_1)(2\mu + \kappa) + 3\kappa] - \frac{8\kappa(\mu + \kappa)}{\lambda + 2\mu + \kappa}\Big\} = 0 \end{aligned}$$

We note that

$$t_{rr}^{(1)} = t_{r\theta}^{(2)}, \qquad t_{rr}^{(2)} = -t_{r\theta}^{(1)} \tag{A.153}$$

It follows from Equations A.135 and A.137 that the boundary conditions reduce to

$$\begin{aligned} &t_{rr}^{(1)} = 0, \qquad t_{r\theta}^{(2)} = 0, \qquad m_{rz}^{(2)} = q_2 \\ &t_{rr}^{(2)} = 0, \qquad t_{r\theta}^{(1)} = 0, \qquad m_{rz}^{(1)} = q_1, \quad \text{for } r = a \end{aligned} \tag{A.154}$$

If we use Equations A.153 and A.151, then the conditions A.154 become

$$N_1 A_2 a - k\gamma a^{-1} A_4 I_2(ka) = 0, \qquad k\gamma A_4 I_1'(ka) - \gamma Q_3 A_2 = q_2$$
$$N_1 B_2 a + k\gamma a^{-1} B_4 I_2(ka) = 0, \qquad k\gamma B_4 I_1'(ka) + \gamma Q_3 B_2 = q_1$$

We find that

$$\begin{aligned} A_4 &= \frac{q_2 N_1 a^2}{k\gamma[N_1 a^2 I_1'(ka) - \gamma Q_3 I_2(ka)]}, \qquad A_2 = \frac{k\gamma}{N_1 a^2} A_4 I_2(ka) \\ B_4 &= \frac{q_1 N_1 a^2}{k\gamma[N_1 a^2 I_1'(ka) - \gamma Q_3 I_2(ka)]}, \qquad B_2 = -\frac{k\gamma}{N_1 a^2} B_4 I_2(ka) \end{aligned} \tag{A.155}$$

Thus, the solution of the problem is given by Equation A.136, where $u^{(\alpha)}$, $v^{(\alpha)}$, $\psi^{(\alpha)}$ are given by Equations A.147, A.148, A.150, and A_2, A_4, B_2, B_4 are defined in Equation A.155.

The plane strain problems for a circular ring-shaped region have been studied by Chiu and Lee [48].

5.7.2 The solution of the problem of extension and bending can be expressed by Equation 5.3.67, where $(u_\alpha^{*(\eta)}, \varphi_3^{*(\eta)})$ is the solution of the problem $\mathfrak{M}^{(\eta)}, (\eta = 1, 2)$, and the constants a_k are given by Equations 5.3.40. We shall study these problems by using the polar coordinates (r, θ). The problem $\mathfrak{M}^{(1)}$ consists in the finding of the functions $u_r^{(1)}, u_\theta^{(1)}$, and $\varphi_3^{(1)}$ which satisfy Equations 5.2.9, 5.2.10, and 5.2.11 with $f_r = f_\theta = 0$, $g_3 = 0$, and the boundary conditions

$$t_{rr} = 0, \qquad t_{r\theta} = 0, \qquad m_{rz} = (\beta + \gamma\nu)\cos\theta \;\text{ for } r = a \tag{A.156}$$

The problem $\mathfrak{M}^{(2)}$ consists in the determination of the functions $u_r^{(2)}, u_\theta^{(2)}$, and $\varphi_3^{(2)}$ which satisfy Equations 5.2.9, 5.2.10, and 5.2.11 in the absence of body loads, and the boundary conditions

$$t_{rr} = 0, \qquad t_{r\theta} = 0, \qquad m_{rz} = -(\beta + \gamma\nu)\sin\theta \;\text{ for } r = a \tag{A.157}$$

The solution of the problem $\mathfrak{M}^{(1)}$ can be obtained from Equations A.136, A.147, A.148, A.150, and A.155 if we take

$$q_1 = 0, \qquad q_2 = \beta + \gamma\nu \tag{A.158}$$

From Equations A.155 and A.158, we obtain

$$A_2 = Z_2, \qquad A_4 = Z_1, \qquad B_2 = B_4 = 0 \tag{A.159}$$

where

$$Z_1 = \frac{N_1 a^2(\beta + \gamma\nu)}{\kappa\gamma[N_1 a^2 I_1'(ka) - \gamma Q_3 I_2(ka)]}, \qquad Z_2 = \frac{k\gamma}{N_1 a^2} Z_1 I_2(ka) \tag{A.160}$$

Thus, the solution of the problem $\mathfrak{M}^{(1)}$ is

$$u_r^{(1)} = u^{(1)}\cos\theta, \qquad u_\theta^{(1)} = v^{(2)}\sin\theta, \qquad \varphi_3^{(1)} = \psi^{(2)}\sin\theta \tag{A.161}$$

where $u^{(1)}, v^{(2)}$, and $\psi^{(2)}$ are given by Equations A.144 and A.147, and the constants A_2 and A_4 are defined by Equations A.159 and A.160.

The solution of the problem $\mathfrak{M}^{(2)}$ can be obtained from Equations A.136, A.147, A.148, A.150, and A.155 if we take

$$q_1 = -(\beta + \gamma\nu), \qquad q_2 = 0 \tag{A.162}$$

From Equations A.135 and A.162, we get

$$A_2 = 0, \qquad A_4 = 0, \qquad B_2 = Z_2, \qquad B_4 = -Z_1 \tag{A.163}$$

The solution of the problem $\mathfrak{M}^{(2)}$ is

$$u_r^{(2)} = u^{(2)}\sin\theta, \qquad u_\theta^{(2)} = v^{(1)}\cos\theta, \qquad \varphi_3^{(2)} = \psi^{(1)}\cos\theta \tag{A.164}$$

where $u^{(2)}, v^{(1)}$, and $\psi^{(1)}$ are given by Equations A.150 with the constants B_2 and B_4 defined by Equations A.163 and A.160.

We note that the divergence of the displacement vector field and the components m_{zr} and $m_{z\theta}$ of the couple stress tensor for the problem $\mathfrak{M}^{(1)}$ are given by

$$\operatorname{div}\mathbf{u} = (3Q_1 - Q_2)Z_2 r\cos\theta, \qquad m_{zr} = \beta[kZ_1 I_1'(kr) - Q_3 Z_2]\sin\theta$$
$$m_{z\theta} = \beta r^{-1}[Z_1 I_1(kr) - Q_3 Z_2 r]\cos\theta \tag{A.165}$$

In the case of the problem $\mathfrak{M}^{(2)}$, we have

$$\operatorname{div}\mathbf{u} = (3Q_1 - Q_2)Z_2 r\sin\theta, \qquad m_{zr} = -\beta[kZ_1 I_1'(kr) - Q_3 Z_2]\cos\theta$$
$$m_{z\theta} = \beta r^{-1}[Z_1 I_1(kr) - Q_3 Z_2 r]\sin\theta \tag{A.166}$$

The functions $(u_\alpha^{*(\rho)}, \varphi_3^{*(\rho)})$ which satisfy the problems $\mathfrak{M}^{(\rho)}, (\rho = 1, 2)$, are given by

$$\begin{aligned}
&u_1^{*(1)} = u^{(1)}\cos^2\theta - v^{(2)}\sin^2\theta, \qquad u_2^{*(1)} = (u^{(1)} + v^{(2)})\sin\theta\cos\theta\\
&u_1^{*(2)} = (u^{(2)} - v^{(1)})\sin\theta\cos\theta, \qquad u_2^{*(2)} = u^{(2)}\sin^2\theta + v^{(1)}\cos^2\theta\\
&\varphi_3^{*(1)} = \psi^{(2)}\sin\theta, \qquad \varphi_3^{*(2)} = \psi^{(1)}\cos\theta
\end{aligned} \tag{A.167}$$

where $u^{(\alpha)}, v^{(\alpha)}$, and $\psi^{(\alpha)}$ are defined in Equations A.161 and A.164. We now can determine D_{ij} from Equations 5.3.41 and 5.3.45. Thus, we obtain

$$\begin{gathered}
x_\alpha^0 = 0, \qquad D_{12} = 0, \qquad D_{\alpha 3} = 0, \qquad D_{33} = \pi a^2 E\\
D_{11} = D_{22} = \frac{1}{4}\pi E a^4 + \pi(2\gamma + \beta\nu + \beta Q_3 Z_2)a^2 - \pi a Z_1 \beta I_1(ka)
\end{gathered} \tag{A.168}$$

where E is introduced in Equation 5.3.45. Here we have used the relations

$$2I_1(kr) + krI_2(kr) = krI_0(kr)$$

$$\int_{\Sigma_1} m_{32}^{(1)}\,da = -\int_{\Sigma_1} m_{31}^{(2)}\,da = -\beta\int_{\Sigma_1}\left[\nu + Q_3 Z_2 - \frac{1}{2}kZ_1 I_0(kr)\right]da$$

It follows from Equations 5.3.40 and A.168 that

$$a_\alpha = \frac{1}{D_{11}}\varepsilon_{\alpha\beta}M_\beta, \qquad a_3 = -\frac{1}{\pi a^2 E}F_3 \tag{A.169}$$

The solution of extension and bending problem has the form 5.3.67 where $u_\alpha^{(\eta)}$ and $\varphi_3^{(\eta)}$, $(\eta = 1, 2)$, are defined in Equation A.167 and the constants a_k are given by Equation A.169. The solution of Saint-Venant's problem for a circular cylinder has been established by Reddy and Venkatasubramanian [188–190].

5.7.3 In the case of the torsion of a Cosserat elastic cylinder, the displacements and the microrotations have the form 5.3.47, where φ and ψ_α satisfy the boundary-value problem 5.3.26, and the constant a_4 is given by

$$a_4 = -M_3/D \tag{A.170}$$

The torsional rigidity D is defined in Equation 5.3.41. We seek the functions φ and ψ_α in the form

$$\varphi = 0, \qquad \psi_\alpha = x_\alpha \Psi(r) \tag{A.171}$$

where Ψ is an unknown function, and $r = (x_1^2 + x_2^2)^{1/2}$. With the help of the relations

$$\psi_{\alpha,\beta} = \Psi\delta_{\alpha\beta} + x_\alpha x_\beta r^{-1}\Psi', \qquad \Delta\Psi_\alpha = x_\alpha\left(\Psi'' + \frac{3}{r}\Psi'\right), \qquad \Psi' = \frac{d\Psi}{dr}$$

we see that Equations $5.3.26_1$ reduce to

$$\Psi'' + \frac{3}{r}\Psi' - s^2\Psi = 0 \tag{A.172}$$

where

$$s^2 = \frac{2k}{\alpha + \beta + \gamma} \tag{A.173}$$

If we introduce the function F by

$$F = r\Psi$$

then Equation A.172 becomes

$$F'' + \frac{1}{r}F' - \left(\frac{1}{r^2} + s^2\right)F = 0 \tag{A.174}$$

The solution of this equation is

$$F = A^* I_1(sr) + B^* K_1(sr) \tag{A.175}$$

where A^* and B^* are arbitrary constants, and I_n and K_n are the modified Bessel functions of order n. From Equations A.171, we get

$$\psi_1 = F\cos\theta, \qquad \psi_2 = F\sin\theta$$

To obtain a solution which is bounded for $r = 0$, we take $B^* = 0$. Thus, we have

$$\varphi = 0, \qquad \psi_\alpha = x_\alpha r^{-1}F, \qquad F = A^* I_1(sr), \qquad u_\alpha = \varepsilon_{\beta\alpha}\tau x_\beta x_3, \qquad u_3 = 0$$

$$\varphi_\alpha = \tau\left(\Psi - \frac{1}{2}\right)x_\alpha = \frac{1}{2}\tau r^{-1}[2A^* I_1(sr) - r]x_\alpha, \qquad \varphi_3 = \tau x_3$$

$$u_r = 0, \qquad u_\theta = \tau z r, \qquad u_z = 0,$$

$$\varphi_r = \varphi_1\cos\theta + \varphi_2\sin\theta = \tau\left[A^* I_1(sr) - \frac{1}{2}r\right]$$

$$\varphi_\theta = 0, \qquad \varphi_z = \tau z \tag{A.176}$$

The conditions 5.3.26$_2$ on the boundary Γ reduce to

$$a(\alpha+\beta+\gamma)\Psi'(a)+(2\alpha+\beta+\gamma)\Psi(a)=\frac{1}{2}(\beta+\gamma)$$

This condition can be written in the form

$$\alpha F(a)+a(\alpha+\beta+\gamma)F'(a)=\frac{1}{2}(\beta+\gamma)a \tag{A.177}$$

Using the relation

$$xI_1'+I_1(x)=xI_0(x)$$

from Equation A.177, we obtain

$$A^*=\frac{a(\beta+\gamma)}{2(\alpha+\beta+\gamma)kI_1(sa)}$$

where

$$k=\frac{asI_0(as)}{I_1(as)}-\frac{\beta+\gamma}{\alpha+\beta+\gamma}$$

From Equations 5.3.41 and A.176, we find that

$$D=\frac{1}{4}\pi a^4(2\mu+\kappa)+\pi a^2(\beta+\gamma)+2\pi\kappa A^*\int_0^a x^2I_1(sx)dx$$

$$+\,2\pi A^*\alpha\int_0^a [sxI_1'(sx)+I_1(sx)]dx$$

With the help of the relations

$$[x^2I_2(x)]'=x^2I_1(x),\qquad xI_1'(x)+I_1(x)=[xI_1(x)]',\qquad I_1(0)=0$$

we obtain

$$D=\frac{1}{4}\pi a^4(2\mu+\kappa)+\pi a^2(\beta+\gamma)+\frac{2}{s}\pi\kappa a^2A^*I_2(as)+2\pi a\alpha A^*I_1(as)$$

The constant a_4 is given by Equation A.170. The torsion problem for a circular cylinder was studied in Refs. 188 and 338.

6.6.1 We use the cylindrical coordinate system (r,θ,z). From Equations 6.5.20, it follows that the solution of the torsion problem has the form

$$u_r=0,\quad u_\theta=\tau rz,\quad u_z=\tau\Phi,\quad \varphi_r=\tau\Phi_r,\quad \varphi_\theta=\tau\Phi_\theta,\quad \varphi_z=\tau z \tag{A.178}$$

where Φ, Φ_r, and Φ_θ are unknown functions of r and θ. Equations 6.5.23 become

$$L_z^{(\rho)}\Lambda=0,\qquad M_r^{(\rho)}\Lambda=\frac{1}{2}(s^{(\rho)})^2r^3,\qquad M_\theta^{(\rho)}\Lambda=0 \text{ on } A_\rho^* \tag{A.179}$$

where $\Lambda = (\Phi, \Phi_r, \Phi_\theta)$ and

$$L_z^{(\rho)}\Lambda = (D_r^2 + D_\theta^2)\Phi + e^{(\rho)} r(D_r + 1)\Phi_\theta - e^{(\rho)} r D_\theta \Phi_r$$

$$M_r^{(\rho)}\Lambda = [D_r^2 + b^{(\rho)} D_\theta^2 - (s^{(\rho)})^2 r^2 - 1]\Phi_r + [(1 - b^{(\rho)})D_r - (1 + b^{(\rho)})] D_\theta \Phi_\theta + \frac{1}{2}(s^{(\rho)})^2 r D_\theta \Phi$$

$$M_\theta^{(\rho)}\Lambda = [b^{(\rho)}(D_r^2 - 1) - r^2 (s^{(\rho)})^2 + D_\theta^2]\Phi_\theta + [(1 - b^{(\rho)})D_r + (1 + b^{(\rho)})] D_\theta \Phi_r - \frac{1}{2} r (s^{(\rho)})^2 D_r \Phi$$

$$D_r = r\frac{d}{dr}, \qquad D_\theta = \frac{d}{d\theta}, \qquad e^{(\rho)} = \kappa^{(\rho)}(\mu^{(\rho)} + \kappa^{(\rho)})^{-1}$$

$$b^{(\rho)} = \gamma^{(\rho)}(\alpha^{(\rho)} + \beta^{(\rho)} + \gamma^{(\rho)})^{-1}, \qquad (s^{(\rho)})^2 = 2\kappa^{(\rho)}(\alpha^{(\rho)} + \beta^{(\rho)} + \gamma^{(\rho)})^{-1}$$

The conditions 6.5.25 and 6.5.27 take the form

$$\begin{gathered}
[\Phi]_1 = [\Phi]_2, \qquad [\Phi_r]_1 = [\Phi_r]_2, \qquad [\Phi_\theta]_1 = [\Phi_\theta]_2 \\
T_z^{(1)}\Lambda - T_z^{(2)}\Lambda = 0, \qquad S_r^{(1)}\Lambda - S_r^{(2)}\Lambda = \alpha^{(2)} - \alpha^{(1)} \\
S_\theta^{(1)}\Lambda - S_\theta^{(2)}\Lambda = 0 \text{ on } \Gamma^* \\
T_z^{(1)}\Lambda = 0, \qquad S_r^{(1)}\Lambda = -\alpha^{(1)}, \qquad S_\theta^{(1)}\Lambda = 0 \text{ on } L
\end{gathered} \tag{A.180}$$

where

$$T_z^{(\rho)}\Lambda = \frac{1}{r}(\mu^{(\rho)} + \kappa^{(\rho)})D_r\Phi + \kappa^{(\rho)}\Phi_\theta$$

$$rS_r^{(\rho)}\Lambda = (\alpha^{(\rho)} + \beta^{(\rho)} + \gamma^{(\rho)})D_r\Phi_r + \alpha^{(\rho)}(D_\theta\Phi_\theta + \Phi_r)$$

$$rS_\theta^{(\rho)}\Lambda = \gamma^{(\rho)} D_r\Phi_\theta + \beta^{(\rho)}(D_\theta\Phi_r - \Phi_\theta)$$

We seek the solution of the problem A.179 and A.180 assuming that Φ, Φ_r, and Φ_θ are functions only of r. Then we obtain

$$\Phi = -e^{(2)} A_1 \int^r I_1(\delta^{(2)} r)dr, \qquad \Phi_r = A_3 I_1(s^{(2)} r) - \frac{1}{2} r$$

$$\Phi_\theta = A_1 I_1(\delta^{(2)} r), \text{ for } 0 \le r \le r_2$$

$$\begin{aligned}
\Phi &= -e^{(1)} \int^r [B_1 I_1(\delta^{(1)} r) + B_4 K_1(\delta^{(1)} r)]dr \\
&\quad + B_5[1 + e^{(1)}\kappa^{(1)}(\gamma^{(1)}\delta^{(1)2})^{-1}]\ln r \\
\Phi_r &= B_3 I_1(s^{(1)} r) + B_6 K_1(s^{(1)} r) - \frac{1}{2} r \\
\Phi_\theta &= B_1 I_1(\delta^{(1)} r) + B_4 K_1(\delta^{(1)} r) - \frac{1}{r}\kappa^{(1)}(\gamma^{(1)}\delta^{(1)2})^{-1} B_5, \text{ for } r_2 \le r \le r_1
\end{aligned} \tag{A.181}$$

where I_n and K_n are the modified Bessel functions of order n, and A_s and B_s are unknown constants, and

$$(\delta^{(\rho)})^2 = (2 - e^{(\rho)})\sigma^{(\rho)}, \qquad \sigma^{(\rho)} = \lambda^{(\rho)}(\gamma^{(\rho)})^{-1}$$

From Equations A.180 and A.181, we find that

$$\begin{aligned} &A_1 = B_1 = B_4 = B_5 = 0, \qquad B_3 = (c_5c_7 - c_4c_8)(c_3c_7 - c_4c_6)^{-1} \\ &B_6 = (c_3c_8 - c_5c_6)(c_3c_7 - c_4c_6)^{-1}, \qquad A_3 = c_1B_3 + c_2B_6 \end{aligned} \tag{A.182}$$

where

$$\begin{aligned} c_1 &= I_1(s^{(1)}r_2)/I_1(s^{(2)}r_2), \qquad c_2 = K_1(s^{(1)}r_2)/I_1(s^{(2)}r_2) \\ c_3 &= c_1[(\alpha^{(2)} + \beta^{(2)} + \gamma^{(2)})I_1'(s^{(2)}r_2) + \alpha^{(2)}I_1(s^{(2)}r_2)(s^{(2)}r_2)^{-1}] \\ &\quad - [(\alpha^{(1)} + \beta^{(1)} + \gamma^{(1)})I_1'(s^{(1)}r_2) + \alpha^{(1)}I_1(s^{(1)}r_2)/(s^{(1)}r_2)]s^{(1)}/s^{(2)} \\ c_4 &= c_2[(\alpha^{(2)} + \beta^{(2)} + \gamma^{(2)})I_1'(s^{(2)}r_2) + \alpha^{(2)}I_1(s^{(2)}r_2)/(s^{(2)}r_2)] \\ &\quad - [(\alpha^{(1)} + \beta^{(1)} + \gamma^{(1)})K_1'(s^{(1)}r_2) + \alpha^{(1)}K_1(s^{(1)}r_2)/(s^{(1)}r_2)]s^{(1)}/s^{(2)} \\ c_5 &= [(\beta^{(2)} + \gamma^{(2)}) - (\beta^{(1)} + \gamma^{(1)})]/(2s^{(2)}) \\ c_6 &= (\alpha^{(1)} + \beta^{(1)} + \gamma^{(1)})I_1'(s^{(1)}r_1) + \alpha^{(1)}I_1(s^{(1)}r_1)/(s^{(1)}r_1) \\ c_7 &= (\alpha^{(1)} + \beta^{(1)} + \gamma^{(1)})K_1'(s^{(1)}r_1) + \alpha^{(1)}K_1(s^{(1)}r_1)/(s^{(1)}r_1) \\ c_8 &= (\beta^{(1)} + \gamma^{(1)})/(2s^{(1)}) \end{aligned}$$

Thus, the solution of the problem A.179 and A.180 is

$$\Phi = 0, \qquad \Phi_r = -\frac{1}{2}r + A_3I_1(s^{(2)}r), \qquad \Phi_\theta = 0, \ \text{ for } 0 \le r \le r_2$$

$$\Phi = 0, \qquad \Phi_r = -\frac{1}{2}r + B_3I_1(s^{(1)}r) + B_6K_1(s^{(1)}r), \qquad \Phi_\theta = 0, \ \text{ for } r_2 \le r \le r_1$$

where A_3, B_3, and B_6 are given by Equation A.182. From Equation 6.5.30, we obtain

$$\begin{aligned} D' &= \frac{1}{4}(2\mu^{(2)} + \kappa^{(2)})\pi r_2^4 + (\beta^{(2)} + \gamma^{(2)})\pi r_2^2 \\ &\quad + \frac{2}{s^{(2)}}\pi\kappa^{(2)}A_3r_2^2I_2(s^{(2)}r_2) + 2\pi\alpha^{(2)}A_3r_2I_1(s^{(2)}r_2) \\ &\quad + \frac{1}{4}(2\mu^{(1)} + \kappa^{(1)})\pi(r_1^4 - r_2^4) + (\beta^{(1)} + \gamma^{(1)})\pi(r_1^2 - r_2^2) \\ &\quad + \{B_3[r_1^2I_2(s^{(1)}r_1) - r_2^2I_2(s^{(1)}r_2)] - B_6[r_1^2K_2(s^{(1)}r_1) \\ &\quad - r_2^2K_2(s^{(1)}r_2)]\}2\pi\kappa^{(1)}/s^{(1)} + 2\pi\alpha^{(1)}\{B_3[r_1I_1(s^{(1)}r_1) - r_2I_1(s^{(1)}r_2)] \\ &\quad + B_6[r_1K_1(s^{(1)}r_1) - r_2K_1(s^{(1)}r_2)]\} \end{aligned}$$

The constant τ is given by Equation 6.5.29.

6.6.2 We consider cylinder B^* for which the cross section Σ_1 is assembly of the domains A_1^* and A_2^*. The solution of extension and bending problem has the form 6.5.16. First, we have to solve the plane strain problems $\mathcal{E}^{(s)}$, $(s = 1, 2, 3)$. We introduce the notations

$$
\begin{aligned}
L_r^{(\rho)} X &= \left(D_r^2 + c^{(\rho)} D_\theta^2 - 1\right) u_r \\
&\quad + [(1 - c^{(\rho)}) D_r - (1 + c^{(\rho)})] D_\theta u_\theta + d^{(\rho)} r D_\theta \varphi_z \\
L_\theta^{(\rho)} X &= [(1 - c^{(\rho)}) D_r + (1 + c^{(\rho)})] D_\theta u_r \\
&\quad + \left[c^{(\rho)} (D_r^2 - 1) + D_\theta^2\right] u_\theta - d^{(\rho)} r D_r \varphi_z \\
M_z^{(\rho)} X &= -\sigma^{(\rho)} r D_\theta u_r + \sigma^{(\rho)} r (D_r + 1) u_\theta + \left(D_r^2 + D_\theta^2 - 2\sigma^{(\rho)} r^2\right) \varphi_z \\
T_r^{(\rho)} X &= [(\lambda^{(\rho)} + 2\mu^{(\rho)} + \kappa^{(\rho)}) D_r u_r + \lambda^{(\rho)} (D_\theta u_\theta + u_r)] r^{-1} \\
T_\theta^{(\rho)} X &= [(\mu^{(\rho)} + \kappa^{(\rho)}) D_r u_\theta + \mu^{(\rho)} (D_\theta u_r - u_\theta)] r^{-1} - \kappa^{(\rho)} \varphi_z \\
S_z^{(\rho)} X &= r^{-1} \gamma^{(\rho)} D_r \varphi_z
\end{aligned}
$$

where $X = (u_r, u_\theta, \varphi_z)$ and

$$
c^{(\rho)} = (\mu^{(\rho)} + \kappa^{(\rho)})(\lambda^{(\rho)} + 2\mu^{(\rho)} + \kappa^{(\rho)})^{-1}, \qquad d^{(\rho)} = \kappa^{(\rho)} (\lambda^{(\rho)} + 2\mu^{(\rho)} + \lambda^{(\rho)})^{-1}
$$

Using these notations, the problems $\mathcal{E}^{(s)}$, $(s = 1, 2, 3)$, become

$$
\begin{aligned}
& L_r^{(\rho)} X^{(1)} = -(1 - 2c^{(\rho)} + d^{(\rho)}) r^2 \cos\theta \\
& L_\theta^{(\rho)} X^{(1)} = (1 - 2c^{(\rho)} + d^{(\rho)}) r^2 \sin\theta, \qquad M_z^{(\rho)} X^{(1)} = 0 \text{ on } A_\rho^* \\
& [X^{(1)}]_1 = [X^{(1)}]_2, \qquad T_r^{(2)} X^{(1)} - T_r^{(1)} X^{(1)} = (\lambda^{(1)} - \lambda^{(2)}) r_2 \cos\theta \\
& T_\theta^{(2)} X^{(1)} = T_\theta^{(1)} X^{(1)}, \qquad S_z^{(2)} X^{(1)} - S_z^{(1)} X^{(1)} = (\beta^{(2)} - \beta^{(1)}) \sin\theta \text{ on } \Gamma^* \\
& T_r^{(1)} X^{(1)} = -\lambda^{(1)} r_1 \cos\theta, \qquad T_\theta^{(1)} X^{(1)} = 0, \qquad S_z^{(1)} X^{(1)} = \beta^{(1)} \sin\theta \text{ on } L
\end{aligned} \tag{A.183}
$$

$$
\begin{aligned}
& L_r^{(\rho)} X^{(2)} = -(1 - 2c^{(\rho)} + d^{(\rho)}) r^2 \sin\theta \\
& L_\theta^{(\rho)} X^{(2)} = -(1 - 2c^{(\rho)} + d^{(\rho)}) r^2 \cos\theta, \qquad M_z^{(\rho)} X^{(2)} = 0 \text{ on } A_\rho^* \\
& [X^{(2)}]_1 = [X^{(2)}]_2, \qquad T_r^{(2)} X^{(2)} - T_r^{(1)} X^{(2)} = (\lambda^{(1)} - \lambda^{(2)}) r_2 \sin\theta \\
& T_\theta^{(2)} X^{(2)} = T_\theta^{(1)} X^{(2)}, \qquad S_z^{(2)} X^{(2)} - S_z^{(1)} X^{(2)} = (\beta^{(1)} - \beta^{(2)}) \cos\theta \text{ on } \Gamma^* \\
& T_r^{(1)} X^{(2)} = -\lambda^{(1)} r_1 \sin\theta, \qquad T_\theta^{(1)} X^{(2)} = 0, \qquad S_z^{(1)} X^{(2)} = -\beta^{(1)} \cos\theta \text{ on } L \\
& L_r^{(\rho)} X^{(3)} = 0, \qquad L_\theta^{(\rho)} X^{(3)} = 0, \qquad M_z^{(\rho)} X^{(3)} = 0, \text{ on } A_\rho^*
\end{aligned} \tag{A.184}
$$

$$
\begin{aligned}
& [X^{(3)}]_1 = [X^{(3)}]_2, \qquad T_r^{(2)} X^{(3)} - T_r^{(1)} X^{(3)} = \lambda^{(1)} - \lambda^{(2)} \\
& T_\theta^{(2)} X^{(3)} = T_\theta^{(1)} X^{(3)}, \qquad S_z^{(2)} X^{(3)} = S_z^{(1)} X^{(3)} \text{ on } \Gamma^* \\
& T_r^{(1)} X^{(3)} = -\lambda^{(1)}, \qquad T_\theta^{(1)} X^{(3)} = 0, \qquad S_z^{(1)} X^{(3)} = 0 \text{ on } L
\end{aligned} \tag{A.185}
$$

where $X^{(s)} = (u_r^{(s)}, u_\theta^{(s)}, \varphi_z^{(s)})$. Let us determine the solutions of these problems. First, we consider the problems $\mathcal{E}^{(\beta)}$, $(\beta = 1, 2)$. We seek the solutions of these problems in the form

$$X^{(\beta)} = (A_1^{(\beta)} \cos\theta + A_2^{(\beta)} \sin\theta, B_1^{(\beta)} \cos\theta + B_2^{(\beta)} \sin\theta, C_1^{(\beta)} \cos\theta + C_2^{(\beta)} \sin\theta) \tag{A.186}$$

where $A_\alpha^{(\beta)}, B_\alpha^{(\beta)}$, and $C_\alpha^{(\beta)}$ are functions of r. From Equations A.183 and A.184, we obtain

$$\begin{aligned} &A_2^{(1)} = B_1^{(1)} = C_1^{(1)} = A_1^{(2)} = B_2^{(2)} = C_2^{(2)} = 0 \\ &A_1^{(1)} = A_2^{(2)} = v_r, \qquad B_2^{(1)} = -B_1^{(2)} = v_\theta, \qquad C_2^{(1)} = -C_1^{(2)} = \psi_z \end{aligned}$$

where v_r, v_θ, and ψ_z satisfy the equations

$$\begin{aligned} &(D_r^2 - 1 - c^{(\rho)})v_r + [(1 - c^{(\rho)})D_r - (1 + c^{(\rho)})]v_\theta + d^{(\rho)} r\psi_z \\ &\quad = -(1 + d^{(\rho)} - 2c^{(\rho)})r^2 \\ &[(1 - c^{(\rho)})D_r + (1 + c^{(\rho)})]v_r - [c^{(\rho)}(D_r^2 - 1) - 1]v_\theta + d^{(\rho)} r D_r \Psi_z \\ &\quad = -(1 + d^{(\rho)} - 2c^{(\rho)})r^2 \\ &(D_r^2 - 1 - 2\sigma^{(\rho)} r^2)\psi_z + \sigma^{(\rho)} r(D_r + 1)v_\theta + \sigma^{(\rho)} r v_r = 0 \text{ on } A_\rho^* \end{aligned} \tag{A.187}$$

and the conditions

$$\begin{aligned} &[Y]_1 = [Y]_2, \qquad T_r^{(2)}Y - T_r^{(1)}Y = (\lambda^{(1)} - \lambda^{(2)})r_2 \\ &T_\theta^{(2)}Y = T_\theta^{(1)}Y, \qquad S_z^{(2)}Y - S_z^{(1)}Y = \beta^{(2)} - \beta^{(1)} \text{ on } r = r_2 \\ &T_r^{(1)}Y = -\lambda^{(1)} r_1, \qquad T_\theta^{(1)}Y = 0, \qquad S_z^{(1)}Y = \beta^{(1)} \text{ on } r = r_1 \end{aligned} \tag{A.188}$$

In the above relations, we have used the notations

$$\begin{aligned} Y &= (v_r, v_\theta, \psi_z) \\ T_r^{(\rho)}Y &= [(\lambda^{(\rho)} + 2\mu^{(\rho)} + \kappa^{(\rho)})D_r v_r + \lambda^{(\rho)}(v_r + v_\theta)]r^{-1} \\ T_\theta^{(\rho)}Y &= [(\mu^{(\rho)} + \kappa^{(\rho)})D_r v_\theta - \mu^{(\rho)}(v_r + v_\theta) - \kappa^{(\rho)} r\psi_z]r^{-1} \\ S_z^{(\rho)}Y &= \gamma^{(\rho)} r^{-1} D_r \psi_z \end{aligned}$$

The general solution of the system A.187 is

$$\begin{aligned} v_r &= C_1 + (2 - 6c^{(2)} + 3d^{(2)})C_2 r^2 \\ &\quad + e^{(2)}(2\delta^{(2)})^{-1} C_3 [I_1(\delta^{(2)} r) - I_0(\delta^{(2)} r)] - \frac{1}{8} r^2 \\ v_\theta &= -C_1 - (6 - 2c^{(2)} + d^{(2)})C_2 r^2 + e^{(2)}(d^{(2)})^{-1} C_3 I_1'(\delta^{(2)} r) - \frac{5}{8} r^2 \\ \psi_z &= -8C_2 r + C_3 I_1(\delta^{(2)} r) - r, \text{ for } 0 \le r \le r_2 \end{aligned}$$

$$
\begin{aligned}
v_r &= D_1 + (2 - 6c^{(1)} + 3d^{(1)})D_2r^2 + e^{(1)}(2\delta^{(1)})^{-1}\{D_3[I_2(\delta^{(1)}r) \\
&\quad - I_0(\delta^{(1)}r) - D_4[K_2(\delta^{(1)}r) - K_0(\delta^{(1)}r)]\} + D_5r^{-2} - \frac{1}{2}d^{(1)}D_6 \\
&\quad + c^{(1)}(2 + 2c^{(1)} - d^{(1)})D_6 \ln r - \frac{1}{8}r^2 \\
v_\theta &= -D_1 - (6 - 2c^{(1)} + d^{(1)})D_2r^2 + e^{(1)}(\delta^{(1)})^{-1}[D_3I_1'(\delta^{(1)}r) \\
&\quad + D_4K_1'(\delta^{(1)}r)] + D_5r^{-2} - \frac{1}{2}[2(1 - c^{(1)})(2c^{(1)} - d^{(1)}) + d^{(1)}]D_6 \\
&\quad - c^{(1)}(2 + 2c^{(1)} - d^{(1)})D_6 \ln r - \frac{5}{8}r^2 \\
\psi_z &= -8D_2r + D_3I_1(\delta^{(1)}r) + D_4K_1(\delta^{(1)}r) - 2c^{(1)}D_6r^{-1}, \quad \text{for } r_2 \leq r \leq r_1
\end{aligned} \tag{A.189}
$$

where C_i, D_i, and D_{3+i} are unknown constants. From Equations A.189 and A.187, we find that

$$
\begin{aligned}
D_1 &= C_1 + 2r_2^2[(2 - 2c^{(2)} + d^{(2)})C_2 - (2 - 2c^{(1)} + d^{(1)})D_2] \\
&\quad - e^{(2)}(2\delta^{(2)})^{-1}[C_3I_0(\delta^{(2)}r_2) - D_3I_0(\delta^{(1)}r_2) + D_4K_0(\delta^{(1)}r_2)] \\
D_2 &= h_1 + h_2D_4 + h_3D_5, \qquad D_3 = h_4 + h_5D_4 + h_6D_5, \qquad D_6 = 0 \\
D_4 &= (g_1g_2 - h_8g_3)(h_7g_1 - h_8h_9)^{-1}, \qquad D_5 = (h_7g_3 - h_9g_2)(h_7g_1 - h_8h_9)^{-1} \\
C_2 &= g_4 + g_5D_2 + g_6D_3 + g_7D_4 + g_8D_5 \\
C_3 &= g_9 + k_1D_2 + k_2D_3 + k_3D_4 + k_4D_5
\end{aligned}
$$

where

$$
\begin{aligned}
h_1 &= \left[-\frac{1}{4}\delta^{(1)}r_1^2I_1'(\delta^{(1)}r_1) - e^{(1)}(\delta^{(1)}\gamma^{(1)})^{-1}(\beta^{(1)} + \gamma^{(1)})I_2(\delta^{(1)}r_1)\right]G_1^{-1} \\
h_2 &= [I_1'(\delta^{(1)}r_1)K_2(\delta^{(1)}r_1) + K_1'(\delta^{(1)}r_1)I_2(\delta^{(1)}r_1)]e^{(1)}G_1^{-1} \\
h_3 &= -2\delta^{(1)}I_1'(\delta^{(1)}r_1)\left(r_1^2G_1\right)^{-1} \\
h_4 &= 2r_1^2[(\gamma^{(1)})^{-1}(\beta^{(1)} + \gamma^{(1)})(2 - 2c^{(1)} + d^{(1)}) - 1]G_1^{-1} \\
h_5 &= [-2(2 - 2c^{(1)} + d^{(1)})\delta^{(1)}r_1^2K_1'(\delta^{(1)}r_1) + 8e^{(1)}(\delta^{(1)})^{-1}K_2(\delta^{(1)}r_1)]G_1^{-1} \\
h_7 &= \eta^{(1)}[K_2(\delta^{(1)}r_2) - h_5I_2(\delta^{(1)}r_2)] \\
&\quad + 2r_2^2[\rho^{(1)}h_2 - \rho^{(2)}Q_1] + \eta^{(2)}Q_2I_2(\delta^{(2)}r_2), \qquad h_6 = -16\left(r_1^2G_1\right)^{-1} \\
h_8 &= -2r_2^{-2} - \eta^{(1)}h_6I_2(\delta^{(1)}r_2) + 2r_2^2(\rho^{(1)}h_3 - \rho^{(2)}Q_3) + \eta^{(2)}Q_4I_2(\delta^{(2)}r_2) \\
h_9 &= -K_1(\delta^{(1)}r_2) + 8r_2h_2 - h_5I_1(\delta^{(1)}r_2) - 8r_2Q_1 + Q_2I_1(\delta^{(2)}r_2) \\
g_1 &= 8r_2h_3 - h_6I_1(\delta^{(1)}r_2) - 8r_2Q_3 + Q_4I_1(\delta^{(2)}r_2) \\
g_2 &= -2r_2^2\rho^{(1)}h_1 + \eta^{(1)}h_4I_2(\delta^{(1)}r_2) + 2r_2^2Q_5\rho^{(2)} - Q_6I_2(\delta^{(2)}r_2)\eta^{(2)} \\
g_3 &= -8r_2h_1 + h_4I_1(\delta^{(1)}r_2) + 8r_2Q_5 - Q_6I_1(\delta^{(2)}r_2)
\end{aligned}
$$

$$g_4 = \Big[(\chi - 1)e^{(2)}(\delta^{(2)}\gamma^{(2)})^{-1}(\beta^{(2)} + \gamma^{(2)})I_2(\delta^{(2)}r_2) + \frac{1}{4}(\omega - 1)\delta^{(2)}r_2^2 I_1'(\delta^{(2)}r_2)\Big] G_2^{-1}$$

$$g_5 = [8\varepsilon\eta^{(2)}I_2(\delta^{(2)}r_2) + 2\omega\rho^{(1)}\delta^{(2)}r_2^2 I_1'(\delta^{(2)}r_2)]G_2^{-1}$$

$$g_6 = [-\varepsilon\eta^{(2)}\delta^{(1)}I_2(\delta^{(2)}r_2)I_1'(\delta^{(1)}r_2) + \omega e^{(1)}\zeta I_2(\delta^{(1)}r_2)I_1'(\delta^{(2)}r_2)]G_2^{-1}$$

$$g_7 = -[\varepsilon e^{(2)}\zeta^{-1}I_2(\delta^{(2)}r_2)K_1'(\delta^{(1)}r_2) + \omega e^{(1)}\zeta K_2(\delta^{(1)}r_2)I_1'(\delta^{(2)}r_2)]G_2^{-1}$$

$$g_8 = 2\omega\delta^{(2)}I_1'(\delta^{(2)}r_2)(r_2^2G_2)^{-1}, \qquad k_1 = 16r_2^2[\omega\rho^{(1)} - \varepsilon\rho^{(2)}]G_2^{-1}$$

$$g_9 = 2r_2^2\big[\omega - 1 + \rho^{(2)}(\beta^{(2)} + \gamma^{(2)})(\gamma^{(2)})^{-1}(1 - \chi)\big]G_2^{-1}$$

$$k_2 = \big[8\omega\eta^{(1)}I_2(\delta^{(1)}r_2) + 2\varepsilon\rho^{(2)}\delta^{(1)}r_2^2 I_1'(\delta^{(1)}r_2)\big]G_2^{-1}$$

$$k_3 = \big[-8\omega\eta^{(1)}K_2(\delta^{(1)}r_2) + 2\varepsilon\rho^{(2)}\delta^{(1)}r_2^2 K_1'(\delta^{(1)}r_2)\big]G_2^{-1}$$

$$k_4 = 16\omega\big(r_2^2G_2\big)^{-1}, \qquad Q_1 = g_7 + h_5g_6 + h_2g_5$$

$$Q_2 = k_3 + h_5k_2 + h_2k_1, \quad Q_3 = g_8 + h_6g_6 + h_3g_5, \quad Q_4 = k_4 + h_6k_2 + h_3k_1$$

$$Q_5 = g_4 + h_1k_5 + h_4g_6, \quad Q_6 = g_9 + h_1k_1 + h_4k_2, \quad \rho^{(\alpha)} = 2 - 2c^{(\alpha)} + d^{(\alpha)}$$

$$\omega = (2\mu^{(1)} + \kappa^{(1)})(2\mu^{(2)} + \kappa^{(2)})^{-1}, \qquad \varepsilon = \gamma^{(1)}(\gamma^{(2)})^{-1}$$

$$\chi = (\beta^{(1)} + \gamma^{(1)})(\beta^{(2)} + \gamma^{(2)})^{-1}, \qquad \zeta = \delta^{(2)}(\delta^{(1)})^{-1}$$

$$\eta^{(\alpha)} = e^{(\alpha)}(\delta^{(\alpha)})^{-1}, \qquad G_1 = 8\eta^{(1)}I_2(\delta^{(1)}r_1) + 2\rho^{(1)}\delta^{(1)}r_1^2 I_1'(\delta^{(1)}r_1)$$

$$G_2 = 8\eta^{(2)}I_2(\delta^{(2)}r_2) + 2\delta^{(2)}r_2^2\rho^{(2)}I_1'(\delta^{(2)}r_2)$$

Therefore, the solutions of the problems $\mathcal{E}^{(\sigma)}$, $(\sigma = 1, 2)$, are

$$\begin{aligned} u_r^{(1)} &= v_r\cos\theta, & u^{(1)} &= v_\theta\sin\theta, & \varphi_z^{(1)} &= \psi_z\sin\theta \\ u_r^{(2)} &= v_r\sin\theta, & u^{(2)} &= -v_\theta\cos\theta, & \varphi_z^{(2)} &= -\psi_z\cos\theta \end{aligned} \tag{A.190}$$

where v_r, v_θ, and ψ_z are given by Equations A.189. The constant C_1 characterizes a rigid translation. Let us consider now the problem $\mathcal{E}^{(3)}$ defined by Equations A.185. From Equation A.185$_1$, we obtain

$$u_r^{(3)} = E_1r, \quad u_\theta^{(3)} = \eta^{(2)}E_2I_1(\delta^{(2)}r), \quad \varphi_z^{(3)} = E_2I_0(\delta^{(2)}r), \text{ for } 0 \le r \le r_2$$

$$u_r^{(3)} = F_1r + F_2r^{-1}, \quad u_\theta^{(3)} = \eta^{(1)}[F_3I_1(\delta^{(1)}r) - F_4K_1(\delta^{(1)}r)] + F_5r^{-1}$$

$$\varphi_z^{(3)} = F_3I_0(\delta^{(1)}r) + F_4K_0(\delta^{(1)}r), \text{ for } r_2 \le r \le r_1$$

where E_α and F_s are unknown constants. If we impose the conditions A.185, we find that

$$E_1 = -\nu^{(1)} + \big[r_2^{-2} + r_1^{-2} - 2\nu^{(1)}r_1^{-2}\big]F_2, \; F_1 = -\nu^{(1)} + (1 - 2\nu^{(1)})r_1^{-2}F_2$$

$$F_2 = (\nu^{(1)} - \nu^{(2)})\big[(2\mu^{(1)} + \kappa^{(1)})\lambda^{(2)}(\nu^{(2)})^{-1}(r_2^{-2} - r_1^{-2}) + r_1^{-2} + r_2^{-2} - 2\nu^{(1)}r_1^{-2}\big]^{-1}$$

$$E_2 = F_3 = F_4 = 0, \qquad \nu^{(\rho)} = \lambda^{(\rho)}(2\lambda^{(\rho)} + 2\mu^{(\rho)} + \kappa^{(\rho)})^{-1}$$

so that the solution of the problem $\mathcal{E}^{(3)}$ is

$$\begin{aligned} &u_r^{(3)} = E_1 r, \qquad u_\theta^{(3)} = 0, \qquad \varphi_z^{(3)} = 0, \quad \text{for } 0 \le r \le r_2 \\ &u_r^{(3)} = F_1 r + F_2 r^{-1}, \qquad u_\theta^{(3)} = 0, \qquad \varphi_z^{(3)} = 0, \quad \text{for } r_2 \le r \le r_1 \end{aligned} \tag{A.191}$$

With the help of Equations 6.5.16, A.190, and A.191, we obtain the solution of the extension and bending problem in the form

$$\begin{aligned} u_r &= \left(-\frac{1}{2}z^2 + v_r\right)(a_1 \cos\theta + a_2 \sin\theta) + a_3 u_r^{(3)} \\ u_\theta &= \left(\frac{1}{2}z^2 + v_\theta\right)(a_1 \sin\theta - a_2 \cos\theta) \\ u_z &= rz(a_1 \cos\theta + a_2 \sin\theta) + a_3 z \\ \varphi_r &= -(a_1 \sin\theta - a_2 \cos\theta)z, \qquad \varphi_\theta = -(a_1 \cos\theta + a_2 \sin\theta)z \\ \varphi_z &= (a_1 \sin\theta - a_2 \cos\theta)\psi_z \end{aligned}$$

where the constants a_s can be determined from the system 6.5.18. From Equations 6.5.19, A.190, and A.191, we find that

$$\begin{aligned} Y_{12} &= Y_{21} = Y_{3\alpha} = Y_{\alpha 3} = 0 \\ Y_{33} &= \pi(\lambda^{(2)} + 2\mu^{(2)} + \kappa^{(2)} + 2\lambda^{(2)} E_1) r_2^2 \\ &\quad + \pi(\lambda^{(1)} + 2\mu^{(1)} + \kappa^{(1)} + 2\lambda^{(1)} F_1)(r_1^2 - r_2^2) \\ Y_{11} &= Y_{22} = \frac{\pi}{4}(2\mu^{(2)} + \kappa^{(2)}) r_2^4 + \pi(\beta^{(2)} + \gamma^{(2)}) r_2^2 \\ &\quad - \pi\beta^{(2)} r_2 \left[C_3 I_1(\delta^{(2)} r_2) - 8 C_2 r_2\right] - 2\pi\lambda^{(2)}(2c^{(2)} - d^{(2)}) r_2^4 C_2 \\ &\quad + \pi\lambda^{(1)} D_5(r_1 - r_2) + \frac{\pi}{4}(2\mu^{(1)} + \kappa^{(1)})(r_1^4 - r_2^4) + \pi(\beta^{(1)} + \gamma^{(1)})(r_1^2 - r_2^2) \\ &\quad - \pi\beta^{(1)}\{D_3\left[r_1 I_1(\delta^{(1)} r_1) - r_2 I_1(\delta^{(1)} r_2)\right] - 8 D_2(r_1^2 - r_2^2) \\ &\quad + D_4\left[r_1 K_1(\delta^{(1)} r_1) - r_2 K_1(\delta^{(1)} r_2)\right]\} - 2\pi\lambda^{(1)}(2c^{(1)} - \delta^{(1)}) D_2(r_1^4 - r_2^4) \end{aligned}$$

so that, from Equation 6.5.18, we obtain

$$a_1 = \frac{1}{Y_{11}} M_2, \qquad a_2 = -\frac{1}{Y_{11}} M_1, \qquad a_3 = -\frac{1}{Y_{33}} F_3$$

Thus, the problem is solved.

Bibliography

[1] M. Abassi, Simple solutions of Saint-Venant's torsion problem by using Tschebycheff polynomials. *Quart. Appl. Math.*, 14(1956), 345–357.

[2] R. Abeyaratne, C.O. Horgan and D.T. Chung, Saint-Venant end effects for incremental plane deformations of incompressible nonlinearly elastic materials. *J. Appl. Mech.*, 52(1985), 847–852.

[3] J. Aboudi, M.J. Pindera and S.M. Arnold, Higher-order theory for functionally graded materials. *Composites, Part B*, 30(1999), 774–832.

[4] E.A. Abramidze, Computer realization of solutions of Almansi-Michell problems for saw-tooth domains (in Russian). *Trudy Vycisl. Tsentra Akad. Nauk Gruzin.*, SSR 21(1981), 5–21.

[5] E.L. Aero and E.V. Kuvshinskii, Fundamental equations of the theory of elastic media with rotationally interacting particles (in Russian). *Fiz. Tverd. Tela*, 2(1960), 1399–1409.

[6] E. Almansi, Sopra la deformazione dei cilindri sollecitati lateralmente. *Atti Accad. Naz. Lincei Rend. Cl. Sci. Fis. Mat. Natur. Ser.*, 5, 10(1901), I: 333–338; II: 400–408.

[7] S.S. Antman, *Nonlinear Problems in Elasticity, Applied Mathematical Sciences*, vol. 107, 2nd ed., Springer, 2004.

[8] T. Ariman and M.J. Zika, On complex potentials in micropolar elasticity. *ZAMM-Z. Angew. Math. Mech.*, 51(1971), 183–188.

[9] N.H. Arutyunyan and B.L. Abramyan, *Torsion of Elastic Bodies* (in Russian). Fizmatghiz, Moscow, 1963.

[10] M.O. Basheleishvili and V.D. Kupradze, New integral equations in the theory of anisotropic elasticity and their applications for solving boundary value problems (in Russian). *Soobsc. Akad. Nauk Gruzin.*, SSR 15(1954), 415–422.

[11] M.O. Basheleishvili, On fundamental solutions of the differential equations of an anisotropic elastic body (in Russian). *Soobsc. Akad. Nauk Gruzin.*, SSR 19(1957), 393–400.

[12] M.O. Basheleishvili, Solution of plane boundary-value problems of statics of anisotropic elastic solids (in Russian). *Trudy Vychisl. Tsentra Akad. Nauk Gruzin.*, SSR 3(1961), 91–105.

[13] M.O. Basheleishvili, Solution of the plane boundary value problems of the statics of anisotropic elastic bodies (in Russian). *Trudy Vychisl. Tsentra Akad. Nauk Gruzin.*, SSR 3(1963), 91–104.

[14] M.O. Basheleishvili, On the basic plane boundary value problems for nonhomogeneous anisotropic elastic body. *Trudy Tibiliss. Gosudarstv. Universiteta*, 111(1966), 117–127.

[15] R. Batra, Saint-Venant's principle in linear elasticity with microstructure. *J. Elasticity*, 13(1983), 165–173.

[16] R.C. Batra and J.S. Yang, Saint-Venant's principle for linear elastic porous materials. *J. Elasticity*, 39(1995), 265–271.

[17] R.C. Batra, F. Dell'Isola and G.C. Ruta, Second-order solution of Saint-Venant's problem for an elastic bar predeformed in flexure. *Int. J. Non-Linear Mech.*, 40(2005), 411–422.

[18] Y. Benveniste and T. Chen, On the Saint-Venant torsion of composite bars with imperfect interfaces. *Proceedings: Mathematical, Physical and Engineering Sciences*, 457(2001), 231–255.

[19] V.L. Berdicevski, On the proof of the Saint-Venant principle for bodies of arbitrary shape (in Russian). *Prikl. Mat. Meh.*, 38(1974), 851–864.

[20] C.A. Berg, The physical meaning of astatic equilibrium in Saint-Venant's principle for linear elasticity. *J. Appl. Mech.*, 36(1969), 392–396.

[21] K. Berglund, Generalization of Saint-Venant's principle to micropolar continua. *Arch. Rational Mech. Anal.*, 64(1977), 317–326.

[22] Y. Biollay, First boundary value problem in elasticity: Bounds for the displacement in Saint-Venant's principle. *ZAMM-Z. Angew. Math. Mech.*, 31(1980), 556–567.

[23] M. Bîrsan, The solution of Saint-Venant's problem in the theory of Cosserat shells. *J. Elasticity*, 74(2004), 185–214.

[24] M. Bîrsan, Saint-Venant's problem for Cosserat shells with voids. *Int. J. Solids Struct.*, 42(2005), 2033–2057.

[25] R. De Boer, Theory of porous media-Past and present. *ZAMM-Z. Angew. Math. Mech.*, 78(1998), 441–466.

[26] B.A. Boley, On a dynamical Saint-Venant principle. *J. Appl. Mech.*, 27(1960), 74–78.

[27] A.P. Boresi and R.J. Schmidt, *Advanced Mechanics of Materials*, John Wiley & Sons, New York, 2002.

[28] C.I. Borş, *Theory of Elasticity for Anisotropic Bodies* (in Romanian). Editura Academiei, Bucureşti, 1970.

[29] J. Boussinesq, Application des potentiels à l'étude de l'équilibre et des mouvements des solides élastiques. Gauthier-Villars, Paris, 1885.

[30] F. Bouyge, I. Iasiuk, S. Boccara and M. Ostoja-Starzewski, A cromechanically based couple-stress model of an elastic orthotropic two-plane composite. *Europ. J. Mech. A: Solids*, 21(2002), 465–481.

[31] S. Breuer and J.J. Roseman, On Saint-Venant's principle in three-dimensional nonlinear elasticity. *Arch. Rational Mech. Anal.*, 63(1977), 191–203.

[32] O. Brulin, Linear micropolar media. In *Mechanics of Micropolar Media* (Edited by O. Brulin and R.K.T. Hsieh), World Scientific, Singapore, 1982, 97–146.

[33] E.J. Brunelle, Stability and vibration of transversely isotropic beams under initial stress. *J. Appl. Mech.*, 39(1972), 819–821.

[34] T.V. Burchuladze, On some generalized potentials for anisotropic elastic bodies (in Russian), *Soobsc. Akad. Nauk Gruzin.*, SSR 23(1959), 135–142.

[35] T.V. Burchuladze, Application of singular integral equations for solving boundary-value problems (in Russian). *Trudy Tbiliskogo Mat. Inst.*, 28(1962), 21–31.

[36] S. Bytner and B. Gambin, Homogenization of Cosserat continuum. *Arch. Mech.*, 38(1986), 289–301.

[37] G. Capriz, Continua with Microstructure. In *Springer Tracts in Natural Philosophy*, vol. 35 (Edited by C. Truesdell), Springer-Verlag, New York, 1989.

[38] D.E. Carlson, Stress functions for couple and dipolar stresses. *Quart. Appl. Math.*, 24(1966), 29–37.

[39] D.E. Carlson, Linear Thermoelasticity, pp. 297–346 of *Flügge's Handbuch der Physik*, vol. VIa/2 (Edited by C. Truesdell), Springer-Verlag, Berlin, Heidelberg, New York, 1972.

[40] A. Carpinteri, *Structural Mechanics: A Unified Approach*, Chapman & Hall, 2000.

[41] A.L. Cauchy, Recherches sur l'équilibre et le mouvement intérieur des corps solids ou fluides, élastiques ou non élastiques. *OEuvres*, 2(1823), 300–304.

[42] P. Chadwick, *Continuum Mechanics. Concise Theory and Problems*, George Allen & Unwin, London, 1976.

[43] Chao Hwei Yuan, On the torsion of non-homogeneous anisotropic elastic cylinders. *Sci. Sinica*, 9(1960), 47–61.

[44] S. Chiriţă, Saint-Venant's problem for nonlocal elastic solids. *An. Şt. Univ. "Al.I. Cuza" Iaşi, s. Matematică*, 24(1978), 147–156.

[45] S. Chiriţă, Saint-Venant's problem for anisotropic circular cylinders. *Acta Mech.*, 34(1979), 243–250.

[46] S. Chiriţă, Some growth and decay estimates for a cylinder made of an elastic material with voids. *Rev. Roum. Math. Pures et Appl.*, 39(1994), 17–26.

[47] S. Chiriţă and R. Quintanilla, On Saint-Venant's principle in linear elastodynamics. *J. Elasticity*, 42(1996), 201–215.

[48] B.M. Chiu and J.D. Lee, On the plane problem in micropolar elasticity. *Int. J. Engng. Sci.*, 11(1973), 997–1012.

[49] I. Choi and C.O. Horgan, Saint-Venant's principle and end effects in anisotropic elasticity. *J. Appl. Mech. Trans.* ASME, Series E, 44(1977), 424–430.

[50] M. Ciarletta and D. Ieşan, *Non-classical Elastic Solids*, Longman Scientific and Technical, Longman Group UK Limited, 1993.

[51] Gh. Gr. Ciobanu, On Almansi-Michell's problem for viscoelastic beams. *An. Şt. Univ. "Al.I. Cuza", s.I a Matematică*, 20(1974), 197–208.

[52] A. Clebsch, *Theorie der Elasticität fester Körper*. B.G. Teubner, Leipzig, 1862.

[53] B.D. Coleman and D. Swigon, Theory of self-contact in Kirchhoff rods with applications to supercoiling of knotted and unknotted DNA plasmids. *Roy. Soc. London A*, 362(2004), 1281–1299.

[54] E. Cosserat and F. Cosserat, *Théorie des corps déformables.* Hermann, Paris, 1909.

[55] R. Courant and D. Hilbert, *Methods of Mathematical Physics*, Vol. I, 1953; Vol. II, Interscience, New York, 1962.

[56] S.C. Cowin, Stress functions for Cosserat elasticity. *Int. J. Solids Struct.*, 6(1970), 389–399.

[57] S.C. Cowin and J.W. Nunziato, Linear elastic materials with voids. *J. Elasticity*, 13(1983), 125–147.

[58] S.C. Cowin, The stresses around a hole in a linear elastic material with voids. *Quart. J. Mech. Appl. Math.*, 36(1983), 247–260.

[59] S. Cowin and P. Puri, The classical pressure vessel problems for linear elastic materials with voids. *J. Elasticity*, 13(1983), 157–163.

[60] S.C. Cowin, The stress around a hole in a linear elastic materials with voids. *Q.J. Mech. Appl. Math.*, 37(1984), 441–465.

[61] I. Crăciun, Potential method in micropolar thermoelasticity. *Rev. Roum. Sci. Techn. Méc. Appl.*, 22(1977), 19–36.

[62] W.A. Day, Generalized torsion: The solution of a problem of Truesdell's. *Arch. Rational Mech. Anal.*, 76(1981), 283–288.

[63] S. De Cicco and L. Nappa, Torsion and flexure of microstretch elastic circular cylinders. *Int. J. Engng. Sci.*, 35(1997), 573–583.

[64] F. Dell'Isola and R.C. Batra, Saint-Venant's problem for porous linear elastic materials. *J. Elasticity*, 47(1971), 73–81.

[65] H. Demiray and E.S. Suhubi, Small torsional oscillations of an initially twisted circular rubber cylinder. *Int. J. Engng. Sci.*, 8(1970), 19–30.

[66] E. Deutsch, On the centre of flexure. *ZAMP* 12(1961), 212–218.

[67] J.B. Diaz and L.E. Payne, Mean value theorems in the theory of elasticity. *Proc. Third. U.S. Natl. Congres Appl. Mech.* (1958), 293–303.

[68] G.Y. Djanelidze, Almansi problem. *Leningrad. Politehn. Inst. Trudy*, 210(1960), 25–38.

[69] S.B. Dong, J.B. Kosmatka and H.C. Lin, On Saint-Venant's problem for an inhomogeneous, anisotropic cylinder-Part III: End effects. *J. Appl. Mech. Trans. ASME*, 68(2001), 392–398.

[70] L. Dragoş, *Principles of Mechanics of Continua* (in Romanian). Ed. Tehnica, Bucureşti, 1983.

[71] L. Dragoş, Fundamental solutions in micropolar elasticity. *Int. J. Engng. Sci.*, 22(1984), 265–275.

[72] G. Duvaut, Élasticité linéaire avec couples de contraintes. Théorèmes d'existence. *J. Mécanique*, 9(1970), 325–333.

[73] J. Dyszlewicz and C. Kolodziej, Selected one-dimensional problems of elastostatics and thermoelastostatics of the hemitropic micropolar medium. *Bull. Acad. Polon. Sci. Ser. Tech.*, 35(1987), 15–25.

[74] J. Dyszlewicz, *Micropolar Theory of Elasticity.* Springer-Verlag, Berlin, Heidelberg, New York, 2004.

[75] F. Erdogan, Fracture mechanics of functionally graded materials. *Compos. Eng.*, 5(1995), 753–770.

[76] J. Ericksen and C. Truesdell, Exact theory of stress and strain in rods and shells. *Arch. Rational Mech. Anal.*, 1(1958), 295–323.

[77] J.L. Ericksen, Equilibrium of bars. *J. Elasticity*, 5(1975), 191–201.

[78] J.L. Ericksen, On the formulation of Saint-Venant's problem. Nonlinear Analysis and Mechanics, Heriot-Watt Symposium vol. 1, 158–186, Pitman, London, 1977.

[79] J.L. Ericksen, On Saint-Venant's problem for thin-walled tubes. *Arch. Rational Mech. Anal.*, 70(1979), 7–12.

[80] J.L. Ericksen, On the status of Saint-Venant's solutions as minimizers of energy. *Int. J. Solids Struct.*, 16(1980), 195–198.

[81] A.C. Eringen and E.S. Suhubi, Nonlinear theory of simple microelastic solids. *Int. J. Engng. Sci.*, 2(1964), I: 189–203; II: 389–404.

[82] A.C. Eringen, *Mechanics of Continua.* John Wiley and Sons, New York, 1967.

[83] A.C. Eringen, Theory of micropolar elasticity. *Fracture*, vol. 2, Academic Press, New York, 1968.

[84] A.C. Eringen and C.B. Kafadar, Polar Fields Theories. In *Continuum Physics* (Edited by A.C. Eringen) vol. IV, Academy Press, New York, 1976.

[85] A.C. Eringen, *Microcontinuum Field Theories I: Foundations and Solids.* Springer-Verlag, New York, Berlin, Heidelberg, 1999.

[86] G. Ferrarese, Sulla compatibilità dei continui alla Cosserat. *Rendiconti di Mat., Serie VI*, 4(1971), 1–23.

[87] G. Fichera, Sulla torsione elastica dei prismi cavi. *Rend. Mat. Ser., 5*, 12(1953) 163–176.

[88] G. Fichera, Existence Theorems in Elasticity, pp. 347–388 of *Flügge's Handbuch der Physick*, vol. VIa/2 (Edited by C. Truesdell), Springer-Verlag, Berlin-Heidelberg-New York, 1972.

[89] G. Fichera, Il principio di Saint-Venant: intuizione dell'ingegnere e rigore del matematico. *Rend. Mat. Ser. VI*, 10(1977), 1–24.

[90] G. Fichera, Remarks on Saint-Venant's principle. *Complex Analysis and its Applications.* I.N. Vekua Anniversary Volume, 543–554, Moscow, 1978.

[91] G. Fichera, Problemi Analitici Nuovi nella Fisica Matematica Classica. Quaderni del Consiglio Nazionale delle Ricerche Gruppo del Consiglio Nazionale di Fisica Matematica, Scuola Tipo-Lito "Istituto Anselmi", Napoli, 1985.

[92] J.N. Flavin, On Knowles' version of Saint-Venant's principle in two-dimensional elastostatics. *Arch. Rational Mech. Anal.*, 53(1974), 366–375.

[93] J.N. Flavin, Another aspect of Saint-Venant's principle in elasticity. *ZAMP*, 29(1978), 328–332.

[94] Y.B. Fu and R.W. Ogden, *Nonlinear Elasticity: Theory and Applications.* Cambridge University Press, Cambridge, 2001.

[95] D.Y. Gao, Nonlinear elastic beam theory with application in contact problem, and variational approaches. *Mech. Res. Comm.*, 23 (1996), 11–17.

[96] D.Y. Gao, Finite deformation beam models and triality theory in dynamical post-buckling analysis, *Int. J. Non-Linear Mech.*, 35(2000), 103–131.

[97] D.Y. Gao and R.W. Ogden, Closed form solutions, extremality and nonsmoothness criteria in a large deformation elasticity problem. *ZAMP* 59(2008), 498–517.

[98] R.D. Gauthier and W.E. Jahsman, Reexamination of micropolar wave experiment. In *Continuum Models of Discrete Systems 4* (Edited by O. Brulin and R.K.T. Hsieh), North-Holland Publ., Amsterdam, 1981, 225–231.

[99] R.D. Gauthier and E.W. Jahsman, A quest for micropolar elastic constants II. *Arch. Mech.*, 33(1981), 717–737.

[100] T.G. Gegelia and R.K. Chichinadze, Basic static problems of elastic micropolar media. *Arch. Mech.*, 28(1976), 89–104.

[101] P. Germain, *Cours de Mécanique des Milieux Continus.* Masson et Cie., Paris, 1973.

[102] G. Geymonat, F. Krasucki and J.J. Marigo, Sur la commutativité des passages à la limite en théorie asymptotique des poutres composites. *C.R. Acad. Sci. Paris*, t.305, Sèrie II(1987), 225–228.

[103] G.M.L. Gladwell, The vibration of cylinders. *Shock and Vibration Digest*, 8(1976), 13–34.

[104] B. Gleeson and J.N. Flavin, A convexity property of inhomogeneous incompressible elastic cylinders. *Mathematical Proceedings of the Royal Irish Academy*, 101 A(2001), 3–6.

[105] J.N. Goodier, An extension of Saint-Venant's principle with applications. *J. Appl. Phys.*, 13(1942), 167–171.

[106] M.A. Goodman and S.C. Cowin, A continuum theory for granular materials. *Arch. Rational Mech. Anal.*, 44(1972), 249–266.

[107] A.E. Green and R.T. Shield, Finite extension and torsion of cylinders. *Phil. Trans. Roy. Soc. Lond.*, A 224(1951), 47–86.

[108] A.E. Green and J.E. Adkins, *Large Elastic Deformations.* Clarendon Press, Oxford, 1960.

[109] A.E. Green and R.S. Rivlin, Simple forces and stress multipoles. *Arch. Rational Mech. Anal.*, 16(1964), 325–353.

[110] A.E. Green and R.S. Rivlin, Multipolar continuum mechanics. *Arch. Rational Mech. Anal.*, 17(1964), 113–147.

[111] A.E. Green, Micro-materials and multipolar continuum mechanics. *Int. J. Engng. Sci.*, 3(1965), 533–537.

[112] A.E. Green and R.S. Rivlin, The relation between director and multipolar theories in continuum mechanics. *ZAMP*, 18(1967), 208–218.

[113] A.E. Green and W. Zerna, *Theoretical Elasticity*, 2nd edition. Clarendon Press, Oxford, 1968.

[114] A. Ya. Grigorenko and G.G. Vlaikov, *Some Problems of Elasticity Theory for Anisotropic Bodies of Cylindrical Form.* Akademperiodika, Kyiv, 2002.

[115] I. Grindei, Flexure of orthotropic beams with multiply-connected section (in Romanian). *St. Cerc. Şt. Fil. Iaşi Acad. R.P.R.*, 13(1962), 383–388.

[116] G. Grioli, *Mathematical Theory of Elastic Equilibrium.* Springer-Verlag, Berlin-Götting-Heidelberg, 1962.

[117] G. Grioli, Microstructure. *Rend. Acc. Naz. dei Lincei, Serie VIII*, 49(1970), I: 261–267, II: 379–385.

[118] G. Grioli, Saint–Venant's theory of the equilibrium of a cylindrical elastic body and exact solutions. In "Il Problema di de Saint-Venant: Aspetti Teorici e Applicativi, Accademia Nazionale dei Lincei, Atti dei Convegni Lincei 140, Roma (1998), 65–76.

[119] M.E. Gurtin, The Linear Theory of Elasticity, pp. 1–295 of *Flügge's Handbuch der Physik*, vol. VIa/2 (Edited by C. Truedell), Springer-Verlag, Berlin-Heidelberg-New York, 1972.

[120] M Haimovici, *Theory of Elasticity* (in Romanian). Editura Didactică şi Pedagogică, Bucureşti, 1969.

[121] R.B. Hetnarski and J. Ignaczak, *Mathematical Theory of Elasticity.* Taylor and Francis, New York, 2004.

[122] T.J. Higgins, A comprehensive review of the Saint-Venant's torsion problem. *Am. J. Phys.*, 10(1942), 248–259.

[123] J.M. Hill, Closed form solutions for small deformations superimposed upon the simultaneous inflation and extension of a cylindrical tube. *J. Elasticity*, 6(1976), 113–123.

[124] S. Hjalmars, Non-linear micropolar theory. In *Mechanics of Micropolar Media* (Edited by O. Brulin and R.K.T. Hsieh), World Scientific, Singapore, 1982, 147–189.

[125] S. Hjalmars, The muscle-relaxed equilibrium of the human spine analyzed by means of an anisotropic beam model. In "*New Problems in Mechanics of Continua*". Proc. of the Third Swedish-Polish Symposion

on Mechanics, Joblonna, Poland, 1982, University of Waterloo Press, Study No. 17, 1983.

[126] I. Hlaváček and M. Hlaváček, On the existence and uniqueness of solution and some variational principles in linear theories of elasticity with couple-stress. Aplikace Matematiky, 14(1969), I: 387–410; II: 411–426.

[127] I. Hlaváček and J. Nečas, On inequalities of Korn's type. *Arch. Rational Mech. Anal.*, 36(1970), 305–311.

[128] C.O. Horgan and J.K. Knowles, The effect of nonlinearity on the principle of Saint-Venant. *J. Elasticity*, 11(1981), 271–292.

[129] C.O. Horgan and J.K. Knowles, Recent developments concerning Saint-Venant's principle. *Adv. Appl. Mech.*, 23(1983), 179–269.

[130] C.O. Horgan, Recent developments concerning Saint-Venant's principle: A second update. *App. Mech. Rev.*, 49(1996), 101–111.

[131] C.O. Horgan, Decay estimates for boundary-value problems in linear and nonlinear continuum mechanics. In *Mathematical Problems in Elasticity* (Edited by R. Russo), *Advances in Mathematics for Applied Sciences*, 38(1996), 47–89.

[132] C.O. Horgan and A.M. Chan, Torsion of functionally graded isotropic linearly elastic bars. *J. Elasticity*, 52(1998/99), 181–199.

[133] C.O. Horgan and R. Quintanilla, Saint-Venant end effects in antiplane shear for functionally graded linearly elastic materials. *Math. Mech. Solids*, 6(2001), 115–132.

[134] C.O. Horgan and J.G. Murphy, Lie group analysis and plane strain bending of cylindrical sectors for compressible nonlinearly elastic materials. *IMA J. Appl. Math.*, 70(2005), 80–91.

[135] G. Horvay, Saint-Venant's principle: A biharmonic eigenvalue problem. *J. Appl. Mech.*, 24(1957), 381–386.

[136] S.C. Hunter, *Mechanics of Continuous Media*, Ellis Horwood Publisher, Chichester, 1976.

[137] C.M. Ieşan, Torsion of inhomogeneous micropolar elastic beams. *Atti. Accad. Sci. Torino, Cl. Sci. Fis. Mat. Natur.*, 18(1973), 657–666.

[138] C.M. Ieşan, Existence theorems in the linear theory of multipolar elasticity. *Aplikace Matematiky*, 18(1973), 385–390.

[139] D. Ieşan, On the generalized plane strain (in Romanian). *St. Cerc. Mat.*, 21(1969), 595–609.

[140] D. Ieşan, Existence theorems in the theory of micropolar elasticity. *Int. J. Engng. Sci.*, 8(1970), 777–791.

[141] D. Ieşan, Torsion of micropolar elastic beams. *Int. J. Engng. Sci.*, 9(1971), 1047–1060.

[142] D. Ieşan, Existence theorems in micropolar elastostatics. *Int. J. Engng. Sci.*, 9(1971), 59–78.

[143] D. Ieşan, On Saint-Venant's problem in micropolar elasticity. *Int. J. Engng. Sci.*, 9(1971), 879–888.

[144] D. Ieşan, Flexure of micropolar elastic beams. *Rev. Roum. Math. Pures et Appl.*, 17(1972), 709–720.

[145] D. Ieşan, On Almansi's problem for elastic cylinders. *Atti Accad. Sci. Ist Bologna Cl. Sci. Fis. Mat. Rend.*, S. XII, 9(1972), 128–139.

[146] D. Ieşan, On the thermal stresses in beams. *J. Eng. Math.*, 6(1972), 155–163.

[147] D. Ieşan, On the second boundary value problem in the linear theory of micropolar elasticity. *Atti Accad. Naz. Lincei Rend. Cl. Sci. Fis. Mat. Natur.*, S. VIII, 55(1973), 456–459.

[148] D. Ieşan, Torsion of anisotropic micropolar elastic cylinders. *ZAMM-Z. Angew. Math. Mech.*, 54(1974), 773–779.

[149] D. Ieşan, Saint-Venant's problem for inhomogeneous bodies. *Int. J. Engng. Sci.*, 14(1976), 353–360.

[150] D. Ieşan, Saint-Venant's problem for inhomogeneous and anisotropic elastic bodies. *J. Elasticity*, 6(1976), 277–294.

[151] D. Ieşan, Saint-Venant's problem for heterogeneous elastic cylinders. *Eng. Trans.*, 24(1976), 289–306.

[152] D. Ieşan, Saint-Venant's problem for heterogeneous anisotropic elastic solids. *Ann. Mat. Pura Appl., (IV)*, 108(1976), 149–159.

[153] D. Ieşan, Saint-Venant's problem for inhomogeneous and anisotropic elastic solids with microstructure. *Arch. Mech.*, 29(1977), 419–430.

[154] D. Ieşan, Saint-Venant's problem for inhomogeneous and anisotropic Cosserat elastic solids. *ZAMM-Z. Angew. Math. Mech.*, 58(1978), 95–99.

[155] D. Ieşan, Deformation of composed Cosserat elastic cylinders. *Rev. Roum. Sci. Techn. Méc. Appl.*, 23(1978), 823–845.

[156] D. Ieşan and S. Chiriţă, Saint-Venant's problem for composite micropolar elastic cylinders. *Int. J. Engng. Sci.*, 17(1979), 573–586.

[157] D. Ieşan, Saint-Venant's problem in micropolar elasticity. In *Mechanics of Micropolar Media* (Edited by O. Brulin and R.K.T. Hsieh), World Scientific, Singapore, 1982, 281–390.

[158] D. Ieşan, Some results in the theory of elastic materials with voids. *J. Elasticity*, 15(1985), 215–224.

[159] D. Ieşan, On Saint-Venant's problem. *Arch. Rational Mech. Anal.*, 91(1986), 363–373.

[160] D. Ieşan, Generalized twist for the torsion of micropolar cylinders. *Meccanica*, 21(1986), 94–96.

[161] D. Ieşan, On the theory of uniformly loaded cylinders. *J. Elasticity*, 16(1986), 375–382.

[162] D. Ieşan, On generalized Saint-Venant's problems. *Int. J. Engng. Sci.*, 24(1986), 849–858.

[163] D. Ieşan, *Saint-Venant's Problem. Lecture Notes in Mathematics.* Vol. 1279, Springer-Verlag, Berlin, Heidelberg, New York, London, Paris, Tokyo, 1987.

[164] D. Ieşan, *Prestressed Bodies*, Longman Scientific and Technical, Longman Group UK Limited, 1989.

[165] D. Ieşan and L. Nappa, On the theory of loaded inhomogeneous cylinders. *Mech. Res. Comm.*, 34(2007), 136–144.

[166] D. Ieşan and R. Quintanilla, On the deformation of inhomogeneous orthotropic elastic cylinders. *European J. Mech. A/Solids*, 26(2007), 999–1015.

[167] D. Ieşan and A. Scalia, On Saint-Venant's principle for microstretch elastic bodies. *Int. J. Engng. Sci.*, 35(1997), 1247–1290.

[168] J. Ignaczak, Tensorial equations of motion for elastic materials with microstructure. In *"Trends in Elasticity and Thermoelasticity"* (Witold Nowacki Anniversary Volume), Wolters-Noordhoff Publishers, Groningen, 1971.

[169] J. Ignaczak, A dynamic version of Saint-Venant's principle in the linear theory of elasticity. *Bull. Acad. Polon. Sci., Sér. Sci. Techn.*, 22(1974), 483–489.

[170] W. Jaunzemis, *Continuum Mechanics.* Macmillan Co., New York, London, 1967.

[171] J.T. Jenkins, Static equilibrium of granular materials. *J. Appl. Mech.*, 42(1975), 603–606.

[172] O.D. Kellogg, *Foundations of Potential Theory.* Springer-Verlag, New York, 1967.

[173] G.M. Khatiashvili, The problem of Almansi for a composite anisotropic cylindrical body (in Russian). *Trudy Vcisl. Tsentra Akad. Nauk Gruzin.*, SSR 4(1963), 29–42.

[174] G.M. Khatiashvili, A method of solving general Almansi-Michell problems for isotropic cylindrical bodies (in Russian). *Trudy Vyčisl. Tsentra Akad. Nauk Gruzin.*, SSR 20(1980), 87–103.

[175] G.M. Khatiashvili, *Almansi-Michell Problems for Homogeneous and Composed Bodies* (in Russian). Izd. Metzniereba, Tbilisi, 1983.

[176] G.M. Khatiashvili, On Saint-Venant's problems for an isotropic two-layered elliptic tube. *Appl. Math. Inform. Mech.*, 9(2004), 72–93.

[177] J.-H. Kim and G.H. Paulino, T-stress in orthotropic functionally graded materials: Lekhnitskii and Stroh formalisms. *Int. J. Fract.*, 126(2004), 345–384.

[178] E. Kirkinis and R.W. Ogden, On extension and torsion of a compressible elastic circular cylinder.*Math. Mech. Solids.*, 7(2002), 373–392.

[179] R.J. Knops and L.E. Payne, Uniqueness theorems in linear elasticity. In *Springer Tracts in Natural Philosophy* (Edited by B.D. Coleman), vol. 19, Springer-Verlag, Berlin-Heidelberg-New York, 1971.

[180] R.J. Knops and L.E. Payne, A Saint-Venant principle for nonlinear elasticity. *Arch. Rational Mech. Anal.*, 81(1983), 1–12.

[181] R.J. Knops and L.E. Payne, Alternative spatial growth and decay for constrained motion in an elastic cylinder. *Math. Mech. Solids*, 10(2005), 281–310.

[182] J.K. Knowles, On Saint-Venant's principle in the two-dimensional linear theory of elasticity. *Arch. Rational Mech. Anal.*, 21(1966), 1–22.

[183] J.K. Knowles and E. Sternberg, On Saint-Venant's principle and the torsion of solids of revolution. *Arch. Rational Mech. Anal.*, 21(1966), 100–120.

[184] J.K. Knowles and C.O. Horgan, On the exponential decay of stress in circular elastic cylinders subject to axisymmetric self-equilibrated end loads. *Int. J. Solids and Struct.*, 5(1969), 33–50.

[185] J.K. Knowles, An energy estimate for the biharmonic equation and its application to Saint-Venant's principle in plane elastostatics. *Indian J. Pure Appl. Math.*, 14(1983), 791–805.

[186] G.V. Kolosov, Über einige Eigenschaften des ebenen Problems der Elastizitäts-theori., *Z. Math. Phys.*, 62(1914), 383–409.

[187] J.B. Kosmatka, H.C. Lin and S.B. Dong, On Saint-Venant's problem for an inhomogeneous anisotropic cylinder-Part II: Cross Sectional Properties. *J. Appl. Mech. Trans. ASME*, 68(2001), 382–391.

[188] G.V. Krishna Reddy and N.K. Venkatasubramanian, Saint-Venant's problem for a micropolar elastic circular cylinder. *Int. J. Engng. Sci.*, 14(1976), 1047–1055.

[189] G.V. Krishna Reddy and N.K. Venkatasubramanian, Flexure of a micropolar elastic circular cylinder. *Proc. Indian Acad. Sci.*, 86A(1977), 575–590.

[190] G.V. Krishna Reddy and N.K. Venkatasubramanian, On the flexural rigidity of a micropolar elastic circular cylinder. *J. Appl. Mech.*, 45(1978), 429–431.

[191] S.A. Kuliev and F.G. Radzhabov, Torsion of prismatic beams reinforced by a round rod. *Izv. Akad. Nauk Azerbaidzhan SSR Ser. Fiz.-Tekhn.* 4(1983), 143–148.

[192] I.A. Kunin, *Elastic Media with Microstructure I, Springer Series in Solid State Sciences 26*, Springer-Verlag, Berlin, 1982.

[193] I.A. Kunin, *Elastic Media with Microstructure II, Springer Series in Solid State Sciences 44*, Springer-Verlag, Berlin, 1983.

[194] V.D. Kupradze, *Method of Potentials in Elasticity* (in Russian). Gosudarstv. Izdat. Fiz., Mat. Lit., Moscow, 1963.

[195] V. Kupradze, Dynamical problems in elasticity. In *Progress in Solid Mechanics*, vol. 3 (Edited by I.N. Sneddon), North-Holland Pub, Amsterdam, 1963.

[196] V.D. Kupradze, T.G. Gegelia, M.O. Bashelishvili and T.V. Burchuladze, *Three Dimensional Problems of Mathematical Theory of Elasticity and Thermoelasticity.* North-Holland Publ., 1979.

[197] G.A. Kutateladze, On the torsion of hollow piecewise homogeneous beams (in Russian). *Dokl. Akad. Nauk*, SSSR 198(1971), 1295–1298.

[198] P. Ladevéze, Principes de Saint-Venant en déplacement et en contrainte pour les poutres droités élastiques semi-infinies. *ZAMP*, 33(1982), 132–139.

[199] P. Ladevéze, Sur le principe de Saint-Venant en elasticité. *J. Méc. Théor. Appl.*, 2(1983), 161–184.

[200] P. Ladevéze, Ph. Sanchez and J.G. Simmonds, Beamlike (Saint-Venant) solutions for fully anisotropic elastic tubes of arbitrary cross section. *Int. J. Solids Struct.*, 41(2004), 1925–1944.

[201] R.S. Lakes, Experiments in Cosserat mechanics. In *Fundamentals of Deformations and Fracture* (Edited by A. Bilby), IUTAM, Cambridge, 1984.

[202] R.S. Lakes, Saint-Venant end effects for materials with negative Poisson's ratios. *J. Appl. Mech.*, 59(1992), 744–746.

[203] S.G. Lekhnitskii, On the torsion of an anisotropic rod by forces distributed over its lateral surface. *Prikl. Mat. Meh.*, 25(1961), 56–67 (in Russian); translated as *J. Appl. Math. Mech.*, 25(1961), 75–92.

[204] S.G. Lekhnitskii, *Theory of Elasticity of an Anisotropic Elastic Body*. Holden-Day, Inc., San Francisco, 1963.

[205] S.G. Lekhnitskii, Elementary solutions of two special problems of equilibrium of nonhomogeneous cylinders. (in Russian), Sb. "Investigations on elasticity and plasticity", Izd. Leningrad Univ., vol. 6, 1967.

[206] S.G. Lekhnitskii, *Anisotropic Plates*. Gordon and Breach Sciences Publishers, New York, 1968.

[207] S.G. Lekhnitskii, Saint-Venant's problem for nonhomogeneous anisotropic beams (in Russian). Sb. "Continuum mechanics and related problems of analysis", Nauka, Moscow, 1972.

[208] E.E. Levi, Sulle equazioni lineari totalmente ellitiche alle derivate parziale. *Rend. Circ. Mat. Palermo*, 24(1907), 275–287.

[209] V.A. Lomakin, *Theory of Nonhomogeneous Elastic Bodies* (in Russian). MGU, Moscow, 1976.

[210] V.A. Lomakin, Problems of generalized theory of elasticity with couple-stresses (in Russian). *Vestnik Mosk. Univ.* S.1, 1(1987), 82–88.

[211] A.E.H. Love, *A Treatise on the Mathematical Theory of Elasticity*, Fourth Edition. Cambridge University Press, New York, 1934.

[212] C.K. Lyons, R.B. Guenther and M.R. Pyles, Considering heterogeneity in a cylindrical section of a tree, *Int. J. Solids and Struct.*, 39(2002), 4665–4675.

[213] O. Maisonneuve, Sur le principe de Saint-Venant. Thèse. Université de Poitiers, 1971.

[214] A. Manolachi, The method of complex potentials in solving two-dimensional problems for media with voids. An. St. Univ. "Al.I. Cuza" Iaşi, sect. I-a. Matematica, 39(1993), 69–94.

[215] K.Z. Markov, On the dilatational theory of elasticity. *ZAMM-Z.Angew. Math. Mech.*, 61(1981), 349–358.

[216] K.Z. Markov, Dilatation theories of elasticity. *New Problems in Mechanics of Continua*. Proc. 3rd Swed.-Pol. Symp. Mech., Jablonna, 1982, SM Stud. Ser. 17(1983), 129–138.

[217] J.E. Marsden and T.J.R. Hugues, *Mathematical Foundations of Elasticity*. Prentice-Hall Inc. Englewood Cliffs, New Jersey, 1983.

[218] M. Matschinskii, Beweis des Saint-Venantschen Prinzip. *ZAMM-Z. Angew. Math. Mech.*, 39(1959), 9–11.

[219] G.A. Maugin, *Material Inhomogeneities in Elasticity*. Chapman and Hall, London, 1993.

[220] S.I. Melnik, Estimates for the Saint-Venant principle (in Russian). *Perm. Gos. Univ. Učen. Zap. Mat.*, 103(1963), 178–180.

[221] J.H. Michell, The theory of uniformly loaded beams. *Quart. J. Math.*, 32(1901), 28–42.

[222] A. Mielke, Saint-Venant's problem and semi-inverse solutions in nonlinear elasticity. *Arch. Rational Mech. Anal.*, 102(1988), 205–229.

[223] S.G. Mikhlin, *The Problem of the Minimum of a Quadratic Functional.* Holden-Day, Inc., San Francisco, 1965.

[224] S.G. Mikhlin, *Multidimensional Singular Integrals and Integral Equations* (in Russian). Nauka, Moskow, 1962.

[225] S.G. Mikhlin, *Numerical Realization of Variational Methods* (in Russian). Moscow, 1966.

[226] L.M. Milne-Thomson, *Antiplane Elastic Systems.* Springer-Verlag, Berlin, Göttingen-Heidelberg, 1962.

[227] R.D. Mindlin and H.F. Tiersten, Effects of couple stresses in linear elasticity. *Arch. Rational Mech. Anal.*, 11(1962), 415–448.

[228] R.D. Mindlin, Microstructure in linear elasticity. *Arch. Rational Mech. Anal.*, 16(1964), 51–78.

[229] R.D. Mindlin and N.N. Eshel, On first strain-gradient theories in linear elasticity. *Int. J. Solids Struct.*, 4(1968), 109–124.

[230] R.D. Mindlin, Galerkin stress functions for non-local theories of elasticity. *Int. J. Solids Struct.*, 8(1972), 1407–1411.

[231] R.D. Mindlin, Solution of Saint-Venant's torsion problem by power series. *Int. J. Solids and Struct.*, 11(1975), 321–328.

[232] R. Mises, On Saint-Venant's principle. *Bull. Amer. Math. Soc.*, 51(1945), 555–562.

[233] M. Mişicu and C. Teodosiu, Solution with the aid of theory of functions of complex variables for nonhomogeneous isotropic body (in Russian). *Prikl. Mat. Mekh.*, 30(1966), 2, 37–53.

[234] M. Mişicu, *Torsion and Flexure* (in Romanian). Editura Academiei, Bucharest, 1973.

[235] G. Moisil, Asupra unui vector analog vectorului lui Galerkin pentru echilibrul corpurilor elastice cu izotropie transversa. *Bul. St. Acad. R.P. Romîne*, 2(1950), 207–216.

[236] R.G. Muncaster, Saint-Venant's problem in nonlinear elasticity: a study of cross-sections. *Nonlinear Analysis and Mechanics*, Heriott-Watt Symposium vol. IV, p. 17–75, Pitman, London, 1979.

[237] R.G. Muncaster, Saint-Venant's problem for slender prisms. *Utilitas Math.*, 23(1983), 75–101.

[238] M.J.P. Musgrave, Stress waves in orthorhombic micropolar media. *Proc. R. Soc. London, Ser.* 415(1988), 163–183.

[239] N.I. Muskhelishvili, Sur le problème de torsion des cylindres élastiques isotropes. *Atti. Accad. Lincei, Ser. 6*, 9(1929), 295–300.

[240] N.I. Muskhelishvili, On the problem of torsion and flexure of elastic beams composed of different materials (in Russian). *Izv. Akad. Nauk*, SSSR, 2(1932), 907–945.

[241] N.I. Muskhelishvili, Some Basic Problems of the Mathematical Theory of Elasticity. Noordhoff, Groningen, 1953.

[242] N.I. Muskhelishvili, Singular Integral Equations (in Russian), Mir, Moscow, 1962.

[243] P.M. Naghdi, The theory of shells and plates. In vol. VIa/2 of *the Handbuch der Physik*, (Edited by C. Truesdell), Springer-Verlag, Berlin, Heidelberg, New York, 1972.

[244] O.I. Napetvaridze, On the boundary value problems of the theory of elasticity with couple-stresses (in Russian). *Seminar of Institute of Applied Mathematics.* Abstract of papers. Tbilisi, 5(1971), 53–67.

[245] L. Nappa, Thermal stresses in microstretch elastic cylinders. *J. Therm. Stresses*, 18(1995), 537–550.

[246] L. Nappa, Decay estimates for micropolar elastic cylinders. *Int. J. Engng. Sci.*, 34(1996), 1601–1609.

[247] L. Nappa and S. Pesce, Thermoelastic deformation of microstretch elastic beams. *J. Therm. Stresses*, 19(1996), 763–777.

[248] J. Nečas, Les méthodes directes en théorie des equations elliptiques. Academia, Prague, 1967.

[249] J. Nečas and I. Hlavacek, *Mathematical Theory of Elastic-Plastic Bodies: An Introduction.* Elsevier Scientific Publishing Company Amsterdam, NY, 1981.

[250] M. Nicolesco, Les fonctions polyharmoniques. Actualites Scientifique et Industrielles 331, Paris, 1936.

[251] I. Nistor, Variational principles for Cosserat body. *Int. J. Nonlinear Mech.*, 37(2002), 565–569.

[252] V.V. Novozhilov and L.I. Slepian, On Saint-Venant's principles in the dynamic of beams, *J. Appl. Math. Mech.*, 29(1965), 293–315.

[253] W. Nowacki, Green functions for micropolar thermoelasticity. *Bull. Acad. Polon. Sci., Sér. Sci. Techn.*, 16(1968), 555–564.

[254] W. Nowacki, *Theory of Micropolar Elasticity.* Springer-Verlag, Wien, New York, 1970.

[255] W. Nowacki, *Theory of Asymmetric Elasticity.* Polish Scientific Publishers, Warszawa and Pergamon Press, Oxford, New York, Paris, Frankfurt, 1986.

[256] J. Nowinski and S. Turski, A study of states of stress in inhomogeneous bodies (in Polish), *Arch. Mech. Stos.*, 5(1953), 397–414.

[257] J.L. Nowinski, *Theory of Thermoelasticity with Applications.* Sijthoff & Noordhoff Int. Publ. Alphen aan Den Rijn, Netherlands, 1978.

[258] J.W. Nunziato and S.C. Cowin, A non-linear theory of elastic materials with voids. *Arch. Rational Mech. Anal.*, 72(1979), 175–201.

[259] O.A. Oleinik and G.A. Yosifian, Boundary value problems for second order elliptic equations in unbounded domains and Saint-Venant's principle. *Ann. della Scuola Norm. Sup. Pisa*, IV, 2(1977), 269–290.

[260] W. Olszak and J. Rychlewski, Nichthomogenitäts Probleme in Elastischen und Vorplastischen Bereich, *Österreichishes Ingenieur Archiv.*, 15(1961), 279–292.

[261] M. Ostoja-Starzewski and D.C. Stahl, Random fiber networks and special elastic orthotropy of paper. *J. Elasticity*, 60(2000), 131–149.

[262] M. Ostoja-Starzewski, Crack patterns in plates with randomly placed holes: A maximum entropy approach. *Mec. Res. Comm.*, 28(2001), 193–198.

[263] M. Ostoja-Starzewski and A. Woods, Spectral finite elements for vibrating rods and beams with random field properties. *J. Sound Vib.*, 268(2003), 779–797.

[264] M. Ostoja-Starzewski, Fracture of brittle microbeams. *J. Appl. Mech.*, 71(2004), 424–427.

[265] H.C. Park and R.S. Lakes, Torsion of a micropolar elastic prism of square cross-section. *Int. J. Solids Struct.*, 23(1987), 485–503.

[266] V.Z. Parton and P.I. Perlin, *Integral Equations in Elasticity* (translated from Russian), Mir, Moscow, 1982.

[267] L.E. Payne, Bounds for solutions of a class of quasilinear elliptic boundary value problems in terms of the torsion function. *Proc. Roy. Soc. Edinburgh. Sect.*, A88(1981), 251–265.

[268] L.E. Payne and G.A. Philippin, Isoperimetric inequalities in the torsion and clamped membrane problems for convex plane domains. *SIAM J. Math. Anal.*, 14(1983), 1154–1162.

[269] L.E. Payne and L.T. Wheeler, On the cross section of minimum stress concentration in the Saint-Venant theory of torsion. *J. Elasticity*, 14(1984), 15–17.

[270] L. Placidi, Solution of Saint-Venant's and Almansi-Michell's problems. Blacksburg, Va: University Libraries, Virginia Polytechnic Institute and State University, 2002.

[271] P. Podio-Guidugli, Saint-Venant formulae for generalized Saint-Venant problems. *Arch. Rational Mech. Anal.*, 81(1983), 13–20.

[272] P. Podio-Guidugli, Some recent results on Saint-Venant problem. In *Il Problema di de Saint-Venant: Aspetti Teorici e Applicativi*, Academia Nazionale dei Lincei, Atti dei Convegni Lincei 140, Roma, 1998, 135–158.

[273] A. Polania, On Saint-Venant's principle in three-dimensional elasticity. *Meccanica*, 11(1976), 98–101.

[274] D.A. Polignone and C.O. Horgan, Pure torsion of compressible non-linearly elastic circular cylinders. *Quart. Appl. Math.*, 49(1991), 591–607.

[275] G. Polya and A. Weinstein, On the torsional rigidity of multiply connected cross-sections. *Ann. of Math.*, 52(1950), 154–163.

[276] P. Puri, Steady-state thermoelastic problems for the half-space with couple stresses. *Arch. Mech. Stos.*, 22(1970), 479–489.

[277] P. Puri and S.C. Cowin, Plane waves in linear elastic materials with voids. *J. Elasticity*, 15(1985), 167–178.

[278] A. Radu, Torsiunea barelor prismatice compuse din mai multe materiale ortotrope. *St. cerc. Şt., Acad, R.P.R., Fil. Iaşi*, 11(1960), 375–379.

[279] A. Radu, Saint-Venant's problem for nonhomogeneous bars (in Romanian), *An.St. Univ. "Al.I. Cuza", Iaşi, s. Matematică*, 12(1966), 415–430.

[280] R.S. Rivlin, Large elastic deformations of isotropic materials. II. Some uniqueness theorems for pure, homogeneous deformations. *Phil. Trans. Roy. Soc. London, Ser. A.* 240(1948), 491–508.

[281] A. Robinson, *Non-Standard Analysis*. North Holland, Amsterdam, 1966.

[282] J.J. Roseman, A pointwise estimate for the stress in a cylinder and its application to Saint-Venant's principle. *Arch. Rational Mech. Anal.*, 21(1966), 23–48.

[283] J.J. Roseman, The principle of Saint-Venant in linear and nonlinear plane elasticity. *Arch. Rational Mech. Anal.*, 26(1967), 142–162.

[284] M.B. Rubin, *Cosserat Theories: Shells, Rods and Points.* Kluwer Academic Publishers, Dordrecht, 2000.

[285] Z.A. Ruhadze, Boundary value problems of elasticity theory for piecewise non-homogenous orthotropic bodies (in Russian). Soobsc. *Akad. Nauk Gruzin.*, SSR 30(1963), 11–18.

[286] A.K. Rukhadze and R.T. Zivzivadze, A generalized Almansi problem for prismatic beams that are composed of various isotropic materials (in Russian). *Soobs. Akad. Nauk Gruzin.*, SSR 103(1981), 41–44.

[287] G. Rusu, Theory of loaded Cosserat elastic cylinders. *An Şt. Univ. "Al.I. Cuza" Iaşi, s. Matematică*, 24(1978), 379–388.

[288] G. Rusu, Existence theorems in elastostatics of materials with voids. An *Şt. Univ. "Al.I. Cuza" Iaşi, s. Matematică*, 30(1985), 193–204.

[289] M. Sadd, *Elasticity: Theory, Applications and Numerics.* Elsevier B.V., 2005.

[290] M.A. Sadowsky and M.A. Goldberg, *Non-Homogeneous Elasticity.* New York, 1958.

[291] A.J.C.B. de Saint-Venant, Mémoire sur la torsion des prismes. *Mémoires présentés par divers savants a l'Academie des Sciences*, 14(1855), 233–560.

[292] A.J.C.B. de Saint-Venant, Mémoire sur la flexion des prismes. *Journal de Mathématiques de Liouville*, Series II, 1(1856), 89–189.

[293] N. Sandru, Le théorème de reciprocité dans l'élasticité asymmétrique (cas dynamique). *Rend. Acad. Lincei, Ser. VIII*, 38(1965), 78–81.

[294] N. Sandru, On some problems of the linear theory of asymmetric elasticity. *Int. J. Engng. Sci.*, 4(1966), 81–94.

[295] U. Saravanan and K.R. Rajagopal, On the role of inhomogeneities in the deformation of elastic bodies. *Math. Mech. Solid.*, 8(2003), 249–376.

[296] V.S. Sarkisyan, On the method of the solution of elasticity theory problem for the inhomogeneous and anisotropic body. *Differential Equations and Their Applications*, Equadiff 5, Proc. 5th Czech. Conf. Bratislava 1981, Teubner-Texte Math. 47(1982), 217–300.

[297] V.S. Sarkisyan, Some modern problems of the nonhomogeneous anisotropic theory of elasticity. In *Continuum Mechanics and Related Problems of Anaylsis* (Tbilisi, 1991), 205–211, 1993.

[298] V.S. Sarkisyan, V.E. Avetikyan and E.Sh. Salekh, On a method for solving the problem of the bending of a nonhomogeneous orthotropic cantilever of rectangular cross-section (in Russian). *Erevan. Gos. Univ. Uchen. Zap. Estestv. Nauki* 179(1993), 15–20.

[299] V.S. Sarkisyan, On the solution of two contact problems for elastic bodies with two heterogeneous finite stringers (in Russian). *Mat. Metodi Fiz-Mekh. Polya*, 46(2003), 114–121.

[300] A. Scalia, Extension, bending and torsion of anisotropic microstretch elastic cylinders. *Math. Mech Solids*, 5(2000), 31–41.

[301] M.R. Scalpato and C.O. Horgan, Saint-Venant decay rates for an isotropic inhomogeneous linearly elastic solid in anti-plane shear. *J. Elasticity*, 48(1997), 145–166.

[302] E. Scarpetta, On the fundamental solutions in micropolar elasticity with voids. *Acta Mech.*, 82(1990), 151–158.

[303] R.D. Schile and R.L. Sierakowski, On the Saint-Venant problem for a non-homogeneous elastic material. *Quart. Appl. Math.*, 23(1965), 19–25.

[304] A.P.S. Selvadurai, *Mechanics of Poroelastic Media.* Kluwer Academic Publishers, Dordrecht, Boston, London, 1996.

[305] D.I. Sherman, The plane strain of anisotropic elastic bodies (in Russian). *Tr. Seism. Inst. Akad. Nauk*, SSSR 86(1938), 1–12.

[306] D.I. Sherman, New solutions of plane strain problems in the theory of anisotropic elastic bodies (in Russian). *Doklady Akad. Nauk*, SSSR 32(1941), 314–315.

[307] D.I. Sherman, On the solution of a plane problem of the theory of elasticity for an isotropic medium (in Russian). *Appl. Math. Mech.* (*Akad. Nauk SSSR Prikl. Mat. Mech.*), 6(1942), 509–514.

[308] D.I. Sherman, Plane deformation in isotropic inhomogeneous media. *Appl. Math. Mech. (Akad. Nauk SSSR Prikl. Mat. Mech.)*, 7(1943), 301–309.

[309] D.I. Sherman, On the problem of plane strain in non-homogeneous media. In *Nonhomogeneity in Elasticity and Plasticity*, Pergamon Press, London, 1959, 3–12.

[310] R.T. Shield, An energy method for certain second-order effects with application to torsion of elastic bars under tension. *J. Appl. Mech.*, 47(1980), 75–81.

[311] R.T. Shield, Extension and torsion of elastic bars with initial twist. *J. Appl. Mech., Trans. ASME*, 49(1982), 779–786.

[312] R.T. Shield and S. Im, Small strain deformations of elastic beams and rods including large deflections. *ZAMP*, 37(1986), 491–513.

[313] I.S. Sokolnikoff, *Mathematical Theory of Elasticity* McGraw-Hill, 1956.

[314] M. Sokolowski, On the statically equivalent loads in Saint-Venant's principle. *Rozprawy Inz.*, 36(1988), 779–791.

[315] L. Solomon, *Élasticité Linéaire.* Masson, Paris, 1968.

[316] E. Soós, Sur le problème de Saint-Venant dans le cas des barres hétérogeneous avec anisotropie cylindrique. *Bull. Math. Soc., Sci. Math. Phys. R.P.R.*, 7(1963), 61–75.

[317] E. Soós, A mixed boundary problem for an aeolotropic nonhomogeneous elastic cylinder. *Rev. Roum. Math. Pures et Appl.*, 9(1964), 981–994.

[318] E. Soós, On Saint-Venant's problem for a nonhomgeneous and anisotropic body (in Romanian). *An. Univ. Bucureşti. Mat. Mech.*, 12(1973), 70–77.

[319] E. Soós, *Discrete and Continuous Models of Solids* (in Romanian). Editura Stiinţifică, Bucureşti, 1974.

[320] R. Stanczak, Green's tensor of mixed boundary value problem of the theory of elasticity and its applications. *Demonstratio Math.*, 2(1970), 145–157.

[321] E. Sternberg, On Saint-Venant's principle, *Quart. Appl. Math.*, 11(1954), 393–402.

[322] E. Sternberg and J.K. Knowles, Minimum energy characterizations of Saint-Venant's solution to the relaxed Saint-Venant problem. *Arch. Rational Mech. Anal.*, 21(1966), 89–107.

[323] E. Sternberg, On Saint-Venant's torsion and the plane problem of elastostatics for multiply connected domains. *Arch. Rational Mech. Anal.*, 85(1984), 295–310.

[324] P.P. Teodorescu, On the plane problem of elasticity of some aeolotropic bodies. *Rev. Roum. Sci. Techn. Méc. Appl.*, 5(1960), 385–399.

[325] P.P. Teodorescu, On the action of concentrated loads in the case of a Cosserat continuum. IUTAM-Symposium 1967, Freudenstadt-Stuttgart, Mechanics of generalized continua, 1968, 120–125.

[326] L. Teresi and A. Tiero, The Saint-Venant problem for beam-like modular trusses. In *Il Problema di de Saint-Venant: Aspetti Teorici e Applicativi*, Accademia Nazionale dei Lincei, Atti dei Convegni Lincei 140, Roma (1998), 135–158.

[327] T.C.T. Ting, *Anisotropic Elasticity: Theory and Applications.* Oxford University Press. Oxford, 1996.

[328] R.A. Toupin, Theories of elasticity with couple-stress. *Arch. Rational Mech. Anal.*, 17(1964), 85–112.

[329] R.A. Toupin, Saint-Venant's principle. *Arch. Rational Mech. Anal.*, 18(1965), 83–96.

[330] R.A. Toupin, Saint-Venant and a matter of principle. *Trans. N Y Acad. Sci.*, 28(1965), 221–232.

[331] C. Truesdell, The rational mechanics of materials-past, present, future. *Appl. Mech. Rev.*, 12(1959), 75–80.

[332] C. Truesdell and R. Toupin, The Classical Field Theories. In vol III/1 of the *Handbuch der Physik* (Edited by S. Flügge). Berlin, Heidelberg, New York, Springer-Verlag, 1960.

[333] C. Truesdel and W. Noll, The Non-Linear Field Theories of Mechanics. In vol III/3 of *the Handbuch der Physik* (Edited by S. Flügge). Berlin, Heidelberg, New York, Springer-Verlag, 1965.

[334] C. Truesdell, The rational mechanics of materials-past, present, future (Corrected and modified reprint of [331]), 225–236 of *Applied Mechanics Surveys*, Spartan Books, 1966.

[335] C. Truesdell, History of Classical Mechanics. Die Naturwissenschaften 63, Part I: to 1800, 53–62; Part II: the 19th and 20th Centuries, 119–130, Springer-Verlag, Berlin, 1976.

[336] C. Truesdell, Some challenges offered to analysis by rational thermomechanics, pp. 495–603 of *Contemporary Developments in Continuum Mechanics and Partial Differential Equations* (Edited by G.M. de la Penha and L.A. Medeiros), North-Holland, Amsterdam, 1978.

[337] G.V. Tsikhistavi, Saint-Venant problems for homogeneous "curvilinearly conic" isotropic bodies with end loading (in Russian). *Trudy Vychisl. Tsentra Akad. Nauk Gruzin.*, SSSR 22(1982), 123–131.

[338] C. Usidus and M. Sokolovski, Torsion of cylindrical shafts made of micropolar material. *Bull. Acad. Polon. Sci., Sér Techn.*, 21(1973), 19–23.

[339] A.I. Uzdalev, Flexure of two-layered anisotropic cylinders (in Russian). *Ing. Sbornik*, 15(1953), 35–42.

[340] I.N. Vekua and A.K. Rukhadze, Torsion problem for a circular cylinder reinforced by a longitudinal circular rod (in Russian). *Izv. Akad. Nauk*, SSSR 3(1933), 1297–1308.

[341] P. Villaggio, *Qualitative Methods in Elasticity.* Kluwer Academic Publishers, Dordrecht, Boston, London, 1996.

[342] W. Voigt, Theoretische Studien über die Elasticitätsverhältnisse der Krystalle. *Abh. Ges. Wiss. Göttingen*, 34(1887), 53–153.

[343] O. Zanaboni, Dimostrazione generale del principio del De Saint-Venant. *Atti Accad Lincei. Rend.*, 25(1937), 117–121.

[344] O. Zanaboni, Valutazione dell'errore massimo cui dà luogo l'applicazione del principio del De Saint-Venant. *Atti Accad. Lincei. Rend.*, 25(1937), 595–601.

[345] A.A. Zelenina and L.M. Zubov, Nonlinear theory of the pure bending of prismatic elastic bodies. *J. Appl. Math. Mech.*, 64(2000), 399–406.

[346] L.M. Zubov, Theory of torsion of prismatic rods under finite deformations (in Russian). *Dokl. Akad. Nauk*, SSSR 270(1983), 827–831.

[347] L.M. Zubov, Semi-inverse solutions of the nonlinear theory of elasticity that lead to two-dimensional boundary value problems. *Dokl. Phys.*, 45(2000), 554–556.

[348] L.M. Zubov, The exact nonlinear theory of tension and torsion of helical springs (in Russian). *Dokl. Akad. Nauk*, 385(2002), 628–630, translation in *Dokl. Phys.*, 47(2002), 623–626.

[349] L.M. Zubov and D.N. Sheidakov, The effect of torsion on the stability of an elastic cylinder under tension. *J. Appl. Math. Mech.*, 69(2005), 49–56.

[350] C.C. Wang and C. Truesdell, *Introduction to Rational Elasticity.* Leyden, Noordhoff, 1973.

[351] K. Washizu, *Variational Methods in Elasticity and Plasticity.* Pergamon, New York, 1968.

[352] M.L. Wenner, On torsion of an elastic cylindrical Cosserat surface. *Int. J. Solids Struct.*, 4(1968), 769–776.

[353] K. Wilmanski and Cz. Wozniak, On geometry of continuous medium with micro-structure. *Arch. Mech. Stos.*, 19(1967), 715–724.

[354] K. Wilmanski, *Thermomechanics of Continua.* Springer, Heidelberg, 1998.

[355] K. Wilmanski, Mass exchange diffusion and large deformations of poroelastic materials. In *Modeling and Mechanics of Granular and Porous Materials* (Edited by G. Capriz, V.N. Ghionna, P. Giovine), Birkhäuser, Basel, 2002, 213–244.

Index